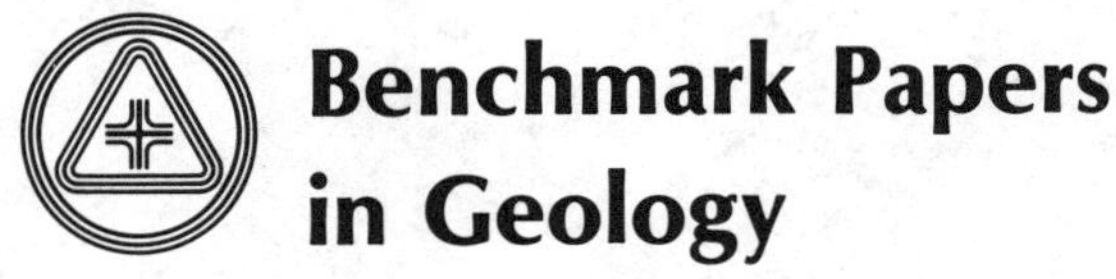

Benchmark Papers in Geology

Series Editor: Rhodes W. Fairbridge
Columbia University

27 PERIGLACIAL PROCESSES / *Cuchlaine A. M. King*
28 LANDFORMS AND GEOMORPHOLOGY: Concepts and History / *Cuchlaine A. M. King*
29 METALLOGENY AND GLOBAL TECTONICS / *Wilfred Walker*
30 HOLOCENE TIDAL SEDIMENTATION / *George deVries Klein*
31 PALEOBIOGEOGRAPHY / *Charles A. Ross*
32 MECHANICS OF THRUST FAULTS AND DÉCOLLEMENT / *Barry Voight*
33 WEST INDIES ISLAND ARCS / *Peter H. Mattson*
34 CRYSTAL FORM AND STRUCTURE / *Cecil J. Schneer*
35 OCEANOGRAPHY: Concepts and History / *Margarget B. Deacon*
36 METEORITE CRATERS / *G. J. H. McCall*
37 STATISTICAL ANALYSIS IN GEOLOGY / *John M. Cubitt and Stephen Henley*
38 AIR PHOTOGRAPHY AND COASTAL PROBLEMS / *Mohamed T. El-Ashry*
39 BEACH PROCESSES AND COASTAL HYDRODYNAMICS / *John S. Fisher and Robert Dolan*
40 DIAGENESIS OF DEEP-SEA BIOGENIC SEDIMENTS / *Gerrit J. van der Lingen*
41 DRAINAGE BASIN MORPHOLOGY / *Stanley A. Schumm*
42 COASTAL SEDIMENTATION / *Donald J. P. Swift and Harold D. Palmer*
43 ANCIENT CONTINENTAL DEPOSITS / *Franklyn B. Van Houten*
44 MINERAL DEPOSITS, CONTINENTAL DRIFT AND PLATE TECTONICS / *J. B. Wright*
45 SEA WATER: Cycles of the Major Elements / *James I. Drever*
46 PALYNOLOGY, PART I: Spores and Pollen / *Marjorie D. Muir and William A. S. Sarjeant*
47 PALYNOLOGY, PART II: Dinoflagellates, Acritarchs, and Other Microfossils / *Marjorie D. Muir and William A. S. Sarjeant*
48 GEOLOGY OF THE PLANET MARS / *Vivien Gornitz*
49 GEOCHEMISTRY OF BISMUTH / *Ernest E. Angino and David T. Long*
50 ASTROBLEMES—CRYPTOEXPLOSION STRUCTURES / *G. J. H. McCall*

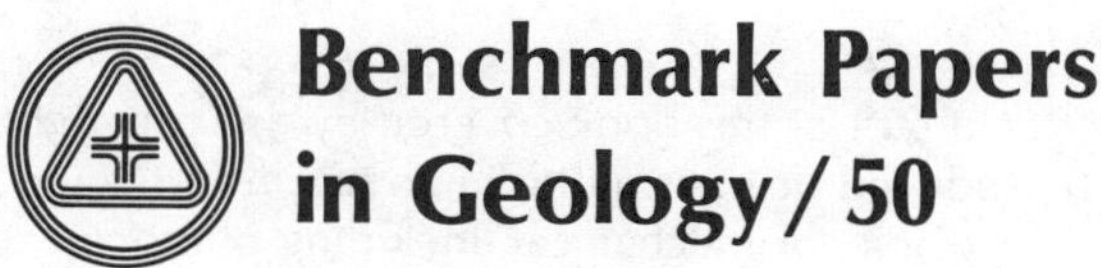

A BENCHMARK® Books Series

ASTROBLEMES— CRYPTOEXPLOSION STRUCTURES

Edited by

G. J. H. McCALL

Fellow, Liverpool University
Department of Geology

STROUDSBURG, PENNSYLVANIA

Benchmark Papers in Geology, Volume 50
Library of Congress Catalog Card Number: 79-10991
ISBN: 0-87933-342-1

81 80 79 1 2 3 4 5
Manufactured in the United States of America.

LIBRARY OF CONGRESS CATALOGING IN PUBLICATION DATA
Main entry under title:
Astroblemes—cryptoexplosion structures.
(Benchmark papers in geology; v. 50)
Includes bibliographical references and indexes.
1. Cryptoexplosion structures—Addresses, essays, lectures. I. McCall, Gerald Joseph Home.
QE613.A87 551 79-10991
ISBN: 0-87933-342-1

Distributed world wide by Academic Press, a subsidiary of Harcourt Brace Jovanovich, Publishers.

SERIES EDITOR'S FOREWORD

The philosophy behind the "Benchmark Papers in Geology" is one of collection, sifting, and rediffusion. Scientific literature today is so vast, so dispersed, and, in the case of old papers, so inaccessible for readers not in the immediate neighborhood of major libraries that much valuable information has been ignored by default. It has become just so difficult, or so time consuming, to search out the key papers in any basic area of research that one can hardly blame a busy person for skimping on some of his or her "homework."

This series of volumes has been devised, therefore, to make a practical contribution to this critical problem. The geologist, perhaps even more than any other scientist, often suffers from twin difficulties—isolation from central library resources and immensely diffused sources of material. New colleges and industrial libraries simply cannot afford to purchase complete runs of all the world's earth science literature. Specialists simply cannot locate reprints or copies of all their principal reference materials. So it is that we are now making a concerted effort to gather into single volumes the critical materials needed to reconstruct the background of any and every major topic of our discipline.

We are interpreting "geology" in its broadest sense: the fundamental science of the planet Earth, its materials, its history, and its dynamics. Because of training in "earthy" materials, we also take in astrogeology, the corresponding aspect of the planetary sciences. Besides the classical core disciplines such as mineralogy, petrology, structure, geomorphology, paleontology, and stratigraphy, we embrace the newer fields of geophysics and geochemistry, applied also to oceanography, geochronology, and paleoecology. We recognize the work of the mining geologists, the petroleum geologists, the hydrologists, and the engineering and environmental geologists. Each specialist needs a working library. We are endeavoring to make the task of compiling such a library a little easier.

Each volume in the series contains an introduction prepared by a specialist (the volume editor)—a "state of the art" opening or a summary of the object and content of the volume. The articles, usually some twenty to fifty reproduced either in their entirety or in significant extracts, are selected in an attempt to cover the field, from the key papers of the last century to fairly recent work. Where the original works are in foreign languages, we have endeavored to locate or commission translations. Geologists, because of their global subject, are often acutely aware of the oneness of our world. The selections cannot, therefore, be restricted to any one country, and whenever possible an attempt is made to scan the world literature.

To each article, or group of kindred articles, some sort of "highlight commentary" is usually supplied by the volume editor. This commentary should serve to bring that article into historical perspective and to emphasize its particular role in the growth of the field. References, or citations, wherever possible, will be reproduced in their entirety—for by this means the observant reader can assess the background material available to that particular author, or, if desired, he or she, too, can double check the earlier sources.

A "benchmark," in surveyor's terminology, is an established point on the ground, recorded on our maps. It is usually anything that is a vantage point, from a modest hill to a mountain peak. From the historical viewpoint, these benchmarks are the bricks of our scientific edifice.

RHODES W. FAIRBRIDGE

PREFACE

In *Astroblemes—Cryptoexplosion Structures* a number of impact structures are described in selected papers or briefly in editorial commentaries in addition to those included in the companion volume, *Meteorite Craters* (G. J. H. McCall, ed., Benchmark Papers in Geology, Vol. 36, Dowden, Hutchinson & Ross, 1977). All these structures at some time or another have been attributed to extraterrestrial agencies. They fall within classes IV, V, and VI in the classification adopted in *Meteorite Craters* and reproduced herein (Tables 1 and 2) with slight additions and amendments. In this volume I deal mainly with the structures of class IV: large, complex, mostly geologically ancient structures, many of which, if they are of extraterrestrial origin, can only be classified as impact-scars—hence the term "astrobleme." Many of the structures are infilled, erosionally-degraded craters, but most lack the physiographic form of a crater, even recognizable in a partly infilled or erosionally-degraded state. In age these structures cover about seven-tenths of the history of the Earth, ranging from Archean to Cenozoic. Most are accompanied by shock deformations of a supposedly unique type, but none has an association with actual meteoric material. A brief portion of the book covers structures of class V, about which more information is needed before any firm conclusion can be drawn concerning their mode of origin. A list is given of those structures of class VI, for which meteoritic origin has been discounted.

In the Introduction, I list the structures and then briefly review the opposing cases advanced for impact and cryptovolcanic origin of these structures. In the first case, the arguments concerning the various shock indicators widely believed to signify impact involvement are briefly considered. In the second case, arguments advanced for cryptovolcanic origin are summarized.

My aim has been to achieve objectivity and to clearly state both cases. Rather than answering questions, the book defines the questions that still need to be answered. I conclude that the origin of terrestrial cryptoexplosion, a real and well-documented geological process, remains obscure, there still being baffling contradictions in the evidence. Despite the use of the term astrobleme in the title of this book, I would urge the continued use of the neutral terms "cryptoexplosion" and "disturbance," rather than this somewhat prejudicial, genetic term.

G. J. H. McCALL

CONTENTS

CONTENTS BY AUTHOR

ASTROBLEMES—CRYPTOEXPLOSION STRUCTURES

INTRODUCTION

The first three classes of impact structures, defined in Table 1 and described in the previous Benchmark volume (McCall, 1977), all have the physiographic form of a crater. This is not so of the deeply buried, highly eroded, or structurally complex manifestations covered in this book. The 71 structures listed in class IVa (Table 2) all display some

Table 1 Classification of meteorite craters and astroblemes used in this volume (modified from Shoemaker and Eggleton, 1961).

Class I	Proved meteorite impact-fragmentation craters and meteorite impact-explosion craters.
Class II	Craters showing shock effects, but no material association of meteorites (all meteorite impact-explosion craters).
Class III	Craters showing the characteristic physiographic form and structure of meteorite craters, but no shock metamorphism, meteoritic material, or bolide association.
Class IV	Structures either deeply eroded or buried, or else of considerable complexity, which do not display an immediately apparent physiographic form of meteorite craters. These include some relatively young but highly complex features, but the majority are of considerable geologic age. The eroded and infilled examples include a number of simple craters, but the majority are of considerable complexity. All are possible meteorite impact-explosion structures.
Class V	Structures concerning which there is too little evidence available at this stage of knowledge to determine origin, and concerning which more information is needed.
Class VI	Structures erroneously attributed to extraterrestial agencies.

Table 2 List of craters and structures that have at some time been attributed to extraterrestrial agencies.*

	Class I
Ia	Odessa, U.S.A.; Haviland, U.S.A.; Dalgaranga, Australia; Boxhole, Australia; Campo del Cielo, Argentina; Kaalijaarv, Estonian S.S.R.; Sikhote-Alin, U.S.S.R.
Ib	Tunguska, U.S.S.R.
Ic	Henbury, Australia; Wabar, Saudi Arabia
Id	Canyon Diablo, U.S.A.; Wolf Creek, Australia
Total = 12	
	Class II
	Bosumtwi, Ghana; Aouelloul, Mauretania; Lonar Lake, India; Monturaqui, Chile; New Quebec, Canada; Pretoria Salt Pan, South Africa; Roter Kamm, South West Africa; Talemzane, Algeria
Total = 8	
	Class III
	Hebron, Canada; Lake Elgh'ytkin, U.S.S.R.; Mora, U.S.A.; Morasko Craters, Poland; Murgab, U.S.S.R.; Patomski, U.S.S.R.; Sayan, U.S.S.R.
Total = 7	
	Class IV
IVa	Agtö, Greenland; Araguainha, Brazil; B. P. Structure, Libya; Beyenchime-Salata, U.S.S.R.; Boltyshka, U.S.S.R.; Brent, Canada; Carswell Lake, Canada; Charlevoix, Canada; Clearwater Lakes, Canada; Crooked Creek, U.S.A.; Decaturville, U.S.A.; Deep Bay, Canada; Des Plaines, U.S.A.; Dycus, U.S.A.; Flynn Creek, U.S.A.; Furnace Creek, U.S.A.; Glasford, U.S.A.; Gosses Bluff, Australia; Gow Lake, Canada; Gusev, U.S.S.R.: Haughton Dome, Canada; Hazel Green Creek, U.S.A.; Hicks Dome, U.S.A.; Holleford, Canada; Howell, U.S.A.; Ilumets, U.S.S.R.; Jeptha Knob, U.S.A.; Kaluga, U.S.S.R.; Kamensk, U.S.S.R.; Kara, U.S.S.R.; Kentland, U.S.A.; Kofels, Austria; Lac Couture, Canada; Lake Dellen, Sweden; Lake Hummeln, Sweden; Lake Janisjarvi, U.S.S.R.; Lake Lappajarvi, Finland; Lake Mien, Sweden; Lake Siljan, Sweden; Liverpool, Australia; Manicouagan, Canada; Manson, U.S.A.; Middlesboro, U.S.A.; Mishinogorsk, U.S.S.R.; Mistastin Lake, Canada; Nicholson Lake, Canada; Oasis Structure, Libya; Obolon, U.S.S.R.; Pilot Lake, Canada; Popigay, U.S.S.R.; Puchez-Katunki, U.S.S.R.; Richat, Mauretania; Rieskessel, Germany; Rochechouart, France; Rose Dome, U.S.A.; Rotmistrovka, U.S.S.R.; Serpent Mound, U.S.A.; Sierra Madera, U.S.A.; Slate Islands, Canada; Steen River, Canada; Steinheim Basin, Germany; St. Martins, Canada; Strangways, Australia; Sudbury, Canada; Tenoumer, Mauretania; Versailles, U.S.A.; Vredefort, South Africa; Wanipitei, Canada; Weaubleau, U.S.A.; Wells Creek Basin, U.S.A.; West Hawk Lake, Canada
Total = 71	

Table 2 Continued

IVb	Kilmichael Dome, U.S.A.; Lac du Bonnet, Canada; Lake Teague, Australia; Sierra da Canghala, Brazil; Willbarger Dome, U.S.A.
Total = 5	
	Class V
	Agheir, Mauretania; Agnak Is., Canada; Al Umchaimin, Iraq; Amguid, Algeria; Amak, Aleutian Is.; Baghdad, Iraq; Butare, Rwanda; Dampier Archipelago, Australia; Dirribindi, Australia; Dzhausov, U.S.S.R.; Eagle Butte, Canada; East Arnhem Land, Australia; East Structure, Kufra, Libya; Elbow, Canada; El Mreiti, Mauretania; Erg Ghech, Algeria; Foum Terouentor, Algeria; Gallouedec, Algeria; Garet et Lefet, Fezzan, Libya; Gourma, Mali; Hagens Fjord, Greenland; Hartney, Canada; Herault, France; Ile Rouleau, Canada; Kakiattukallak Lake, Canada; Kalkkop, South Africa; Lac Bouchet, France; Labynkyr, U.S.S.R.; Lac La Moinerie, Canada; Lake Michikamau, Canada; La Sauvetat, France; Limestone Mountain Disturbance, U.S.A.; Matam, Senegal; Mecatina, Canada; Meen Lake, Canada; Mejouada, Mauretania; Menihek Lakes, Canada; Paris Lakes, France; Ramgarh, India; Savonoski, Alaska; Semsiyat, Mauretania; Skeleton Lake, Canada; St. Magnus Bay, Shetland Is., Scotland; Tademait, Algeria; Temimichat, Mauretania; Tindouf, Morocco; Tvaren Bay, Sweden; Zhamanshin, U.S.S.R.
Total = 48	
	Class VI
	Bay of Campeche, Mexico; Bishoftu, Ethiopia; Bushveld, South Africa; Carolina Bays, U.S.A.; Colluma, Bolivia; Crestone, U.S.A.; Duckwater, U.S.A.; Dzioua, Algeria; Ellef Ringnes Is., Canada; Eyre Peninsula, Australia; Falkland Is. Arc; Freetown Complex, Sierra Leone; Franktown, Canada; Glover Bluff, U.S.A.; Great Dike, Rhodesia; Gulf of St. Lawrence, Canada; Gwarkuh, Baluchistan; Hungarian Plateau; Hudsons Bay, Canada; Keeley Lake, Canada; Konder, U.S.S.R.; Laguna Guatavita, Columbia; Macamic Lake, Canada; Malha, Sudan; Melville Is., Canada; Michlifen, Morocco; Mt. Doreen, Australia; Nebiewale, Ghana; Panamint, U.S.A.; Piute, U.S.A.; St. Georges Bay, Argentina; Sault-au-Cochons, Canada; Sea of Japan; Socotra; Tibesti, Chad; Tiffin, U.S.A.; Tierra del Fuego; Ubehebe, U.S.A.; Upheaval Dome, U.S.A.; Ungava Bay, Canada; Weddell Sea, Antarctica; Wilkes Land, Antarctica; Wiltshire, England; Winkler, U.S.A.; Witwatersrand Basin, South Africa
Total = 45	

*Since its publication in *Meteorite Craters* (McCall, 1977), this table has been revised in accordance with more recent information primarily from the USSR, which has also required revision of the classification of the Ilumets and Puchez-Katunki structures. The Sayan crater was incorrectly listed previously (McCall, 1977, pp. 4, 352) due to an error. New information has also required revision of the classification of the Elbow and Winkler structures.

form of shock metamorphism or brecciation, but the remaining five in class IVb do not. All the terrestrial impact structures of significant dimensions that have at some time been attributed to meteorite impact are listed in Table 2.

The structures of class IV range in diameter from 1 to more than 100 km. It has been found convenient to subdivide the large number of structures in this class under geographic groups, treated in separate chapters:

- I Europe (excluding European U.S.S.R.)
- II European Russia and Asia
- III Canada and Greenland
- IV Central United States
- V Africa
- VI Australia
- VII South America

Five morphological subclasses are erected for class IVa:

- IVa/1 Simple structures
- IVa/2 Complex structures with central uplift
- IVa/3 Complex structures with both central uplift and enclosing ring depression
- IVa/4 Complex structures with only a central depression
- IVa/5 Complex, irregular structures

Some structures do not fit readily into the classification (Table 1). The Richat structure in Mauretania is quite young and highly complex: it does not fit well into any class and is considered here, for convenience, with the structures of class IV in Africa. The Tenoumer structure, also in Africa, has the form of a quite youthful crater while displaying some structural complexity in the form of ring dikes: though it could possibly be referred to class II, the complication of ring dikes serves to ally it to the structures of class IVa.

THE CASE FOR METEORITIC ORIGIN OF LARGE CRYPTOEXPLOSION STRUCTURES

Research on meteorite craters proceeded on orthodox lines up to the late 1950s, being mainly concerned with impact fragmentation craters and the Arizona Canyon Diablo Crater. Prior to this, certain original thinkers had advanced meteoritic hypotheses for certain large cryptic structures: Vredefort (Daly, 1947); the cryptic structures of the central United States (Boon and Albritton, 1937; Dietz, 1946, 1947); the Rieskessel (Werner, 1904; Stutzer, 1936); Bosumtwi (McClaren, 1931); and the Pretoria Salt Pan (Rohleder, 1933). It was not until the dawn of the Space Age that Dietz (1959, 1961, 1964) focused attention on this potential field of research. Numerous studies were published in the ensuing decade. The ample funds available from NASA sources were directed more toward establishing the existence of ancient impact scars on the terrestrial surface rather than toward a totally objective investigation of the phenomenon of cryptoexplosion. But despite the consequent loss of objectivity, this program advanced the study of cryptoexplosion beyond the initial statements of theory (Bucher, 1936, cryptovolcanic; Baldwin, 1949, impact explosion) to a situation in which an entire new family of cryptoexplosion structures has been added to the geologic record. A profusion of data is available concerning their field relationships, petrography, mineralogy, shock deformations, geochemistry, geophysics, and geochronology. It is this data that is included in this volume. Occurrences studied have ranged from the immense, ancient Vredefort and Sudbury complexes to quite simple recent craters (McCall, 1977). Some older and larger structures are complicated by association with magmatic activity of an apparently normal mantle-derived type (for example, large Paleozoic to Mesozoic structures in Canada and the Proterozoic Sudbury structure). This also appears to be the case in some quite young structures in Mauretania. Peculiar glasses, glassy breccias, and glassy lavas are associated with a number of cryptoexplosion structures, both potassic and sodic alkaline types. These add to the complexity and the mystery of cryptoexplosion. Are they forms of impactite—impact-generated melts derived by crustal fusion—or are they secondarily triggered mantle-melting derivitives? Are the magmatic rocks not true endogenic eruptives and are the glassy phases not products of some obscure form of endogenic fluidization related to a primary highly reactive gas phase? Immense structures such as Sierra Madera and Gosses Bluff display a complete absence of such magmatic associations despite explosive deformation and uplift of immense scale.

These diverse structures are all linked by a common factor: peculiar, if not unique, shock deformations and effects. Their formation certainly involved intense and virtually instantaneous explosive events related to central-point foci, and situated, in most cases, at no great depth in the crust. That they are of extraterrestrial origin is, however, by no means established with certainty. The structures of class I are virtually established as meteoritic (there are possible reservations concerning the Wolf Creek Crater, which is unproven). Beyond class I, the status of structures of classes II to V is probable or possible, depending on the weight of the evidence. None are of proved extraterrestrial origin.

The case for extraterrestrial origin is stated succinctly by Guy-Bray (Paper 20): "the reason for belief in impact origin is simple: *Shock Metamorphism.* All the changes described are attributed to the passage of high-pressure shock waves: the only known method of producing such high-pressure shock waves is the hypervelocity impact of a large meteorite." Various indicators have been defined, which are supposed to provide a sort of litmus test of meteorite impact-induced explosive shock. They include shatter cones, stishovite, coesite, maskelynite, other thetomorphic glasses, fusion glasses, shock-lamellation in mineral grains, and certain characteristic types of breccia (suevite etc.). Calibration of these indicators has been developed by laboratory experiment, and by comparison with artificial nuclear and chemical explosions. Models of impact explosion have been erected based both on experiment and on field observation at a number of cryptoexplosion sites. Much of the theory of zonation of shock effects and distribution of melts and breccias in meteorite impact explosion has been derived from such field observations (including Canyon Diablo Crater, Brent, and the Rieskessel). Type examples have been erected and used for purposes of analogy. It is important to bear in mind that only the Canyon Diablo Crater in Arizona is proved to be of extraterrestrial origin and even though this unique crater has contributed much to theory, it is inadequate to supply comparative data applicable to impact explosions of quite different scale and type and within quite different host rocks. In this, there is a possible inherent weakness in the theory of impact-explosion cratering. This weakness is illustrated by the case of the Brent Crater, Canada, which has been interpreted as endogenic by Currie (1971). Currie's views are considered seriously by many earth scientists, and it is not unlikely that this supposed type-example of a meteorite impact-explosion crater is not of extraterrestrial origin at all. The possibility of entertaining an *impact-volcanic* hypothesis is a real one. This would resolve the anomaly in the case of structures, usually large ones in which magmatic rocks are present, of apparently normal magmatic derivation. Triggering of partial melting in the mantle may be invoked—a possibility that has been explored by French (1972) and Dence (Paper 21). French has recognized that even in the optimum case of the immense Sudbury structure there are severe constraints on such a process. Apparently normal magmatic components are present in much smaller structures where the constraints must be greater. Could these constraints be fatal to the impact-volcanic hypothesis? Both derivation of such magmatic rocks by large-scale crustal melting and contamination of primary magmas unrelated to the impact explosion by products of shock-induced crustal melting likewise come up against insuperable theoretical difficulties.

THE SUGGESTED UNIQUE INDICATORS OF IMPACT-EXPLOSION

Shatter Cones

Shatter cones are conical fracture surfaces with striae that fan outward in horsetail packets. When complete, they form striated cup and cone structures. The striae are sharp grooves between intervening sharp, rounded ridges, or the reverse on negative mold facies (Dietz, 1972). Small parasitic half-cones shingle the face of the master cone. Length ranges from <1 cm to >12 m. Cones develop best in dense aphanitic rocks, especially carbonate rocks. They also develop well in sandstone, quartzite, chert, shale, basalt, granite, and gneiss. They appear to be distinct from cone-in-cone structure (Amstutz, 1965). Shatter cones are covered by a number of papers by Dietz (1959, 1960, 1961, 1963a, b, 1964, 1968, 1972) which amplify earlier observations by Branca and Fraas (1905) and Bucher (1936). Dietz concludes that they represent an unique type of fracture created on a boundary between a plastic and elastic shock wave. The horsetail represents the movement of the plastic domain relative to the elastic (Johnson and Talbot, 1964). The cones appear to be formed by tensional release in an axially aysmmetrical field, following compression by an expanding shock wave (Manton, 1965). They are created by the intersection of an elastic precursor shock wave with some inhomogeneity, such as a pebble or nodule present at the apex (for example, as described from Subury by Guy-Bray et al., Paper 23). Susceptibility of a particular rock type to shatter coning is a function of its Hugoniot curve. Dietz believes that shock pressures between 20 and 80 kilobars are required for cones to form; but Currie (Paper 16) cited experiments by Shoemaker et al. (1962) in support of a much lower pressure requirement of only 5 kilobars. Shatter cones display uniformity of size in any one stratum. Apical angle is usually 90°, but acute angles may indicate higher shock pressures. Distribution is symmetrical around "ground zero" and apices are mostly directed toward the explosion focus, though reflection can produce the converse. In small structures, shatter cones are restricted to central uplifts; but in larger ones, they occur up to 35 km from the focus.

Neither Currie (Paper 16 and 1968) nor Zimmerman (1971) accepted that shatter cones are unique to meteorite impact-explosion structures. Certainly occurrences at Crooked Creek and Decaturville (Paper 25) are particularly difficult to reconcile with such a restriction. The relationship of shatter cones to actual meteorite material is established, but does not necessarily endorse such restriction. A crude example was described by Dietz (1963b) from the Arizona crater, but it could be a form of slickensides; Dietz (1972) records cones in association with meteoric iron at Kaalijaarv, Estonia; and Krinov (1971) records shatter cones of crude form in a small impact-fragmentation crater containing meteoric iron at Sikhote-Alin, Siberia. This last occurrence seems to confirm that immense shock pressures are not necessary for these cones to form, and to suggest that they can form under condition of impact fragmentation without explosion. Shatter cones have been produced in nuclear cratering experiments (Bunch and Quaide, 1968).

High Pressure Polymorphs of Silica

Under dynamic compression the framework silicates quartz and felspar undergo transition into dense phases with silicon and aluminum in coordination with six oxygen atoms (Stöffler, 1971a). Quartz (zero-pressure density = 2.65) undergoes irreversible transition to stishovite (4.28) in natural shocked rocks (Chao et al., 1962) and in recovery shock experiments (Milton and di Carli, 1963). An additional polymorph, coesite (2.93), may be formed in dynamic compression; and this has been used as an indicator for recognition of impact structures (Cohen et al., 1961). Synthesis of these polymorphs was achieved by Stishov and Popova (1961) and Coes (1953). They were subsequently found in nature at the Canyon Diablo Crater (Chao et al., 1960, 1962). Stishovite is only known from there and the Rieskessel (Paper 1), but coesite is known from a number of meteorite craters and cryptoexplosion structures. It has also been artificially produced in craters formed in experiments involving nuclear and TNT explosion. Full details of these polymorphs are given by Stöffler (1971a). Coesite and stishovite are supposed only to result from immense shock pressures, such as are believed to be unattainable in nature except in impact explosion. There has, however, been some argument about their field of stability. It has been suggested, that, given raised temperature, they can be formed at much lower pressures. It has also been suggested that they can form in zones of intense tectonic cataclasis, although they have proved difficult to locate in mylonite zones. Despite this, some occurrences such as those at Lake Mien (Paper 2), Lake Lappajarvi (Paper 4), and Agtö Island (p. **113**) could be in sites of extreme endogenic rather than exogenic deformation. Other occurrences, such as those at Brent (Paper 7) and Richat (Part V), are in structures of debatable origin—the second is not considered to be of extraterrestrial origin by Dietz et al. (1969) despite the recorded occurrence of coesite there. Though stishovite and coesite are undoubtedly indicators of shock and probably of extreme shock pressures, such shock and pressures could result from extreme geometrical confinement and multiplication effects (Jamieson, 1963; Nicolaysen, 1972) or from chemical explosion (Currie, Paper 16).

Shock Lamellations

The planar features known as shock lamellations in mineral grains have been described by Short (1966a,b), Chao (1967, 1968), Bunch et al. (1968), Horz (1968), Dence (1968), and Robertson et al. (1968). Stöffler has erected a series of progressive stages of shock metamorphism, correlating them with a pressure/temperature scale (Table 3). Dence subdivided the range equivalent to stages 0 and I in Stöffler's scale into 5 shock zones, based on the crystallography of successively developed sets of lamellae in quartz. French (1968b) also considered the progressive development of shock metamorphism, presenting the sequence

Stage 0 Crushing and granulation identical with that produced in natural metamorphism—no distinctive shock features
Stage 1 Planar features (shock lamellae) develop in quartz and felspar
Stage 2 Planar features in quartz—recrystallization in felspar
Stage 3 Recrystallization in quartz—extensive plastic deformation and recrystallization of felspar
Stage 4 Eutectic melting of contacts of deformed quartz and felspar grains
Stage 5 Concurrent melting of all constituent minerals to form a physically and chemically homogeneous glass
Stage 6 Melting, mixing, and accretion of glass with rock and mineral fragments to form heterogeneous, inclusion-bearing, and flow-banded glassy bodies

Bunch (1968) concluded that "the common occurrence in explosion craters of complex deformation bands, kink bands, strain induced triclinicity of alkali felspars, deformation twins with secondary slip, deformation lamellae, multiple planar fractures and fracture in felspars, all of which are associated with unique shock produced microstructures in quartz, suggest a high stress level rarely observed in rocks from tectonites."

Planar features in quartz may be deformation lamellae or cleavage fractures, but, regardless of their exact nature, appear to possess their own unique intrinsic properties, which have not been reproduced either in static experiments on quartz or in quartz from other natural occurrences (Carter, 1965). Basal deformation lamellae in quartz only, in Carter's view, could have been produced by shock waves from an impacting extraterrestrial body.

Table 3 Comparison of various classifications of nonporous quartzofelspathic rocks from natural-impact craters and from nuclear-explosion craters (after Stöffler, 1971b)

Shock effects in framework silicates	Shock stages[a]	Shock categories[b]	Pressure (Kb)	Postshock temp. (°C)	Shock zones[c]	Shock categories[d]	Shock categories[e]	Peak pressure[f] (approx. Kb)	Peak postshock temp.[f] (approx. °C)
Fractured quartz and felspar	0	1	100	81	i		B	100	100
Diaplectic quartz and felspar	I	2 3	200 380	168 382	ii iii iv v	i ?	C	350	300
Diaplectic quartz and felspar glasses	II	4	450	1080		ii	D	450	900
Fused felspar (vesiculated glass)	III	5 6	510	1600	vi?	iii iv	E	550-600	1300-1500
Inhomogeneous rock glasses	IV	7	600	2580	"melt zone"	v	F	> 800	> 3000
Silicate vapour	V								

[a]Stöffler, 1965, 1966, 1971b; von Englehardt and Stöffler, 1968.
[b]Chao, 1968.
[c]Dence, 1968.
[d]James, 1969.
[e]Short, 1969.
[f]Approximate.

Evidence of a circumstantial nature suggesting that shock lamellations are not restricted to impact processes was advanced by Greenwood (1967) who reported basal 0001 and other lamellae in tectonically deformed quartz from the Star Mine, Idaho. French (1968) questioned the conclusions of Greenwood; but Greenwood (1968) seems to have met this criticism, pointing out that Carter (1965) reported similar lamellations in tectonically deformed rocks. Greenwood suggested that the negative evidence for the unique status of quartz lamellae is tenuous, and urged more study—only 22 occurrences had been studied in detail up to that time.

Bunch, et al. compared natural lamellations with those produced in the Sedan nuclear cratering experiment. He invoked pressures of more than 150 kilobars to account for the deformations at the Rieskessel (Paper 1), Clearwater Lakes (Paper 14), and Manicouagan (Papers 15 and 16).

Impactites and Impact Melts

Vesicular impactite glasses produced from target rocks were described from Wabar, Arabia, and Henbury, Australia by Spencer (1933); and from Canyon Diablo, Arizona, by Nininger (1954). These are unquestionably impactites. Unlike tektites, such glass contains undigested, shocked fragments of the target rocks. They also contain traces of meteoric iron in the form of spheroids. The subject of impactite glasses is covered by Dence (1971). In large, recent impact craters, isolated glassy bodies or slag-like bombs are found. Many large cryptoexplosion structures are associated with similar developments or more ambiguous glass or recrystallized glass masses contained in mixed breccias. The latter commonly contain shocked and comminuted country rock. Further evidence of impact origin may be found in the form of nickel-iron spherules or geochemical anomalies in such glasses (as at El Aouelloul, Mauretania, and Monturaqui, Chile [McCall, 1977] and at the Rieskessel, Germany [Paper 1]). Such occurrences strengthen the case for impact origin without being conclusive, for native nickel-iron is known from terrestrial rocks (McCall, 1973, p. 235). Common associations of cryptoexplosion structures with peridotitic, picritic, and alkaline igneous rocks also engenders some doubt, for such rocks are likely to be associated with magnesium enrichment and magnesium/nickel geochemical anomalies. Nickel is a not uncommon geochemical trace element, and is not uncommon in laterite. Geochemical anomalies involving nickel are less valid evidence than the actual presence of nickel-iron. The process whereby grains are fused by shock with accompanying vaporization, and fusion glasses are produced, has been covered by Stöffler (1971b). Such fusion may be only partial; but the process, if carried to completion, produces an impact melt. Melting of a particular phase may be selective, depending on mutual melting point differences between phases, but eventually mixed glasses will be formed. Flow structures develop with more complete melting. Such effects, due to explosive shock whatever its origin, are difficult to distinguish from melting in thermal metamorphism or fluidization processes of endogenic origin. Both fusion glasses and pseudotachylite manifestations (dikes and veins of fine, dark fluidized rock crush with or without minor of magmatic components or secondary fusion glass) are common features of cryptoexplosion structures, but by no means unique to them.

Diaplectic Glasses

Diaplectic glasses appear to be peculiar to reversion from corresponding high pressure phases. They are solid-state glasses and have been artificially produced in both nuclear and shock-recovery experiments. In nature, they are characteristic of terrestrial, lunar, and meteoric-shocked rock, including rock from impact sites. Maskelynite is one such glass, derived from plagioclase felspar and well known from stony meteorites. The maskelynite from Clearwater Lakes and Manicouagan (Papers 14, 16) reproduces exactly the phenomena as the famous Shergotty shocked achondrite meteorite (McCall, 1973, pp. 145, 174). Maskelynite may require shock pressures in excess of 50 kilobars to form, but Currie (Papers 13 and 16) questions this, believing that maskelynite is neither an indicator of extreme shock pressure nor exclusively of extraterrestrially originated shock. Heated maskelynite reverts to crystalline plagioclase, with only slight preservation of lattice disorder. In this it is unlike a liquid-stage glass. Stöffler and Hornemann (1972) concluded that pressures in the order of 300 to 500 kilobars are required for thetomorphic glasses to form, values much higher than those suggested by Bunch et al. (1967) for lamellation to develop in quartz.

Breccias

Various types of monomict and polymict breccia—breccia composed of country rock more or less in situ but consisting of fragmented and rotated blocks; transported breccias; breccias injected, but not ejected into the air; and breccias so ejected (fall-back or throw-out breccias)—have been described from cryptoexplosion structures. Complex associations of melt and breccia fragments have been termed "suevite," "tagamite," and "karnaite." The theory of distribution of such breccias in the process of impact explosion is covered by von Engelhardt (1971).

In the case of the Brent Crater (Paper 7), distribution of the breccias is well established, but their origin remains controversial. Impact theory leans heavily on this occurrence, investigated in detail by drilling. Reconstructions based on analogy with Brent have been advanced for other structures. There is some doubt whether the glassy breccias formed in a neatly ordered manner, layer on layer, are suggested in such reconstructions, which generally involve the piercing of the layered array of breccias by a central uplift. One model involves projection upward from the focus of a spray of suevite—a model that fits the case of the Rieskessel but is not applicable to more eroded structures such as Manicouagan, where the suevite manifestly emanated, not from a central focus, but from dispersed-sheet, dike, ring-dike, and pipe-like feeders. Genetic connotation has been afforded to suevite and other such breccia types, but it is really only the internal features such as lamellations and thetomorphic glasses that are unique, in any sense, not the breccias, which are otherwise not intrinsically different from breccias of endogenic origin.

THE ALTERNATIVE CASE FOR ENDOGENIC ORIGIN OF CRYPTOEXPLOSION STRUCTURES

Guy-Bray (Paper 20) stated "If it can be shown that some as yet undocumented form of volcanism can reproduce features of shock metamorphism, then a real alternative to impact will be available. In the meantime it has been argued that there is a certain lack of intellectual satisfaction in a hypothetical cryptovolcanic process that must mimic so exactly the evolution of a meteorite crater." Dence (1968) remarked that "the contention that volcanism can generate strong shocks has no supporting evidence." French (1968, pp. 9, 10) makes similar observations. In these statements lies the entire *corpus delecti* of the controversy. We are concerned here with philosophy of science. Those engaged in the exact sciences operate first and foremost by theory and experiment; but geology is not an exact science; the geologist operates first and foremost from existing physical evidence and follows the Darwinian approach of multiple hypotheses to reconcile the facts observed. Since exact quantification or simulation of natural geologic processes by theory or experiment is unattainable, the geologist is aware that his limited view of the problem may lead him to a conclusion at variance with the dictates of theory and experiment. Nevertheless, his conclusion may be the correct one and fails to mesh with theory and experiment because unknown facts or influences are involved. In the case of cryptoexplosion (an ideal geologic subject for experimental simulation because it is instantaneous and thus eliminates the problem of simulating passage of immense spans of time), the geologist is not obliged to reject the contrary circumstantial evidence to be found—anomalous igneous rock associations, anomalous clusterings, anomalous groupings with endogenic structures incorporating explosive phases, or anomalous alignment along endogenic tectonic lineaments—simply because there is no fully defined endogenic process capable of generating the shock deformations. It is open to him to invoke some obscure agency such as chemical explosion as an alternative to impact. It is correct to state that no such process is fully defined, but the lack of such definition does not warrent rejection *de facto* of an obscure endogenic process. A quarter of a century ago, the ideas of Wegener, du Toit, and others concerning continental drift were held to ridicule because physicists could find no mechanism capable of effecting such immense lateral displacements. Yet it has lately been demonstrated that global strains are systematized in such a way that such movements must be accepted by reasonable scientists, though the geodynamic causative mechanism is still unknown. There is a direct parallel between this case and the case of cryptoexplosion.

It boils down to "Who bears the onus of proof?" (Paper 27). Neither is it reasonable, on the basis of the above philosophical argument, to reject an impact origin for cryptoexplosion—or to exclude endogenic origin. We must continue to entertain multiple hypotheses.

There is no direct evidence of meteoric involvement in any cryptic structure of class III, IV, or V, and very meagre evidence, if any, in all but two of the structures of class II (Monturaqui, Aouelloul; McCall, 1977). Likewise, the above summary of the indicators of impact involvement shows that whereas they may be strong indirect evidence of impact involvement, there is no *certainty* that these "indicators" cannot be formed by endogenic agencies. There is much circumstantial evidence suggesting that, however unlikely it may appear, this may be the case—that a number of cryptoexplosion structures are endogenic. This evidence is summarized below and covered in more detail in the text following this introduction.

Anomalies in Cryptoexplosion Structures

Evidence from the Kansas-Missouri-Illinois Axis (Paper 25). It is difficult to avoid the conclusion that the entire linear chain of cryptoexplosion structures is endogenic, controlled by a deep-seated master fracture in the basement. The repeated occurrence of the highly explosive magma types of the kimberlite-meimechite kindred along this axis is difficult to dismiss as a fortuitous coincidence (as is the recent discovery of kimberlite very close to the Middlesboro structure). Yet Crooked Creek and Decaturville are cryptoexplosion structures that show the supposed unique shock deformations and are components of this linear chain.

Tectonic Lineament Control in Swabia, South Africa, and Canada. Bucher (1963) remarked on the apparent control of distribution by a tectonic lineament in the first two cases. He also noted the association on the lineament of the rare melilitic and carbonatitic explosive magmatic developments in the first case. Currie (1965b) has suggested a similar tectonic control in Canada, and even claimed to have used it to predict new discoveries (Currie, 1968). There is some reason to suspect that in both Canada and the United States, cryptoexplosion structures are strung out along weakly developed anorogenic zones, which could have a physiographic reflection in upwarps, downwarps, or linear faults (McCall, 1963). However, there could be a preferred clustering of ancient cryptoexplosion structures on upwarps simply because they provide optimum situations for exposure, in contrast to downwarps which are likely to be mantled by young sediments or superficial cover. The sort of linear control recognized in the Kansas-Missouri-Illinois axis, in Swabia, and in South Africa is, however, only explicable by invoking endogenic causes for the cryptoexplosion. The grouping in the first case along a narrow master fracture and its intersection with intermittent, oblique second-order fractures could surely never be achieved by random meteorite strikes?

Anomalies in the Mauretanian Structures. Coesite is recorded from Richat (Part V), a structure rejected by Dietz et al. (1969) and by Short and Bunch (1968). Although it surely must exist, Dietz could not find the coesite-bearing breccia. The Mauretanian structures also appear as if strung out along a tectonic lineament, which could include some ring-dike developments and developments of other igneous rocks of apparently normal magmatic origin. If there is a structural lineament control, however, it is much less well defined than in the case of the Kansas-Missouri-Illinois line. The depth of two of the craters, of supposed impact origin, is anomalous in terms of impact explosion theory, and difficult to explain: one has a ring-dyke association of what appear to be normal magmatic igneous rocks; Richat has an association with developments of a rare, highly alkaline analcimolite igneous rock.

Deeply Eroded Structures. Certain of the deeply eroded cryptoexplosion structures in Sweden and the Agtö structure in Greenland display linear rather than circular patterns of fault and mylonite zones. A newly recorded occurrence on the Kola peninsula (Part II) also does not appear to have any association with a circular cryptoexplosion structure. The possibility that such structures are the deep roots of endogenic cryptoexplosion centers cannot be discounted. It is a fact that diatremes, terrestrial magmatic explosion pipes, lose their circularity and become linear in depth.

Anomalies in the Vredefort and Sudbury Structures. The focal situation of the Vredefort structure within the Witwatersrand Basin, which was formed long prior to the cryptoexplosive event and over a protracted period, is anomalous in terms of possible impact origin, being only explicable by invoking the remote eventuality of a hit right in the bullseye. The shatter cones at Vredefort appear to be developed in late intrusions, and there is evidence suggesting that there was more than one age of shatter coning. There is strong evidence of integration of Vredefort in a linear chain of igneous complexes, the remainder of which, including the Bushveld, Trompsberg, and the Great Dike occurrences, cannot conceivably be of meteoritic origin. At Sudbury, the

Onaping Formation, Chelmsford Formation, and the Irruptive itself are all anomalous: the vast volumes of igneous rocks outpoured cannot be explained in terms of any impact-magmatic theory (a weakness as significant as that attributed by Guy-Bray to endogenic cryptoexplosion—and, if one structure carrying all the unique shock manifestations is found wanting, the whole house of cards tumbles); there are also anomalies in the structural history of the Sudbury basin. The suggestion of multiphase development in the case of Vredefort (mentioned above) is to be matched in several other structures, the best example being the Rieskessel, where Bucher (1963) recognized two stages of development.

Central Uplifts. As in the case of lunar craters, the central uplift has always been a very difficult structural element to explain in cryptoexplosion structures. Central uplifts are typical of large structures, but occur in structures down to one or two kilometers diameter, for example the Steinheim Basin (Paper 2). Delayed isostatic rebound cannot be seriously entertained as an explanation for these uplifts: they appear to be an integral part of the catastrophic deformation, for the most part, contemporaneous and inseparable from the shock deformation. They range from immense diapir-like, inward and upward squeezes to punched up cylinders of crystalline rocks and angular, fault-bounded crystalline massifs. The latter could be structures formed much later than the shock deformation. The origin of such uplifts is critical to the origin of cryptoexplosion structures. Prompt rebound (cometary) fits in better with the observed facts and with experiment, but seems to be only applicable to large structures. Could cometary impact really produce structures as small as the Steinheim Basin? There is room for more specialist study of the uplifts at Flynn Creek, Gosses Bluff, Sierra Madera, Wells Creek, and Kentland from the viewpoint of modern structural geology.

Resemblance to Caldera, Cauldron, and Ring-Dyke Phenomena. The structural patterns of cryptoexplosion structures are in many cases not unlike those of ring complexes (roots of central volcanoes, which commonly involve cauldron subsidence) and surface calderas. This resemblance is particularly well marked at Vredefort, Clearwater Lakes, Manicouagan, Charlevoix, Mistastin Lake, Nicholson Lake, Lake St. Martin, Siljan, Richat, and Tenoumer. Features such as ring-dikes, pseudotachylite, granophyre screens, pecular silicic intrusive breccias, and ash-flow-like associations are shared.

Igneous Rock Associations. Rocks of the alkali basalt, picrite, melilitite, meimechite-kimberlite, and carbonatite associations appear with surprising regularity (for such rocks, of mantle provenance, are of rare terrestrial occurrence) in close association with cryptoexplosion structures—scattered haphazardly through cratons or strung along linear anorogenic zones traversing the cratons. They are erupted explosively due to a release of volatiles as they approach the surface and typically form diatremes.

Potassic alkaline rocks form a second group that reappears repeatedly in cryptoexplosion structures, particularly the large Canadian occurrences. They could be produced by melting of country rocks consequent on impact explosion, but only a highly selective mixture of country rocks could produce these surprisingly uniform doreite magmas. Alternatively, they could be either simple magmatic products of differentiation, or fluidized products of mixture of primary magma with melted country rock, or direct products of melting of country rock by endogenic agencies (in which case the compositional objection above again applies). There is some evidence from Canadian occurrences that the doreites have evolved from magma chamber differentiation. Potash alkaline rocks of magmatic origin (mantle derived) are characteristically erupted in small volumes and in a highly explosive manner.

There is need for a systematic investigation of the world wide distribution of cryptoexplosion structures alongside the distribution of all known igneous rocks of the above kindreds, together with fuller petrographic study and trace-element study of such rocks. Modern trace-element techniques should be able to discriminate between crustal melt components and primary magmatic components. Such study should include intrusive breccias, fluidized systems, and the suevite kindred. Another possibly profitable line of research is a time-based comparison of cryptoexplosion events and worldwide eruption of these rare magmatic suites. The lack of any record of the unique shock deformations in manifestly endogenic kimberlite and carbonatite associations is surprising, but the recent discovery of these shock effects in the Kola Peninsula (a classic province of alkaline central magmatism) suggests that more rigorous, systematic, and objective study of such associations could reveal that they *are* represented. These associations are certainly coupled with point-focus explosions of immense-scale.

Alteration Zones and Alkaline Metamorphism. The alteration zones and alkaline metamorphism recognized at Brent (Paper 7) and several other Canadian structures, including the New Quebec Crater (Currie, 1965a), have been attributed to the reactivation of ground water by heat produced in the impact. Currie (1971), on the other hand, invoked fenitization, a metosomatic effect operative at the margins of carbonatite intrusions (an alnoite dike certainly establishes that carbonatite is present at Brent, fortuitously or otherwise). Currie in various writings suggests that the alteration zones are incompatible with meteoritic origin.

Global Unbalance of Distribution. Lack of exploration of certain areas of the land surface of the globe must contribute to the unbalance observed in the distribution of cryptoexplosion structures in terms of expectation based on astronomical considerations, of more or less random global scattering. However, this cannot account for the entire lopsidedness of their distribution. Thirty-nine class IVa structures occur in Canada, Greenland, and the United States; thirteen occur in Russia; and there are only nineteen throughout the rest of the world. If, indeed, these structures are astroblemes, and since recent extensive mineral exploration all over Australia should have revealed any large structures on that continent, their paucity in Asia and Australia is anomalous. This distributional unbalance calls to mind the lopsided distribution of the lunar maria, never satisfactorily explained in terms of impact.

Arguments Against Endogenic Origin

Three features of cryptoexplosion structures that have been advanced against endogenic origin are (1) lack of feeders to the bodies of rocks of magmatic aspect, (2) waning of deformation in depth, and (3) isolation from other magmatic rocks or sites of volcanism. None of these is really valid. Highly mobile effusions rarely display feeders connecting them with their source in the crust. Both impact explosion and volcanic explosion will produce shock effects that wane in depth below the explosion focus. Magmatic explosions occur close to the surface, due to release of a gas phase at a critical pressure drop threshold as the magma approaches the surface. Point foci of such explosions will be at no great depth, like impact-explosion foci; and the patterns of the two will be very similar. Isolation is characteristic of centers of the rarer explosive types of volcanism that occur within cratonic areas. The fact that cryptoexplosion centers are commonly isolated in the same way is a reasonable argument for impact origin, but not a conclusive one. And the fact that cryptoexplosion is virtually restricted to cratonic areas of the globe is a possible counterargument.

The possibility of using LANDSAT imagery to test whether cryptoexplosion structures have a preferred spatial relationship to endogenic (tectonic) lineaments should not be overlooked, and such a relationship would be impossible to reconcile with impact theory. Another interesting line of research would be to look at the relationship of the very large isolated structures such as Sierra Madera, Gosses Bluff, Sudbury, and the Rieskessel to major preexisting sedimentary basins. The focal relationship to such a basin in the case of Vredefort is well established; it may be that other such relationships have been overlooked.

The requirement of an endogenic process capable of substituting for impact explosion in cryptoexplosion structures is stated by French (1968, p. 9). It must

(1) generate pressures of hundreds of kilobars near the surface of the earth;
(2) contain such pressures over areas of 10 square miles until they can be suddenly released as a shock wave;
(3) operate in confined spaces at shallow depth;
(4) release total energy in the order of 10^{24} kilobars (possibly up to 10^{28}-10^{30} kilobars);
(5) develop temperatures of 1500°C;
(6) involve rapid cooling and quenching; and
(7) produce these effects without any or with only sparse development of magma.

Item 2 above is questionable. It seems to ignore the possibility of adiabatic expansion with the sudden release of a gas phase as a magma approaches the surface and the possibility that the new gas phase in such a process might be a highly reactive one such as hydrogen. There would, in such a case, be no requirement for prolonged storage of high pressures. There have probably been much greater energy involvements in endogenic volcanic explosions than the well-known value estimated for Bezymyanny in Siberia (Gorshkov, 1969). The point-focus explosions in the carbonatite diatremes, productive of great arrays of closely spaced cone sheets, could well be of another, higher order of magnitude. Man, in his brief span on earth, may never have experienced the earth's more immense volcanic manifestations. We may be living in a geologic age of quiet in the eruptive context. Neither in impact explosion nor in volcanic explosion is there a requirement for prolonged storage of high pressures.

CONCLUSION

The above paragraphs show that there are arguments in favor of both impact-explosion and cryptovolcanic explosion as explanation of cryptoexplosion structures. It may be that somewhere in the record there is an insensible passage from exogenic to endogenic structures, but it is very difficult to separate two distinct groups, and rather we see a continuum of apparently congenetic structures. The present evidence is conflicting, theory and experiment favoring impact and circumstantial relationships favoring cryptovolcanism. An impact-explosion mechanism capable of producing cryptoexplosion structures is known and understood, but it could never have produced the immense-scale magmatic associations at Sudbury or the distribution pattern of the Illinois-

Missouri-Kansas axis. The magmatic associations strongly favor cryptovolcanism, but no process is known that could, endogenically, produce the recorded cryptoexplosion structures; though, there are intimations of what the nature of such a process might be.

A recent book on the disappearance of Tsar Nicholas II and his family concludes "The file on the Tsar remains open." The file on cryptoexplosion remains open and scientists will be well advised to adopt objective and nongenetic descriptive methods in future discussions of the subject, until the origin of cryptoexplosion in geology is discovered. The editor is advised (personal communication, A. J. A. Janse) that the neutral term "disturbance" is now in use in America, in accord with this thinking.

REFERENCES

Amstutz, G. C. 1965. Tectonic and Petrographic Observations on Polygonal Structures in Missouri. *New York Acad. Sci. Annals,* **123**(article 2):876–894.

Baldwin, R. B. 1949. *The Face of the Moon.* Chicago: University of Chicago Press.

Boon, J. D., and C. C. Albritton. 1937. Meteorite Scars in Ancient Rocks. *Field and Lab.* **5**:53–64.

Branca, W., and E. Fraas. 1905. Das kryptovulkanische Becken von Steinheim. *K. Preuss. Akad. Wiss. Abh.,* pp. 1–64, Berlin.

Bucher, W. H. 1936. Cryptovolcanic Structures in the United States. *Internat. Geol. Cong., 16th, Proc.* **2**:1055–1084.

Bucher, W. H. 1963. Cryptoexplosion Structures Caused from Within or Without the Earth (Astroblemes or Geoblemes?). *Am. Jour. Sci.,* **261**:567–649.

Bunch, T. E. 1968. Some Characteristics of Selected Minerals from Craters. In French and Short, 1968, pp. 413–432.

Bunch, T. E., and W. L. Quaide. 1968. Shatter Cones in the Danny Boy Nuclear Crater. In French and Short, 1968, p. 285.

Bunch, T. E.; A. J. Cohen; and M. R. Dence. 1968. Shock Induced Structural Disorder in Plagioclase and Quartz. In French and Short, 1968, pp. 509–518.

Bunch, T. E.; M. R. Dence; and A. J. Cohen. 1967. Natural Terrestrial Maskelynite. *Am. Mineralogist,* **52**:721–733.

Carter, N. L. 1965. Basal Quartz Deformation Lamellae—A Criterion for Recognition of Impactites. *Am. Jour. Sci.* **263**:786–806.

Chao, E. C. T. 1967. Shock Effects in Certain Rock Forming Minerals. *Science* **156**: 192–202.

Chao, E. C. T. 1968. Pressure and Temperature Histories of Impact Metamorphosed Rocks Based on Petrographic Observations. In French and Short, 1968, pp. 135–158.

Chao, E. C. T., J. J. Fahey, J. Littler, and D. J. Milton. 1962. Stishovite, SiO_2, a Very High Pressure New Mineral from Meteor Crater, Arizona. *Jour. Geophys. Research* **67**:419.

Chao, E. C. T.; E. M. Shoemaker; and B. M. Madsen. 1960. First Natural Occurrence of Coesite. *Science* **132**:220.

Coes, L. 1953. A New Dense Crystalline Silica. *Science* **118**:131.

Cohen, A. J., T. E. Bunch, and A. M. Reid. 1961. Coesite Discoveries Establish Cryptovolcanics as Fossil Meteorite Craters. *Science* **134**:1624–1625.

Currie, K. L. 1965a. The Geology of the New Quebec Crater. *Canadian Jour. Earth Sci.* **2**:141–160.

Currie, K. L. 1965b. Possible Analogues of Lunar Craters in the Canadian Shield. *New York Acad. Sci. Annals,* **123**(article 2):915–940.

Currie, K. L. 1968. Mistastin Lake, Labrador, a New Canadian Crater. *Nature* **222**: 776–777.

Currie, K. L. 1971. Origin of Igneous Rocks Associated with Shock Metamorphism as Suggested by Geochemical Investigations of Canadian Craters. *Jour. Geophys. Research* **76**:5575–5585.

Daly, R. A. 1947. The Vredefort Ring Structure of South Africa. *Jour. Geol.* **55**: 125–145.

Dence, M. R. 1968. Shock Zoning at Canadian Craters: Petrography and Structural Implications. In French and Short, 1968, pp. 169–184.

Dence, M. R. 1971. Impact Melts. *Jour. Geophys. Research* **76**:5552–5565.

Dietz, R. S. 1946. Geological Structures Possibly Related to Lunar Craters. *Popular Astron.* **54**:1–3.

Dietz, R. S. 1947. Meteoritic Impact Suggested by Orientation of Shatter Cones at the Kentland, Indiana, Disturbance. *Science* **105**:42–43.

Dietz, R. S. 1959. Shatter Cones in Cryptoexplosion Structures (Meteorite Impact?). *Jour. Geol.* **67**:496–507.

Dietz, R. S. 1960. Meteorite Impact Suggested by Shatter Cones in Rock. *Science,* **131**:1781–1784.

Dietz, R. S. 1961. Vredefort Ring Structure: Meteorite Impact Scar? *Jour. Geol.* **69**:499–516.

Dietz, R. S. 1963a. Astroblemes, Ancient Meteorite Impact Structures on Earth. In *The Solar System, Vol IV, Moon, Meteorites and Comets,* G. P. Kuiper, and B. Middlehurst, eds. Chicago: University of Chicago Press, pp. 285–300.

Dietz, R. S. 1963b. Cryptoexplosion Structures: A Discussion. *Am. Jour. Sci.* **261**:650–664.

Dietz, R. S. 1964. Sudbury Structure as an Astrobleme. *Jour. Geol.* **72**:412–434.

Dietz, R. S. 1968. Shatter Cones in Cryptoexplosion Structures. In French and Short, 1968, pp. 267–285.

Dietz, R. S. 1972. Shatter Cones (Shock Fractures) in Astroblemes. *Internat. Geol. Cong., 24th, Proc.* **15**:112–118.

Dietz, R. S.; R. Fudali; and W. A. Cassidy. 1969. Richat and Semsiyat Domes (Mauretania), Not Astroblemes. *Geol. Soc. America Bull.* **80**:1367–1372.

Engelhardt, W. von. 1971. Detrital Impact Formations. *Jour. Geophys. Research,* **76**:5566–5574.

Engelhardt, W. von, and D. Stöffler. 1968. Stages in Shock Metamorphism of Crystalline Rocks in the Ries Basin, Germany. In French and Short, 1968, pp. 159–168.

French, B. M. 1968. Shock Metamorphism as a Geological Process. In French and Short, 1968, pp. 1–17.

French, B. M. 1972. Production of Deep Melting by Large Scale Meteorite Impacts: The Sudbury Structure, Canada. *Internat. Geol. Cong. 24th, Proc.* **15**:125–132.

French, B. M., and N. M. Short, eds. 1968. *Shock Metamorphism of Natural Material.* Baltimore: Mono Book Corp.

Gorshkov, G. S. 1969. Gigantic Eruption of the Volcano Bezymyanny. *Bull. Volcano.* **20**:77–109.

Greenwood, W. R. 1967. Deformation Lamellae in Quartz. *Science* **158**:1180.

Greenwood, W. R. 1968. Deformation Lamellae in Quartz. *Science* **160**:907.

Horz, F. 1968. Statistical Measurements of Deformation Structures and Refractive Indices in Experimentally Shock-Loaded Quartz. In French and Short, 1968, pp. 243–254.

James, O. B. 1969. Shock and Thermal Metamorphism of Basalt Nuclear Explosion, Nevada Test Site. *Science* **166**:1615.

Jamieson, J. C. 1963. Possible Occurrence of Exceedingly High Pressures in Geological Processes. *Geol. Soc. America Bull.* **74**:1067–1070.

Johnson, G., and R. Talbot. 1964. Theoretical Study of Shock Wave Origin of Shatter Cones. Thesis, Wright-Patterson Air Force Base, Ohio.

Krinov, E. L. 1971. New Studies of the Sikhot-Alin Iron Meteorite Shower. *Meteoritics* **6**:27–138.

McCall, G. J. H. 1963. Are Cryptovolcanic Structures Due to Meteoric Impact? *Nature* **201**:251-254.

McCall, G. J. H. 1973. *Meteorites and Their Origins.* New York: Halstead Press.

McCall, G. J. H. 1977. *Meteorite Craters.* Benchmark Papers in Geology, vol. 36. Stroudsburg, Pa.: Dowden, Hutchinson & Ross.

McLaren, M. 1931. Lake Bosumtwi, Ashanti. *Geog. Jour.* **78**:270–276.

Manton, W. I. 1965. The Orientation and Origin of Shatter Cones in the Vredefort Ring. *New York Acad. Sci. Annals,* **123**(article 2):1017–1049.

Milton, D. J., and P. S. di Carli. 1963. Maskelynite Formation by Explosive Shock. *Science* **140**:670–671.

Nicolaysen, L. O. 1972. North American Cryptoexplosion Structures Interpreted as Diapirs which Obtain Relief from Strong Natural Confinement. *Geol. Soc. America Mem. 132,* pp. 605–620.

Nininger, H. H. 1954. Impactite Slag at Barringer Crater. *Am. Jour. Sci.* **252**:277-290.

Robertson, P. B.; M. R. Dence; and M. A. Vos. 1968. Deformation in Rock Forming Minerals from Canadian Craters. In French and Short, 1968, pp. 433–452.

Rohleder, H. P. T. 1933. The Steinheim Basin and Pretoria Salt Pan. *Geol. Mag.* **70**: 489–498.

Shoemaker, E. M., and R. E. Eggleton. 1964. Re-examination of the Stratigraphy and Structure of Sierra Madera, Texas. In *Astrogeologic Studies Annual Progress Report, Aug. 25, 1962–July 1, 1963, U.S. Geol. Survey Open File Report,* Part B, pp. 98–106.

Shoemaker, E. M.; R. E. Gault; and R. V. Lugn. 1962. Shelter Cones Formed by High Speed Impact in Dolomite. *U.S. Geol. Survey Prof. Paper 477,* pp. 424–426.

Short, N. M. 1966a. Effect of Shock Pressures from a Nuclear Explosion on Mechanical and Optical Properties of a Granodiorite. *Jour. Geophys. Research* **71**: 1195–1216.

Short, N. M. 1966b. Shock Processes in Geology. *Jour. Geol. Education* **14**:149–166.

Short, N. M. 1969. *Progressive Shock Metamorphism of Quartzite Ejecta from Sedan Nuclear Explosion Crater,* Goddard Preprint No. X622-69-537. Greenbelt, Md.: Goddard Space Flight Center.

Short, N. M., and T. E. Bunch. 1968. A World-Wide Inventory of Features Characteristic of Rocks Associated with Presumed Meteorite Impact Craters. In French and Short, 1968, pp. 255–266.

Spencer, L. J. 1933. Meteoritic Iron and Silica Glass from the Meteorite Craters of Henbury, Central Australia; and Wabar, Arabia. *Mineralog. Mag.* **23**:387–404.

Stishov, S. M., and S. V. Popova. 1961. New Dense Polymorphic Modification of Silica. *Geokhimiya* **10**:837–839.

Stöffler, D. 1965. Anzeichen besonderer mechanischer Beanspruchung an Mineralien der Kristallineinschlüsse des Suevits (Stosswellenmetamorphose). *Neues Jahrb. Mineralogie Monatsh.* **9-11**:350.

Stöffler, D. 1966. Zones of Impact Metamorphism in the Crystalline Rocks of the Nordlinger Ries Crater. *Contr. Mineralogy and Petrology* **16**:51.

Stöffler, D. 1971a. Coesite and Stishovite in Shocked Crystalline Rocks. *Jour. Geophys. Research* **76**:5474–5488.

Stöffler, D. 1971b. Progressive Metamorphism and Classification of Shocked and Brecciated Rocks from Impact Craters. *Jour. Geophys. Research* **76**:5541–5551.

Stöffler, D., and U. Hornemann. 1972. Quartz and Felspar Glasses Produced by Natural and Experimental Shock. *Meteoritics* **7**:371–394.

Stutzer, O. 1936. Die Talweitung von Kofels im Otztal (Tirol) als Meteorkrater. *Deutsch. Geol. Gesell. Zeitschr.* **88**:523–524.

Werner, E. 1904. Das Ries in der Schwabischen-Frankischen Alb. *Bl. Schwab. Albvieren* **16**:153–167.

Zimmerman, R. A. 1971. Formation of Shatter Cones in the Steinheim Basin. *Neues Jahrb. Mineralogie Monatsh.* **1**:19–25.

Part I
STRUCTURES OF CLASS IV IN EUROPE

Editor's Comments on Papers 1 Through 4

1 DENNIS
Ries Structure, Southern Germany, A Review

2 ENGELHARDT
Impact Structures in Europe

3 KRAUT and FRENCH
The Rochechouart Meteorite Impact Structure, France: Preliminary Geological Results

4 SVENSSON
Lappajärvi Structure, Finland: Morphology of an Eroded Impact Structure

Although the large number of cryptoexplosion structures now known in North America has tended to overshadow the European structures in recent years, three of the latter were historically the first to attract attention and to be attributed to impact. These are the Kofels structure, the Rieskessel, and the Steinheim Basin. Fifteen more are now known, one in France, five in Scandinavia, and ten in European Russia. The latter are considered in Part II, alongside those in Asiatic Russia. They are, however, listed in Table 4, which includes all known cryptoexplosion structures in Europe.

RIESKESSEL

No less than 57 papers dealing with the Rieskessel are listed by Freeberg (1966). It is difficult to do justice to such a vast production of literature. The most notable work is the study by Shoemaker and Chao (1961), which deals specifically with the impactite phenomena. The review paper by Dennis (Paper 1) is concise, eclectic, and up to date; valuable additional observations are included in Paper 2 by von Engelhardt, which covers all the structures in Europe.

The Rieskessel is a 22 km diameter crateriform structure in South

Table 4 Known cryptoexplosion structures in Europe

Paper no.	*Name*	*Country*	*Diameter* (km)	*Class*	*Age*
1, 2	Rieskessel	Germany	45/140[a]	IVa/4	Miocene
2	Steinheim Basin	Germany	2.5	IVa/2	Miocene
2	Kofels	Austria	4	IVa/5	Recent
3	Rochechouart	France	15	IVa/2	Jurassic
2	Lake Hummeln	Sweden	1.3	IVa/5	Early Cambrian
2	Lake Mien	Sweden	5	IVa/5	Precambrian?[b]
2	Lake Dellen	Sweden	15	IVa/1?	Precambrian?[b]
2	Lake Siljan	Sweden	40	IVa/3	Post-Ordovician
2, 4	Lake Lappajarvi	Finland	10	IVa/5	Precambrian?[b]
2, 5	Lake Janisjarvi	U.S.S.R.	10	IVa/5	Precambrian?[b]
5	Puchezh-Katunki	U.S.S.R.	80	IVa/2	Lower Jurassic
5	Boltyshka	U.S.S.R.	25	IVa/2	Lower Cretaceous
5	Rotmistrovka	U.S.S.R.	2	IVa/1	Cretaceous
5	Kaluga	U.S.S.R.	15	IVa/1	Upper Devonian
5	Mishinogorsk	U.S.S.R.	2.5	IVa/1	Late Palaeozoic
5	Ilumets	U.S.S.R.	5	IVa/1	Early Palaeozoic
5	Kamensk	U.S.S.R.	25	IVa/2	Paleogene
5	Gusev	U.S.S.R.	8	IVa/1	Paleogene
	Obolon	U.S.S.R.	12	IVa/2	Jurassic

[a]Structure/ejectamenta.
[b]Though these structures appear to be geologically ancient no upper limit can be placed on their age.

Germany, surrounded by at least two outer rings. It occurs within the Swabian volcanic province, which includes the older Kaiserstuhl and Hegau phases as well as the younger Urach tuffisite pipes. The Steinheim Basin is closely associated with it, and all form the Swabian lineament, the two cryptoexplosion structures being sited at the northeast end. The Urach pipes are of a melilitic alkaline picrite basalt; the Kaiserstuhl, where the lineament meets the Rhine Graben, is a carbonatite complex. Von Engelhardt (Paper 2) has recorded the interesting fact that a single block of ejectamenta has been recovered 140 km from the focus. The asymmetry of the structure is not entirely of erosional origin, and von Engelhardt attributes much of it to oblique, low-angle impact from the northwest.

Despite the coincidence of fission track ages, the *moldavite* tektites are of markedly different composition and character from the Rieskessel glasses, and it is extremely difficult to support congenesis (Chao, 1963).

The Rieskessel, like the Brent Crater in Canada (Paper 7), is critical to the astrobleme theory. Certain relationships are more in accord with endogenic than exogenic origin. Bucher (1963) advanced arguments in favor of an endogenic origin, based on the relationship of the Rieskessel, Steinheim Basin, and Urach "vulkanembryonnen" to the regional Swabian tectonic lineament and its junction with another major northwest trending lineament. Geophysical evidence of an underlying dense magmatic body, evidence of hot-spring activity in the central uplift at Steinheim, the close relationship to diatremes of kimberlitic (i.e., highly explosive) affinities, and evidence of multistage development derived from the sedimentary infill have all been advanced to discount meteoritic origin. The Rieskessel remains controversial, though widely believed to be of meteoritic origin.

STEINHEIM BASIN

The Steinheim Basin was described by Bucher (1963), who advanced a cryptovolcanic explanation, following Branca and Fraas (1905), who coined the term. Von Engelhardt (Paper 2) favored impact origin, and believed that the evidence in the fossil record favors contemporaneiety with the larger Rieskessel structure nearby.

The Steinheim Basin is a ring-shaped, 2.5 km diameter depression with its floor situated 80 m above the level of the plateau of the Swabian Alps. There is a low central hill within the basin, the Klosterberg, which is 40 m high and partly enveloped in hot-spring sinter. Disturbance of the rocks is intense in the basin, but lacking outside. The older rocks have been carried up, by some 150 m displacement in the central hill, as disordered blocks with steep dips, a confused kneading of heterogeneous units. The structure has been dated by the infill of Miocene lake beds, which dip inward at gentle to moderate angles, the dips being a combination of initial dips and sliding during sedimentation. A ring-shaped collapse combined with uplift was suggested by Bucher, who cited evidence that the hill formed slightly later than the shallow basin (Kranz and Gottschick, 1925). Intense brecciation of Upper Jurassic limestone occurs along the outer edge of the disturbance, testifying to the explosive nature of the originating process, the beds being broken and tilted in diverse directions, some being vertical. Innumerable fractures and intense shattering, which has obliterated all traces of original bedding, are features of these rocks. There is no trace of volcanic material, but hot springs issued at the center and eastern edge of the structure. Their products, aragonite, opal and dolomite, and their physiological effect on fresh water snails, were repeated twice in the stratigraphic sequence of lake sediments.

KOFELS

Von Engelhardt (Paper 2) also favors impact origin for the Kofels structure, which has baffled geologists for more than a century. Situated in Otzdal, 60 km south of Innsbruck, it is a basin 4 km across cut in the west side of a glacial valley 1500 m deep. Though comparatively recent, it lacks the physiographic form of an explosion crater. Fission-track data suggest an age similar to that of the Arizona crater. Intensely fractured granite gneiss in giant blocks, and vesicular pumice-like glass in a fissure in the gneiss (forming a dike), occur here. The glass is uniform or contrastingly banded and contains rock fragments and also fragments of dark breccia. Quartz grains with lechatelierite, felspars with rims of felspar glass, diaplectic quartz and felspar glass, and shock lamellations in quartz are all recorded (Storzer et al., 1971; Milton, 1965). The glass appears to be derived by incomplete fusion of the country rock, during an episode of high temperature and rapid fusion according to Kurat and Richter (1968), who found the chemistry to be compatible with such an origin. Volcanic interpretations have been advanced (Hammer, 1923, 1924; Ampferer, 1929; Pichler, 1863); Preuss (1971) invoked an immense landslip producing pseudotachylite; impact origin has been favored by Seuss (1936), Stutzer (1936), and Kurat and Richter (1968) as well as von Engelhardt.

ROCHECHOUART

The Rochechouart structure is discussed by Kraut and French (Paper 3) and by von Engelhardt (Paper 2). It is an ancient structure 15 km in diameter situated 50 km west of Limoges, France. It has no topographic expression to indicate the presence of a circular structure and lies amid gently rolling farmland. Set in a basement of Hercynian granite, gneiss, schist, porphyry, and microgranite dikes, overlain locally by Triassic and Cretaceous sediments, the Rochechouart structure includes shocked and brecciated autochthonous basement rocks (monomict breccias, in situ); vesicular flow-textured melt rocks; red welded breccia carrying shocked enclaves of basement rocks set in a welded fragmental matrix; polymict lithic breccia consisting of shock metamorphosed and shatter-coned fragments, but lacking glass; and a highest unit of glassy, poorly sorted breccia of shocked basement enclaves and heterogeneous fladen set in a suevite matrix. The present ground surface is believed to coincide with the crater floor and the structure is believed to possess a central uplift of autochthonous breccia 4 km in diameter. The shock lamellations in the quartz close to the focus include those of type D (Robertson et al., 1968). Kraut and French believe that the structure is an eroded meteorite crater of Jurassic age.

SWEDISH STRUCTURES

Von Engelhardt has described the four Swedish structures in Paper 2. Lake Hummeln has also been described by Svensson (1968b). Situated 15 km northwest of Okarshamn, it does not have circular shape, being an elongated depression of rounded form, measuring 1300 X 1000 m. It is set amid Gothian (Precambrian) granite. Cambrian sandstone and Ordovician limestone are represented among boulders on the southern shore of the lake. These are believed to be glacially dredged from the bottom. Breccias consisting of granite fragments outcrop on the lake shore, grading outwards into undisturbed rock. Joints of both radial and tangential orientation relative to the depression show small displacements, and breccias occur in zones parallel to fault planes. Petrographic shock features are not recognized. An early Cambrian age is suggested. Von Engelhardt and Svensson believe it to be a meteoritic structure.

Lake Mien (Paper 2) has also been described by Stanfors (1969). Situated 30 km north of Karlshamn, it is a fluvioglacial depression with few outcrops. Granite is exposed on the north shore, and peculiar rocks, discovered among boulders south of the lake, were originally supposed to indicate a rhyolite volcano. These acid rocks were later found to contain coesite and to resemble suevite (Svensson and Wickman, 1965). The structure is rhombic, not circular. The basement rocks are Precambrian granites and gneisses, cut by diabase dikes trending north-northeast. The shape of the lake appears to be controlled by linear faults. Rhyolite boulders include:

- black, glassy, fragile, obsidian-like rocks, with sparse inclusions
- black to grey glassy to finely crystalline lava-like rocks, with inclusions of felspar, quartz, and Archean rock fragments
- tuff-like breccias composed of fragments of the same character, albite, and pieces of the lava-like rock, all set in a porous brown to yellow matrix
- breccias of Archean gneiss and granite fragments, cemented by finely crystalline quartz and devoid of lava fragments.

Drilling has revealed a 6–8 m thick zone of lava-like rock, vesicular with zeolite infillings and localized flow texture in the matrix. The tuff-like breccia resembles suevite in its upper part, and carries glassy black fladen (Paper 1). The lower part consists of basement fragments set in a brown earthy matrix. There are no shatter cones. The structure has been subjected to deep erosion. It is situated 80 km north of the only known younger volcanic center in Fennoscandia.

According to von Engelhardt (Paper 2) there are suevite bombs aerodynamically shaped. The glass in them displays fluidal texture. The coesite occurs in shocked fragments within the suevite. Planar deformations in quartz, diaplectic quartz, and heterogeneous glass are all recognized. Von Engelhardt believes that the original crater has been modified and enlarged by erosion. He favors meteoritic origin, as does Svensson (1969); but Stanfors (1969) did not favor any particular origin, considering that the structure might be cryptovolcanic or meteoritic.

LAKE DELLEN

Lake Dellen, situated 300 km north of Stockholm, consists of two lakes with a narrow neck between them (Paper 2). It is a circular, deeply eroded structure, 15 km in diameter. The basement is Precambrian gneiss and granodiorite. Dellenites, defined as acid lavas (Svenonius, 1895), outcrop on the north shore. More recently, rocks of suevite character are reported, with aerodynamically shaped bombs. The suevite contains fragments of devitrified glass, gneiss, quartz, and mineral grains. The aphanitic melt rocks display fluidal texture and contain rock fragments. The bombs are composed of brownish devitrified glass, displaying many schlieren and including vesicles and mineral fragments (mainly quartz with planar shock lamellations; Svensson, 1968a). The age of the structure, which has only a crude circularity, could be anything from Precambrian upward.

LAKE SILJAN

Lake Siljan is situated 270 km north of Stockholm. It is a deeply eroded ring complex 40 km across. The ring, partly defined by faults, is outlined by outcrops of Ordovician rocks resting in a partly disturbed state on an older surface of weathered Precambrian granite. The sedimentary layers are locally uplifed and overturned. Granite forms a circular central massif. The country around the structure is underlain by granite porphyry, Jötnian sandstone, and older granites and migmatites. The ring has been interpreted as a large caldera (Asklund, 1936; Rutten, 1966), but impact origin was suggested by Fredriksson and Wickman

(1963), and is favored by von Engelhardt. Shatter cones are recorded from the central granite close to the focus (Dietz, 1970), and also quartz displaying shock lamellations. The Ordovician rocks are involved in the brecciation and are predisturbance—the disturbance is therefore post-Ordovician. Stam (1967) has suggested that this structure is far more complex than has been realized and deserves much more detailed study.

LAKE LAPPAJARVI

The last structure is in Finland and is described by Svennson in Paper 4, as well as by von Engelhardt in Paper 2. Lake Lappajarvi is situated 320 km north of Helsinki in a flat terrain of drumlins and fluvioglacial deposits. The structure, 10 km in diameter, is deeply eroded. The country rocks are Precambrian granites, grandiorites, and gneisses. Holmberg (1858) recorded a peculiar rock form here, "karnäite," believed to be pyroclastic and mainly of agglomerate to tuff character, pure lava being rare. Eskola (1921) reported rhyodacitic composition. Quartz in fragments in this rock displays shock lamellation (Svensson, 1968b). Coesite, maskelynite, and diaplectic glass were discovered in boulders that originated in this structure (Lehtinen, 1970). The karnäite layer has a roughly circular outcrop 6 km in diameter: there are two varieties, one containing microlites of sanidine, plagioclase, cordierite, orthopyroxene, and quartz set in a glassy groundmass (Lehtinen, 1970), the other a hard lava-like rock containing bed-rock fragments. Suevite (fall-out breccia) is also recorded, containing fragments of bed rock, mineral grains, and fluidal glass altered to zeolite and devitrified to quartz, which shows planar deformations. Diaplectic glass and felspar glass are also recorded. The presence of shock lamellations in quartz that is a devitrification product is highly anomalous, suggesting, as it does, a period of time (to allow vitrification) between original formation of the glass and occurrence of the shock. Coesite occurs in the suevite breccias and in granite fragments from weakly shocked breccias, devoid of glass. Dence (1968) regarded this structure as meteoritic, as did Svensson and von Engelhardt. If it is an impact-explosion structure, the focus would have to be in the northern part of the lake. Oja (1978) provides a recent account and illustrations of this structure.

REFERENCES

Ampferer, O. 1929. Die geologischen Hypothisen uber die Formung de Talraums zwischen Umhausen un Langenfeld im Otztale. *Akad. Wiss. Wien., Math.-Naturw. Kl. Sitz. Ber.* abt. 1, **148**:401.

Asklund, B. 1936. Diskussioninlagg med anledning av P. Thorslunds foredrag "Nagra stratigrafiska och tektoniska ron fran Siljansromradet." *Geol. Fören. Stockholm Förh.* **58**:385-387.

Branca, W., and E. Fraas. 1905. Das Kryptovulkanische Becken von Steinheim. *K. Preuss. Akad. Wiss. Abh.*, pp. 1-64, Berlin.

Bucher, W. H. 1963. Cryptoexplosion Structures Caused from Within or Without the Earth (Astroblemes or Geoblemes?). *Am. Jour. Sci.* **261**:567-649.

Chao, E. C. T. 1963. The Petrographic and Chemical Characteristics of Tektites. In *Tektites,* J. A. O'Keefe, ed. Chicago: University of Chicago Press, pp. 51-94.

Dence, M. R. 1968. Shock Zoning at Canadian Craters: Petrography and Structural Implications. In *Shock Metamorphism of Natural Materials,* B. M. French and N. M. Short, eds. Baltimore, Md.: Mono Book Corp., pp. 169-184.

Dietz, R. S. 1970. Siljan Ring, Sweden. *Meteoritics* **5**:192.

Eskola, P. 1921. On Volcanic Rocks in Lake Janisjarvi in Eastern Finland. *Finlande Comm. Géol. Bull.* **55**:1-13.

Fredriksson, K., and F. E. Wickman. 1963. Meteoriter. *Svensk Naturvetenskap* **16**: 121-157.

Freeberg, J. H. 1966, 1969. Terrestrial Impact Structures: A Bibliography. *U.S. Geol. Survey Bull. 1220,* 91 pp., *1320,* 39 pp.

Hammer, W. 1923. Uber das vorkommen jungvulkanische gesteine im Ötztale, (Tirol), und ihr alter. *Akad. Wiss. Wien. Math.-Naturw. Kl. Sitz. Ber.* abt. 1, **102**:329-342.

Hammer, W. 1924. Ein vorkommen jungvulkanischens Gesteins in der Tiroler Zentral-Alpen. *Zeitschr. Vulkanol.* **8**:238-243.

Holmberg, H. J. 1858. Materialer till Finlands geognosi: Bidrag till Finlands Naturekannedom etnografi och statitik. *Finska Vetenskapssocieteten* **4**:118.

Kranz, W., and F. Gottschick. 1925. Zur Tekonic des Steinheimer Beckens. *Deutsch. Geol. Gesell. Zeitschr.* **77**:37-65.

Kurat, G., and W. Richter. 1968. Ein alkali feldspatglas im Impaktit von Kofels, Tirol. *Naturwissenschaften* **55**:490.

Lehtinen, M. 1970. New Evidence for an Impact Origin of Lake Lappajarvi, Western Finland. *Geol. Soc. Finland Bull.* **42**:89-93.

Milton, D. J. 1965. Fused Rock from Kofels, Tyrol. *Tschermaks Mineralog. u. Petrog. Mitt.* **9**:86-94.

Oja, H. 1978. *Tulipalloja Taivaala.* Helsinki: Tahtitieteellinen Yhdistys Ursary., 208 pp.

Pichler, A. 1863. Zur geognosie Tirols, II, Die Vulkanischen Reste von Kofels. *Jahrb. K. K. Geol. Reichsanstalt Wien* **13**:591-593.

Preuss, E. 1971. Uber den Bimstein von Kofels/Tirol. *Fortschr. Mineralogie* **49**:70.

Robertson, P. B.; M. R. Dence; and M. A. Vos. 1968. Deformation in Rock Forming Minerals from Canadian Craters. In *Shock Metamorphism of Natural Materials,* B. M. French and N. M. Short, eds. Baltimore, Md.: Mono Book Corp., pp. 433-452.

Rutten, M. M. 1966. The Siljan Ring of the Palaeozoic, Central Sweden: A Posthumous Ring Complex of a Late Precambrian Dala Porphyries Caldera. *Geologie en Mijnbouw* **45**:125-136.

Shoemaker, E. M., and E. C. T. Chao. 1961. New Evidence for Impact Origin of the Ries Basin. *Jour. Geophys. Research* **66**:3371-3388.

Stam, J. C. 1967. On the Geology and Tectonics of the Lake Siljan Area, Central Sweden. *Geologie en Mijnbouw* **46**:467-481.

Stanfors, R. 1969. Lake Mien, An Astrobleme or Volcanotectonic Structure. *Geol. Fören. Stockholm Förh.* **91**:173-186.

Storzer, D., P. Horn, and B. Kleinmann. 1971. The Age and Origin of the Fused Rock from the Kofels Structure, Austria (abstr.) *Meteoritics* **6**:319-320.

Stutzer, O. 1936. Die Talweitung von Kofels im Otztal (Tirol) als Meteorkrater. *Deutsch. Geol. Gesell. Zeitschr.* **88**:523-524.

Suess, F. E. 1936. Der Meteor-Krater von Kofels bei Umhausen im Otztale, Tirol. *Neues Jahrb. Mineralogie Beil.* abt. A, **72**:98-155.

Svenonius, F. 1895. In Praktikst geologiska undersokiningar inom Gefleborgs lan. (Blomberg, A.), *Sveriges Geol. Undersikning Ser. Ac.* **5**:54-59.

Svensson, N. B. 1969. Lake Mien, Southern Sweden, A Possible Astrobleme. *Geol. Fören. Stockholm Förh.* **91**:73-86.

Svensson, N. B. 1968a. The Dellen Lakes, A Probable Meteorite Impact in Central Sweden. *Geol. Fören. Stockholm Förh.* **90**:309-316.

Svensson, N. B. 1968b. Lake Lappajarvi, Central Finland: A Possible Meteorite Impact Structure. *Nature* **217**:438.

Svensson, N. B., and F. E. Wickman. 1965. Coesite from Lake Mien, Southern Sweden. *Nature* **205**:1202-1203.

1

Reprinted from *Jour. Geophys. Research* **76**:5394–5406 (1971)

Ries Structure, Southern Germany, A Review[1]

John G. Dennis

Department of Geology
California State College, Long Beach 90801

In southern Germany a prominent north-northwest–facing scarp is interrupted by a roughly circular depression known as the Ries basin. Surface topography and geophysical investigations reveal a complicated crater geometry consisting of an inner zone roughly 8 km in diameter with only modest surface expression, followed outward by a concentric zone of flat to hummocky relief, which itself is bordered by the above-mentioned scarp to the south, and by a less prominent rim to the north. The diameter of the entire depression is 22–24 km. The present crater floor is a flat plain as a result of crater filling by fallback breccias as well as fresh-water lake and alluvial deposits. The distribution, structure, texture, composition, and mechanism of deposition of the various ejecta are important clues for reconstructing the cratering mechanism. The undisturbed crystalline basement of the Ries originally was overlain by a sequence of Mesozoic sedimentary rocks, roughly 600 meters thick. The ejecta consist of a variety of sedimentary breccias (*Bunte Trümmermassen*) and crystalline breccias (monomict breccias, polymict breccias, suevite) derived from these preexisting rocks. Volumetrically the sedimentary breccias dominate, forming a discontinuous ejecta blanket especially in the south. Most crystalline breccias are confined to the crater itself or to the proximity of the rim, though sporadic patches occur as far away as one crater radius. Ejecta block size decreases with increasing distance from the crater. In general, where discernible the pre-Ries stratigraphic sequence tends to be reversed within the ejecta blanket. Suevite is the single most important breccia derived from basement rocks. It contains a variety of crystalline rock types in all stages of shock metamorphism. Its most characteristic components are aerodynamically shaped glass bombs. Detailed petrographic investigations indicate peak shock pressures in excess of 600 kb and 2000°C. Suevite is responsible for strong local negative magnetic anomalies. Oriented samples show reversed remanent magnetism. Gravity measurements have revealed a strong negative anomaly corresponding to about 7×10^{6} g mass deficiency. The Ries event has been dated by K-Ar and fission track techniques at 14.8 ± 0.7 m.y. and 14.0 ± 0.6 m.y., respectively, and these dates are compatible with stratigraphic evidence (Upper Tortonian).

The amount and variety of geological and geophysical evidence actually and potentially available for reconstruction of a cratering event makes the Ries an outstanding geological structure. The ejecta deposits in particular are unique in their structural state, composition, extent, and accessibility. The Ries has been and continues to be the subject of extensive field and laboratory investigations, most recently documented in a symposium [*Preuss and Schmidt-Kaler*, 1969; *Stöffler*, 1971; *von Engelhardt*, 1971]. The present review intends to make the most important results of Ries investigations accessible to a wider readership.

[1] Lunar Science Institute Contribution 52.

Regional Geological Setting

The Ries basin is located on the southern rim of the Südwestdeutsche Gross-Scholle [*Carlé*, 1955]. This platform, built on Variscan (Paleozoic and older) basement is bounded by the Rhine graben to the west, and by the Franconian line, a fracture zone along the western border of the Bohemian massif (Moldanubicum) to the east (Figure 1). Volcanic areas of Oligocene to Pliocene age mark the margin of the platform to the north and southwest. Along its entire southern rim the platform dips gently south toward the Molasse basin, which borders the Alps. Table 1 shows a representative stratigraphic succession of the platform cover in the Ries area.

The south-southeast dipping Malmian limestone forms a cuesta-type mountain chain, the

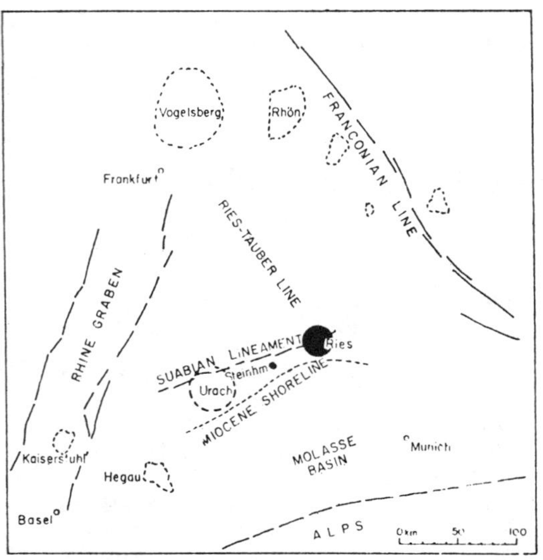

Fig. 1. Tectonic setting of the Ries. Shading indicates Miocene volcanic areas. Based on *Dehm* [1962], Fig. 2.

Alb, with a north-northwest-facing scarp line that generally follows the Malm-Dogger contact as shown on Figure 2. A fracture zone known as the Suabian lineament [*Seibold*, 1951] follows the Alb scarp, and the hinge line of the Molasse trough is parallel with the same trend farther south (Figure 1).

Advocates of volcanic origin of the Ries have drawn attention to volcanic activity in southwest Germany within a comparatively narrow time span that would have terminated with the Ries event [e.g. *Dehm*, 1962; *Bucher*, 1963]. The Kaiserstuhl and Hegau volcanism to the southwest is oldest, followed by the Urach tuffisite pipes, some of which come close to the age of the Ries, *Dehm* [1962], *Carlé* [in *von Engelhardt*, 1962], and *Bucher* [1963] also attach significance to the straight alignment of the Ries and Steinheim basins with the Urach volcanic area (Figure 1). The Kaiserstuhl and Hegau areas are approximately along the same trend (Figure 1). The volcanic products are confined to phonolites, melilite basalts, and alcali olivine basalts and thus differ significantly from Ries products in their bulk chemistry.

LOCAL GEOLOGICAL SETTING

'Ries' is the name of a flat, circular topographic depression some 22–24 km in diameter that locally takes the place of the Alb scarp and its foothills (Figure 2). The Ries plain is

TABLE 1. Stratigraphic Succession in the Ries Area

System	Series	Lithology	Approximate Thickness, meters
Jurassic	Malmian (Weissjura)	Thick bedded to massive (reefy) white limestone; thin-bedded and platy intercalations near the top.	200
		Well-bedded, marly in part.	100
	Dogger (Braunjura)	Oolithic limestones, sandstones, shales, calcareous sandstones, and oolithic iron ore beds.	50
		Bluish gray shales (Opalinus shales) with ironstone geodes.	90
	Lias (Schwarzjura)	Marls, shales, bituminous shales, phosporitic limestones, and sandstones.	Approx. 30
Triassic (Upper and Middle)	Keuper	Red marls.	25
		Sandstones, some shales.	Approx. 220
		Shales, sandstones.	Approx. 220
	Muschelkalk	Calcareous sandstones.	50
Lower Triassic and Permian		Continental facies, transgressive on basement (uncertain).	0–200
Basement (Moldanubicum)		Predominantly granitic and granodioritic plutonic rocks, gneisses, calc-silicate rocks, and amphibolites.	

Simplified after *Schmidt-Kaler* [1969] and R. Hüttner (personal communication, 1971).

at an average elevation of 420 meters above sea level; the surrounding hills reach elevations of over 600 meters to the south, where they are in Malmian limestones of the Alb, and about 500 meters to the north, where they are sculptured mainly in softer middle and lower Jurassic rocks. It is noteworthy that the Alb scarp has retreated very little since the cratering event and that at least part of the drainage network of the area has become re-established along pre-Ries valleys [*Hüttner,* 1958].

The basin rim, though evident in general outline, cannot be unequivocally defined in detail. One convention takes the 450-meter elevation contour as the outline of the Ries perimeter (see Figure 7). This outline is clearly somewhat polygonal and reflects certain known fracture systems [*von Engelhardt,* 1967*a*].

The Ries basin is filled by Tortonian and Sarmatian (late Miocene) lake deposits, with marl and clay uppermost, that are covered in part by alluvial and Pleistocene periglacial deposits. The maximum known thickness of crater fill is over 300 meters [*Reich and Horrix,* 1955]. Some elevations that were shoals in the former Ries crater lake are capped by fresh water limestones.

The present physiographic crater floor is truly flat only in its inner region. A broken alignment of low hills, few exceeding 50 meters in height above the plain, passes through Langenmühle, Nördlingen, and the Wennenberg (Figure 2) in the shape of a horseshoe open to the north-northeast and marks the 'inner crater.' Between this inner loop of low hills and the Ries perimeter, the basin surface is more irregular and is even hilly in part; these topographic irregularities are due to deposits of very unusual breccias (Figure 2).

RIES EJECTA

Sedimentary breccias (Bunte Trümmermassen). These breccias include material derived from every rock type known to have existed in the site of the Ries prior to the cratering event, from Variscan basement to Tertiary deposits; however, their predominant (99%) and characteristic components are derived from the Mesozoic sedimentary sequence (Table 1). On the basis of block size, two categories can be distinguished: (1) 'Bunte Breccie,' the dominant type of the sedimentary breccias; and (2) allochthonous, giant blocks of Malmian limestone large enough to be mapped separately. These blocks are true klippes [*Dennis,* 1967]. The distinction of category (2) is one of convenience only and depends entirely on the scale of the map. *Hüttner* [1969] suggests 25 meters as minimum diameter for maps 1:25,000. No klippes are shown separately on Figure 2.

'Bunte Breccie,' by definition [*Hüttner,* 1969], is the name given to all those components of the sedimentary breccias that cannot be mapped separately as klippes. On the mesoscopic scale the matrix is clayey, locally more or less sandy; it consists of material from the more readily disaggregated rocks of the succession (Lower and Middle Jurassic and Triassic). Where Bunte Breccie overlies soft or weathered rocks, the rocks have been incorporated in the matrix [*Hüttner,* 1969].

This breccia is extremely heterogeneous, both in rock type and in clast size. Malmian limestone is ubiquitous; other lithologies appear to be more irregularly distributed. Size distribution is irregular for any given outcrop, but larger sizes diminish in frequency away from the Ries [*Hüttner,* 1958, 1969]. W. Schneider [in *von Engelhardt et al.,* 1969] has analyzed the mineral composition of samples of Bunte Breccie and concludes that stratigraphically older constituents tend to be statistically more common in the upper portions of Bunte Breccie. Schneider similarly demonstrates that older constituents predominate near the crater rim, and that proportions of younger constituents increase away from the Ries. This suggests inversion of the primary succession. [See *Gault et al.,* 1968.]

On a mesoscopic scale the matrix occupies a larger volume than the included fragments in most exposures; the heterogeneous distribution of fragments, as well as poor sorting of grain sizes in the matrix [W. Schneider, in *von Engelhardt et al.,* 1969], suggests intensive and extensive mixing. Some individual fragments of the softer lithologies are deformed in fluidal style, but no preferred alignment has been discovered, except in part at contacts with autochthonous bedrock and at the top, where de-

Fig. 2. *(Opposite)* Geological outline map of the Ries. Based on geological map of the Ries [*Hüttner et al.,* 1969].

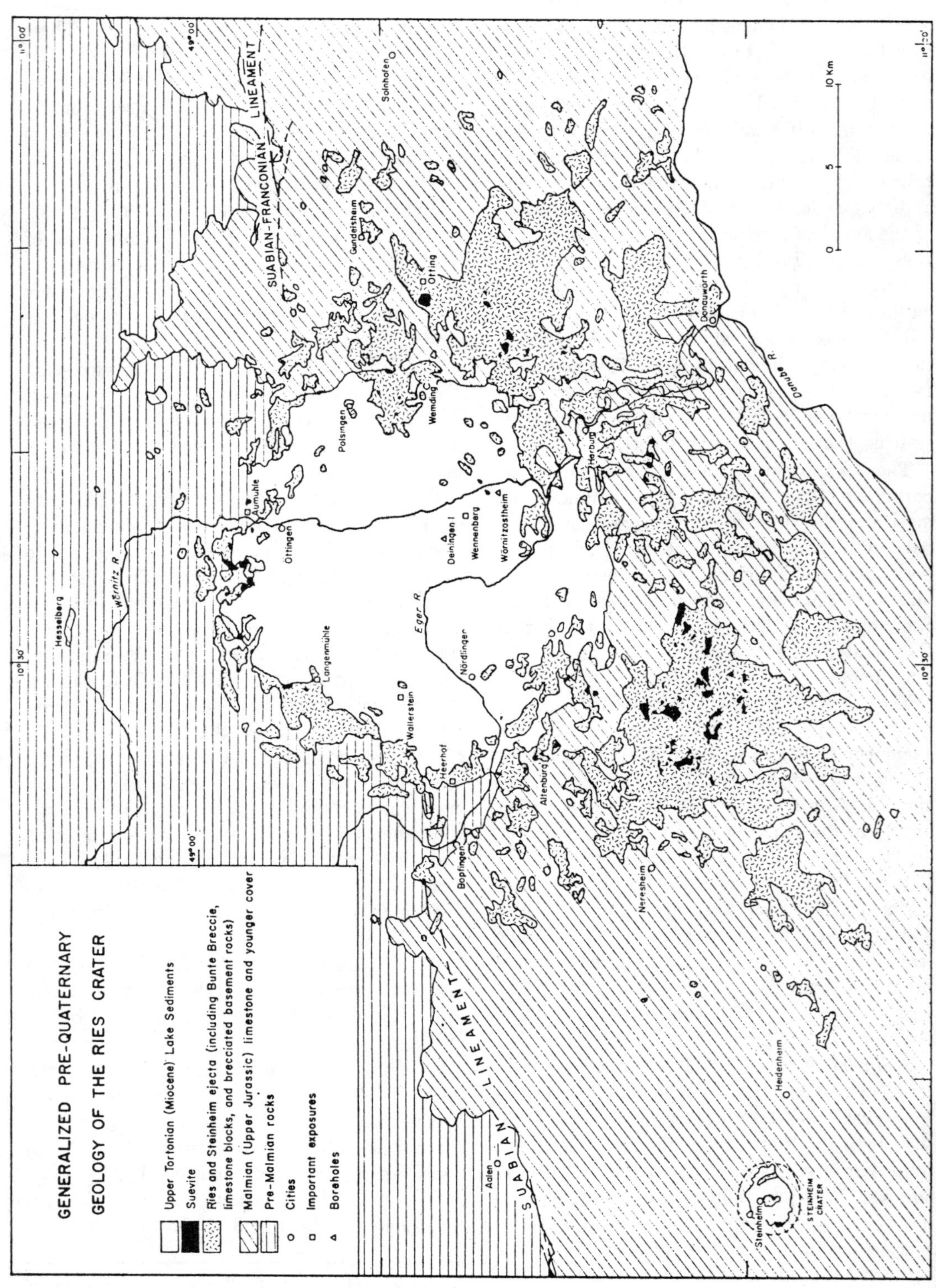
GENERALIZED PRE-QUATERNARY
GEOLOGY OF THE RIES CRATER
Upper Tortonian (Miocene) Lake Sediments
Suevite
Ries and Steinheim ejecta (including Bunte Breccie, limestone blocks, and brecciated basement rocks)
Malmian (Upper Jurassic) limestone and younger cover
Pre-Malmian rocks
Cities
Important exposures
Boreholes
SUABIAN-FRANCONIAN LINEAMENT
SUABIAN LINEAMENT
Solnhofen
Gundelsheim
Otting
Wemding
Polsingen
Aumühle
Öttingen
Deiningen I
Wennenberg
Wörnitzostheim
Harburg
Donauworth
Danube R.
Wörnitz R.
Hesselberg
Eger R.
Nördlingen
Langenmühle
Wallerstein
Heerhof
Altenburg
Bopfingen
Neresheim
Aalen
Heidenheim
Steinheim
STEINHEIM CRATER
0
5
10 Km
11°00'
10°30'
49°00'

formation appears to be due to periglacial mass movement [*Hüttner,* 1969].

Where Bunte Breccie directly overlies Malmian limestone, the bedrock surface is commonly polished and grooved, giving an appearance not unlike that of a glacially abraded surface. Statistically, these grooves are radially oriented with respect to the Ries center (Figure 3) [*Wagner,* 1964]. Grooved and polished surfaces also occur within Bunte Breccie on surfaces of allochthonous blocks.

Hüttner [1969] and *Dressler et al.* [1969] have reported structures that could be interpreted as weakly developed shatter cones from allochthonous Malmian inclusions in Bunte Breccie. W. von Engelhardt (oral communication, 1970) has found well-developed shatter cones in a granite basement fragment within Bunte Breccie at Ronheim, near Harburg (Figure 2).

Klippes are predominantly of Malmian limestone, but a few consist of older rocks. Most form discordant topographic elevations. They measure from 25 to over 1000 meters in length. Most of the larger klippes are deformed to varying degrees. In the more massive blocks, mainly of upper Malm, deformation is by fracture. This has resulted in two types of brittle

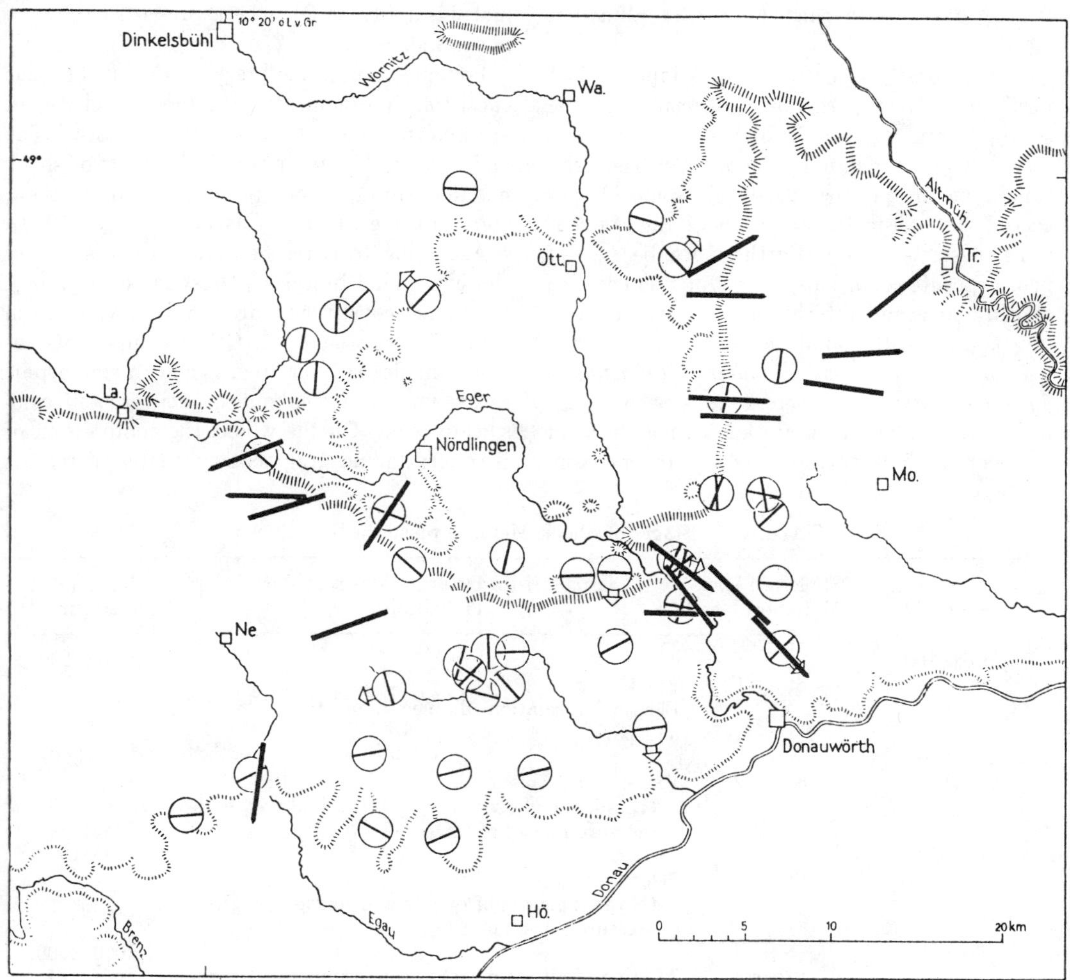

Fig. 3. Movement traces on polished bedrock surfaces, and orientation of deformation structures in klippes. Black bars, movement direction determined on polished surfaces; pointed end indicates sense of movement where determined. Circles, deformation structures in klippes; diametric line indicates *B* and *δ* axes; arrow indicates sense of displacement of entire klippe. Hachuring, escarpments. From *Hüttner* [1969], plate 7, after *Wagner* [1964].

failure: (1) fracturing along surfaces, irregular in spacing and orientation, but locally systematic; and (2) 'mortar structure' [*Hölder*, 1942], in which fragments of limestone, centimeters or less in size, are set in finely ground material of identical composition. Mortar structure is uncommon in blocks less than 25 meters in length [*Hüttner*, 1969].

In klippes made up of bedded limestone, deformation can be by folding, in part. Fracturing is less intense and is controlled by bedding, with the most prominent fracture sets normal to bedding surfaces. Both fold axes and prominent bedding/fracture intersections tend to be oriented preferentially at small angles or parallel to the Ries crater rim (Figure 3) [*Wagner*, 1964].

The greatest concentration of klippes is just inside the Ries perimeter, especially in the south, where the rim is highest. That is one reason why, as mentioned before, the basin rim cannot be unequivocally defined: some klippes exceed the 450-meter elevation of the conventional basin boundary. Furthermore, because of poor exposure conditions, the distinction between allochthonous and autochthonous limestones cannot be made with confidence.

Crystalline fragmented material (Kristalline Trümmermassen). Coherent masses of fragmented crystalline rock are known mainly from the loop of hills surrounding the inner basin. They may be blocks 100 meters and more across. They form what is known as the 'inner rampart' of the Ries (Figure 7). Its rampart-like structure in depth has been revealed by the geophysical work of *Reich and Horrix* [1955]. Some of these masses are nearly undeformed; others are fractured to varying degrees but display no signs of shock deformation.

Genuine crystalline breccias are of two kinds, monomict and polymict [*von Engelhardt*, 1971], Monomict crystalline breccia consists of finely fractured material of one single-rock type (mainly granites or gneisses). Some occur as fissure fillings in crystalline or sedimentary fragmented material. Indications of shock metamorphism (Table 2), where present, do not go beyond stage 1.

Polymict crystalline breccias, that is, breccias consisting of fragments of different rock types, are known from patchy occurrences both inside and outside the crater rim. Indications of shock metamorphism, where present, are of stages 1 and 2 (Table 2) [*von Engelhardt et al.*, 1969].

According to a recent petrographic study of the crystalline breccias [*Dressler et al.*, 1969], their chief constituents are granites, which form over 75% of the total. Other components include mafic igneous rocks and metamorphic rocks, mainly amphibolites, gneisses, and calc-silicate rocks. On the whole, the composition of Ries crystalline fragmentary material reflects

TABLE 2. Stages of Shock Metamorphism in Ries Rocks

Pressure, kb	Stage of Shock Metamorphism	Characteristic Deformations and Phase Transitions	Residual Temperature, °C
Approx. 100			Approx. 100
	Stage 1	Fracturing Plastic deformation (diaplectic quartz and feldspar)	
250–300			200–300
	Stage 2	Phase transitions (Diaplectic glasses of quartz and feldspar, high-pressure phases of SiO_2)	
500–550			1200–1500
	Stage 3	Selective melting (Normal glasses of quartz and feldspar, high-pressure phases of SiO_2)	
600–650			2000–3000
	Stage 4	Melting of all main rock-forming minerals (Inhomogeneous rock melts, glass bombs)	
Approx. 1000			Approx. 5000
		Volatilization	

After *von Engelhardt and Stöffler* [1968]; also see *Stöffler* [1971].

the composition of the nearest exposed Variscan basement at the western border of the Bohemian massif.

Suevite. This very unusual breccia (typically 5–10 meters thick) mesoscopically is seen to have a fine-grained matrix, greenish-gray when fresh, which makes up some 80% of the rock. Crystalline rock fragments account for between 2 and 7%. Between 0.1 and 1% are sedimentary rock fragments, and the strikingly characteristic glass bombs make up 11 to 18% of the rock [*Ackermann,* 1958]. The exact percentages vary with the different exposures. The recognizable clastic constituents of the matrix resemble components of basement rocks. F. Schüle [cited by *von Engelhardt et al.,* 1969] has undertaken a petrographic examination of the matrix. He has found microscopic glass wisps (but no shards!), no different in any respect other than size from the glass bombs described below. X-ray examination of matrix material has revealed that its finest fraction consists predominantly of a clay of the montmorillonite group, which appears to be secondary [*von Engelhardt et al.,* 1969].

Most of the glass bombs, also known as *Fladen* or *Flädle* ('pancake') have a characteristic flattened, aerodynamic shape. They show no plastic deformation, suggesting complete solidification before impact (Figure 4). They have a recognizable preferred alignment in that 60% of the longest axes are inclined at less than 30% to the horizontal. Statistically, teardrop-shaped bombs point with their thicker end (flight front) away from the Ries center. A common size range is 2 to 12 cm, but smaller and larger sizes are also quite common. (E. Preuss, written communication, 1970, mentions a bomb 120 cm long.) A number of bombs enclose fragments of basement rock. A detailed study of the Ries bombs is by *Hörz,* [1965].

The glass bombs have been classified according to their cooling history into three types by *von Engelhardt* [1967*b*]. Type 1 shows no recrystallization and was evidently chilled rapidly. Type 3 is totally recrystallized. Type 2 shows incipient recrystallization. Where the original top or bottom of suevite is exposed, the areas are characterized by chilled zones exclusively

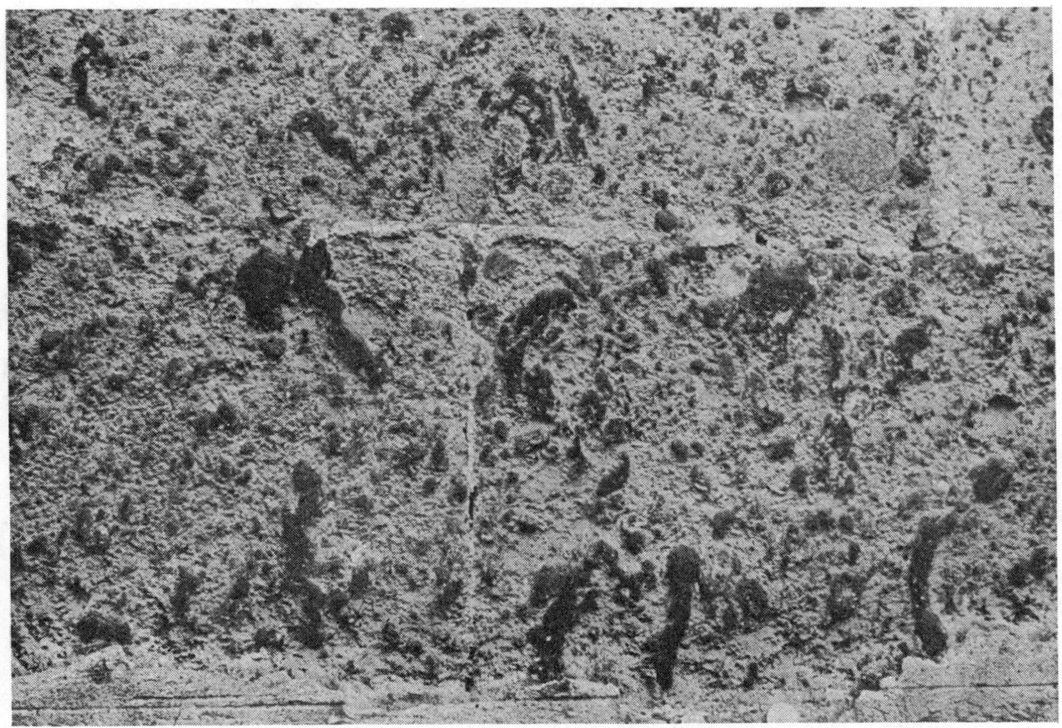

Fig. 4. Suevite exposed in wall of St. George's Church, Nördlingen. Long side of photograph, 75 cm. Black indicates Ries bombs, displaying preferred orientation. White indicates highly shocked crystalline fragments.

containing type 1 bombs. Drilling at Wörnitzostheim (Figure 2) has revealed a complete 84-meter section of suevite overlying brecciated granite. The lowermost 17 meters and the uppermost 6 meters of that section are chilled zones containing type 1 bombs. The suevite here is overlain by lake sediments. A complete petrographic account of that section is by *Förstner* [1967]. The existence of chilled zones at the bottom and top of suevite, together with other evidence, indicates that the suevite was deposited in one single episode.

On a microscopic scale, most type 1 bombs are inhomogeneous and abound in schlieren. The bombs contain angular inclusions of quartz, feldspar, and other mineral (and rock) fragments. Mafic minerals are extremely rare. Type 3 bombs occur between the chilled zones. They are readily recognized by the dull luster of fresh fractures. Minerals formed as a result of devitrification include mainly pyroxene and plagioclase. Pyroxene tends to form as very unusual hairlike or wormlike crystals. Secondary quartz and montmorillonite are also present [*Förstner*, 1967; *von Engelhardt et al.*, 1969].

All Ries bombs, and especially those of type 1, are remarkably uniform in over-all chemical composition, which corresponds to that of a certain gneiss. In view of the wide range of crystalline basement rocks, such a uniformity of composition of the chilled bombs suggests a well-defined source material within the crater [*von Engelhardt*, 1967*b*].

Hörz [1965] has drawn attention to the unusually low $Fe^{+3}:Fe^{+2}$ ratio in suevite bombs, indicating temperatures in the melt in excess of 1500°C. *El Goresy* [1964] has made a detailed study of the opaque minerals in glass bombs. On the basis of partially molten ilmenite and rutile he arrived at temperatures above 1700°C. Theoretical considerations, based on equation-of-state data [*Wackerle*, 1962], require temperatures in excess of 2000°C.

Indications of all stages of shock metamorphism, as outlined in Table 2, have been observed throughout the suevite. The glass bombs represent stage 4. Quartz and feldspar are highly sensitive indicators of stages 1, 2, and 3. Stage 2 is most common among affected crystalline fragments in suevite [*von Engelhardt et al.*, 1969; *Stöffler*, 1971; *Chao*, 1968]. The ubiquitous presence of shock metamorphic features including high-pressure phases (see Stöffler) are the most convincing evidence for impact origin. Accordingly, suevite would be an impact breccia derived from local crystalline bedrock. A volcanic origin as postulated in the past [see *Schnell*, 1926] is no longer tenable.

Absolute age determinations performed on suevite glasses have yielded an age for the Ries event itself. The K-Ar method resulted in 14.8 ± 0.7 m.y. [*Gentner et al.*, 1963] and the fission-track method yielded an age of 14.0 ± 0.6 m.y. [*Gentner et al.*, 1967].

A strikingly different type of 'suevite' is exposed near Polsingen (Figure 2). Its characteristic features are a reddish to grayish appearance, higher density, greater strength, and a vesicular groundmass that suggests wholesale melting. Under the microscope the groundmass, unlike that of normal suevite, is not montmorillonitic; it consists largely of a network of tiny feldspar laths. The red coloration is due to hematite [*von Engelhardt et al.*, 1969]. In contrast to all other suevites, no aerodynamically shaped bombs are present, indicating flowage rather than ejection along ballistic trajectories. This observation, as well as its location on the crater rim, seems to agree with characteristics of impact melts [*Dence*, 1971].

Spatial Relationships of the Ries Crater and its Ejecta

The near-surface filling of the inner crater zone consists of sediments of the former Ries lake, overlain by Pleistocene and recent deposits and underlain by suevite. Some topographic elevations underlain by the 'crystalline rampart,' which evidently were shoals in the Ries lake, are capped by calcareous tufa. Beyond the rampart, in the outer crater zone, the filling consists mainly of basement and Triassic components of the Bunte Breccie, overlain by sporadic patches of suevite. Toward the crater rim Jurassic components become more numerous, and the crater wall itself is lined with a discontinuous ring of Malm klippes, especially in the south. The Ries lake sediments between the crater wall and the 'crystalline rampart' vary in thickness but are commonly thin. Freshwater limestones, mostly algal, have formed along the outer crater rim.

The patchiness of the ejecta blanket appears to be due, in large though unknown measure,

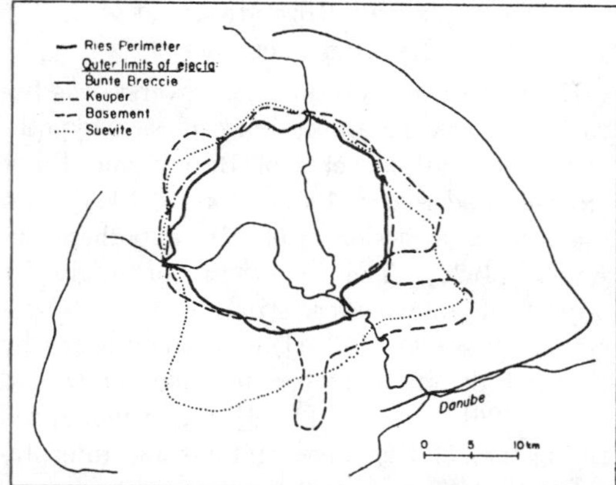

Fig. 5. Maximum lateral extent, as observed today, of individual stratigraphic units within various Ries breccias. Bunte Breccie, outline of all sedimentary breccias and thus entire ejecta blanket; Keuper, extent of Triassic components within Bunte Breccie; basement, polymict and monomict crystalline breccias as well as moderately, but mostly unshocked, crystalline inclusions in Bunte Breccie; suevite, highly shocked basement. After *von Engelhardt et al.* [1969].

to post-Ries erosion [*Hüttner and Wagner*, 1965]. The extreme sharpness of the contact between Bunte Breccie and suevite is not yet fully understood. It may reflect different modes of travel. The present limits of ejecta distribution are shown in Figure 5. Total thickness of the ejecta apron varies considerably and depends on relief. Average thickness is about 20 to 24 meters.

Geophysical Investigations

The classical refraction work by *Reich and Horrix* [1955] revealed an 'inner crater,' now filled with lake sediments to a maximum depth of 350 meters, and two subsidiary depressions. A reflection profile by *Angenheister and Pohl* [1969] indicates details of the internal structure of a crater segment (Figure 6). Preliminary results of new refraction profiles (see Angenheister and Pohl) suggest that the surface of autochthonous basement under the Ries crater is at least 1 to 1.4 km deeper than the basement outside the crater.

Pohl and Angenheister [1969] have measured the total intensity of the geomagnetic field near the Ries (Figure 7). The strong negative anomalies within the 'crystalline rampart' are conspicuous. A local, strong negative anomaly was also encountered near Wörnitzostheim (Figure 2) and was subsequently drilled. The core displayed suevite at 20 meters, underlain in downward succession by fragmented granite, Triassic sedimentary rock, and Jurassic sedimentary rock in reversed, somewhat incomplete stratigraphic sequence [*Förstner*, 1967]. These units are evidently part of an ejecta blanket between the inner and outer crater. Core drilling at Deiningen (Figure 2) revealed suevite at a depth of 330 meters, overlain by lake sediments [*Mosebach*, 1964; *Förstner*, 1967].

Pohl and Angenheister [1969] found that the average *total* magnetization of suevite (about $+50 \cdot 10^{-5}$ oe) is more than an order of magnitude greater than that of unaltered basement rocks. Measurements on oriented suevite specimens from quarries and drill holes have revealed that the direction of remanent magnetization is reversed ($i = -57°$, $d = 194°$). The strong magnetization of the suevite suggests crystallization of new magnetite during the melting

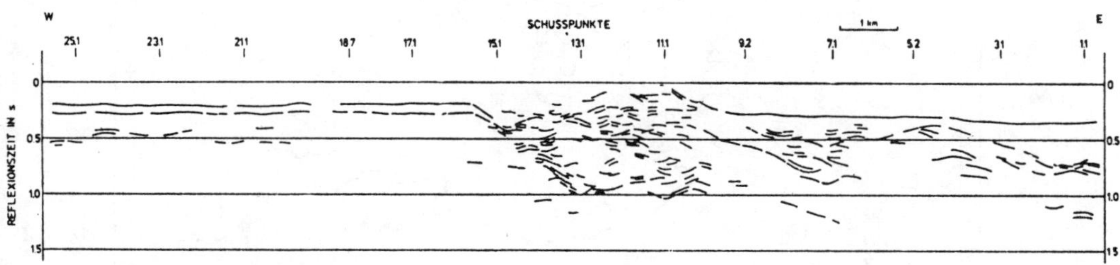

Fig. 6. Seismic reflection profile through western half of the crater. Eastern end of profile is located at approximate center of Ries. Vertical scale shows travel time in seconds adjusted to 400 meters above sea level. Numerals designate shot points. Note well-defined break at point 15.1 between undisturbed and disturbed zones. Some reflections between shot points 11.1 and 15.1 may be caused by lateral reflections. From *Angenheister and Pohl* [1969].

process, requiring local temperatures of at least 1400°C. Cooling past the Curie point (580°C) must have taken place shortly after the suevite was deposited, as indicated by the extremely constant direction of the magnetic field in samples from vastly different localities.

Clearly, the suevite can account for all magnetic anomalies. On the basis of the present total magnetic field, it is postulated that the immediate floor of the inner zone is made up of suevite varying in thickness between 50 and 500 meters.

Though the regional gravity field is rather complicated, *Jung et al.* [1969] could clearly estabish a negative anomaly at the Ries crater. Figure 8 illustrates the residual local field. Abrupt steepening of the gradient indicates a distinct inner crater, corroborating the results of the seismic and magnetic work. The total negative anomaly is estimated to correspond to a mass deficiency of 7×10^{16} g.

ORIGIN OF THE RIES STRUCTURE AND RELATED PROBLEMS

Early twentieth century controversy was between the volcanic uplift and explosion hypothesis with multiple vents of Branco and Fraas [*Branco and Fraas,* 1901; *Branco,* 1903], and the central-explosion ('blast') hypothesis of *Kranz* [1922]. The hypothesis of origin by impact of an extraterrestrial body, first suggested by *Werner* [1904], was strengthened by the first discovery of coesite [*Shoemaker and Chao,* 1961]. *Dehm* [1969] has summarized the history of Ries investigations and interpretations.

It is clear from the nature and distribution of the ejecta that they are derived locally. *David* [1969] argues that much of the suevite must have originated at depths greater than 1.4 km. The sedimentary breccias originated at shallower levels, and at least a large part of

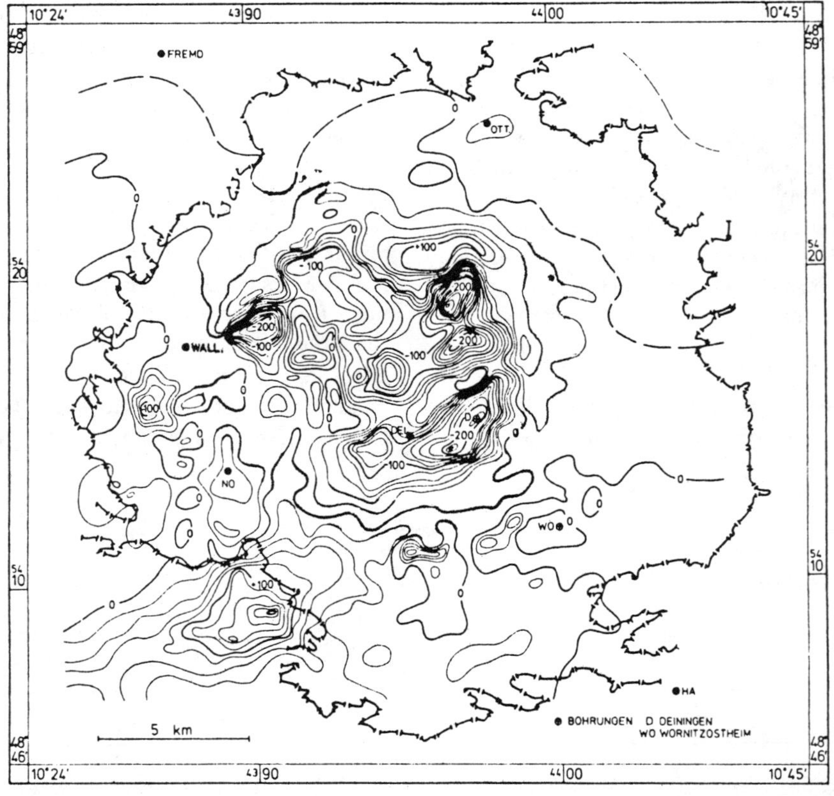

Fig. 7. Contours of total magnetic intensity anomalies in 10^{-5} oe. Shading indicates 'rampart' of fragmented crystalline basement outlining inner crater. Broken line shows 450-meter topographic contour. Cross circles mark locations of boreholes. From *Pohl and Angenheister* [1969].

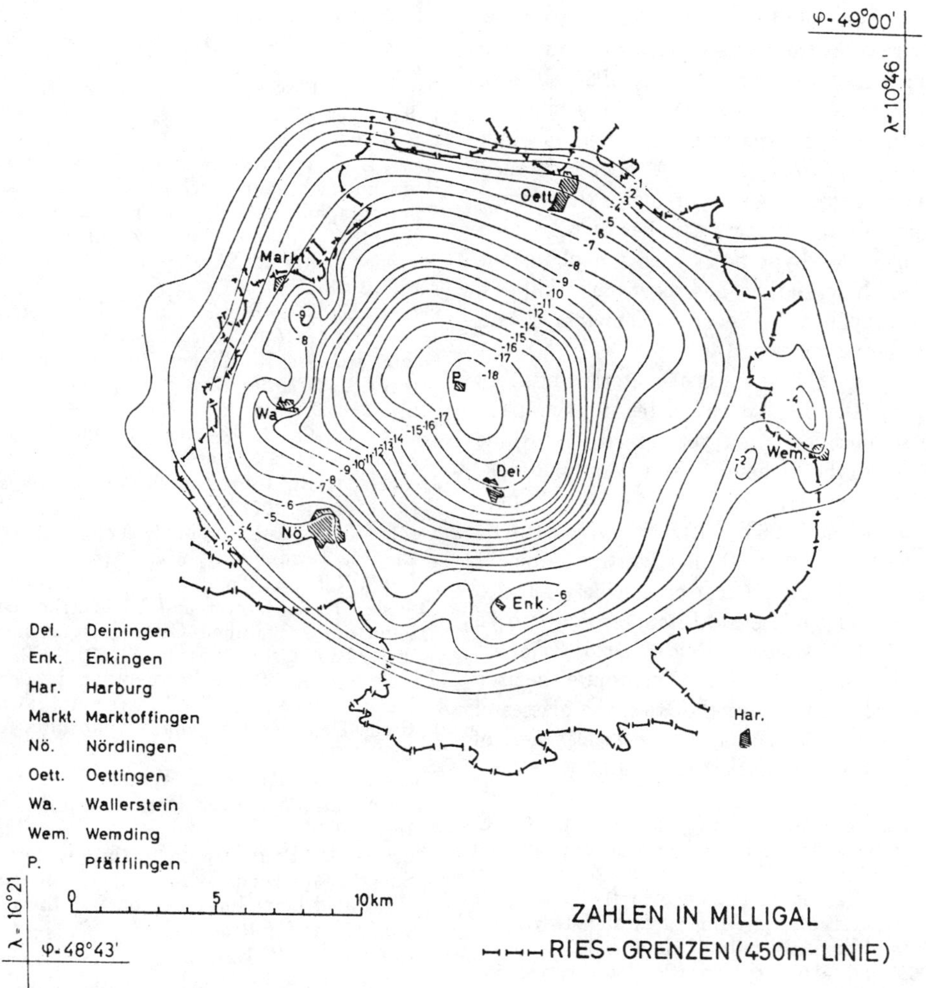

Fig. 8. Residual gravity field at Ries crater adjusted to 400 meters above sea level. Contours in milligals. From *Jung et al.* [1969].

them traveled along the ground, as evidenced by bedrock striations. The distribution and extent of breccias derived from different stratigraphic levels, as well as their inversed stratigraphy in the ejecta blanket, is compatible with experimental impact cratering [*Gault et al.*, 1968] and high-yield explosive cratering (*Oberbeck* [1971]; D. J. Roddy, unpublished data, 1971).

Petrographic evidence from crystalline breccias, and especially from suevite, indicates that unusually high transient temperatures and pressures must have prevailed. The evidence for abnormally high pressures and temperatures of short duration, as well as the over-all energy balance, cannot be reconciled with any known near-surface geological phenomena of internal origin. The impact of an extraterrestrial body (meteorite or comet head) offers the only available explanation compatible with present evidence.

Many important questions remain to be answered. These include the configuration of the basement under the Ries. A deep drill hole is one of the first current scientific priorities. Another problem to be resolved is that of the manner of transportation and emplacement of the various breccias. Although the suevites probably were ejected along ballistic trajectories, field evidence indicates that large parts of the sedimentary breccias must have moved along the ground or close to it. Recent observations of base surges accompanying volcanic eruptions and nuclear explosions [*Moore*, 1967; *Fisher and Waters*, 1970] introduce the possibility of at least partial transport by a fluidized medium [*Waters and Fisher*, 1971]. Although detailed work to test this hypothesis remains to be done, it might help in solving a number of remaining questions. The hypothesis might explain, for instance, how large heterogeneous klippes, rapidly propelled for several kilometers, retain their essential cohesion.

Striking age coincidences of materials possibly related to the Ries event noted in the past, whose significance is not yet agreed on, include (1) that of the Bohemian and Moravian tektites (moldavites), dated at 14.7 ± 0.7 m.y. [*Gentner et al.*, 1963]; (2) that of 'Reuter's blocks,' small (20 to 40 cm) weathered blocks of Malmian limestone scattered widely along a Pliocene land surface at distances of up to 65 km from the Ries center [*Reuter*, 1926; *Schaefer*, 1957; *Herold*, 1969]; (3) a 'bentonite' horizon in the Molasse south of the Ries, the glasses of which have yielded a fission-track age of 14.6 ± 0.8 m.y. [*Gentner et al.*, 1969].

Acknowledgments. I am indebted to G. C. Amstutz for much help and encouragement, and to R. Hüttner, E. Preuss, and R. Zimmermann for very enlightening guidance in the field while I was a guest of the Mineralogical-Petrographic Institute of the University of Heidelberg, Germany in 1968 and 1969. I also wish to express my appreciation for extensive helpful discussions with G. Angenheister, R. Dehm, W. von Engelhardt, F. Hörz, J. Pohl, and D. Stöffler. H. Vidal kindly supplied copy for Figures 3, 6, 7, and 8. Figures 1, 2, and 5 were prepared by T. Edmundson.

Publication was supported by NASA contract NSR-09-051-001.

REFERENCES

Ackermann, W., Geologisch-petrographische Untersuchungen im Ries, *Geol. Jahrb., 75,* 135, 1958.

Angenheister, G., and J. Pohl, Die seismischen Messungen im Ries von 1948–1969, *Geol. Bavarica, 61,* 304, 1969.

Branco, W., Das vulcanische Vorries und seine Beziehungen zum vulcanischen Riese bei Nördlingen, *Abhandl. Preuss. Akad. Wiss., Berlin, 1902,* 132 pp., 1903.

Branco, W., and E. Fraas, Das vulcanische Ries bei Nördlingen in seiner Bedeutung für Fragen der allgemeinen Geologie, *Abhandl. Preuss. Akad. Wiss., Berlin, 1901,* 169 pp., 1901.

Bucher, W. H., Cryptoexplosion structures caused from without or within the earth? ('Astroblemes' or 'Geoblemes'), *Amer. J. Sci., 261,* 597, 1963.

Carlé, W., Bau und Entwicklung der südwestdeutschen Grosscholle, *Geol. Jahrb., Beih., 16,* 1, 1955.

Chao, E. C. T., Pressure and temperature histories of impact metamorphosed rocks—based on petrographic observations, in *Shock Metamorphism of Natural Materials,* edited by B. M. French and N. M. Short, p. 135, Mono, Baltimore Md., 1968.

David, E., Das Ries-Ereignis als physikalischer Vorgang, *Geol. Bavarica, 61,* 350, 1969.

Dehm, R., Das Nördlinger Ries und die Meteortheorie, *Mitt. Staatssammlung Palaeontol. Hist. Geol., 2,* 69, Munich, 1962.

Dehm, R., Geschichte der Riesforschung, *Geol. Bavarica, 61,* 25, 1969.

Dence, M. R., Impact melts, *J. Geophys. Res., 76,* this issue, 1971.

Dennis, J. G., International Tectonic Dictionary, English Terminology, *Mem. Amer. Ass. Petrol. Geol., 7,* 97 pp., 1967.

Dressler, B., G. Graup, and K. Matzke, Die Gesteine des kristallinen Grundgebirges im Nördlinger Ries, *Geol. Bavarica, 61,* 201, 1969.

El Goresy, A., Die Erzmineralien in den Ries- und Bosumtwi-Krater-Gläsern und ihre genetische Deutung, *Geochim. Cosmochim. Acta, 28,* 1881, 1964.

Fisher, R. V., and A. C. Waters, Base surge bed forms in maar volcanoes, *Amer. J. Sci., 268,* 157, 1970.

Förstner, U., Petrographische Untersuchungen des Suevits aus den Bohrungen Deiningen und Wörnitzostheim im Ries von Nördlingen, *Contrib. Mineral. Petrol., 15,* 281, 1967.

Gault, D. E., W. L. Quaide, and V. R. Oberbeck, Impact cratering mechanics and structures, in *Shock Metamorphism of Natural Materials,* edited by B. M. French and N. M. Short, p. 87, Mono, Baltimore, Md., 1968.

Gentner, W., and G. A. Wagner, Alterbestimmungen an Riesgläsern und Moldaviten, *Geol. Bavarica, 61,* 296, 1969.

Gentner, W., H. J. Lippolt, and O. A. Schaeffer, Argonbestimmungen an Kaliummineralien, 11, Die Kalium-Argon-Alter der Gläser des nördlinger Rieses und der böhmisch-mährischen Tektite, *Geochim. Cosmochim. Acta, 27,* 191, 1963.

Gentner, W., B. Kleinmann, and G. A. Wagner, New K/Ar and fission track ages of impact glasses and tektites, *Earth Planet. Sci. Lett., 2,* 83, 1967.

Herold, R., Eine Malmkalk-Trümmermasse in der Oberen Süsswassermolasse Niederbayerns, *Geol. Bavarica, 61,* 413, 1969.

Hölder, H., Geologische Untersuchungen in der Umgebung von Lauchheim, *Neues Jahrb. Mineral. Geol. Palaeontol., suppl. vol. B, 86,* 315, 1942.

Hörz, F., Untersuchungen an Riesgläsern, *Beitr. Mineral. Petrogr., 11,* 621. 1965.

Hüttner, R., Geologische Untersuchungen im SW-Vorries auf Blatt Neresheim und Wittlingen, thesis, University of Tübingen, 1958.

Hüttner, R., Bunte Trümmermassen und Suevit, *Geol. Bavarica, 61,* 142, 1969.

Hüttner, R., and G. H. Wagner, Bericht über Bohrungen in Suevittuffen des württembergischen Riesgebietes, *Jahresh. Geol. Landesamtes Baden-Wuerttemberg, 7,* 223, 1965.

Hüttner, R., H. Schmidt-Kaler, and W. Treibs, Geological map of the Ries and surroundings, 1:100,000, in *Das Ries; Geologie, Geophysik, und Genese eines Kraters, Geol. Bavarica, 61,* edited by E. Preuss and H. Schmidt-Kaler, Bayerisches Geologisches Landesamt, Munich, 1969.

Jung, K., H. Schaaf, and H.-G. Kahle, Ergebnisse gravimetrischer Messungen im Ries, *Geol. Bavarica, 61,* 337, 1969.

Kranz, W., Der geologische Aufbau und Werdegang des nördlinger Rieses, *Rieser Heimatbuch, 1,* 25, 1922.

Mosebach, R., Das nördlinger Ries, vulkanischer Explosionskrater oder Einschlagstelle eines Grossmeteoriten?, *Ber. Oberhess. Ges. Natur Heilk. Giessen, Naturw. Abt., 33,* 165, 1964.

Moore, J. G., Base surge in recent volcanic eruptions, *Bull. Volcanol., 30,* 337, 1967.

Oberbeck, V. R., Laboratory simulation of impact cratering with high explosives, *J. Geophys Res., 76,* this issue, 1971.

Pohl, J., and G. Angenheister, Anomalien des Erdmagnetfeldes und Magnetisierung der Gesteine im nördlinger Ries, *Geol. Bavarica, 61,* 327, 1969.

Preuss, E., and H. Schmidt-Kaler, eds., *Das Ries, Geologie, Geophysik, und Genese eines Kraters, Geol. Bavarica, 61,* 478 pp., Bayerisches Geologisches Landesamt, Munich, 1969.

Reich, H., and W. Horrix, Geophysikalische Untersuchungen im Ries und Vorries und deren geologische Deutung, *Geol. Jahrb., Beih., 19,* 119 pp., 1955.

Reuter, L., Die Verbreitung jurassischer Kalkblöcke aus dem Ries im südbayerischen Diluvialgebiet, *Jahresber. Mitt. Oberrheinischen Geol. Ver., Neue Folge (1925), 14,* 191, 1926.

Schaefer, I., *Erläuterungen zur geologischen Karte von Augsburg und Umgebung, 1:50,000,* 92 pp., Bayerisches Geologisches Landesamt, Munich, 1957.

Schmidt-Kaler, H., Versuch einer Profildarstellung für das Rieszentrum vor der Kraterbildung, *Geol. Bavarica, 61,* 38, 1969.

Schnell, T., Der bayerische Trass und seine Entstehung, *Jahresber. Mitt. Oberrheinischen Geol. Ver., Neue Folge (1925), 14,* 222, 1926.

Seibold, E., Das schwäbische Lineament zwischen Fildergraben und Ries, *Neues Jahrb. Geol. Palaeont., Abhandl., 93,* 285, 1951.

Shoemaker, E. M., and E. C. T. Chao, New evidence for the impact origin of the Ries basin, Bavaria, Germany, *J. Geophys. Res., 66,* 3371, 1961.

Stöffler, D., Progressive metamorphism and classification of shocked and brecciated crystalline rocks at impact craters, *J. Geophys. Res., 76,* this issue, 1971.

von Engelhardt, W., Mineralogische Bemerkungen zu der neuentfachten Diskussion um das nördlinger Ries, *Jahresh. Verein Vaterländ, Naturkunde Württemberg, 117,* 18–23, 1962.

von Engelhardt, W., Neue Beobachtungen im nördlinger Ries, *Geol. Rundsch., 57,* 165, 1967*a*.

von Engelhardt, W., Chemical composition of Ries glass bombs, *Geochim. Cosmochim. Acta, 31,* 1677, 1967*b*.

von Engelhardt, W., Detrital impact formation, *J. Geophys. Res., 76,* this issue, 1971.

von Engelhardt, W., and D. Stöffler, Stages of shock metamorphism in the crystalline rocks of the Ries basin, Germany, in *Shock Metamorphism of Natural Materials,* edited by B. M. French and N. M. Short, p. 159, Mono, Baltimore, Md., 1968.

von Engelhardt, W., D. Stöffler, and W. Schneider, Petrologische Untersuchungen im Ries, *Geol. Bavarica, 61,* 229, 1969.

Wackerle, J., Shock compression of quartz, *J. Appl. Phys., 33,* 922, 1962.

Wagner, G. H., Kleintektonische Untersuchungen im Gebiet des Nördlinger Rieses, *Geol. Jahrb., 81,* 519, 1964.

Waters, A., and R. V. Fisher, Base surges and their deposits: Capelinhos and Taal volcanoes, *J. Geophys. Res., 76,* this issue, 1971.

Werner, E., Das Ries in der schwäbisch-fränkischen Alb, *Blätter des schwäbischen Albvereins, 16,* 153, 1904.

(Received January 7, 1971;
revised March 11, 1971.)

2

Reprinted from *Internat. Geol. Cong., 24th, Proc.* **15**:90–111 (1972)

Impact Structures in Europe

W. VON ENGELHARDT,
F.R. Germany

ABSTRACT

For ten probable impact craters in Europe, the geological setting, morphological features, textural peculiarities, impact rocks and shock phenomena are described. These structures are: Mien, Hummeln, Siljan and Dellen in Sweden; Lappajärvi in Finland; Jänisjärvi in USSR; Rochechouart in France; Steinheim and Ries in Germany; Köfels in Austria.

INTRODUCTION

TABLE 1 describes the main features of structures in Europe which are considered to have been produced by impacts of extraterrestrial bodies (meteorites or comets) large enough to reach the surface of the earth with planetary velocities. It is the scope of this paper to summarize for each of these structures the general topographic and geological setting and all observations, so far known, from which an impact origin can be deduced. The arguments include morphological and structural features, textural peculiarities like shatter cones and intense brecciation, breccias containing shocked rocks and minerals, glassy ejecta (bombs), and impact melt rocks which are differentiated from volcanic formations by texture and mode of occurrence. Most important are effects of shock waves of high peak pressure, such as typical deformation structures in minerals, diaplectic glasses, high-pressure phases, and partial and complete fusion. Another purpose of this paper is to give a full account of the literature on these structures. In all cases where the number of publications is too abundant, the reader is referred to reviews and recent articles from which the references to publications not quoted in the reference list of this article may be obtained.

The breccias produced by impact and rocks of the individual structures will be described using a general terminology proposed by Dence (1971) dealing with impact melt rocks and the author (Engelhardt, 1971) on impact breccias. Shock phenomena in rocks are classified using the progressive stages I - IV of shock metamorphism as defined by Stöffler (1971).

The structures of Table 1 are arranged according to their visible sizes. It appears that clear indications of shock can only be expected in craters larger than about 100 m in diameter. The formation of central uplifts seems to depend not only on the crater size, but also on the nature of the bedrock and perhaps the kind of impacting body. The impact formations observable at one particular site depend on the degree of erosion. At young craters like the Ries, which are only little modified by erosion, surficial ejecta blankets outside the crater can be studied, but the breccias and melts filling the crater cavity and the shape of the crater itself are hidden unless they are disclosed by drill holes. On the other hand,

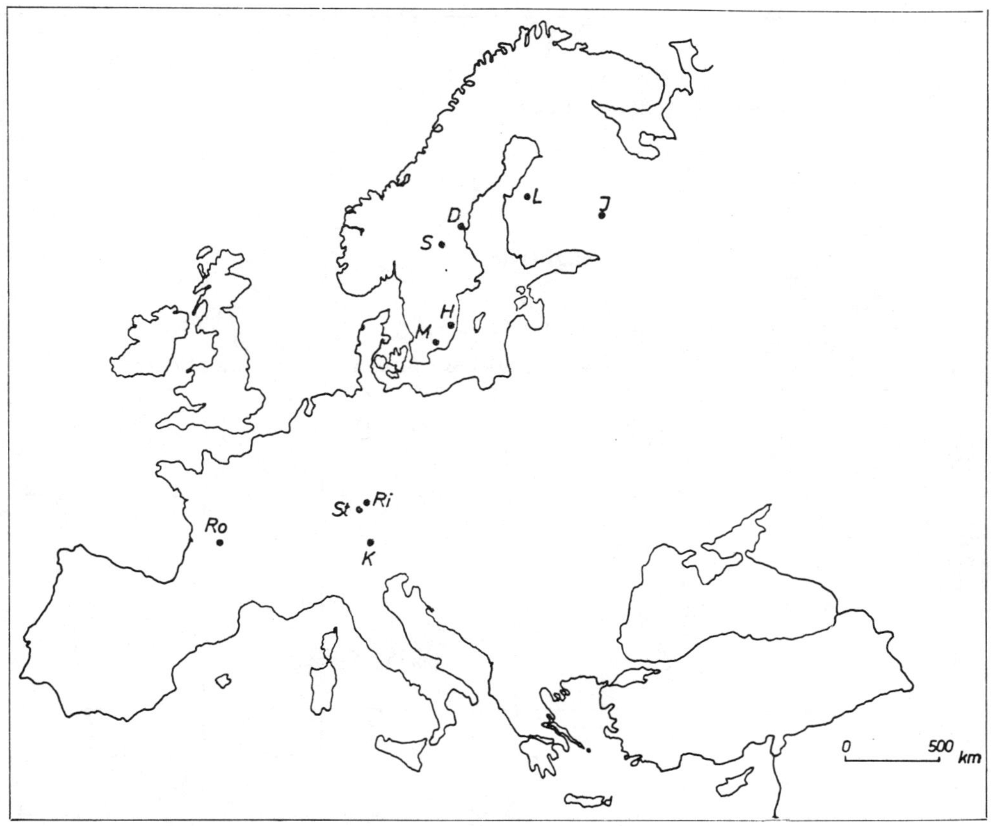

Figure 1 — Impact craters in Europe.

D = Dellen	M = Mien
J = Jänisjärvi	Ri = Ries
H = Hummeln	Ro = Rochechouart
K = Köfels	S = Siljan
L = Lappajärvi	St = Steinheim

craters like those on the Baltic shield, which were excavated by ice, or the nearly completely eroded structure of Rochechouart, reveal materials which were deposited within the crater or even at its bottom.

Figure 1 shows the location of the ten impact sites currently known in Europe. Their distribution corresponds to the geological history of the continent. Most craters are accumulated on the oldest part of Europe, the Baltic shield, built up of Precambrian metamorphic rocks and intrusives. It seems unlikely that impact structures survived the Jotnian (Algonkian) period of strong denudation, so it can be assumed that the impacts of Mien, Hummeln, Siljan, Dellen, Lappajärvi and Jänisjärvi were all formed in post-Jotnian times (i.e., within the last 1000 m.y.). Within the Caledonian areas of Norway and the British islands no impact structures have been found, as yet. The only impact site on a Hercynian basement is the deeply eroded structure of Rochechouart, situated on the Plateau Central, France. It has an age of approximately 200 m.y. The Ries and Steinheim impacts occurred 14.8 million years ago. The crater of Köfels is the only impact site of post-Pleistocene age (8000 years).

TABLE 1 — Impact Structures in Europe

Name Location	Diameter of the visible structure (km)	Age (years	slightly (s) conside-rably (c) extremely deeply eroded (d)	Morphology	Central elevation	Bedrock
Humeln, Sweden 57°22′N 16°15′E	1	6.10^8	c	lake	—	granite
Steinheim, Germany 48°02′N 10°01′E	3,5	15.10^6	s	circular depression	hill	lime-stone
Köfels, Austria 47°13′N 10°58′E	4	8.10^3	s	circular cavity filled by landslide	—	gneiss
Mien, Sweden 56°25′N 14°52′E	5	?	c	lake	island	granite gneiss
Lappajärvi, Finland 63°10′N 23°40′E	10	?	c	lake	island	granite, gneiss
Jänisjärvi, Karelia, USSR 61°58′N 30°55′E	10	?	c	lake	island	mica-schist
Dellen, Sweden 61°55′N 16°32′E	15	?	c	lake	pen-insula	gneiss, grano-diorite
Rochechouart, France 45°49′N 0,5°E	15	2.10^8	d	—	—	gneiss, schist, granite
Ries, Germany 48°53′N 10°37′E	24	$14{,}8.10^6$	s	circular depression	ring uplift ?	sedi-ments, gneiss, granite
Siljan, Sweden 61°05′N 15°00′E	40	post Ordo-vician	d	annular lake	granite massif	granite, lime-stone

Shatter cones	Detrital impact formations: W: Weakly shocked breccias S: Suevite a: in situ b: fallback o: fallout	Glassy ejecta (bombs)	Impact melt rocks	Planar deform-ation structures	Dia-plectic glass	Coesite(C) Stisho-vite(St)
—	Wa	—	—	—	—	—
+	Wb Wo	—	—	+	—	—
—	Wb?	—	+	+	+	—
—	Wa Wb Sb	+	+	+	+	C
—	Wb Sb	—	+	+	+	C
—	?	—	+	?	?	?
—	Sb	+	+	+	—	—
+	Wa Wb Sb	+	+	+	+	—
+	Wb Wo Sb So	+	+	+	+	C St
+	W	—	—	+	—	—

It may be asked in which regions of Europe additional impact sites could be expected to exist. Six craters are known on the southern part of the Baltic shield which, south of 63° latitude, occupies an area of about 2.1 x 10^5 km^2. They were formed during the last 1 x 10^9 years. No impact structures are known, as yet, on the Baltic shield north of 63° latitude. Because this part of the shield represents an area of about 3.0 x 10^5 km^2, it may be suggested that another six or even seven impact sites do exist within the area of northern Sweden, northern Finland, northern Karelia and Kola peninsula. Other regions of Europe where the occurrence of impact sites appears most probable are the Precambrian shield of the Ukraine, the Caledonian regions of Norway and the British islands and some of the Hercynian massifs, like those of central France, northwestern France and Bohemia.

Figure 2 gives sizes and shapes of the structures described in detail in the following paragraphs.

IMPACT STRUCTURES ON THE BALTIC SHIELD

In 1963, Fredriksson and Wickman indicated five structures in Sweden which they regarded as possible impact sites. Since then, for four of them (Mien, Hummeln, Siljan, Dellen), this assumption has been confirmed or become very plausible. No additional information exists about the structures of Tvären, a circular depression, 2 km in diameter, in a fjord 80 km south of Stockholm. Its bottom is 80 m below sea level, whereas the average water depth of the fjord is 30-35 m. A block of Lower or Middle Ordovician limestone has been dredged from this hole.

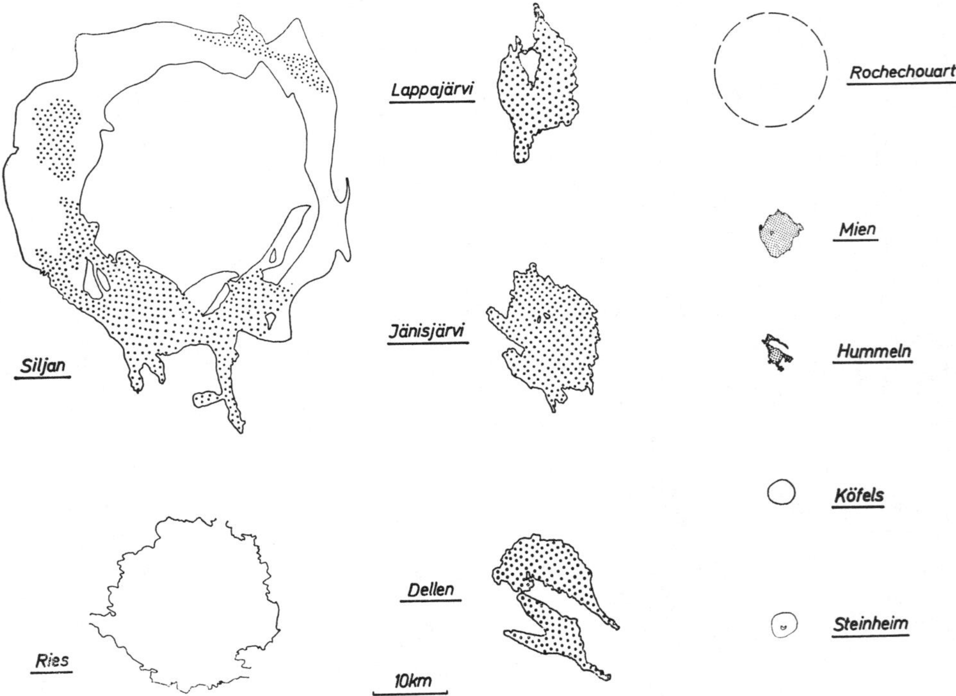

FIGURE 2 — Morphology of European impact sites. Dotted areas are water.

Lake Mien, southeastern Sweden, is 30 km north of the town of Karlshamn. The lake has a more rhombic than circular form, about 5 km in average diameter. In the northwestern part of the lake, the island of Ramsö forms a steep elevation between the shallow eastern and the much deeper western half of the lake. The greatest depth in the eastern half is 43 m.

Bedrocks of the region are granites, gneissic granites and gneisses of Gothian and Dalslandian age, intersected by numerous diabase dikes. The area around Lake Mien is covered by glacial deposits. A large esker extends from the southern end of the lake, in a south-southwestern direction, to the Baltic Sea (direction of ice movement).

On the island of Ramsö a grey to black aphanitic lava-like rock occurs with pale yellow inclusions and many vesicles. As seen under the microscope, this rock shows, besides vesicles, rounded inclusions of recrystallized bedrock and sandstone in a devitrified matrix. Devitrification products of the matrix are trichitic microcrystallites and small laths of feldspar, some with skeletal terminations, pyroxene and opaque minerals. The feldspar laths are randomly oriented or form spherulitic textures.

Two drill holes on Ramsö (Stanfors, 1969) have revealed a layer of the lava-like rock, 20-25 m in thickness under a 3-5-m-thick cover of moraine. The content of bedrock fragments increases with increasing depth. Below the lava-rock a layer follows, about 2 m thick, which is similar to the suevite of the Ries crater. The tuff-like breccia contains, in a grey matrix, glassy black lumps of irregular form and rock fragments. The breccia grades in its lower parts into a mixture of angular fragments of different gneissic and granitic rocks and broken pieces of the lava-like rock. Both boreholes terminated in a breccia composed of angular unaltered bedrock fragments, cemented by quartz. This breccia grades into fractured gneissic granite.

Within the esker south of the lake occur fragments of the Ramsö lava rock, breccias and glass bombs. The breccias are suevites containing rock and mineral fragments in all stages of shock metamorphism and fragments or lumps of fluidal, vesicular glass. Coesite was found (Svensson and Wickman, 1965; Svensson, 1969) in shocked rock fragments from this breccia. The bombs are mostly flat, apparently shaped by aerodynamical forces. They consist of a very heterogeneous, mostly devitrified, glass with vesicles and schlieren, rock and mineral fragments, many of them with clear indications of shock. Most mineral fragments are quartz; mafic minerals are absent or very rare. Both breccias and bombs are so similar in appearance to suevite and bombs (Fladen) from the Ries crater that they cannot be distinguished megascopically.

Holst (1890), who first investigated the peculiar Mien rocks, regarded them as volcanics. Högbohm (1910) suggested an impact origin of the Mien structure and compared it with Meteor Crater, Arizona. Fredriksson and Wickman (1963) listed the Mien structure among the possible astrobleme structures of Sweden. Bergquist (1954) believed that the Mien structure was formed by material from the moon which fell down on the surface of the earth. Stanfors (1969) gave a description of the morphology of Mien Lake and a report about the drill-hole profiles on Ramsö. Svensson (1969) described the Mien structure as a possible impact crater.

Shock indications in minerals, such as planar deformation structures in quartz (Engelhardt and Stöffler, 1965), diaplectic glasses of quartz and feldspar, coesite, heterogeneous glasses and glass bombs prove that the Mien structure was formed by impact. The lava-like rock of Ramsö, apparently, represents an impact melt. Suevite breccias and bombs occurring in the esker are parts of the original crater filling which was removed by glacial erosion.

Stanfors (1969) pointed out that the thick layer of lava-like rock on Ramsö is against the impact origin of the structure, because layers of impact melt have only been observed in much larger craters. However, it must be considered that the original crater diameter may have been appreciably reduced by erosion.

The impact occurred sometime between Precambrian and Pleistocene times.

Lake Hummeln, Småland, southern Sweden, is 15 km northwest of the city of Oskarsham. The lake contains a depression of a somewhat elongated, rounded form with a flat bottom, 61 m deep at its deepest point. The axes of the depression are 1300 and 1000 m long, respectively. The surrounding bedrock consists of porphyritic granite of Gothian age. For a long time boulders of Cambrian sandstone and Ordovician limestone have been known to occur along the southern shores of the lake (Hisinger, 1825; Linnarsen, 1878; Svedmark, 1892, 1904, and others; see Svensson, 1966). The limestone occurs as big boulders on and near the surface of a narrow moraine ridge stretching northwest- southeast from the southeastern shore (direction of the ice movement during the last glaciation). Cambrian sandstone boulders are abundant at the southeastern shore. To the north, west and east of the lake no limestone boulders and only a few small sandstone boulders have been found. Apparently, the Paleozoic sediments originally filled the depression and were excavated by glacial erosion.

Breccias occur in outcrops on the southern shore. They consist of angular fragments of the granite, up to 50 cm in size, and of its minerals. Radially from the center of the depression, and within a distance of 12 - 15 m, the breccia gradually passes over into undisturbed bedrock. Small movements have taken place within the bedrock along two main joint directions, one radial to the center and the other tangential to it. Parallel to the tangential faults, bands with large fragments alternate with other bands of heavily crushed bedrock. Wedges of breccia penetrate beyond these bands radially from the center into fissures of the bedrock. Thin sections of the breccias show only heavily crushed rock and mineral fragments down to the finest grain sizes. Petrographic shock features are absent.

Although, so far, no direct evidence for high shock pressures is known from this locality the abnormally deep depression, the breccias and the faults make it most probable that the structure was produced by the impact of a small extraterrestrial body. The impact may have occurred in early Cambrian times.

The impact origin of the Hummeln structure was first suggested by Wickman in 1961 (personal communication to Svensson) and later established by Svensson (1966).

Siljan Ring, central Sweden, 270 km northwest of Stockholm. A recent description of the geology of the area was given by Hjelmqvist (1966). North of Lake Siljan, Paleozoic rocks form a conspicuous ring-shaped belt, about 40 km in diameter, which is partly defined by faults. The sediments rest, partly undisturbed on the older surface. The bottom layers of the Paleozoic deposits are represented by the Ordovician Obolus conglomerate, which is found in the eastern and southern parts of the ring-shaped belt. At Sjurberg, in the south, the conglomerate rests directly on the weathered granite surface and contains pebbles of granite and other Precambrian rocks. A considerable part of the Paleozoic belt is occupied by Ordovician limestone, sometimes alternating with layers of shale. The Ordovician is overlain by Silurian shales and the Ora sandstone, the uppermost part of the Silurian.

By the violent movements to which the region has been subjected, the Paleozoic layers have frequently been uplifted and even overturned. The central part of the structure is occupied by a large granite massif of nearly circular outline. It consists in the northern half of the reddish, medium- to coarse-grained, acid Siljan granite, in the southern half of the coarse-grained, greyish red, quartz-poor Järna granite. The area around the Siljan Ring structure consists of Precambrian rocks. It is composed, west of the structure, of porphyries and granites of sub-Jotnian age, overlain by Jotnian sandstone. East of the structure, older migmatites and granites, of Svionian age, prevail.

Different attempts have been made to interpret the formation of the Siljan Ring. According to one of these the ring is a relic of a big caldera (Asklund, 1936; Rutten, 1966). Another interpretation explains the ring-shaped depression as the result of undirected stretching of an inhomogeneous material (Cloos, 1933). Fredriksson and Wickman (1963) hold that the ring structure has originated by a large impact.

The impact hypothesis is supported by the occurrence of peculiar breccias which have been found at the boundary between granite and Ordovician limestone, in the south of the structure. Connected with the breccias occur dark veins, 10-20 cm in thickness, which thin into fine veinlets less than 1 cm thick. Asklund (1936) interpreted the veins as diabase and the breccias as volcanic explosion breccias. At Stumsnäs, in the outermost south of the structure, the breccia zone between granite and Ordovician limestone is about 20 m thick. In some parts the breccia contains fragments of granite in a grey-green groundmass (Hjelmqvist, 1966). Recently shatter cones were discovered by Svenson and Robertson (Dietz, 1970) in granite, near the center of the structure. Svenson (1971) found quartz with planar deformation structures in granite samples from the central cupola. Predominant are structures parallel to $(10\bar{1}3)$; other orientations are: $(10\bar{1}2)$, $(10\bar{1}1)$, $(12\bar{3}1)$.

Because Ordovician limestone is involved in the brecciation, the structure has been formed in post-Ordovician, but pre-Pleistocene, time.

Dellen Lakes, Central Sweden, 300 km north-northwest of Stockholm. The two lakes (Norra and Södra Dellen) are separated by the peninsula of Norrbonäset. Both lakes together form a depression with an approximately circular outline, 14 - 16 km in diameter. The bedrock of the surrounding peneplain consists of gneissose granodiorites of Svionian age.

Svenonius (1895) gave the name dellenite to a rock which he found on the northern shore of Norrbonäset. Boulders of dellenite occur also in the direction of ice movement southeast of Lake Dellen (Lundquist, 1963). Tryggvason (1940) gave a petrographic description and a chemical analysis of dellenite boulders which were found at the coast of Uppland near Roslagen and Alvkarleby 200 m south-southeast of Dellen. The Dellen structure was thought to be of volcanic origin and a description of the Dellen rocks as volcanics was given by Redaelli (1957).

The dellenite rocks comprise suevites, impact melt and glass bombs. The suevite breccia contains rock (gneiss, quartzite) and mineral fragments with and without shock indications and devitrified glass fragments. The impact melt is represented by aphanitic grey to greenish rocks of a fluidal texture, containing fragments of rocks and minerals in a more or less devitrified glass matrix. Quartz is by far the most abundant fragmental mineral; mafic minerals are absent or very rare. Devitrification products of the groundmass are laths of feldspar and pyroxene, arranged either irregularly or radiating in all directions. The bombs are aerodynamically shaped. Their forms are similar or even identical to those of

Lake Mien and Ries crater bombs. They consist of brownish devitrified glass with many schlieren of different color, vesicles and mineral fragments, predominantly quartz. Many quartz grains contain planar deformation structures. Svensson (1968) identified orientations parallel to $(10\bar{1}3)$, $(10\bar{1}2)$, $(21\bar{1}2)$, $(10\bar{1}1)$ and (0001). Most frequent are orientations parallel to $(10\bar{1}3)$.

Högbom (1910) compared the Dellen structure with Meteor Crater in Arizona and suggested first an origin by impact. Bergqvist (1954) believed that the Dellen structure was formed by material which fell from the moon. Fredriksson and Wickman (1963) mentioned Dellen Lakes among the five geological structures in Sweden which they regarded as possible astroblemes. Svensson (1968) supported the interpretation of Dellen Lakes as an impact crater by the observation of shock-induced quartz deformations.

Shock deformation features in mineral and rock fragments, suevite breccia, impact melt and glassy bombs prove that the Dellen structure was produced by impact. Moving land-ice eroded the crater-filling breccias and destroyed the morphology of the crater. Therefore, the size of the structure is not exactly known. Svensson (1968) estimated its size to be of the order of 8-10 km. The original crater was certainly larger. The impact occurred sometime between the Precambrian and the Pleistocene.

Lake Lappajärvi, west-central Finland, 320 km northwest of Helsinki. The shallow, nearly elliptical depression of the lake has a long axis, in a north-south direction, of about 17 km and a short one of about 10 km. The average depth of the lake is between 5 and 10 m. Two narrow bottom valleys, up to 30 m deep, occur close to the eastern and western shore lines. Another one stretches northeast-southwest and divides the lake bottom into two shallow platforms, a northern one with the island of Kärnasaari, and a southern one. The topography of the lake bottom was formed by glacial erosion which excavated softer materials filling the depression west, south and north of the island, whereas the harder rocks of Kärnasaari prevented the northern platform from being eroded (Svensson 1971). The bedrock of the area is composed of Precambrian granodiorite, biotite plagioclase gneiss and granite. The surface is generally covered by till.

On the island of Kärnasaari, Holmberg (1858) found a peculiar rock type, later named kärnäite (Berghell in Saksela, 1949). Kärnäite boulders have been transported by ice movement as far away as about 320 km south-southeast from Lappajärvi. They do not occur north of Kärnasaari island (Laitakari, 1942; Mölder, 1948; Saksela, 1949; Kulonpalo (1969). Field investigations by Lehtinen (1970) disclosed, besides kärnäite, several other types of impact rocks. Lehtinen distinguished two types of kärnäite (I and II) and two breccias, one with glass fragments (suevite) and another without glass (weakly shocked breccia). Kärnäite I contains, in a glassy groundmass, microlites of sanidine, plagioclase, orthopyroxene, cordierite and fragments of quartz. Fragments of biotite and feldspar are almost decomposed. Kärnäite II, the most common type, is a lava-like hard rock of bluish-black color, splitting into sharp-edged pieces. The crystalline fine-grained groundmass contains fragments of bedrock and minerals in various stages of shock metamorphism. The glassy Kärnäite I grades continuously into Kärnäite II. Both rocks are regarded as impact melts.

Suevite breccia, named fallout breccia by Lehtinen (1970), is a grey-green or brownish rock which contains bedrock and mineral fragments and fragments of fluidal glass. The glass is very heterogeneous, mostly altered to zeolites or devitrified, and contains abundant fragments of quartz, quartz with planar deformation structures, diaplectic quartz glass and a few fragments of diaplectic and fused feldspar glass. Weakly shocked breccias, named impact breccias by Lehtinen (1970), are represented by rocks composed of angular, sometimes large fragments of bedrock and mineral fragments. These breccias do not contain glass. Indications of shock (minerals with shock-deformation structures, diaplectic glasses) are frequent.

The following indications have been found in rock and mineral inclusions in kärnäite and breccias: planar deformation structures in quartz parallel to $(10\bar{1}3)$, $(10\bar{1}2)$, $(21\bar{3}1)$, $(10\bar{1}1)$, $(10\bar{1}0)$, (0001) and $(51\bar{6}1)$ (Svensson, 1968; Lehtinen, 1970); reduced refractive indices of quartz with planar structures; diaplectic quartz glass; diaplectic plagioclase with isotropic lamellae and planar deformation structures; microcline with deformation lamellae and reduced birefringence; diaplectic plagioclase glass; biotite with kinkbands; transformation of biotite into an oxidized brown mass; kinkbands in graphite. Coesite was found in several specimens from suevite breccia and in a granite fragment from weakly shocked breccia (Lehtinen, 1970).

In the bedrocks of the area, Lehtinen (1970) found average contents of 6 ppm Co and 15 ppm Ni (Co/Ni=0,4). In kärnäite and breccias he found average contents of 4 ppm Co and 370 ppm Ni (Co/Ni=0.01). These results may indicate an admixture of meteoritic material.

In the general gravity map of Finland (Bouguer anomaly), Lake Lappajärvi basin occupies a negative anomaly of 5 - 10 milligals (Lehtinen, 1970).

The occurrence of shock-induced deformations, transformations and fusion clearly prove that the Lappajärvi structure was formed by impact. The crater was deeply eroded and kärnäite of Kärnasaari most likely represents an impact melt layer near the bottom of the crater (Svensson 1971). From morphology the diameter of the present structure may be estimated to be about 10 km. The impact occurred sometime between Precambrian and Pleistocene times.

Lake Jänisjärvi, Karelia, USSR, north of Lake Ladoga, 250 km north of Leningrad. A publication by Eskola (1921) about peculiar rocks from this locality makes it most probable that the lake occupies an old impact site. Lake Jänisjärvi has an elliptical form, with a long axis of 17 km and a short one of 11 km. The greatest depth is 50 m. Two small islands (West- and East Selkäsaari) are situated in the northwestern part of the lake. Bedrocks of the area are micaschists of variable composition.

On the Selkäsaari islands, Eskola (1921) found rocks which he interpreted as volcanics. At the north shore of East Selkäsaari a compact grey lava-like rock with one horizontal and two vertical joint planes rises to about 10 m above the lake level. It contains fragments of micaschist, quartzitic schist and lava in an amorphous groundmass with microlites of plagioclase, hypersthene, biotite and ellipsoidal amygdules. Toward the western shore of East Selkäsaari the rock plain sinks, gradually passing into agglomerates containing mostly fragments of the lava-like material, micaschist fragments and amygdules. At the southern shore lava-like rocks associated with agglomerates show very distinct fluctuation structures. Exposures at West Selkäsaari show mostly lava-like rocks with fluctuation structures.

Eskola (1921) showed that the chemical composition of the Jänisjärvi rock is little in accordance with that of normal igneous rocks (Table 2). Due to its high content in alumina and alkalies it has the composition of an argillaceous sediment which must be almost exactly the same as that of the sedimentogeneous crystalline schists of the region. Eskola therefore concluded that the Jänisjärvi rock was formed by a large-scale assimilation of country rocks by a volcanic lava. Eskola draws attention to the close similarity of

Lake Jänisjärvi to Mien, Dellen and Lappajärvi. The rocks from these localities agree in having a somewhat uncommon composition as compared with normal volcanics (high in alkalies, especially potassium, and lime). All these rocks are aphanitic lava-like rocks, associated with agglomerates or breccias, and all occur in the midst of round open lakes, strikingly different in shape from almost all the other lakes on the Fennoscandian shield.

These analogies make it very likely that the rock of Selkäsaari represents an impact melt and that the whole Jänisjärvi structure was formed by impact.

TABLE 2 — Chemical Composition of Impact Melt Rocks From Craters on the Baltic Shield (Eskola, 1921)

	I	II	III	IV
SiO_2	68.55	68.36	67.20	59.63
Al_2O_3	14.73	13.24	14.65	18.11
Fe_2O_3	0.37	1.29	—	1 12
FeO	2.56	3.39	3.24	5·67
MnO	0.73	0.27	—	0·14
MgO	1.39	1.15	1.03	2·15
CaO	2.83	2.51	2.78	3·06
Na_2O	2.88	2.05	3.28	2·17
K_2O	4.10	5.34	4.28	3·75
TiO_2	—	—	0.63	0.97
P_2O_5	—	—	0.13	0.17
H_2O	2.34	2.63	1.10	3..68
	100.48	100.23	99.51	100.62

I: Ramsö island, Lake Mien (Analyst H. Santesson)
II: Norrbonäset peninsula, Dellen Lakes (Analyst H. Santesson)
III: Kärnasaari island, Lake Lappajärvi (Analyst E. Mäkinen)
IV: Lake Jänisjärvi (Analyst P. Eskola)

IMPACT STRUCTURES IN CENTRAL EUROPE

Rochechouart, south-central France, 35 km west of Limoges. Outcrops of unusual breccias in the region around Rochechouart have attracted the attention of geologists for over 100 years. They were regarded as of volcanic origin. [For the older literature see Kraut (1969)]. Recent investigations have shown that the rocks contain abundant indications of shock (Kraut, 1967, 1969a and b; Kraut *et al.*, 1969; Kraut and French, 1971).

The rocks of the region are gneisses, granites, schists, porphyries and microgranite dikes belonging to a basement complex of Hercynian age. Farther west the basement is overlain by a series of Triassic and Cretaceous sediments.

The outcrops of impact formations occupy an area within a circle about 15 km in diameter. The rocks can be classified into four main types.

(1) Weakly shocked and brecciated authochthonous basement rocks are widely distributed in several small outcrops. The breccias are generally monomict. All gradations occur between fragmental breccias and extensively shattered but unfragmented rocks. Petrographic shock features are rare. West of Valette, exposed basement rocks exhibit planar deformation structures in quartz. In the same area shatter cones occur. The orientation of one set of cone axes suggests that the center of the structure is located in this region.

(2) Polymict weakly shocked breccias (lithic breccia of Kraut and French, 1971). The best exposure is the cliff underlying the château of Rochechouart, where the breccia is exposed for a thickness of 30 m. It is composed of fragments of different basement rocks ranging in size up to several decimeters. Observed shock features are planar deformation structures in quartz and kink bands in biotite. No glassy material is present. Shatter-coned fragments have been found at Rochechouart. Breccias of this kind are widely distributed. Their upper and lower contacts have not been established and thicknesses are unknown. Breccias of Rochechouart type are considered to have composed the central portion of the crater fill. They are, apparently, overlain by the suevite breccia of Chassenon.

(3) Strongly shocked breccias (or suevites) occur at Chassenon (glassy breccia of Kraut and French, 1971) and at Montoume (red welded breccia of Kraut and French, 1971). At Chassenon the buff to greenish breccia (10 m exposed) contains rock fragments without shock indications and fragments in all stages of shock metamorphism (I - IV) up to complete melting. Products of shock-induced melting are elongated or flat bodies of a heterogeneous glass including vesicles and mineral fragments. The glass is in part devitrified. The glass bodies appear to be identical to the glassbombs (Fladen) of the Ries crater. The suevite of Montoume (10 m exposed) apparently directly overlies the basement rocks. It is a hard, massive, dark red rock which exhibits crude jointing. It contains unshocked and shocked basement rocks in stages I - IV of shock metamorphism, up to several decimeters in size, and elongated bodies of a fluidal, heterogeneous and devitrified glass. The matrix is composed of fine fragments enclosed in devitrified, glassy material. Many of the shocked fragments are recrystallized and show indications of higher temperatures than the suevite of Chassenon.

(4) Impact melt. At Babaudus and Fonteceverane in the central part of the structure a massive, buff-colored crystalline rock overlies shocked and slightly brecciated basement rocks. It contains vesicles, basement rock and mineral fragments in a groundmass of chloritized glass with small laths of pyroxene and plagioclase. Some fragments show shock features of stages I - IV of shock metamorphism. This rock is considered to be a layer of impact melt representing the deepest unit of the crater fill.

The structure of Rochechouart represents a very deeply eroded impact crater. From the distribution of impact rocks and shock metamorphism, Kraut and French (1971) concluded that a central uplift, about 4 km in diameter, was situated in the area of Valette-Babaudus-Fonteceverane. For the diameter of the original crater a minimum value of 15 km was estimated.

Preliminary K-Ar age determinations of the impact melt rock have given ages of 154 and 173 million years, respectively, which should be regarded as a minimum value due to the alteration of the rocks.

Ries, southern Germany, around the city of Nördlingen, is 100 km northwest of Munich (Fig. 3). The circular depression, 24 km in diameter, has been the object of many geological investigations, for more than a hundred years. Older theories assumed that the structure and its peculiar breccias were produced by volcanic forces, or even by glaciation. The impact theory, tentatively expressed in 1904 by Werner and in 1936 by Stutzer, was first confirmed by the detection of coesite in suevite by Shoemaker and Chao (1961). Later physical, geophysical, geological and petrological investigations provided many further contributions to the knowledge of this crater and its formations, formed by an impact 14.8 million years ago. A theoretical approach to describing the Ries impact as a physical process was published by David (1969). Reviews in the older literature about the Ries were given by Löffler (1925), Kranz (1925-1952), Preuss (1964, 1969) and Dehm (1969). Several articles, representing the knowledge up to 1969, are contained in the volume "Das Ries", edited by Preuss and Schmidt-Kaler (1969). A summarizing review was published by Dennis (1971). The following remarks can only deal with the facts of general importance.

Nearly the whole Ries area has been geologically mapped at the 1:25,000 scale. For the published and unpublished sheets see Preuss (1964), Hüttner, Schmidt-Kaler and Treibs (1969). The mapping was done by several people with different theoretical backgrounds, over a long period of time (since 1925), and before the impact origin of the crater was recognized. Therefore, the maps, although representing extremely careful and reliable observations,

FIGURE 3 — Ries crater, Germany.

The outermost contour defines the area around the crater covered by multicolored breccia, consisting of fragmental sedimentary rocks. Hatched areas are rich in crystalline rocks excavated from the basement as large uniform blocks or monomict and polymict breccias. Black spots are occurrences of suevite. The morphological crater rim is represented by the 500-m contour line in the southern half and the 460-m contour in the northern half of the structure, respectively. The horseshoe-shaped ring of hills surrounding the central crater is indicated by a dashed line.

D = drill hole of Deiningen
W = drill hole of Wörnitzostheim
A — B = seismic profile, Figure 4.

do not provide all the information which is desirable from the point of view of the impact theory. A geological map (scale 1:100,000) of the whole Ries area was recently published by Hüttner, Schmidt-Kaler and Treibs (1969). It represents a synopsis of the information available to 1969.

In the Ries area the crystalline basement (gneisses, granites, amphibolites of Hercynian age) is overlain by a sequence of Mesozoic sediments, roughly 700 m thick. The sequence consists of about 300 m of Upper Jurassic limestone, about 150 m of Middle and Lower Jurassic sandstones and shales, about 250 m of Upper to Middle Triassic (Keuper) marls, sandstones and shales and Lower Triassic or even Permian sediments of variable thickness. The sediments dip gently to the south. The hard Upper Jurassic limestone forms an escarpment, the Swabian Alb, striking west-southwest. At the time of the impact the escarpment crossed the present Ries plain close to its center.

The flat bottom of the Ries depression has an average elevation of 420 m above sea level. The surrounding hills reach elevations of over 600 m in the south, where they are built up by Upper Jurassic limestone, but only 500 m in the north, where they consist of softer Middle Jurassic, Lower Jurassic and Keuper rocks. The basin rim has a somewhat polygonal outline, probably due to pre-existing fracture systems. The central part of the basin forms a flat plain of circular outline, about 10 km in diameter. It is surrounded by a horseshoe-shaped ring of hills up to 50 m in height which is open to the north. Between this inner ring of hills and the outer rim is a marginal zone of hummocky and irregular relief. The interior of the basin is filled by Tortonian and Sarmatian (Upper Miocene) lake sediments, covered in part by alluvial and Pleistocene deposits. Upper Miocene fresh-water limestones occur on top of the hills of the horseshoe ring and along the inner slope of the outer rim. At a distance of a few km (at most 5 km), the present rim is surrounded by a watershed. Two rivers enter the basin, the Wörnitz from the south and the Eger from the southwest. The Wörnitz river, after joining the Eger, leaves the basin through a narrow outlet in the southeast. Inlets and outlets represent old valleys already existing before the impact.

Johnson and Vand (1967) carried out an harmonic analysis of the Ries crater morphology based on 1:25,000 topographic maps. A main rim was found, 24 km in diameter, and two secondary ones, 34 and 45 km in diameter, respectively. No geological or other field evidence is known for ground movements at such great distances outside the crater. The contours of the rims obtained in this analysis are slightly elliptical. The major axis has a general north-northwest — south-southeast direction. Masses excavated by the impact and raised from the crater floor occur in the horseshoe ring, within the hummocky zone between this ring and the outer rim and as an ejecta apron outside the crater. Impact breccias have also been found in boreholes below the lake sediments in the central crater.

Within and around the Ries crater the following main types of impact formations can be distinguished and mapped:

(1) Blocks of shattered but uniform masses of Mesozoic sediments, sometimes more than 1 km in size. Predominant are blocks and intensely fractured masses of Jurassic limestones. Such masses occur within the marginal zone of the crater and outside the rim, within the region occupied by multicolored breccia. The participation of lower members of the stratigraphic column diminishes with increasing distance from the crater (Hüttner, 1969). Smaller fragments of Upper Jurassic limestone also occur south of the Danube as boulders at the base of Pleistocene deposits (Reutersche Blöcke) and em-

bedded in Upper Miocene fluviatile sediments which have the same age as the Ries event. Most of the known occurrences in Miocene sediments are located 50-70 km south and southeast of the Ries center. An isolated mass of heavily fractured limestone, about 2 m in diameter, has been found 140 km southeast of the crater (Hüttner, 1969; Herold, 1969; Engelhardt, unpublished observations).

(2) Multicolored breccia (Bunte Breccie) is a weakly shocked polymict breccia composed of all kinds of sedimentary rocks in variable proportions, with a small admixture of material from the crystalline basement. The components range in size from large blocks to the finest grains. Multicolored breccia occurs within the marginal zone of the crater and outside of it. Around the crater the multicolored breccia forms an apron extending from the rim up to 26 km to the southwest and up to 24 km to the south-southeast, but only 3-4 km to the north. In the south the extension of the breccia is interrupted by the Danube valley. The thickness is very variable; the average may be 20-24 m, but maximum values up to 80 m have been observed in the southeast. The amount of older sediments and of crystalline material in the breccia diminishes with increasing distance from the center. Indications of shock metamorphism in the multicolored breccia are a few quartz grains with planar deformation structures and some fragments of crystalline rocks in stages I and II of shock metamorphism. The multicolored breccia was moved along the ground, as can be concluded from striations on the surface of underlying limestone pointing to the crater centre, and from strong interaction and mixing with local unconsolidated Miocene sediments in the south (Hüttner, 1969; Schneider *in* Engelhardt *et al.*, 1969, 1971).

(3) Fragmental material from the crystalline basement occurs as uniform blocks of granite or gneiss, up to some 100 m in size, or as monomict and polymict weakly to moderately shocked breccias. The larger blocks are situated predominantly within the horseshoe ring. Crystalline breccias occur at the inner ring, within the marginal zone and outside the crater, predominantly in two rays extending to the south and southeast. Monomict breccias consist of fragments of one crystalline rock and show only minor shock effects (up to stage I of shock metamorphism). Polymict breccias are mixtures of different gneisses, granites and amphibolites. Shock effects include stages I and II of shock metamorphism, but no melting. Coesite and stishovite have been found in these breccias (Hüttner, 1969; Dressler, Graup and Matzke, 1969; Abadian, 1971; Graup, 1971; Stöffler *in* Engelhardt *et al.*, 1969).

(4) Suevite is a strongly shocked breccia consisting of crystalline rock fragments, a few sedimentary rocks, mineral fragments (predominantly quartz), glass fragments and glass bombs in a groundmass of montmorillonite. Crystalline rock and mineral fragments occur in all stages of shock metamorphism up to complete melting. Coesite and stishovite have been found in shocked rock fragments. Characteristic constituents are aerodynamically shaped bombs (Fladen) consisting of a heterogeneous schlieren-rich, mostly devitrified glass with inclusions of many mineral fragments (predominantly quartz), lechatellerite and vesicles. The chemical composition of the glass of all suevite occurrences is very uniform, corresponding to a gneiss of granitic norm (Hörz, 1965; Engelhardt and Hörz, 1965; Engelhardt, 1967). The Ni-contents seem to be higher than those of the crystalline source rocks (Staehle, 1970). As can be concluded from the remanent magnetization, the average temperature of the suevite at the time of its deposition was above the Curie temperature of magnetite, i.e. $>580°C$. Vertical vents formed by escaping vapor or gases

and early hydrothermal alterations show that the suevite mass included a certain amount of volatiles.

Suevite occurs (1) as a blanket underlying the lake sediments in the central basin (fall-back breccia) and (2) at the surface as isolated spots in the marginal zone and outside the crater, especially to the south and east (fallout breccia), up to 20 km from the crater rim. In most cases fallout suevite is underlain by multicolored breccia. The occurrences at the surface are remnants of an originally more or less continuous blanket of variable thickness. The present spot-like occurrences may represent thicker suevite accumulations trapped in local pockets of the hummocky multicolored breccia relief. Suevite was deposited as the last of all masses ejected from the crater. As can be concluded from structures at the lower contact (Schneider, 1971), the suevite mass, as a hot density current, overflowed the crater rim and rolled over the first-formed deposits of multicolored breccia. Because the shapes of the bombs indicate their rapid flight through the air it is probable that the suevite current developed from the breakdown of an eruption column ejected more or less vertically from the crater center. Suevite breccias below the lake sediments represent that part of the material which landed within the crater. Although suevite deposits outside the crater never show laminated or dune-like structures their mode of formation may be somewhat similar to that of volcanic base surge deposits as described by Waters and Fisher (1971). (Hüttner, 1969; Engelhardt, 1969, and in unpublished records).

(5) Impact melt. Red vesicular rocks, occurring in two small outcrops near the eastern rim (Polsingen, Amerbach), may represent impact melts. The rocks show a fluidal texture and contain many crystalline fragments in a groundmass consisting of feldspar laths, pyroxene, some cristobalite and hematite. They may be outlying tongues of impact melt rock sheets which occur deeper in the crater (Engelhardt, 1969; Dence, 1971).

Shatter cones are very rare in the Ries. Some not very well developed cones in a brecciated limestone from a crystalline breccia and some small ones in crystalline fragments have been found near the eastern rim. A large granite block with many well-developed shatter cones has been found in multicolored breccia at the southeastern rim (Dressler, Graup and Matzke, 1969; Graup, 1970, unpublished observations; Engelhardt, 1970, unpublished observations).

Results of gravity measurements carried out in the Ries area have been summarized by Jung, Schaaf and Kahle (1969) and Kahle (1969). The residual field obtained by subtraction of the regional field from the measured gravity values shows a centrosymmetrical negative anomaly the center of which (−18 milligal) is located at the topographical Ries center. By integrating the gravity anomaly, a total deficient mass is obtained between 6.8×10^{16} and 7.7×10^{16} g.

The Ries crater shows negative magnetic anomalies which are mainly located within the inner crater. The measurement of the magnetization of suevite has shown that the anomaly is due to a suevite layer of varying thickness, up to 400 m, in the central crater (Pohl and Angenheister, 1969).

Refractive seismic experiments from 1948 to 1952 and recent reflection profiles have revealed the complicated structure of the Ries crater. A profile, based on reflection data from outside the crater through the western rim and ending in the center (18 km), is shown in Figure 4. An inner crater, 10 km in diameter, is surrounded by the inner horseshoe rim. Within the inner crater Tertiary sediments of about 350 m maximum thickness have been

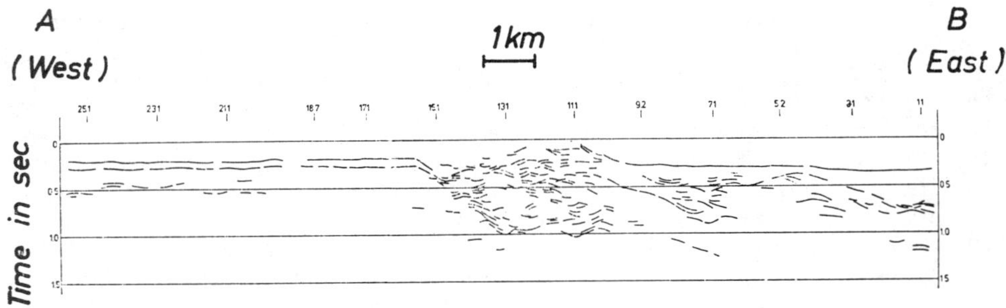

FIGURE 4 — Ries crater. Seismic reflection profile from outside the crater to the center. For location of the profile, see Figure 3.

deposited after the Ries event. Below these sediments lies a layer of suevite of varying thickness, up to 300-400 m. The layer underlying the suevite is a breccia of unknown composition. The inner ring is probably composed of fractured crystalline rocks. The marginal zone between the ring and the outer rim is mostly filled with various breccias. The western rim is clearly indicated by the abrupt disappearance of the reflections in the undisturbed Mesozoic layers and from the surface of the basement (Angenheister and Pohl, 1969).

Two drill holes have penetrated the lake sediments. The hole at Deiningen, approximately 1 km south of the topographic center, has revealed suevite underlying a sequence of 310 m of lake sediments. The hole at Wörnitzostheim, located in the southeast of the basin, in the marginal zone, close to the inner ring, has revealed 20 m of lake sediments, underlain by 80 m of suevite. Below the suevite lies 1.3 m of granite and multicolored breccia composed of Keuper and Middle Jurassic, in an inverted sequence (Förstner, 1967).

The distribution of ejecta masses is not centrosymmetrical around the Ries crater. Multicolored breccia, crystalline breccias and suevite occupy larger areas south and southeast of the crater than to the north and northwest. To some degree, the present distribution may be influenced by erosion which was more intense in the north than to the south. However, the present asymmetry cannot entirely be attributed to erosion alone. Undoubtedly, more brecciated material was ejected and deposited to the southeast and south than to the north and northeast. Asymmetrical features can also be recognized in the morphology of the crater. The horseshoe ring of crystalline material raised to the surface is open to the north or northwest and Johnson and Vand (1967) showed that the topographic crater shows an elliptical contour, the long axis pointing to the southeast. It is probable that these asymmetries originate from an oblique incidence of the impacting body which came in at a low angle from the north-northwest.

Age determinations (K-Ar) of glassy bombs from several suevite occurrences gave 14.8 ± 0.7 m.y.). These values have been confirmed by fission track measurements. The same age was found for the moldavites and bentonitic glass tuffs which occur near Landshut, Bavaria, 90 km southeast of the Ries center, in Upper Miocene sediments. It was suggested that both moldavites and glass tuffs were produced by the Ries impact. Several theories have been proposed about source material and mode of transportation, such as melting of surface material and ejection of jets early in the impact process, or on condensation from a silicate vapor. The chemical composition of moldatives and Bavarian glass tuffs is markedly different from that of Ries glass bombs (Cohen, 1961, 1963; Gentner, Lippolt and Schaeffer, 1963; Gentner and Wagner, 1969; David, 1969; Engelhardt and Hörz, 1965; Engelhardt, 1967).

Steinheim basin, southern Germany, is 65 km west-southwest of Stuttgart and 40 km west of the center of the Ries. The Steinheim basin has been investigated for a long time by geologists. A review of the older literature was given by Kranz (1924). Summaries of the present knowledge were published by Groschopf and Reiff (1966, 1969).

The nearly circular depression has a diameter of about 3.5 km. The bottom is 100-120 m lower than the surrounding peneplain of the Swabian Alb, which consists of Upper Jurassic limestone. In the center of the basin a crescent-shaped central uplift (Steinhirt-Kosterberg) rises to an height of 50-55 m above the basin floor. The depression has two outlets into the Stuben valley in the south, one to the south and a broader one to the southeast. The basin is filled by Upper Miocene (Sarmatian) lake sediments which are covered by Pleistocene and alluvial deposits. Lake sediments also occur at the slopes surrounding the basin and around the central uplift.

The central uplift consists of a chaotic mixture of brecciated and fractured Upper Jurassic limestones and Middle Jurassic clays and sandstones which have been raised from their original positions for up to 200-300 m. A recent borehole has revealed Upper Triassic (Keuper) rocks in a deeper level of the uplift. Shatter cones are frequent in Upper Jurassic limestone fragments of the uplift. Minor amounts of breccias consisting of Upper Jurassic limestone fragments and traces of Middle Jurassic rocks have been reported to occur at the southeastern and eastern slopes of the rim. Outside the basin, in the southeast, west and northwest, a few blocks of Upper Jurassic limestone have been found which may have been ejected from the crater. Outcrops in the southern and southwestern rim show the Upper Jurassic limestone slightly dipping into the basin, local brecciation and indications of overthrusting.

Boreholes in the basin have shown that at a depth of about 40 m below the surface the lake sediments are underlain by a weakly shocked fallback breccia consisting up Upper Jurassic limestone, Middle Jurassic limestone and Middle Jurassic clays and sandstones in a fine-grained greyish matrix. Middle Jurassic sandstone fragments and loose quartz grains from this breccia show planar deformation structures, 60% parallel to (0001) and 30% parallel to (10$\bar{1}$1) (Engelhardt, Bertsch, Stöffler, Groschopf and Reiff, 1967). The breccia has a thickness of 40 to 50 m and is underlain by slightly fractured Upper Jurassic limestone. Sediments, breccias and brecciated bedrock are indicated by a slight negative gravity anomaly which amounts to about 1 milligal in the center of the basin (Jensch, 1965; Müller, Ansorge, Emter and Greiner, 1969). The magnetic field shows a small positive anomaly 3 km south-southeast of the center (Müller *et al.*, 1969).

The topography of the basin, the central uplift with shatter cones, the weakly shocked fallback breccia and brecciation phenomena in the rim indicate that the Steinheim basin was formed by an impact and not by some kind of endogenic force as was assumed by earlier authors. Fossils in the lake sediments of Steinheim are of Sarmatian age and of the same kind as occur in the lake sediments of the Ries. Both impacts probably occurred at the same time. It seems most likely that they both were produced by one single event.

This event may have delivered several impacting bodies, besides the two largest ones which produced the Steinheim and Ries craters. Illies (1969) supposed a shallow depression in Upper Jurassic limestone at Pfahldorf, 58 km east-northeast of the Ries center, to be an impact structure belonging to the Ries-Steinheim event. The elliptical depression, 2.5 to 2 km in diameter, is partially filled by Upper Miocene fresh-water limestone. A group of depressions north of Kehlheim, 100 km east of the Ries, which are filled by breccias and Upper Miocene deposits, are supposed by Rutte (1971, personal communication) to be of impact origin. Rauser, Steinbrunn and Storzer (1971) suggested an impact site of Ries age near Weissenburg, 23 km northeast of the Ries center, on account of a slight culmination in Lower Jurassic beds and planar deformation structures in quartz.

Köfels, Ötztal, Austria, 60 km southwest of Innsbruck. The Ötztal is a typical U-shaped glacial valley with a flat bottom, approximately 1.5 km wide, and sharp flanks that rise to 1500 m. South of Umhausen a semicircular basin, about 4 km in diameter, is cut into the western flank of the valley. This embayment is occupied by a mass of coarse rubble and giant blocks which form a barrier, called Maurach, across the Ötztal. The river draining the valley has cut a narrow gorge, 4 km long, through the Maurach. The village of Köfels is situated in the northwestern part of the basin, 400 m above the bottom of the Ötztal. The Maurach mass as exposed in the gorge and around Köfels is composed of intensely fractured country rocks which consist in the Köfels area of granitic gneiss. The Maurach overlies gravel and fine sediments of the Ötztal.

Vesicular pumice-like glasses have been known from Köfels for more than a hundred years (Pichler, 1863; Hammer, 1923, 1924; Ampferer, 1939). They occur north of Köfels as a dike-like body, about 10 cm wide, at least 5 m deep and more than 10 m long, which fills a fault fissure in the granitic gneiss. The fault strikes about east-west, in a direction tangential to the outline of the basin. Another dike filled with glass, up to 4 cm wide, has been found in the gneiss east of Köfels (Heissel, 1965). Scattered pieces of the same glass have been found elsewhere near Köfels.

The Köfels glass exhibits varying degrees of vesicularity. Some glasses resemble volcanic pumice and contain so many bubbles, 0.5 to 5 mm in diameter, that the specimens float in water. In some of these samples the glass is uniform; others exhibit flow textures of glasses varying in color and refractive index. Dense glasses with few small vesicles are fluidal mixtures of various light and brown glassy materials. The glasses are not devitrified, except for some dark-colored portions which contain minute microlites. All glasses are very rich in inclusions of fragmental material, mostly quartz and some feldspar. Also, rock fragments occur and small pieces of dark breccias. Petrographic shock features are generally absent. Rims of lechatellerite around quartz grains, of feldspar glass around fragmental feldspar and melted biotite indicate a very high temperature of the melt. Recently, Storzer, Horn and Kleinmann (1971) reported, from a glass excavated at a deeper level of the Köfels vein, the occurrence of planar features in quartz and diaplectic quartz and feldspar glasses. Kurat and Richter (1968, 1969 and 1972) have shown that the chemical composition of the Köfels glass may be explained by short and incomplete fusion of the country rock at very high temperatures followed by very rapid cooling.

Pichler (1963), Hammer (1923, 1924) and Ampferer (1939) interpreted the Köfels structure and the glass as being produced by a volcanic event, although no young volcanism is known within the Alps. An impact origin was first suggested by Suess (1937) and Stutzer (1936) and later emphasized by Milton (1966) and Kurat and Richter (1968, 1969). Milton (1966) interpreted the Köfels glass as a veinlet of impact melt. Preuss (1971) explained it as a pseudotachylite formed by a large landslide. From glacial-geological observations, Senarclens-Grancy (1956) concluded that the Köfels event took place about 8150 B.C. The fission-track age of the glass was determined by Storzer, Horn and Kleinmann (1971) to be 8000 ± 6000 years.

REFERENCES

Abadian, M., 1971. (In Press). Petrographie Stoss wellen metamorphose und Entstehung polymikter kristalliner Breccien im Nördlinger Ries. Thesis, Univ. Tübingen, Contr. Mineral. and Petrol.

Ampferer, O., 1939. Die geologischen Hypothesen über die Formung des Talraums zwischen Umhausen und Langenfeld im Ötzale. Sitz. Ber. Akad. Wiss. Wien, Math.-nat. Kl. Abt. 1, 148, p. 401.

Angenheister, G., and Pohl, J., 1969. Die seismischen Messungen im Ries. Geologica Bavarica 61, p. 304-326.

Asklund, B., 1936. Diskussionsinlägg med anledning av P. Thorslunds föredrag "Nagra stratigrafiska och tektoniska rön fran Siljansomradet". Geol. För. Stockholm Förh. 58.

Bergquist, N. O., 1954. The moon puzzle. Copenhagen.

Cohen, A. J., 1961. A semi-quantitative asteroid impact hypothesis of tektite origin. J. Geophys. Res. 66, p. 2521.

———, 1963. Asteroid — or comet — impact hypothesis of tektite origin: the moldavite strewn field. Tektites (ed. by J. A. O'Keefe), p. 189-211, Univ. Chicago Press, Chicago.

David, E., 1969. Das Ries-Ereignis als physikalischer Vorgang. Geol. Bavarica 61, p. 350-378.

Dehm, R., 1969. Geschichte der Riesforschung. Geologica Bavarica 61, p. 25-35.

Dence, M. R., 1971. Impact melts. J. Geophys. Res. 76, p. 5552-5565.

Dennis, J. G., 1971. Ries structure, Southern Germany, a review. J. Geophys. Res. 76, p. 5394-5406.

Dietz, R., 1970. Siljan ring, Sweden. Meteoritics 5, p. 192.

Dressler, B., Graup, G., and Matzke, K., 1969. Die Gesteine des kristallinen Grundgebirges im Ries. Geol. Bavarica 61, p. 201-228.

Engelhardt, W. v., 1967. Chemical composition of Ries glass bombs. Geochim. Cosmochim. Acta, 31, p. 1677-1689.

———, 1971. Detrital impact formations. J. Geophys. Res. 76, p. 5566-5574.

Engelhardt, W. v., and Stöffler, D., 1963. Spaltflächen in Quarz als Anzeichen für Einschläge grosser Meteoriten. Naturwissenschaften 52, p. 489-490.

Engelhardt, W. v., and Hörz, F., 1965. Riesgläser und Moldavite. Geochim. Cosmochim. Acta 29, p. 609-620.

Engelhardt, W. v. Bertsch, W., Stöffler, D., Groschopf, D., and Reiff, W., 1967. Anzeichen für den meteoritischen Ursprung des Beckens von Steinheim. Naturwissenschaften 54, p. 198-199.

Engelhardt, W. v., Stöffler, D., and Schneider, W., 1969. Petrologische Untersuchungen im Ries. Geol. Bavarica 61, p. 229-295.

Eskola, P., 1921. On volcanic rocks in Lake Jänisjärvi in Eastern Finland. Bull. Comm. Géol. Finlande 55, p. 1-13.

Förstner, U., 1967. Petrographische Untersuchungen des Suevits aus den Bohrungen Deiningen und Wörnitzostheim im Ries von Nördlingen. Contr. Mineral. Petrol. 15, p. 281-308.

Fredriksson, K., and Wickman, F. E., 1963. Meteoriter. Svensk Naturvetenskap 16, p. 121-157.

Gentner, W., Lippolt, H. J., and Schaeffer, O. A., 1963. Argonbestimmungen an Kaliummineralien XI: Die Kalium-Argon-Alter der Gläser des Nördlinger Rieses und der böhmisch-mährischen Tektite. Geochim. Cosmochim. Acta 27, p. 191-200.

Gentner, W., and Wagner, G. A., 1969. Altersbestimmungen an Riesgläsern und Moldaviten. Geol. Bavarica 61, p. 296-303.

Graup, G., 1971. Kristalline Schollen des Nördlinger Ries. Ph.D. thesis, Tübingen. (In preparation).

Groschopf, P., and Reiff, W., 1966. Ergebnisse neuer Untersuchungen im Steinheimer Becken (Würtemberg). Jh. Ver. vaterl. Naturkde. Würtemberg 121, p. 155-168.

———, 1969. Das Steinheimer Becken. Ein Vergleich mit dem Ries. Geol. Bavarica 61, p. 400-412.

Hammer, W., 1923. Über das Vorkommen jungvulkanischer Gesteine im Ötztal (Tirol) und ihr Alter. Sitz. Ber. Akad. Wiss. Wien, Math.-nat. Kl, 102, p. 329-342.

———, 1924. Ein Vorkommen jungvulkanischen Gesteins in den Tiroler Zentralalpen. Z. Vulkanol. 8, p. 238-243.
Heissel, W., 1965. Das "Bimssteinvorkommen" von Köfels im Ötztal (Tirol). N. Jb. Mineral. Mh. 1965, p. 257-367.
Herold, R., 1969. Eine Malmkalk-Trümmermasse in der Oberen Süsswassermolasse Niederbayerns. Geol. Bavarica 61, p. 413-427.
Hisinger, W., 1825. Underrättelse om lager af petrifaktförande kalksten pa Humlenäs i Calmar län. Kongl. Vetenskapacademiens Handlingar.
Hjelmqvist, S., 1966. Beskrivning till berggrundskarta över Kopparbergs län. Sver. Geol. Unders. Ser. Ca. No. 40.
Högbom, A. G., 1910. Geol. Fören. Stockholm Förh. 32, p. 482.
Holmberg, H. J., 1858. Materialier Finlands geognosi. Bidrag till Finlands naturkännedom, etnografi och statistik. Finska Vetens. Soc., 4 Häftet, Helsinki, 118.
Holst, N. O., 1890. Ryoliten vid sjön Mien. Sveriges Geologiska Undersökning, Ser. C. No. 110.
Hörz, F., 1965. Untersuchungen an Riesgläsern. Beitr. Mineral. Petrogr. 11, p. 621-661.
Hüttner, R., 1969. Bunte Trümmermassen und Suevit. Geol. Bavarica 61, p. 142-200.
Hüttner, R., Schmidt-Kaler, H., and Treibs, W., 1969. Geologische Übersichtskarte des Ries. Geol. Bavarica 61, p. 451-454 und Beilage.
Illies, H., 1969. Nördlinger Ries, Steinheimer Becken, Pfahldorfer Becken und die Moldavite: strukturelle Zusammenhänge einer Impakt-Gruppe. Oberrhein. geol. Abh. 18, p. 1-31.
Jensch, A., 1965. Geophysikalische Untersuchungen im Steinheimer Becken. N. Jb. Mineral. Mh. 1965, p. 279-284.
Johnson, G. G., and Vand, V., 1967. Application of a Fourier data smoothing technique to the meteorite crater Ries Kessel. J. Geophys. Res. 72, p. 1741-1750.
Jung, K., Schaaf, H., and Kahle, H. G., 1969. Ergebnisse gravimetrischer Messungen im Ries. Geol. Bavarica 61, p. 337-342.
Kahle, H. G., 1969. Abschätzung der Störungsmasse im Nördlinger Ries. Z. Geophys. 35, p. 317-345.
Kranz, W., 1924. Das Steinheimer Becken. *In*: Begleitworte zur geognostischen Spezialkarte von Württemberg. Atlasblatt Heidenheim, 2. Aufl. Württemb. Statist. Landesamt Stuggart, 138 p.
———, 1925-1952. Beiträge zum Nördlinger Ries-Problem. 3.-7. und 9. Fortsetzung: Zbl. Mineral. (B) 1925, 199-201; 1927, 403-407; 1934, 262-267; 1937, 215-216; N. Jb. Mineral., Mh. 1945-1948, 336-339; N. Jb. Geol. Paläontol., Mh. 1952, 49-50.
Kraut, F., 1969a. Sur la présence de cônes de pression ('shatter cones') dans les brèches et roches éruptives de la région de Rochechouart. C.R.H. Acad. Sci., Ser. D, 269, p. 1486.
———, 1969b. Über ein neues Impaktit-Vorkommen im Gebiete von Rochechouart-Chassenon (Departements Haute-Vienne und Charente), Frankreich. Geol. Bavarica 61, p. 428-450.
Kraut, F., Short, N. M., and French, B. M., 1969. Preliminary report on a probable meteorite impact structure near Chassenon, France (Abs.). Meteoritics 4, p. 490.
Kraut, F., and French, B. M., 1971. The Rochechouart meteorite impact structure, France: Preliminary geological results. J. Geophys. Res. 76, p. 5407-5413.
Kulonpalo, M., 1969. Kärnäite blocks in central and south Finland or the largest block fan in Finland. Geologi, 21, p. 80-81.
Kurat, G., and Richter, W., 1968. Ein Alkalifeldspatglas im Impaktit von Köfels. Tirol. Naturwissenschaften 55, p. 490.
———, 1969. On the "pumice" from the Köfels crater, Austria. Meteoritics 4, p. 192.
———, 1972. Impaktite von Köfels, Tirol. Tschermaks Min. Petr. Mitt. 17, p. 23-45.
Laitakari, A., 1942. Suomen Geologinen Yleiskartta, Lehti B 3 Vaasa, 39, 64 p. Suom. Geol. Toimikunta, Helsinki.
Lehtinen, M., 1970. New evidence for an impact origin of Lake Lappajärvi, Western Finland. Bull. Geol. Soc. Finland 42, p. 89-93.
Linnarson, G., 1878. De paleozoiska bildningarna vid Humlenäs i Smaland. Sveriges Geologiska Undersökning Ser. C. No. 28.
Löffler, R., 1925. Beiträge zur Riesentstehungshypothese. Mitt. oberrh. geol. Ver., N.F. 14, p. 75-83.
Lundqvist, G., 1963. Beskrivning till jovdartskarta över Gävleborgs län. Sveriges Geologiska Undersökning Ser. Ca. No. 42.
Milton, D. J., 1965. Fused rock from Köfels, Tyrol. Tschermaks Mineralog. petrogr. Mitt. 9, p. 86-94.
Mölder, K., 1948. Die Verbreitung de Dacitblöcke in der Moräne in der Umgebung des Sees Lappajärvi. Bull. Comm. Géol. Finlande 142, p. 45.

Müller, St., Ansorge, J., Emter, D., and Greiner, G., 1969. Geomagnetische Messungen im Gebiet des Steinheimer Beckens. Oberrhein. geol. Abh. 18, p. 33-46.
Pichler, A., 1863. Zur Geognosie Tirols II. Die vulkanischen Reste von Köfels. Jahrb. K. K. Geol. Reichsanstalt Wien 13, p. 591-593.
Pohl, J., and Angenheister, G., 1969. Anomalien des Erdmagnetfeldes und Magnetisierung der Gesteine im Nördlinger Ries. Geol. Bavarica 61, p. 327-336.
Preuss, E., 1964. Das Ries und die Meteoritentheorie. Fortschr. Mineral. 41, p. 271-312.
———, 1969. Einführung in die Ries-Forschung. Geol. Bavarica 61, p. 12-24.
———, 1971. Uber den Bimsstein von Köfels/Tirol. Fortschr. Mineral. 49, Beiheft 1, 70.
Preuss, E., and Schmidt-Kaler, H. (*Editors*), 1969. Das Ries. Geol. Bavarica 61, 478 p.
Rauser, P., and Steinbrunn, F., 1971. Evidence for a triplet cratering event in the Ries area formed by fission of a single meteorite under the earth's tidal forces (Abstract). Meteoritics 6, p. 304.
Redaelli, L. L., 1957. A petrological investigation in Lake Norra Dellen by means of frog-man equipment. Sver. Geol. Unders. Ser. C. No. 548.
Rutten, M. G., 1966. The Siljan ring of Paleozoic. Central Sweden: a posthumous ring complex of a late Precambrian Dala porphyries caldera. Geol. en Mijnbouw 45, p. 125-136.
Saksela, M., 1949. Das pyroklastische Gestein von Lappajärvi und seine Verbreitung als Geschiebe. C.R. Soc. géol. Finlande No. 22, p. 19-30.
Schneider, W., 1971. Petrologische Untersuchungen der Bunten Breccie im Nördlinger Ries. N. Jb. Miner. Abh. 114, 136-180.
Senarclens-Grancy, W., 1956. Zur Glazialgeologie des Ötztals und seiner Umgebung. Mitt. Geol. Ges. Wien 49, p. 256.
Shoemaker, E. M., and Chao, E. T. C., 1961. New evidence for the impact origin of the Ries basin, Bavaria, Germany. J. Geophys. Res. 66, p. 3371-3378.
Staehle, V., 1970. Nickel und Kobalt in Gesteinen des Nördlinger Ries. Contr. Mineral. and Petrol. 28, 72-88.
Stanfors, R., 1969. Lake Mien — an astrobleme or a volcanic-tectonic structure. Geol. Fören. Stockholm Förh. 91, p. 73-86.
Stöffler, D., 1971. Progressive metamorphism and classification of shocked and brecciated crystalline rocks at impact craters. J. Geophys. Res. 76, p. 5541-5551.
Storzer, D. Horn, P., and Kleinmann, B., 1971. The age and the origin of the fused rock from the Köfels structure, Austria (Abstract). Meteoritics 6, p. 319-320.
Stutzer, O., 1936. Die Talweitung von Köfels im Ötztal (Tirol) als Meteorkrater. Z. Deut. Geol. Ges. 88, p. 523-524.
———, 1936. Meteor Krater und Nördlinger Ries. Z. Deut. Geol. Ges. 88, 510-523.
Suess, F. E., 1936. Der Meteor-Krater von Köfels bei Umhausen im Ötztale, Tirol. N. Jb. Mineral. Beil. Bd. 72, Abt. A, p. 98-155.
Svedmark, E., 1892. Mötet den 7 april 1892. Geol. Fören. Stockholm Förh. 14.
———, 1904. Beskrivning till kartbladet Oskarshamn. Sver. Geol. Unders. Ser. Ac., No. 5.
Svenonius, F., *in* Blomberg, A., 1895. Praktiskt geologiska undersökningar inom Gefleborgs län. Sver. Geol. Unders., Ser. C., No. 152, p. 54-59.
Svensson, N. B., 1966. Lake Hummeln, a possible astrobleme in Southern Sweden. 1. The bottom topography. Sver. Geol. Unders. Ser. C., No. 608.
———, 1968. The Dellen lakes, a probable meteorite impact in Central Sweden. Geol. Fören. Stockholm Förh. 90, p. 309-316.
———, 1968. Lake Lappajärvi, central Finland: a possible meteorite impact structure. Nature 217, p. 438.
———, 1969. Lake Mien, southern Sweden — a possible astrobleme. Geol. Fören. Stockholm Förh. 91, 101-110.
———, 1969. Meteorite impact craters on the Scandinavian Precambrian basement. Meteoritics 4, p. 208.
———, 1971. Lappajärvi structure, Finland: morphology of an eroded impact structure. J. Geophys. Res. 76, p. 5382-5386.
———, 1971. Probable meteorite impact crater in Central Sweden. Nature Physical Sci. 229, p. 90-92.
Svensson, N. B., and Wickman, F. E., 1965. Coesite from Lake Mien in Central Sweden. Nature 205, 1202-1203.
Tryggvason, T., 1940. Zwei erratische Blöcke von der Küste Upplands. Bull. Geol. Inst. Upsala 28, 175-190.
Waters, A., and Fisher, R. V., 1971. Base surges and their deposits: Capelinhos and Taal volcanoes. J. Geophys. Res. 76, 5596-5614.
Werner, E., 1904. Das Ries in der schwäbisch-fränkischen Alb. Bl. schwäb. Albverein 16, p. 153-167.

3

Reprinted from *Jour. Geophys. Research* **76**:5407-5413 (1971)

The Rochechouart Meteorite Impact Structure, France: Preliminary Geological Results[1]

FRANÇOIS KRAUT

Mineralogie, Muséum Nationale d'Histoire Naturelle Paris, 5e, France

BEVAN M. FRENCH

Planetology Branch, NASA, Goddard Space Flight Center Greenbelt, Maryland 20771

The Rochechouart meteorite impact structure is located in south-central France, about 50 km west of the city of Limoges and near the town of Rochechouart (45.49°N latitude, 0.50°E longitude). No topographic expression of a circular feature is apparent in the region, which consists of gently rolling farm and forest land with a maximum relief of about 100 meters. The rocks underlying the region are a basement complex of Hercynian age that includes gneisses, granites, schists, porphyries, and microgranite dikes. Further west, near the city of La Rochefoucauld, this basement is unconformably overlain by a series of sediments of Triassic to Cretaceous age. (For detailed discussions of the geology, see *Kraut* [1969*b*] and references cited there.)

[1] Lunar Science Institute Contribution 35.

The remnants of the impact structure itself consist of scattered outcrops of unusual breccias that have attracted attention for over 100 years [*Kraut*, 1969*b*]. Recently, it was recognized that the rocks contain shock-metamorphic effects regarded as diagnostic for meteorite impact [*Kraut*, 1967, 1969*a*, *b*; *Kraut et al.*, 1969]. The results reported here are based on a joint field investigation of the structure in the summer of 1969 [*Kraut*, 1969*a*, *b*].

STRATIGRAPHY AND ROCK TYPES

The present ground surface lies at approximately the level of the original crater floor, exposing scattered outcrops of a variety of crater-related rock units. These units represent both the fragmental crater fill and the shocked autochthonous rocks below the crater floor (Figures 1, 2). Although no one locality shows the complete section, it is possible to reconstruct the

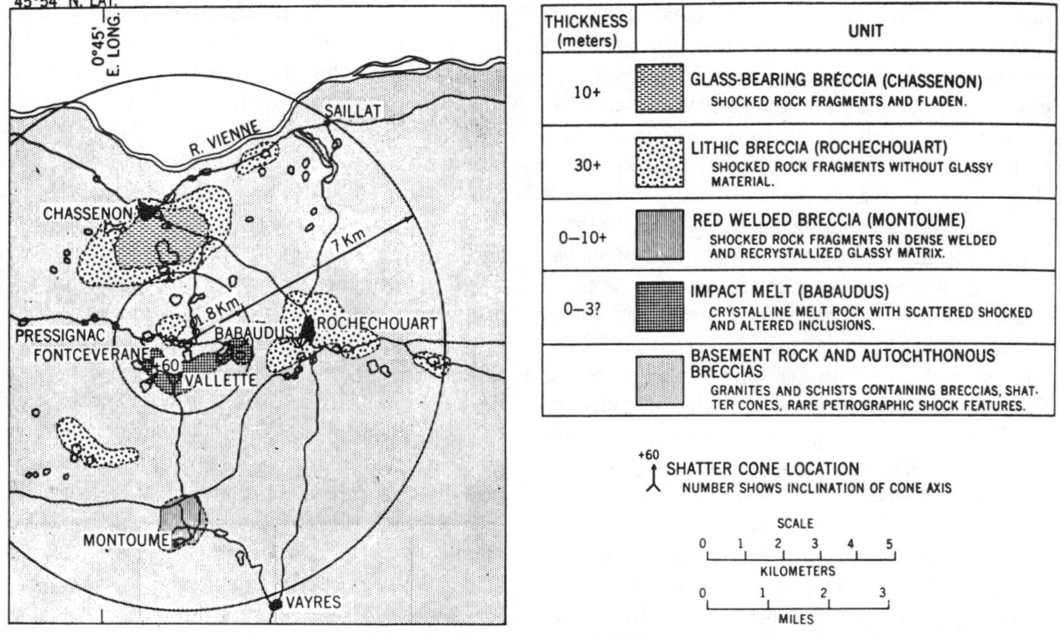

Fig. 1. Preliminary geological map of the Rochechouart structure, France, showing remnants of impact-produced rock units from the original crater fill. Small outlined areas are actual outcrops; remainder of distribution is inferred.

original stratigraphic column by analogy with rock types from other impact craters, particularly Brent and Clearwater Lakes, Canada [*Dence*, 1968].

Five distinct impact-related rock units can be distinguished at Rochechouart, the shocked autochthonous rocks below the original crater floor and four units from the original crater fill. These units have been combined into a stratigraphic column as follows, from the base upward:

1. *Shocked and brecciated autochthonous basement rocks.* Distinctive shock-metamorphic features, together with breccias of possible impact origin, are found in the rocks of the crystalline basement which were below the original crater floor. The distribution of the breccias is uncertain because of the limited field work and the scarcity of outcrops. Where found, the breccias are generally monomict and composed of fragments of the host rock. Rotation and movement of the constituent fragments vary in different outcrops, and gradations exist between clearly fragmental breccias and extensively shattered but unfragmented rock. Petrographic shock features are generally not found in these breccias, and their formation by impact cannot therefore be definitely established. However, they strongly resemble breccias found in similar locations in the Brent crater, Canada [*Dence*, 1968, p. 172].

Definitely shock-metamorphosed basement rocks are exposed in the central part of the Rochechouart structure near the villages of Babaudus, Vallette, and Fontceverane (Figure 1). Shater cones, developed in both schists and microgranite, were discovered in 1969 about 1 km west of Vallette [*Kraut*, 1969*a*] (see Figure 1). The orientation of one set of cone axes [see *Manton*, 1965; *Howard and Offield*, 1968] suggests that the center of the structure is located in this region but does not provide an exact location. In this same area, exposed basement rocks exhibit well-developed planar features (shock lamellae) in quartz that show a range of distinctive fabrics identical to those at other impact structures (see below and Figure 3).

2. *Impact melt.* The rock type believed to be the lowest unit of the crater fill is a thin, discontinuous, buff-colored, massive, crystalline

melt rock that overlies the shocked basement rocks in the central part of the Rochechouart structure (Figure 1). The contact is well exposed at Babaudus and Fontceverane, and the melt rock, which is probably less than two meters thick, rests directly on shocked and slightly brecciated basement rocks. The rock is vesicular, often heterogeneous, and sometimes exhibits local flow structure. Inclusions of shock-metamorphosed basement rocks and mineral fragments are common; these are often found as reddish bands or layers intermixed with buff-colored melt. In thin section, the melt rock is fine grained and consists of laths of pyroxene and plagioclase feldspar in a matrix of chloritized glass [*Kraut*, 1969*b*, pp. 442–444]. Vesicles are often lined or filled with secondary iron-rich chlorite.

This unit of melt rock strongly resembles impact melt formed by fusion of target rocks at other impact craters, especially Brent and Clearwater Lakes, Canada [*Dence*, 1968] and Tenoumer, Mauritania [*French et al.*, 1970]. The value of $^{87}Sr/^{86}Sr$ determined by P. M. Hurley on a sample of the Babaudus impact melt was 0.744, too high for a primary magma and strongly indicative of derivation from the underlying basement rocks [*Hurley*, 1969, p. 30]. Preliminary K-Ar age determinations on the same material give ages of 154 and 173 m.y., both ± 8 m.y. (J. B. Hartung, personal communication, 1969); in view of the observable alteration of the melt rock, these numbers should be considered as minimum values only.

3. *Red welded breccia.* A massive unit composed of large shocked inclusions of basement rock in a welded fragmental matrix is exposed in a quarry near the village of Montoume in the southern part of the structure (Figure 1), where it forms a cliff about 10 meters high and apparently lies directly on basement rock without any intervening impact melt. As exposed in the quarry, the welded breccia is a massive dark-red unit that exhibits crude jointing and contains shocked and altered inclusions of various basement rocks up to several decimeters in size. The matrix is composed of finer fragments enclosed in a slightly devitrified glassy material that contains numerous small crystals of hematite. Welding of the material has been sufficiently strong that joints and other fractures pass through the larger inclusions.

The inclusions in the welded breccia show a range of strong shock-metamorphic effects, including extensive development of planar features in quartz, partial isotropization of feldspar, oxidation of biotite, and incipient selective melting [*Kraut*, 1969*b*, pp. 441–442]. Many of the shocked inclusions are annealed, recrystallized, and altered, probably from thermal

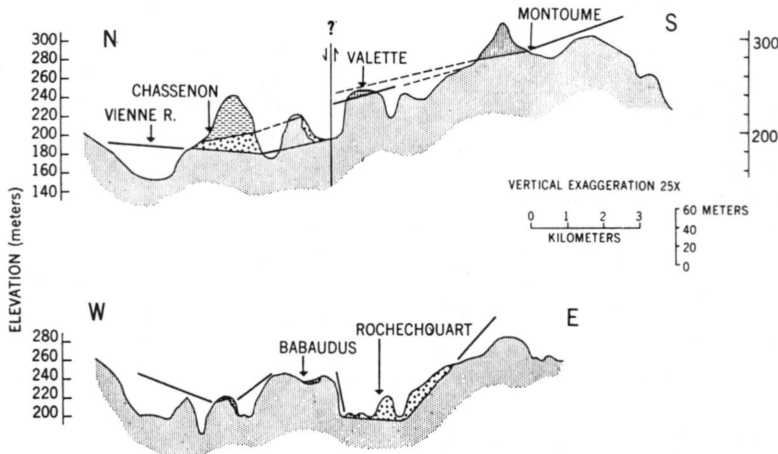

Fig. 2. Schematic N–S and E–W cross sections through the center of the Rochechouart structure, France. Formation symbols are the same as in Figure 1. Heavy line is the contact between original crater fill and shocked autochthonous basement rocks, which corresponds to the original crater floor. Note that the E–W cross section, despite vertical exaggeration, appears roughly symmetrical, whereas the N–S cross section clearly indicates uplift of the southern part of the structure relative to the northern. The fault shown is inferred.

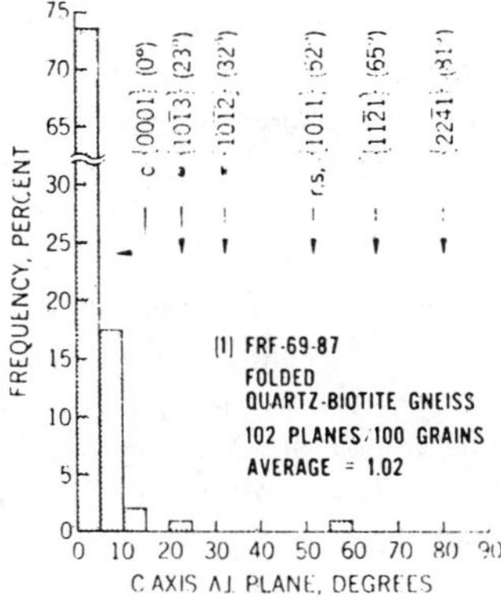

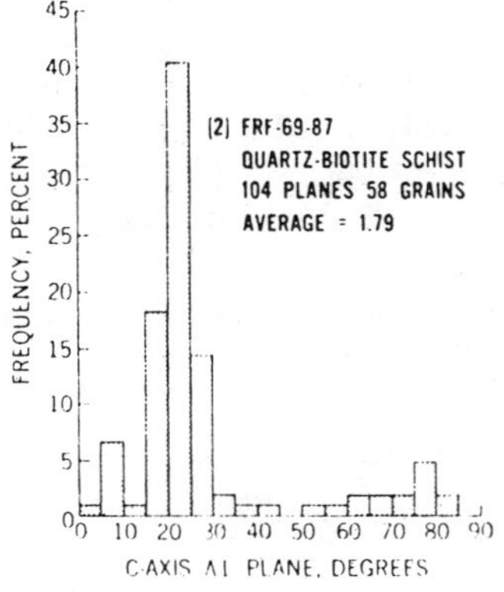

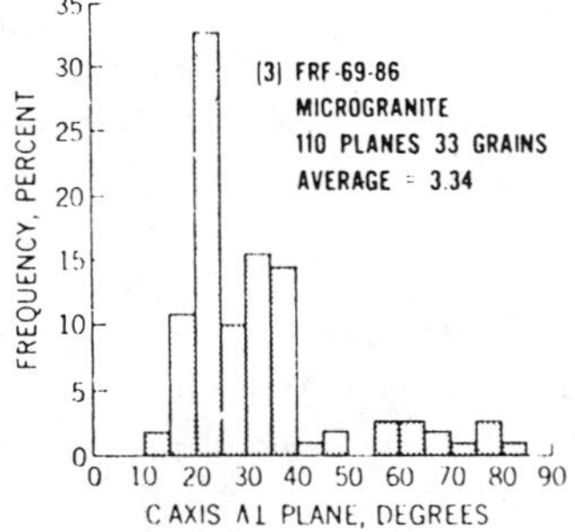

Fig. 3. Histograms showing orientations of planar features in quartz grains of specimens collected from shocked autochthonous basement rocks in the central part of the Rochechouart structure, France. Abscissa is the angle between the quartz *c* axis and the plane normal; ordinate is frequency in per cent. Plotting interval is 5°, estimated uncertainty is ±4°. The strong concentration of planes at angles corresponding to specific crystal planes (listed at top) is typical of shock-metamorphic features. The diagrams duplicate fabrics described from Canadian craters [*Robertson et al.*, 1968]. Note the strong concentration of basal planes (type A, *top*), the strong $\omega\{10\bar{1}3\}$ concentration (type B, *center*), and the combined $\omega\{10\bar{1}3\}$-$\pi\{10\bar{1}2\}$ concentrations (type D, *bottom*).

effects associated with compaction and welding of the breccia.

The welded breccia at Montoume is very similar to rocks observed at other impact craters, particularly to a welded breccia at Nicholson Lake, Canada [*Dence et al.*, 1968, pp. 353–355].

4. *Lithic breccia.* The term *lithic breccia* designates a unit of crater fill composed of diverse rock fragments, many of which show shock-metamorphic features, and lacking glassy material. The best exposure is in the cliff underlying the chateau of Rochechouart, where the lithic breccia forms a massive unit about 30 meters thick. The unit is polymict and poorly sorted; fragments range in size up to several decimeters and represent all basement rocks exposed in the region. Petrographic shock effects, especially planar features in quartz and kink bands in biotite and muscovite, are common [*Kraut*, 1969*b*, pp. 444–446], and shatter-coned fragments have been found in the unit at Rochechouart [*Kraut*, 1969*a*].

The lithic breccia is more widely distributed than is the impact melt or the welded breccia (Figure 1). Its upper and lower contacts have not been established, and its thickness is uncertain. It is considered to have composed the central portion of the section of crater fill and to be similar to the 'weakly shocked' and 'moderately shocked' breccias preserved in the Brent crater, Canada [*Dence*, 1968, pp. 172–173]. It is apparently overlain by the more friable glass-bearing breccias exposed in the neighborhood of Chassenon.

5. *Glassy breccia* (= suevite). The highest unit of crater fill recognized at Rochechouart is a poorly sorted buff to greenish breccia containing both shock-metamorphosed basement rock fragments and heterogeneous bodies composed of altered glass and rock fragments which appear identical to the *Fladen* from the Ries basin, Germany [*Hörz*, 1965; *von Engelhardt et al.*, 1969]. This unit is best exposed in the vicinity of Chassenon (Figure 1), where a minimum thickness of about 10 meters is exposed in Gallo-Roman quarries. Except for more extensive alteration and chloritization of glassy material, this unit appears identical to the suevite from the Ries basin [*Kraut*, 1967, 1969*b*]. Shock effects are common in the rock fragments and include planar features in quartz

and feldspar, kink-banding in micas, partial to complete isotropization of feldspar, incipient selective melting, and the formation of heterogeneous glassy bodies by extreme melting. The total thickness of the unit is unknown, and the location of the deposits at Chassenon with respect to the original crater rim is not yet known.

Shock-Metamorphic Features

Despite the relatively great age of the Rochechouart structure and the alteration of many of the impact-produced rocks, a wide variety of distinctive shock-metamorphic effects are present throughout the structure [*Kraut*, 1967, 1969*a*, *b*; *Kraut et al.*, 1969]. Because of the similarity between these features at Rochechouart and those observed at numerous other proved impact structures, the origin of Rochechouart by meteorite impact is considered established, despite the scattered outcrops and the absence of any observable circular topographic feature.

One result of the 1969 field study was the recognition that outcrops of weakly to moderately shocked autochthonous basement rocks are associated with scattered outcrops of impact melt in the central part of the structure (Figure 1). The similar association of these features in better-exposed structures [*Robertson*, 1967; *Dence*, 1968] indicates that this area may represent the central uplift of the original Rochechouart crater.

The orientation of the planar features (shock lamellae) within the host quartz grains was measured in three samples collected from different locations in this central area. Standard Universal stage techniques were used to determine the angle between the *c* axis and the normal to the plane [*Robertson et al.*, 1968; *von Engelhardt and Bertsch*, 1969; *French et al.*, 1970]. The results are presented in the conventional histogram plot (Figure 3).

The Rochechouart specimens clearly exhibit typical shock-metamorphic fabrics and show strong concentrations of plane normals at specific angles that correspond to unique planes in the quartz crystal. More important, the three specimens show different orientation patterns, all of which are identical to patterns observed at numerous other impact craters, and in particular to patterns observed in drill core from beneath the Brent crater, Canada [*Dence*, 1968; *Robertson et al.*, 1968]. The strong concentration of basal planes in specimen FRF-69-79 (Figure 3) is identical to the type A pattern from Canadian craters [*Robertson et al.*, 1968]. Higher shock levels are indicated in the other two specimens; the development of the $\omega\{1013\}$ peak at 23° indicates the Type B fabric, whereas the occurrence of the $\pi\{1012\}$ peak at 32° in the third Rochechouart specimen is typical for the type D fabric (compare with *Robertson et al.* [1968], p. 440–442).

These results are extremely important for interpretation of the Rochechouart structure. First, the fabric exhibited by the planar features (shock lamellae) in quartz provides additional evidence for the impact origin of the Rochechouart structure and is consistent with the wide variety of shock-metamorphic effects already observed.

Second, the occurrence of the intense type D fabric indicates that the basement rocks at this locality were close to both the center of the structure and the original crater floor. At the Brent crater, Canada, this fabric is restricted to rocks less than about 60 meters below the base of the melt layer, which apparently marks the original crater floor [*Dence*, 1968, pp. 172–174]. At Rochechouart, material showing the type D fabric was collected from a 17-meter-deep well on the property of M. Quichaud at Le Bouchet, a small farm located about midway between Babaudus and Fontceverane (Figure 1). This point, which is also consistent with the one shatter-cone orientation available, can be considered provisionally to be the center of the Rochechouart structure.

Third, the occurrence of a range of shock fabrics in quartz collected from the basement rocks in the central area indicates the possibility of mapping different shock zones in the basement to give a better location for the center of the structure [see *Robertson*, 1967].

Geological Structure

The original circular Rochechouart crater has been largely removed by erosion, and a topographic circular feature is no longer detectable. However, the distribution of impact-produced rocks and shock-metamorphic effects makes possible a reasonable reconstruction of the original crater.

All presently preserved outcrops of impact melt and breccias from the original crater fill can be enclosed by a circle about 15 km in diameter (Figure 1). Although it is possible that some of the suevitelike glassy breccias at Chassenon may have been deposited beyond the original crater rim, the value of 15 km appears reasonable for the minimum original diameter of the crater. The association of shatter cones, shocked autochthonous basement rocks, and impact melt in the central part of the structure suggests further that the area of Vallette-Babaudus-Fontceverane represents a central uplift about 4 km in diameter (Figure 1), a feature that should be expected in a crater of this size [*Dence et al.*, 1968].

Schematic cross sections drawn E–W and N–S through the center of the structure (Figure 2) strongly indicate deformation of the crater after its formation such that the southern part has been raised relative to the northern part. Allowing for the sparse outcrops and the high vertical exaggeration, the relations shown in the E–W cross section appear to represent an originally symmetrical crater profile. In contrast, the N–S cross section clearly shows that the crater floor has been tilted up at the south end, so that the welded breccia at Montoume, presumably low in the original crater fill, has been elevated above the stratigraphically higher lithic and glassy breccias of Rochechouart and Chassenon. This deformation could be due to faulting or to regional tilting, possibly accompanying the episode of later faulting that has affected the Mesozoic sediments to the west near La Rochefoucauld.

Conclusions

The Rochechouart impact structure is the first to have been recognized by the discovery of shock-metamorphic effects in an area where no individual structure is topographically visible. Despite erosion and excavation of the structure to a level approximating the original crater floor, information derived from the study of other impact craters allows a reasonable reconstruction of the Rochechouart structure. The original crater is considered to have been approximately 15 km in diameter, with a central uplift about 4 km in diameter that still preserves highly-shocked basement rocks overlain by impact melt.

The rocks of the Rochechouart structure exhibit a wide variety of shock-metamorphic effects which establish its origin through meteorite impact. Shock-metamorphic effects observed include shatter cones, planar features (shock lamellae) in quartz, partial to complete isotropization of feldspar, kink-banding in muscovite and biotite, formation of heterogeneous glassy bodies (*Fladen*), and development of an impact-melt unit by fusion of the basement rock.

Despite its extensive erosion, the Rochechouart structure holds considerable promise for future study because of the wide variety of impact-related rock types and the easy accessibility of subsurface shock zones by drilling. More detailed geological mapping of the structure, including mapping of the petrographic shock zones in the central uplift, will allow more exact estimation of the original dimensions of the crater. Further age dating of both basement rocks and impact melt is needed for better estimates of the age of the structure and to establish whether the crater was formed before, during, or after the deposition of the Mesozoic sediments now preserved to the west of Rochechouart.

Acknowledgment. Publication was, in part, supported by NASA contract NSR 09-051-001.

References

Dence, M. R., Shock zoning at Canadian craters: Petrography and structural implications, in *Shock Metamorphism of Natural Materials*, edited by B. M. French and N. M. Short, p. 169, Mono, Baltimore, Md., 1968.

Dence, M. R., M. J. S. Innes, and P. B. Robertson, Recent geological and geophysical studies of Canadian craters, in *Shock Metamorphism of Natural Materials*, edited by B. M. French and N. M. Short, p. 339, Mono, Baltimore, Md., 1968.

French, B. M., J. B. Hartung, N. M. Short, and R. S. Dietz, Tenoumer crater, Mauritania: Age and petrologic evidence for origin by meteorite impact, *J. Geophys. Res.*, *75*, 4396, 1970.

Hörz, F., Untersuchungen an Riesgläsern, *Beitr. Mineral. Petrogr.*, *11*, 621, 1965.

Howard, K. A., and T. W. Offield, Shatter cones at Sierra Madera, Texas, *Science*, *162*, 261, 1968.

Hurley, P. M., Rb-Sr isotopic investigation of crater rocks of possible impact origin, in *Variations in Isotopic Abundances of Strontium, Calcium, and Argon and Related Topics*, *Doc. MIT-1381-17* (Annual Progress Report for 1969, U.S.A.E.C. contract At(30-1)-1381), p. 27, MIT

Press, Cambridge, Mass., 1969.

Kraut, F., Sur l'origine des clivages du quartz dans les brèches 'volcaniques' de la région de Rochechouart, *C. R. H. Acad. Sci., Ser. D, 264,* 2609, 1967.

Kraut, F., Sur la présence de cônes de pression ('shatter cones') dans les brèches et roches éruptives de la région de Rochechouart, *C. R. H. Acad. Sci., Ser. D, 269,* 1486, 1969*a*.

Kraut, F., Über ein neues Impaktit-Vorkommen im Gebiete von Rochechouart-Chassenon (Départements Haut-Vienne und Charente, Frankreich), *Geol. Bavarica, 61,* 428, 1969*b*.

Kraut, F., N. M. Short, and B. M. French, Preliminary report on a probable meteorite impact structure near Chassenon, France (abstract), *Meteoritics, 4,* 490, 1969.

Manton, W. I., The orientation and origin of shatter cones in the Vredefort Ring, *Ann. N.Y. Acad. Sci., 123,* 1017, 1965.

Robertson, P. B., La Malbaie structure, Quebec—a Paleozoic meteorite impact site, *Meteoritics, 4,* 89, 1967.

Robertson, P. B., M. R. Dence, and M. A. Vos, Deformation in rock-forming minerals from Canadian craters, in *Shock Metamorphism of Natural Materials,* edited by B. M. French and N. M. Short, p. 433, Mono, Baltimore, Md., 1968.

von Engelhardt, W., and W. Bertsch, Shock induced planar deformation structures in quartz from the Ries crater, Germany, *Contrib. Mineral Petrol., Berlin, 20,* 203, 1969.

von Engelhardt, W., D. Stöffler, and W. Schneider, Petrologische Untersuchungen im Ries, *Geol. Bavarica, 61,* 229, 1969.

(Received January 5, 1971;
revised February 25, 1971.)

4

Reprinted from *Jour. Geophys. Research* **76**:5382–5386 (1971)

Lappajärvi Structure, Finland: Morphology of an Eroded Impact Structure[1]

NILS-BERTIL SVENSSON[2]

Norrlandsfonden, S-951 00 Luleå, Sweden

Within the last ten years, data accumulated from several distinct lines of research have produced strong evidence for the occurrence of large meteorite impacts on the surface of the earth through geologic time [*French,* 1968]. Current tabulations consider more than 50 structures to have originated from such events [*French,* 1969].

During a large meteorite impact, a relatively small volume of rock near the impacting body will be subjected to shock pressures in excess of 500 kb, sufficient to produce melting of the rock. Such impact-melted material has been found in several forms, including individual bodies composed of mixtures of glass and rock and mineral fragments, glass-rich breccias containing numerous shocked rock fragments, and relatively thick and uniform layers resembling sills, associated with the breccias but composed of completely crystalline rock with few or no inclusions.

Dence [1968] has suggested that these impact melt layers are material melted close to the penetrating meteorite. Their location should indicate the approximate depth of penetration of the meteorite into the target rock. The layers may be as much as several hundred meters thick and range in character from medium-grained to aphanitic. Their characteristics include: (1) chemical composition similar to that of the underlaying target rock; (2) considerable secondary hydrothermal alteration and zeolitization; (3) absence of phenocrysts; (4) absence of flow structure, indicating crystallization virtually in place; (5) generally fine grain with numerous quench textures in the glassy varieties; (6) presence of occasional shocked inclusions or lechatelierite derived from fused quartz grains; (7) association with distinctively shocked and shock-melted breccias.

[1] Lunar Science Institute Contribution 62.

[2] On leave from the Geological Survey of Sweden, S-104 05 Stockholm, Sweden.

DESCRIPTION OF THE LAPPAJÄRVI STRUCTURE

Location and morphology of the Lappajärvi area. Lake Lappajärvi (63°10′N, 23°40′E) is situated about 200 miles northwest of Helsinki (Figure 1). The area was glaciated during the last glaciation period.

The topography of the Lappajärvi area is characterized by its flatness [*Mölder and Salmi,* 1955]. The surface is generally covered by till, the depressions of which are filled with peat bogs, and only small and scattered outcrops of bedrock occur. Drumlins are the most discernible till form. The glaciofluvial deposits are of minor importance and include some areas of gravel, sand, and silt.

Odenwall [1934] sounded Lake Lappajärvi. The depth contours have been transferred onto the map (Figure 1). Two narrow bottom valleys occur close to the eastern and western shorelines of the lake. Another one stretches NE–SW and divides the lake bottom into two shallower platforms, a northern one with the island of Kärnäsaari and a southern one. Several small islands also occur. They have not been marked on the general map.

Geology. The bedrock in the Lappajärvi area

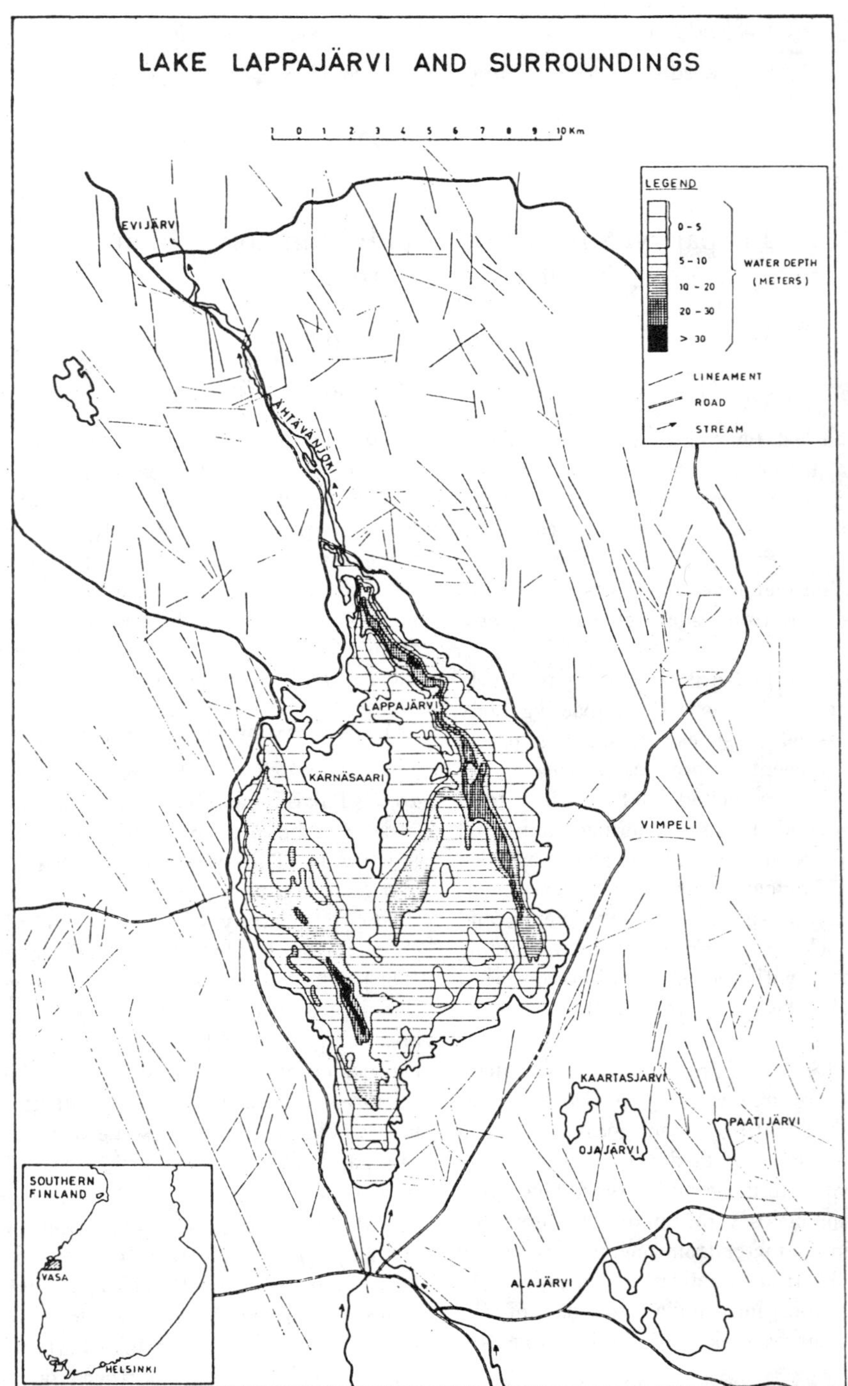

Fig. 1. Map of Lake Lappajärvi and surroundings with photogeological lines and depth contours. The approximate location of the geological map sheet Vasa with Lappajärvi is evident from the survey map of southern Finland.

is composed of Precambrian rocks, including granodiorite, biotite plagioclase gneiss, and granite [*Laitakari,* 1942]. On the island of Kärnäsaari, (Figure 1), *Holmberg,* [1858] found a peculiar rock type later named kärnäite, (H. Berghell, in *Saksela* [1949]). It has a variable appearance and was thought to be composed of pyroclastic material, agglomerates, and tuffs. Pure lava types were recorded to be of secondary importance (see Saksela). The fragments range in size from several centimeters to sand granules. They are made up of material from the bedrock surrounding the lake. The matrix has an aphanitic texture. The composition of kärnäite is rhyodacitic according to a chemical analysis [*Eskola,* 1921, analysis by E. Mäkinen].

I have examined fragments in boulders of kärnäite originating from Lake Lappajärvi [*Svensson,* 1968*a*] and have recorded the occurrence of planar features in quartz grains following the crystallographic planes including $(10\bar{1}3)$ or $(01\bar{1}3)$, $(10\bar{1}2)$ or $(01\bar{1}2)$, $(21\bar{3}1)$ or $(12\bar{3}1)$, and $(10\bar{1}1)$ or $(01\bar{1}1)$. Such planar features have been developed as a consequence of shock waves during laboratory experiments at pressures of at least 100 kb [*Hörz,* 1968; *Müller and Défourneaux,* 1968]. They are regarded as strong criteria for a meteorite impact-event [*Bunch and Cohen,* 1963, 1964; *von Engelhardt and Stöffler,* 1965]. I [*Svensson,* 1968*a*] concluded that the Lappajärvi structure most probably originated from such an incident. *Lehtinen* [1970] strongly confirmed this opinion by reporting such shock-metamorphosed minerals as coesite, maskelynite, and diaplectic quartz in breccia boulders originating from the structure. The kärnäite thus most probably correspond to an impact-melt layer according to *Dence* [1968]. This rock is younger than the surrounding Precambrian bedrock. *Laitakari* [1942], *Mölder* [1948] and *Saksela* [1949] studied the distribution pattern of kärnäite boulders that had been transported from the structure during the last glaciation. They concluded that kärnäite most probably can be found all over the northern platform as far to the north as about the village of Lappajärvi. This means that the meteorite impact crater is located in the northern part of the lake.

The crystalline bedrock, including granites and gneisses, is by comparison with most other rock types very resistant to ice erosion. However, certain types of kärnäite boulders have been found as far away as about 200 miles south-southeast of Lappajärvi, transported there by the ice [*Kulonpalo,* 1969]. This rock type thus also seems to have offered great resistance to mechanical disintegration. The same holds true for the kärnäite in situ. In this connection it is noteworthy that the northern of the two platforms in Lappajärvi, composed of kärnäite, is shallower than the southern, where basement rock types seem to occur on the islands [*Laitakari,* 1942]. Indeed, the depth of the southern platform is probably representative of what is normal for a lake of Lappajärvi size in this basement area.

A study of available topographic maps indicates that no significant height differences exist between the Kärnäsaari island and the neighborhood around Lake Lappajärvi. Thus the melted bedrock material in the crater seems to be at least as capable of resisting decomposition and erosion as the surrounding bedrock material that has been eroded to approximately the same level.

The narrow valleys on the bottom of the lake follow weaker parts of the bedrock that have been excavated through the glacial ice erosion. They encircle and separate the area that is covered by kärnäite from the surrounding normal Precambrian rock types. The impact-melt layer in the Lappajärvi structure is roughly circular and has a diameter of 5–6 km (Figure 1).

DISCUSSION AND CONCLUSIONS

The lineament pattern. I have studied the lineament pattern of the Lappajärvi area from stereopairs of black and white aerial photographs at the scale of about 1 : 65,000. No field checking was done. The aim of the study was to see if the orientation of the bottom valleys that follow weaker parts of the bedrock and surround the Kärnäsaari platform bore any relationship to the general lineament pattern of the area (Figure 1).

The general trend of the lineament pattern in the area is NNW–SSE and parallels the ice-movement direction. A secondary lineament system stretches about NE–SW. Ice-erosion was greatest along lines of weakness parallel to the direction of movement and least perpendicular to this direction. Indeed, in the latter case, sedimentation of material might even have occurred. This might very well explain the predominance of the NNW-SSE general trend of the lineament pattern over the NE-SW trend. Some glaciomorphological features such as drumlins might further have been mistaken for true lineament lines and strengthened this impression.

Bottom valleys not consistent with a tectonic origin. The river Ähtävänjoki (Figure 1) most probably follows an important fracture zone that parallels the NNW-SSE lineament trend. The zone probably coincides with the narrow bottom valley along the eastern shoreline of Lake Lappajärvi but is not so pronounced further southeastward. There it can be followed with interruptions and is, for example, represented along the western shores of the lakes Kaaratasjärvi and Alajärvi. Thus the eastern bottom valley in the lake might very well be interpreted as having been formed by erosion of some of the weaker parts of the bedrock along the Ähtävänjoki tectonic zone.

The curved bottom valley along the western shore of Lappajärvi is somewhat broader than the eastern one. This probably is due to the fact that a part of the zone of kärnäite rock has been transported away by the ice. The southern part of the valley might be interpreted as representing another zone of weakness parallel to that of the Ähtävänjoki river. The central and northern parts of the valley, however, do not fit with such an orientation nor with the general northeastern strike of the secondary lineament.

A tectonic origin for the more easily eroded bedrock along the western bottom valley thus does not seem to be very likely. The valley to the southeast of the Kärnäsaari platform might be regarded as part of the NE-SW lineament system. However, it has been excavated to a considerable extent by the ice in spite of the fact that it is situated perpendicular to the ice-movement direction. The valley must thus represent an important weak zone in the bedrock. However, it does not extend either to the northeast or the southwest of the lake. This should most probably be the case if the valley's orientation was tectonically controlled.

Thus it does not seem to be very likely that the valleys to the west and southeast of the Kärnäsaari platform have a tectonic origin although this might be the case with the eastern bottom valley along the Ähtävänjoki line.

Bottom valleys consistent with meteorite impact. This interpretation of the morphology of the Lappajärvi structure is based on a comparison with the Brent crater, which has most probably been caused by a meteorite impact [*Dence*, 1965; 1968]. The stratigraphy in the central part of Brent crater includes, from the top to the bottom, (1) sedimentary fill; (2) irregular melt zones with coarse-grained, well-mixed, vesicular fragment breccias and lenses of less deformed breccias; (3) breccias with rarely occurring melted fragments; (4) melt zone; and (5) fractured gneisses. Along the marginal parts of the crater, melted layers are insignificant or lacking.

The Lappajärvi crater has been eroded to such an extent that a horizon corresponding to zone (4) in the Brent crater now appears on the top of the remaining structure. For example, this layer will act like a diabase sill and will protect the underlying breccias from being eroded.

By comparison with the circumstances in the Brent crater, the breccias in the marginal parts of the Lappajärvi structure are thought to have been left uncovered by any melt zone. The ice erosion attacked these parts of the crater much more effectively than either the cap of melt zones or the surrounding Precambrian bedrock. The marginal parts of the structure thus now appear as valleys (Figure 1), except on the northern and a part of the southern sides, where, because of the NNW-SSE movement of the ice, deposition of material appears to have occurred, particularly to the north.

Similarity to other impact structures. Boulders resembling the kärnäite boulders in Lappajärvi have been reported from Lake Mien, southern Sweden [*Holst*, 1890] and the Dellen lakes (F. Svenonius, in *Blomberg* [1895]). These structures were later interpreted as being caused by meteorite impacts [*Svensson and Wickman*, 1965; *Svensson*, 1968*b*] and seem also to be capped by irregular melt zones.

Stanfors [1969] found narrow valleys along the shorelines of the Mien structure. He interpreted them as being caused by erosion along tectonic zones. It appears, however, that not all of them can be satisfactorily explained in this way. Instead, they might very well be interpreted in the same way as the valleys on the bottom of Lake Lappajärvi. F. Svenonius [*Blomberg*, 1895] reported that the central part of the north Dellen lake was comparatively shallow. Along the shoreline, however, deeper troughs appear to occur. The topography of the Mien and the Dellen structures seem to resemble the topography of the Lappajärvi crater.

CONCLUSIONS

Meteorite craters that have zones with considerable amounts of melted bedrock will be easily eroded down to bedrock levels. The margins of the craters that are not protected will be further eroded, giving rise to troughs or valleys. Such a stage most probably exists for relatively long periods of time. This type of crater can probably be found randomly all over the world, whereas smaller craters lacking melted layers must very soon be covered by sediments to survive erosion.

Acknowledgment. This publication was, in part, supported by NASA contract 09-051-001.

REFERENCES

Blomberg, A., Praktiskt geologiska undersökningar inom Gefleborgs län, *Sveriges Geol. Undersokn., C 152,* 54–59, 1895.

Bunch, T. E., and A. J. Cohen, Coesite and shocked quartz from Holleford crater, Ontario, Canada, *Science, 142,* 379–381, 1963.

Bunch, T. E., and A. J. Cohen, Shock deformation of quartz from two meteorite craters, *Bull. Geol. Soc. Amer., 75,* 1263–1266, 1964.

Dence, M. R., The extraterrestrial origin of Canadian craters, in *Symposium on Geological Problems in Lunar Research, Trans. N. Y. Acad. Sci., 123,* 941–969, 1965.

Dence, M. R., Shock zoning at Canadian craters: Petrography and structural implications, in *Shock Metamorphism of Natural Materials,* edited by B. M. French and N. M. Short, p. 169, Mono, Baltimore, Md., 1968.

Eskola, P., On volcanic necks in Lake Jänisjärvi in Eastern Finland, *Bull. Comm. Géol. Finlande, 55,* 1–13, 1921.

French, B. M., Shock metamorphism as a geological process, in *Shock Metamorphism of Natural Materials,* edited by B. M. French and N. M. Short, p. 1, Mono, Baltimore, Md., 1968.

French, B. M., Possible relations between meteorite impact and igneous petrogenesis as indicated by the Sudbury structure, Ontario, Canada, *NASA Doc. X-644-69-371,* 50 pp., 1969.

Holmberg, H. J., Materialier till Finlands geognosi: Bidrag till Finlands naturkännedom, etnografi och statistik, *Finska Vetenskapssocieteten, no. 4,* Helsinki, Finland, 1858.

Holst, N. O., Ryoliten vid sjön Mien, *Sveriges Geol. Undersökn., Ser. C, 110,* 3–50, 1890.

Hörz, F., Statistical measurements of deformation structures and refractive indices in experimentally shock loaded quartz, in *Shock Metamorphism of Natural Materials,* edited by B. M. French and N. M. Short, p. 243, Mono, Baltimore, Md., 1968.

Kulonpalo, M., Kärnäiittilohkareita keski—ja eteläsuomessa eli suomen pisin lohkareviuhka (Kärnäite blocks in central and south Finland or the longest block fan in Finland): *Geologi, 21,* 80–81, 1969.

Laitakari, A., *Suomen Geologinen Yleiskartta, Lehti B 3 Vaasa, 39,* 64 pp., Suomen Geologinen Toimikunta, Helsinki, Finland, 1942.

Lehtinen, M., New evidence for an impact origin of Lake Lappajärvi, western Finland, *Bull. Geol. Soc. Finland, 42,* 89–93, 1970.

Mölder, K., Die Verbreitung der Dacitblöcke in der Moräne in der Umgebung des Sees Lappajärvi, *Bull. Comm. Géol. Finlande, 142,* 45, 1948.

Mölder, K., and Salmi, M., *The General Geological Map of Finland, Sheet B 3 Vaasa,* Geologinen Tutkimuslaitos, Helsinki, Finland, 1955.

Müller, W. F., and M. Défourneaux, Deformationsstrukturen in Quarz als Indikator für Stosswellen: Eine Experimentelle Untersuchung an Quarz-Einkristallen, *Z. Geophys., 34,* 483–504, 1968.

Odenwall, E., Lake Lappajärvi. Bathymetric chart. Notes on thermal conditions, *Hydrografisen Toimiston Tiedonantoja 6,* pp. 1–24, Helsinki, Finland, 1934.

Saksela, M., Das pyroklastische Gestein von Lappajärvi und seine Verbreitung als Geschiebe, *C. R. Soc. Geol. Finlande, 22,* 19–30, 1949.

Stanfors, R., Lake Mien—an astrobleme or a volcanotectonic structure, *Geol. Fören. Stockholm Förh., 91,* 73–86, 1969.

Svensson, N. B., Lake Lappajärvi, central Finland: A possible meteorite impact structure, *Nature, 217,* 438, 1968*a*.

Svensson, N. B., The Dellen lakes, a probable meteorite impact in central Sweden, *Geol. Fören. Stockholm Förh., 90,* 314–315, 1968*b*.

Svensson, N. B., and F. E. Wickman, Coesite from Lake Mien, southern Sweden, *Nature, 205,* 1202–1203, 1965.

von Engelhardt, W., and D. Stöffler, Spaltflächen im Quarz als Anzeichen für Einschläge grosser Meteoriten, *Naturwiss, 52,* 489–490, 1965.

Part II
STRUCTURES OF CLASS IV IN EUROPEAN AND ASIATIC RUSSIA

Editor's Comments on Papers 5 and 6

5 MASAYTIS
Astroblemes in the U.S.S.R.

6 MASAYTIS, MIKHAYLOV, and SELIVANOVSKAYA
The Popigay Meteorite Crater

The nine structures in the European part of Russia are considered together with three in Asia, which are the only definite identifications of cryptoexplosion structures throughout the entire continent, though the Ramgarh structure in India (Crawford, 1972; Part VIII) could be another. These structures are listed in Table 5.

Table 5 Structures of class IV in European Russia and Siberia

Paper no.	*Name*	*Country*	*Diameter* (km)	*Class*	*Age*
		Europe			
2, 5	Janisjarvi		17	IVa/2	Post-Proterozoic
5	Puchezh-Katunki		80	IVa/2	Jurassic
5	Boltyshka		25	IVa/2	Cretaceous
5	Rotmistrovka		2	IVa/1	Cretaceous
5	Kaluga		15	IVa/1	Devonian
5	Mishinogorsk		2.5	IVa/1	Paleozoic
5	Ilumets*		5	IVa/1	Paleozoic
5	Kamensk		25	IVa/2	Paleogene
5	Gusev		3	IVa/1	Paleogene
—	Obolon**		12	IVa/2	Jurassic
		Asia			
5, 6	Popigay		100	IVa/4	Paleogene
5	Kara		50	IVa/2	Cenozoic
5	Beyenchime-Salata		8	IVa/2	Cenozoic

*Classified as class III previously (McCall, 1977, pp. 4, 352).
The Obolon structure was described by A. A. Val'tev, Ye. P. Gurov, and V. A. Ryabenko in *Dokl. Akad. Nauk S.S.S.R.* **232:170-173 (1977). A twin funnel structure, probably with a central uplift, this is associated with shock lamellations,

LAKE JANISJARVI

All the twelve structures described by Masaytis in Paper 5 are of class IVa. The Lake Janisjarvi structure is also described by von Engelhardt in Paper 2. Situated 250 km north of Leningrad, this lake is elliptical (17 X 11 km), and up to 50 m deep. The country rocks are Precambrian mica schists. On an island within the lake, Eskola (1921) recorded grey lava-like volcanic rocks, with inclusions of schist and lava fragments set in an amorphous matrix containing amygdules and microlites of plagioclase, hypersthene, and biotite. Agglomerates were recorded as outcropping on the east shore of the island. Chemical analyses reveal the lavas to have an unusually high alumina and alkali content, their composition being equivalent to that of argillite. Eskola suggested widespread assimilation of country rock. Their peculiar chemistry resembles that of "lavas" from Lake Mien, Lake Dellen, and Lake Lappajarvi, all showing high alkali content (especially potassium) and also high lime. The fact that these unusual lavas occur around open lakes, which contrast with other Fennoscandian glacial lakes, is considered by von Engelhardt to favor impact origin.

POPIGAY

Popigay is covered by Masaytis in Paper 5 and in more detail by Masaytis, Mikhaylov, and Selivanovskaya in Paper 6. It is an immense structure, 100 km in diameter, situated on the north-central Siberian platform. It has been supposed to be a volcano-tectonic depression filled with Cretaceous to early Cenozoic lavas and tuffs of intermediate composition, but Masaytis et al. (Paper 6) draw analogy to the Rieskessel and consider it to be of impact origin. There can be little doubt that it is broadly analogous to the Rieskessel, whatever the origin of Popigay and the Rieskessel. There are two confocal funnels, respectively 75 and 100 km in diameter, and a semicircular rim uplift zone composed of Archean gneiss of the Anabar shield. In the outer funnel, thrusting outward from the funnel over deformed sediments (which are in situ) has occurred. Authigenic breccias of gneiss and associated pseudotachylites; fusion of breccia fragments; shatter cones; shock deformations of mineral grains, including lamellations in quartz; and isotropization are recorded. Ejected blocks up to 15 m in diameter and allogenic breccia extend as far as 70 km from the focus, that is, 40 km from the edge of the inner funnel. Formations previously termed tuffs are now regarded as megabreccias. They consist of a variety of rocks derived from both basement and platform cover, cemented by friable sand/clay material within the confines of the basin (Masaytis et al., 1972). Glasses, pre-

---

diaplectic glasses, and magnetic particles in allogenic breccias. Taenite and kamacite have been identified diffractometrically in the latter.

viously termed andesite porphyries, are now considered to be impactites, analogous to suevite. Lechatelierite and maskelynite are recorded. Dikes, necks, and irregular projections of locally porous fragmental glasses and partly crystallized massive glasses (tagamites) are exposed within the inner crater. They mainly cut allogenic breccias as discontinuous sheets fed by them overlie these breccias. The tagamites are claimed to have the overall composition of the shield rocks, but details of their major and trace-element chemistry do not seem to entirely endorse this claim, and there is some doubt about the manner in which the average values have been derived. These "vitroclastic impactites" (sic) are said to show some resemblance to ignimbrites (cf. Stevenson, 1972). They show a lack of sprinklings of plagioclase and a low Fe_2O_3/ FeO ratio (0.3–0.5). Chemical analyses from Popigay are given in Table 6.

The Popigay meteorite is considered to have driven to a depth of 1.5 km beneath the surface and on explosion to have liberated energy in the order of 10^{30} ergs within the Archean basement. The scale of this

Table 6 Average chemical composition of impactites and rocks of crystalline shields (evaluated at 100% without water and volatiles

	1	*2*	*3*	*4*	*5*
SiO_2	64.5	64.8	64.4	64.4	65.5
TiO_2	.8	.9	.6	.6	.8
Al_2O_3	14.9	14.1	16.6	15.5	15.7
Fe_2O_3	1.8	2.6	2.4	1.8	.8
FeO	5.7	5.2	3.1	2.8	3.8
MnO	.1	.1	.1	.1	–
MgO	3.7	3.4	2.3	2.0	2.7
CaO	3.7	4.4	4.2	3.8	3.6
Na_2O	1.9	1.8	3.1	3.5	3.2
K_2O	2.7	2.6	3.0	3.3	3.7
P_2O_5	.2	.1	.2	.2	.3
Number of analyses	15	16	39	–	24

Source: From Masaytis et al., 1972.
1 Average composition of tagamites of Popigay basin
2 same for suevite
3 Average composition, rocks of Anabarskii shield (Rabken, 1959)
4 Average composition of crystalline shields (Poldervaart)
5 Average composition of suevite, Nördlinger Ries (Hörz, 1965)

structure exceeds any other known cryptoexplosion structure except the early Proterozoic Sudbury and Vredefort structures. Masaytis et al. consider that the immense scale of the cryptoexplosion structure accounts for the prolific pseudovolcanic expressions of Popigay.

OTHER CLASS IV RUSSIAN STRUCTURES

The other structures described by Masaytis from Russia have only recently become known to western science. They include large and complex structures with central uplifts, ring deformations, and complex deformation patterns (Puchez-Katunki, 80 km diameter; Kara, 50 km; Boltyshka, 25 km; Kamensk, 25 km; Kaluga, 15 km). All are associated with suevite and massive "impactite." Shatter cones are present at Puchez-Katunki, Kara, and Kamensk. Diaplectic minerals and glasses are associated with all but Kamensk. Surprisingly, none of the Russian structures of large size has yielded high-pressure silica polymorphs, except Popigay. Of the smaller Russian structures, Beyenchime-Salata (8 km) and Mishinogorsk (2.5 km) have yielded shatter cones, and both the latter and the Rotmistrovka structure (2 km) are associated with diaplectic glasses. Characteristically, as elsewhere, the smaller structures are less complex and do not possess central uplifts.

The spectrum of cryptoexplosion in Russia is a very similar to that in Canada; though, as yet, the available descriptive literature is much less detailed for the Russian structures and photographic illustrations are virtually unavailable.

The petrography of this group and the chemistry of the "igneous rocks" calls for much more descriptive study. The tagamite and suevite of Popigay appears less potassic than the doreites of the large Canadian structures, but the rocks of Janisjarvi appear to be similar.

A recent paper by Nechayeva (1975) reports shock effects of unique type associated with supposed astroblemes in alkali granites of the Kola peninsula in northwest Russia. No definite cryptoexplosion structures have been outlined. This occurrence is intriguing, for the Kola peninsula is a classic igneous province of alkaline igneous intrusions of central type associated with explosion vents (diatremes).

REFERENCES

Crawford, A. R. 1972. Possible Impact Structure in India. *Nature* **237**:96.

Eskola, P. 1921. On Volcanic Rocks in Lake Janisjarvi in Eastern Finland. *Finlande Comm. Géol. Bull.* **55**:1-13.

Horz, F. 1965. Untersuchungen an Riesglasern. *Contr. Mineralogy and Petrology* **11**:621.

McCall, G. J. H., ed. 1977. *Meteorite Craters.* Benchmark Papers in Geology, vol. 36. Stroudsburg, Pa.: Dowden, Hutchinson & Ross.

Masaytis, V. L.; M. S. Mikhaylov; and T. V. Selivanovskaya. 1972. Popigay Basin—an Explosion Meteorite Crater. *Meteoritics* 7:39–45.

Nechayeva, I. A. 1975. Shock Effects in Rocks of the Central Part of the Kola Peninsula. *Akad. Nauk SSSR Doklady* **223**:1227–1230.

Rabken, M. I. 1959. Geology and Petrology of the Anabar Crystalline Shield. (Reference incomplete as provided by Masaytis et al., 1972.)

Stevenson, J. S. 1972. The Onaping Ash-Flow Sheet, Sudbury, Ontario. *Geol. Assoc. Canada Spec. Paper 10,* pp. 41–48.

5

Reprinted from *Internat. Geology Rev.* **18**:1249-1258 (1976)

Astroblemes in the USSR

V.L. Masaytis

The territory of the Soviet Union has only twice been subjected to grandiose cosmic bombardment within living memory. In 1908, in the basin of the Podkamennaya Tunguska River, at a height of 5-10 km above the earth's surface, a gigantic explosion of a cosmic body occurred, assumed to have been a small comet. This explosion caused no geologic distrubances, and no remains of the exploded body have been found (7). Four decades later, in 1947, an iron meteorite shower fell in Primor'ye, which resulted in over two hundred small impact craters (up to 26 m in diameter), funnels, and holes. The total weight of the fallen meteoritic material has been estimated at several tens of tons (20).

The detailed study of the traces of these phenomena, continuing up to the present time, and also the seven small explosion and impact craters (diameter up to 110 m) known for several decades in Kaaliyarv in Estonia (1), has provided much new information. The suggestion has been made that the Ilumetsa craters in southeastern Estonia (diameter of the largest, 80 m) also have a meteorite origin, but remains of meteorites have not so far been found here (2).

Many investigators have repeatedly pointed out that ancient explosion meteorite craters (astroblemes) may be found in the Soviet Union (4, 8, 23). But only recently, in different regions of the USSR, has it been possible to identify reliable traces of cosmic catastrophes of the geological past, considerably exceeding in scale and traces those recorded above (fig. 1). The diameter of a number of such astroblmes reaches several tens of kilometers.

The theory of formation of meteorite explosion craters was developed more than two decades ago (21, 22, 29). However, only the study of the impact metamorphism of rocks and minerals, caused by ultra-high velocity impacts and explosions, has enabled us to make wide use of this theory for diagnosing such structures (28), including those in the USSR.

The table presents generalized data on astroblemes identified up to the end of 1973. Their origin has for a long time been in dispute, although many have been studied in depth by geophysical methods and by drilling. Features of impact metamorphism, established in the rocks forming the structures, indicate their extraterrestrial origin (3, 6, 12, 13, 14, 15). In particular, this applies to the Puchezh-Katunki structure (24), the Yanis'yarvi structures (28), and certain others.

The characteristics of the astroblemes presented below (except for Popigay and Yanis'-yarvi, where special observations have been made, and the results of which have been partly published) and the maps of their construction have been based on an interpretation of geologic and geophysical data, sections through drill holes, and maps of these items (9, 10, 11, 16, 17, 19). We have also used information kindly supplied by G. M. Karpov (on the Rotmistrovka structure), M. V. Mikhaylov (on the Beyenchime-Salata strcture), and Ye. M. Movshovich (on the Kamensk and Gusev structures). The impact-metamorphosed rocks and impactites have been described on the basis of the author's petrographic observations[1].

The Popigay astrobleme (fig. 2a) occurs on the northern edge of the Anabar shield, whose crystalline rocks are overlain by Proterozoic and Cambrian quartzites, dolomites, and limestones, and also Permian sandstones and siltstones enclosing dolerite sills. In some places, friable Cretaceous sands and sandstones with coal lenses have been preserved overlying these rocks. In spite of the fact that some 30 million years have passed since the time of formation of the crater, according to radiometric age determinations on the impactites, the rounded depression (up to 200-400 m), about 70 km in diameter, has been partially filled with Quaternary sands and conglomerates, all still well expressed in the relief.

The geologic and geophysical data indicate the complicated construction of the astrobleme. In the inner funnel of the crater, about 72 km across, there is a ring uplift of gneissic authigenic breccia, 45 km in diameter (amplitude up to 1.5-2 km). The breccia crops out in the western and northwestern sectors, and in the remaining part the rise has been concealed below the overlying rocks. The gneisses, which possess features of impact metamorphism (shatter cones and diaplectic minerals and glasses), have been cut by veins of pseudotachylites and impactites. The inner funnel has been filled almost exclusively with suevites, in which stratiform and discordant bodies of massive

Translated from Astroblemy na territorii SSSR, Sovetskaya Geologiya, 1975, no. 11, p. 52-64. The author is with the All-Union Scientific Research Institute of Geology (VSEGEI).

[1] At the same time as this report was being prepared for publication, a number of investigators published data supporting the cosmic origin of the Puchezh-Katunki, Boltyshka, Il'inets, and Kamensk struc tures, and the Zhamanshin structure.

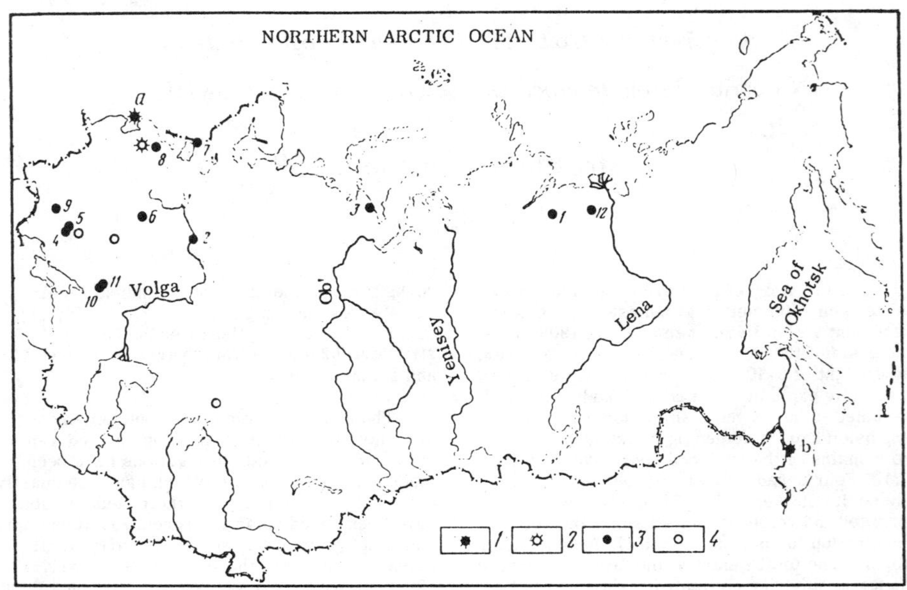

FIGURE 1. Meteorite craters and astroblemes of the USSR.

1 - meteorite craters: a) Kaaliyarv, b) Sikhote-Alin'; 2 - assumed meteorite craters; 3 - astroblemes: 1) Popigay, 2) Puchezh-Katunki; 3) Kara; 4) Boltyshka; 5) Rotmistrovka, 6) Kaluga, 7) Yanis'yarvi, 8) Mishinogorsk, 9) Il'inets, 10) Kamensk, 11) Gusev, 12) Beyenchime-Salata; 4 assumed astroblemes.

Astroblemes of the USSR

Name of structure	Diameter, km	Method of study		Morphology of structure			Features of impact metamorphism							Geologic age
		Drilling	Geophysical investigations	Basin, swell (includ. buried forms)	Central and annular uplift	Ring deformations	Explosion breccias	Suevites	Massive impacties	Shatter cones	Diaplectic minerals	Diaplectic glasses	High-pressure phases	
Popigay	100	+	+	+	+	+	+	+	+	—	+	+	+	Pg_3
Puchezh-Katunki	80	+	+	+*	+	+	+	+	+		+	+		J_1
Kara	50	+	+	+	+	+	+	+	+	+	+			Kz
Boltyshka	25	+	+	+*	+		+		+	—	+	+		Cr_1
Rotmistrovka	2	+		+*			+				+			Cr_1
Kaluga	15	+	+	+*		+	+	+			+	+		D_3
Yanis'yarvi	15		+	+	+		+	+	+	+	+	+		Pt_3
Mishinogorsk	2,5	+	+			+	+			+	+	+		Pz_3?
Il'inets	5	+	+				+	+	+		+			Pz_1
Kamensk	25	+	+	+*	+		+	+		+				Pg_1
Gusev	3	+	+	+*			+							Pg_1
Beyenchime-Salata	8			+		+	+			+				Kz?

*Presence of buried structures.

impactites (tagamites*) up to several tens of meters thick have been included. The latter also form floor veins below the suevites directly on the gneisses of the ring uplift. The thickness of the suevites in the central part of the crater probably reaches 2-2. 5 km.

The flat outer rim embraces almost the entire ring of the inner rim up to 25-30 km wide, broken in places only in the southwest, where the crater has seemingly been superimposed directly on the crystalline rocks of the shield. The sedimentary rocks that form the margin of the outer rim have been intensely deformed (the deformations weaken toward the side of the crater); centrifugal overthrusts, klippen (up to 1-2 km across), faults, including radial types, and various anticlines and synclines, in general parallel to the margins, have been observed. The throw of the vertical displacements along the overthrusts is measured in a few hundred meters, and that of the horizontal displacements, up to 10-15 km and more. The deformed rocks are overlain by an allogenic block breccia, rising from below the suevites; its visible thickness is up to 150 m. This breccia, along with impactites, also fills the ploughed-up tongues cut into the edge of the outer rim. They are distributed radially from the side of the inner rim, and are up to 5-7 km across and up to 15 km in length.

The allogenic breccia (fig. 3) consists of fragments and blocks of various sedimentary and crystalline, partly impact-metamorphosed rocks (they are fractions of a meter up to a few hundred meters in size) and a friable coptoclastic (from the Greek "kopto", to collide, to break by impact) matrix, developing during crushing and powdering of the Permian and Cretaceous sandstones, sands, and clays; fragments of glass have sometimes been recorded. The coptoclastites, almost devoid of large fragments of rocks, lie above the suevites in the central part of the crater.

The greatest interest centers around the impactites, similar in chemical composition to the gneisses, and especially the massive impactite-tagamites (see fig. 3), which consist of partially crystallized glass, enclosing up to 25% volume percent of fragments of various rocks (mainly impact-metamorphosed and fused gneisses) and their minerals.

*Tagamite - a variety of impactite rocks named after the Tagamy ridge in the Popigay crater. "... mas sive, sometimes porous, glassy or slightly crystallized rocks ... partly saturated with the fused fragments of gneisses and their minerals ... occur as dikes, necks and irregular injection stratal bodies within the suevites and allogenic breccia, but mainly in the form of thick (up to 60 m) discontinuous sheets on the surface of the allogenic breccia and the suevites .... They have no analogs among the igneous and metamorphic rocks." (Masaytis et al., 1972, Internat. Geology Rev., v. 14, no. 4, p. 329.) -IGR Editor.

The vitroclastic impactites (suevites) (analogous to the rocks of the Ries crater) consist of fragments and bombs of similar glass, cemented by the same finely crushed glass with particles of the above-named rocks and minerals. The impactites contain fragments of diaplectic minerals and glasses, and also various high-pressure phases. In addition, spherules of sulfides with a large nickel content are present.

The Popigay crater, owing to its large diameter, complex internal structure, good preservation of the ejected material and the variety of its lithological types, and also the large scale of impact fusion and certain features of the mineral transformations, is unique, and its study may clarify the processes of formation of such morphostructures. Some features of the mechanism of its origin can at present only be reconstructed in a general form.

The cosmic body penetrated the sedimentary cover and was completely brought to a standstill in the crystalline basement, where the main portion of its energy was released, which controlled the explosion (up to 10^{30} ergs). Judging by the depth of penetration of the body (about 2. 5 km) it must have possessed a high density. The sedimentary cover in the central zone of the crater was crushed and thrown out by the wave; in the marginal zone, it was deformed and displaced by the masses of block breccias being shifted along the surface. Along the periphery of the crater, the rocks were crumpled as a result of alternating gentle undulate annular fluctuations of small amplitude (up to 100 m), and also movements of the frontal parts of the overthrusts, and the ploughing-up and crushing action of the slickenside breccias and falling blocks. The height of the annular ring of ejected material was probably insignificant.

The impact melting developing in the epicenter of the explosion as a result of fusion of the gneisses spread radially with great rapidity, forming annular ridges, and farther from the center, streams, flows, etc., which covered the major portion of the floor of the crater. In the marginal zone these streams covered over the decelerated movement of the mass of breccias. The spreading melt was almost simultaneously buried by "nuées ardentes," which gave rise to the suevites, and later, by clastic material also, which had steeper trajectories during ejection and was colder. As a result of gas lift, and probably uplifts of the crater basement, the impact melt was partially squeezed into the upper horizons of suevites and coptoclastites and possible onto the surface also. The origin of the asymmetric annular uplift of the basement evidently began at the moment of explosion and continued as a result of recoil even after the filling of the crater. It may be suggested that the transformation of the various rocks and minerals could have been achieved at an impact pressure of 1 megabar and a temperature of

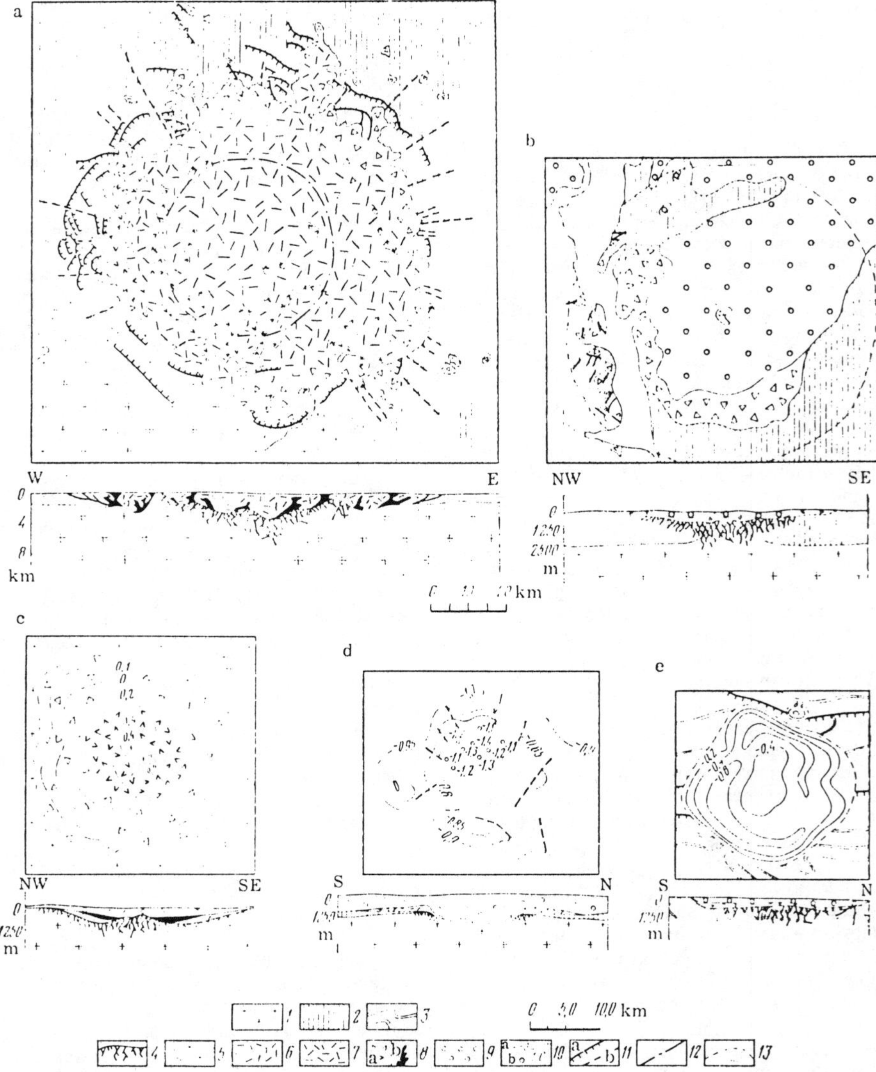

FIGURE 2. Geologic structure and sections of the Popigay (a), Puchezh-Katunki (b), Boltyshka (c), Kaluga (d), and Kamensk and Gusev (e) astroblemes.

1 - Archean and Lower Proterozoic crystalline rocks; 2 - Lower Proterozoic, Paleozoic, and Mesozoic sedimentary rocks, lying horizontally; 3 - the same rocks, folded; 4 - cataclased and brecciated rocks (authigenic breccia); 5 - allogenic breccia; 6 - allogenic breccia and suevites; 7 - suevites; 8 - massive impactites (a - on maps, b - in sections); 9 - Paleozoic, Mesozoic, and Cenozoic sedimentary rocks, overlying breccias and impactites; 10 - structure contours of the base of the craters (a) and their markers (b) based on geophysical drilling data; 11 - overthrusts (a) and other faults (b); 12 - axis of annular uplift; 13 - contour of assumed distribution of impactites and allogenic breccia.

FIGURE 3. Outcrop of allogenic breccia consisting of giant blocks of various sedimentary and metamorphic rocks. The breccia is overlain by massive impactites with columnar parting (Popigay astrobleme).

up to 1800-2000°, although at the epicenter of the explosion, where vaporization of the impacted body and the terrestrial rocks took place, both the pressure and temperature were probably substantially higher.

The Puchezh-Katunki astrobleme (see fig. 2b), located to the north of Gor'kiy, has the form of a sloping funnel (depth about 600 m) in sedimentary deposits resting horizontally on the crystalline basement, and with a thickness of about 2 km. It has been filled with an allogenic breccia (crushed and folded Proterozoic, Paleozoic, and Triassic rocks) enclosing large torn-off fragments (up to several hundreds of meters across) and blocks of the same rocks, and occasionally gneiss fragments; Along the periphery, the allogenic breccia rests on deformed Permian and Lower Triassic deposits, with the surface nature of the deformations (folds, faults, etc.) being clearly identified. The allogenic breccia has been covered by Middle Jurassic and Cretaceous sediments, up to 400 m thick, forming a trough. In the center of the funnel is an uplift of gneiss authigenic breccia, about 10 km across; along the sides, it passes into brecciated and crumpled Paleozoic rocks (authigenic breccia of Paleozoic rocks). It may be assumed that in the center of the structure, the crystalline authigenic breccia passes at depth into undeformed or weakly deformed rocks of the basement. The amplitude of the central uplift (the so-called Vorotilovo rise) is about 2 km.

The authigenic breccia consists of brecciated and cataclased gneisses, amphibolites, in places transformed into a finely crushed mass, enclosing fragments of rocks, lenses, and sectors of porous fluidal polymineralic glass. In the gneisses and amphibolites occur diaplectic quartz, plagioclase, and biotite, and a diaplectic plagioclase glass. The polymineralic glasses contain fragments of different minerals and have been partially crystallized, with the formation of andesine microlites, and in places have been converted into montmorillonite, chlorite, and calcite; glasses in the form of fragments are also found in the allogenic breccia, which on the basis of structure and composition approximates in this case to suevite. In spite of the large diameter of the structure, impactites are insignificantly developed here. The radiometric age of the glasses is about 183 ± 3 m. y., which agrees with the geologic data.

The Kara astrobleme, almost completely covered by Quaternary deposits up to 100 m thick, is located in the north of Pay-Khoy in the delta of the Kara River among folded Paleozoic (including Permian) terrigenous and carbonate sequences, intruded by diabases and striking northwesterly. In the center and along the periphery of the basin weakly defined in the relief, an allogenic breccia and suevites, more than 400 m thick, rise from below the friable deposits. The breccias contain fragments and blocks of Paleozoic rocks, and also sandstones with a Late Cretaceous fauna, not found in solid outcrop. Shatter cones are observed in the deformed sequences that underlie the allogenic breccia (fig. 4).

The suevites, transitional into the allogenic breccia, consist of particles of polymineralic glasses and fragments of quartzites, sandstones, carbonaceous shales, limestones, and their component minerals. They sometimes enclose small bodies of massive impactites, usually

FIGURE 4. Shatter cones in sandstones (Kara astrobleme).

porous and weakly crystallized. In the center of the structure is an uplift of brecciated and cataclased Paleozoic rocks (authigenic breccia), up to 7-8 km across near the surface. In the samples with traces of impact metamorphism, collected from the allogenic and authigenic breccias, we have found diaplectic quartz, plagioclase, biotite, pyroxene, amphibole, and also diaplectic plagioclase glass. In chemical composition, the massive impactites correspond to shales, and mullite is present in the hyaline groundmass in the form of microlites.

The Boltyshka astrobleme (see fig. 2c) is located on the Ukrainian shield, 30 km to the north of Kirovograd. The base of the buried crater lies approximately 1 km below the surface of the Precambrian basement, which rests outside the structure at a depth of 70-100 m below friable Cenozoic deposits. Below the level of the latter, the basin of the crater has been filled with Cretaceous (?) — Paleogene lacustrine sediments with seams of oil shales. In the center of the crater is an uplift of crushed cataclased and partially fused granites (authigenic breccia), 2 × 4 km in size, with a relative height of about 500 m above the base.

The allogenic breccia which rests everywhere on the authigenic breccia consists mainly of fragments of granites and encloses lenses and sectors of polymineralic, partly crystallized glasses. It is distributed over a radius of 12-13 km from the center, and gradually wedges out; its maximum thickness is about 200 m. On the margins of the structure, this friable breccia encloses individual blocks of granite. On the breccia, in the form of a ring with an outer diameter of about 7-8 km, are massive impactites, the maximum thickness of which is little more than 200 m. The massive impactites contain a large number of blocks, fragments, and small particles of crystalline rocks. The glass of the impactites is in places completely crystallized with the formation of an aggregate of plagioclase, hypersthene, alkaline feldspars, and quartz. The mineral fragments in the impactites and in the breccia frequently have a diaplectic aspect (quartz, plagioclase, potassium feldspar, and biotite), and diaplectic plagioclase and K-feldspar glasses have been found. Part of the diaplectic glass has been recrystallized.

The radiometric age of the impactites (in chemical composition, they correspond to potassic granites of the base of the crater) is about 70 m. y. In spite of the comparatively small diameter of the structure, the scale of melting of the crystalline rocks is extremely significant, which is probably explained by their occurrence on the surface at the moment of impact and the comparatively strong opposition to compression.

The Rotmistrovka astrobleme occurs 30 km to the north-northwest of the Boltyshka astrobleme and is extremely similar to the latter in geological construction. The crater has been filled with Cretaceous sediments, below which lie breccias with glass and cataclased granite-gneisses, containing diaplectic quartz, plagioclase, biotite, and pseudotachylitic glass, partially zeolitized and montmorillonitized; it is about 300 m deep. It may be suggested that the Rotmistrovka and Boltyshka astroblemes were close in time of formation.

The Kaluga astrobleme (see fig. 2d), located at the town of Kaluga, has been buried below Upper Devonian and Lower Carboniferous strata about 800 m thick. It forms a rounded basin a few hundred meters deep, traceable in the crystalline basement and in the horizontally overlying sandstones and clays of the Valday series of the Upper Proterozoic, and the sandstones, siltstones, and marls of the Middle Devonian that have been intensely deformed along its periphery. The allogenic breccia with fragments of sedimentary and crystalline rocks forms a swell rising 300 m above the floor of the basin and 150-200 m above the original surface of the Middle Devonian sediments. On the side away from the swell, the allogenic breccia overlies undeformed rocks of the Valday series and the Middle Devonian.

The basin has been filled with breccias and suevites, not less than 90 m thick. The radiometric age of the glasses is about 400-430 m. y., which is probably somewhat underestimated. In the fragments and blocks of schists and granite-gneisses from the breccias, which have normally been crushed and cataclased, diaplectic quartz, plagioclase, and biotite, and sometimes diaplectic glass after plagioclase (maskelynite) have been observed. The suevites consist of fragments of various polymineralic glasses, including fluidal, vesicular, and other types. The glasses contain fragments of minerals, including diaplectic forms, that also occur in the matrix of the suevites.

FIGURE 5. Shatter cones in metasiltstones (Yanis'yarvi astrobleme).

The Kaluga astrobleme is interesting in that it is one of the most deeply buried occurrences, revealed by geophysical investigations and by drilling. The crater can be traced in the structure of the overlying Paleozoic sediments, many hundreds of meters thick, which agrees with the experimentally established estimate of the ratio of the depth of the original basin to the thickness of the overlying sequences (about 1:3), under which the basin may still be traced on the surface.

The Yanis'yarvi astrobleme lies in the southeastern part of the Baltic shield, to the north of Lake Ladoga. This is the oldest of the astroblemes known in the USSR: the radiometric age of the impactites is about 700 m.y. The structure has been substantially transformed as a result of glacial erosion. The circular basin in Proterozoic schists (13 × 17 km) has been flooded by the waters of Lake Yanis'yarvi. Almost at its center, are three islands, which consist of impactites and breccias. The latter are also known on the southwestern margin.

The breccias consist of fragments and blocks of schists, pegmatites, and quartz, cemented by the same, but more finely crushed rocks. Locally in the breccias, crystallized impactite glass has been recorded in the form of lumps and irregular fragments. There are shatter cones in the breccia (fig. 5). The massive impactites, containing numerous fragments of schists, rest on top of the breccias in the form of a sheet-like body. The rocks are finely crystalline and consist of plagioclase, hypersthene, amphibole, a small amount of quartz, ore minerals, glass, chlorite, etc.

The massive impactites correspond precisely in chemical composition to that of the schists. Microscope investigations indicate the presence of fragments of diaplectic quartz (the predominant orientation of the planar elements in the quartz is $\{10\bar{1}3\}$) and recrystallized maskelynite in the impactites and breccias, and also recrystallized quartz fusion glasses.

The Mishinogorsk astrobleme is located 25 km southeast of Gdov, in horizontal Paleozoic sequences. The structure is rounded in plan and is surrounded by a ring of deformed Middle Devonian, Ordovician, Cambrian, and upper Proterozoic sedimentary rocks, dipping away from the astrobleme at varying angles, with all of them broken by faults. The crystalline basement outside the crater is 500 m below the surface. The cup-shaped basin in the rocks of the basement and cover, about 700 m deep, has been filled with a gigantic allogenic breccia. In the center of the structure, the breccia rests directly on cataclased granite-gneisses; the blocks and fragments of dolomites and limestones which compose it frequently have a mortar structure. The minerals of the crystalline rocks (quartz and biotite) have been intensely deformed, and in part have a diaplectic nature; the presence of a diaplectic plagioclase glass has been recorded. In places, fusion glasses have been observed in the breccia, including fluidal and plicated types, with inclusions of fragments of diaplectic minerals. The gneiss and schist fragments frequently contain small shatter cones.

The Il'inets astrobleme, lying 45 km southeast of Vinnitsa, is a deeply eroded structure: under a thin cover of Neogene sands and clays, only a lens (up to 200-250 m) of suevites and allogenic breccia has been preserved, consisting of fragments of granites, gneisses, and other Precambrian rocks of the Ukrainian shield. This lens, about 5 km in diameter, is rounded in plan and is underlain by brecciated granites. The suevites contain sectors and fragments of vesicular glasses and impactites of acid composition, in turn enclosing fragments of crystalline rocks. The granites and gneisses possess features of impact metamorphism; quartz (with the planar elements oriented predominantly along $\{10\bar{1}3\}$) and diaplectic feldspars and biotite are also present. The early Paleozoic age of the structure (490-500 m.y.) has been determined on the basis of radiometric dating of the impactites, and also arising from the fact that they are overlain by Devonian clays; fragments of lower Paleozoic siltstones are present in the breccia.

Rocks similar in outward appearance to the impactites described and also breccias have been found within six small (up to 1 km across) sectors over a distance of up to 30 km to the west and southwest of the Il'inets lens of impactites and breccias. These rocks are in places underlain by cataclased granites. If during further investigations we can successfully establish the impact origin of these rocks and their time equivalence, then this will indicate that there was a group of seven craters here, and the Il'inets crater was the largest of them.

The Kamensk structure (see fig. 2e) lies near Kamensk-Shakhtinskiy, in the area of distribution of folded Carboniferous, Permian, and

Lower Triassic deposits (sandstones, limestones, shales, and marls), with sublatitudinal strike. The crater has been buried below a 200-300 m sequence of horizontal Danian-Paleocene marls, also developed beyond its boundaries. It has been filled with a breccia of blocks and fragments (up to 10 m and more) of sedimentary rocks of Carboniferous, Permian, Triassic, and Late Cretaceous age, with a sand-clay matrix, containing small fragments of the same rocks (coptoclastites). In the zone of the gentle central uplift, the thickness of the blocky breccia reaches 500-600 m, and the amplitude of uplift is about 400 m. Like almost the entire base of the crater, it has been formed of brecciated but not displaced Carboniferous rocks. The assumed thickness of this breccia is up to 1.5-2 km.

The Gusev structure (see fig. 2e), located 1 km northeast of the Kamensk structure, is an ellipsoidal basin about 400 m deep, filled with a breccia like that described above. Although no features of impact metamorphism have yet been found in either the Kamensk or the Gusev structures, data on the deep-seated construction, obtained as a result of drilling, most probably indicate their extraterrestrial origin. Evidently, both of these astroblemes developed during the fall of a single cosmic body, which broke up near the earth's surface into two uneven parts. The outlines of the structures, somewhat differing from the present-day rounded forms, have been controlled by the superimposition of subsequent tectonic movements and by uneven erosion.

The Beyenchime-Salata astrobleme was discovered on the watershed of the Olenëk and Bur rivers. Here is a horizontal sequence of Cambrian carbonate and terrigenous rocks below which are upper Proterozoic limestones and dolomites. The circular basin in the relief has a relative depth of up to 100 m, and in its inner part, weakly cemented polymict breccias enclosing fragments of Proterozoic (?), Cambrian, and Permian rocks (the latter have been preserved only outside the crater) rise from below the cover of drift.

In the annular swell, which consists of uplifted and mixed blocks of sandstones, limestones, and dolomites, an overthrust of older rocks onto younger types has been recorded along a system of centrifugal thrusts; radial joints have also been observed. The rocks have a mortar structure, and well-defined shatter cones have been found in the dolomites. The preservation of the morphostructure, similar to that of the Popigay astrobleme, located 350 km to the west, suggests that it is Cenozoic in age.

In the USSR, in addition to those described above, there are a number of other circular structures, the origin of which is so far unclear (5, 18). No features of impact metamorphism have yet been established in their component rocks, but the features of the morphology, construction, and composition of the infilling rocks suggests an extraterrestrial origin for these structures. Here are a few examples.

In the Zhamanshin reserve, to the north of the Aral Sea, there is an isometric shallow basin up to 15 km in diameter. Breccias have been found here, consisting of blocks of Proterozoic and Paleozoic rocks, and slags and glasses overlie horizontal Paleogene deposits; the radiometric age of the glasses is about 4-5 m. y. Young volcanic manifestations are not known in this region, so that there are grounds for suggesting that the slags and glasses are of impact origin.

The buried Obolon basin, circular in plan, located 90 km north-northeast of the Boltyshka astrobleme, may also be regarded as a probable impact-explosion structure. It is about 12 km in diameter, and about 900 m deep. It is possible that the isometric Kursk basin, 5 km in diameter, lying near the city of Kursk, is also a similar buried structure. In its central part, there is an uplift of crushed crystalline basement with an amplitude of up to 250 m. The structure, which has been traced in Devonian and Carboniferous deposits, is concealed below a 200-m sequence of Jurassic and Cretaceous strata.

So far, there are no geologic and petrographic data available that would enable us to consider the El'gytygyn and Labynkyr structures and the Patom crater as cosmogenic objects (4). The assumed Kondër "crater" is an intrusion of the central type, sculptured by erosion, consisting of ultrabasic and alkaline-earth igneous rocks.

The discovery and study of the astroblemes in the USSR, begun only recently, have already developed rapidly. As of today, most of them have been identified in well-studied regions of the European part of the Soviet Union. In this respect, a number of important data have been obtained on the morphology and structure of the astroblemes discovered, and on the composition and impact metamorphism of the infilling rocks. The available information now enables us mutually to compare these astroblemes, and also to compare their features with those of analogous structures in other countries (27, 28).

The data on the time of formation of the astroblemes in the USSR, in spite of the approximate nature of a number of estimates, all support the pattern revealed in dating the craters of other regions of the globe (27): a significant portion of them has an age up to 100 m.y., which is in many ways explained by the better preservation of the younger craters and the comparative ease of their discovery.

In contrast to the astroblemes in Canada, located mainly on the shield, the great majority of the astroblemes discovered in the USSR occur within the ancient plates or regions of Paleozoic folding, which were stable during the Cenozoic. Therefore, in many of the astroblemes in the USSR, distinguished by their good preservation, not only lenses of authigenic breccia are available for sutdy, but also allogenic breccia and impactites. However, even the comparatively young Cenozoic (or late Mesozoic in part) astroblemes are practically devoid of a rapidly eroded swell, and sometimes only the annular uplift of deformed rocks along the periphery of the crater remains.

The good preservation of a number of astroblemes in the USSR makes it possible to compare their morphology with the relatively young craters of the Moon. For example, the construction of the Popigay crater is similar to that of the 200 km Tsiolkovsky crater on the Moon. In this respect, a comparison of the features of the rocks that form the morphostructures on both planets is of particular importance. The available data demonstrate that, in spite of the difference in the composition of terrestrial and lunar breccias, these and others are mutually alike both in fabric features and in types of transformation of the rocks and minerals (25), which gives grounds for proposing the hypothesis on the similarity in the conditions of their origin.

The suevites of the Popigay, Kara, and other astroblemes have enabled us to study certain properties of the fluidized high-temperature three-phase system gas — melt particles — rock and mineral particles, which is distributed radially from the centers of the craters in the form of "nuées ardentes." The absence of any post-impact igneous rocks in the large astroblemes of the USSR, where impactites have been widely developed, most strongly indicates that the intense meteorite impact was still inadequate for the manifestation of trigger volcanism (if it is indeed in general possible under terrestrial conditions). There are numerous other problems, which require study, associated with the processes of impact-explosion crater formation and their subsequent effect on the formation of the earth's crust. It may be hoped that a further detailed study of the astroblemes found in the USSR will contribute to the development of this new direction in geologic and planetologic investigations.

REFERENCES

1. Alloe, A.O., 1963, HISTORY OF STUDY OF THE KAALA METEORITE CRATERS: EstSSR Inst. Geol. Trudy, vyp, 11, p. 25-34.

2. ______ 1963, NEW DATA ON STRUCTURE OF THE ILUMETSA CRATERS: AN EstSSR Inst. Geol. Trudy, vyp. 11, p. 35-43.

3. Val'ter, A.A., and Ryabenko, V.A., 1973, PETROGRAPHIC FREATURES OF IMPACT-METEORITE ORIGIN OF THE IL'INETS STRUCTURE (VINNITSA REGION): Geol. Zhur., v. 33, no. 6. p. 142-144.

4. Zotkin, I.T., and Tsvetkov, V.I., 1970, THE SEARCH FOR METEORITE CRATERS ON THE EARTH: Astronom. Vestnik, v. 4, no. 1, p. 55-65.

5. Kiryukhin, L.G., Florenskiy, P.V., and Sobolev, Yu.S., 1969, THE ZHAMANSHIN ENIGMA: Priroda, no. 3, p. 70-72

6. Masaytis, V.L., Raykhlin, A.I., Reshetnyak, N.B., Selivanovskaya, T.V., and Shitov, V.A., 1974, COESITE FROM THE POPIGAY CRATER (FIRST DISCOVERY IN THE USSR). Vses. Miner. Obshch. Zapiski, ch. 103, vyp. 1, p. 122-127.

7. Krinov, Ye.L., 1949, THE TUNGUSSKA METEORITE: Izd-vo AN SSSR, Moscow.

8. ______ 1962, METEORITE CRATERS ON THE EARTH'S SURFACE: Meteoritika, vyp. 22, p. 3-30.

9. Logvin, E.I., 1968, HYDROTHERMAL MINERALIZATION OF VOLCANOGENIC ROCKS IN THE SOB RIVER BASIN. In RIPHEAN VOLCANISM AND METAL OCCURRENCES IN WESTERN PART OF UKRAINIAN SHIELD: p. 132-142, Izd-vo Naukova Dumka, Kiev.

10. Malakhovskiy, D.B., and Buslovich, A.L., 1966. NEW DATA ON ORIGIN OF THE GDOV DISLOCATIONS. In DATA ON GEOLOGY AND MINERAL RESOURCES OF THE NORTHWEST OF THE RSFSR: vyp. 5, p. 125-134, Izd-vo Nedra, Leningrad.

11. Malovitskiy, Ya.P., 1959, ORIGIN OF NORTH DONETS "AGGLOMERATE". In PROBLEMS OF GEOLOGY OF OIL AND GAS: Mosk. Inst. Narodn. Khozyaystva i Gornoy Promyshlennosti Trudy, p. 217-228, Gosgeoltekhizdat, Moscow.

12. Masaytis, V.L., [no date given], GEOLOGICAL CONSEQUENCES OF FALLS OF CRATER-FORMING METEORITES. Izd-vo Nedra, Leningrad, 18 pp,

13. ______ 1974, SOME ANCIENT METEORITE CRATERS IN THE USSR: Meteoritika, vyp. 33, p. 64-68.

14. ______ Mikhaylov, M.V., and Selivanovskaya, T.V., 1971, POPIGAY BASIN — AN EXPLOSION METEORITE CRATER: AN SSSR Doklady, v. 197, no. 6, p. 1390-1393.

15. Masaytis, V.L., Mikhaylov, M.V. and Selivanovskaya, T.V., THE POPIGAY METEORITE CRATER: Sovetskaya Geologiya, no. 6, p. 143-147

16. Nikol'skiy, A.P., 1969, ORIGIN OF MAGMA AND TECTONIC BASIN OF THE BOLTYSHKA VOLCANO. In VOLCANISM AND FORMATION OF DEPOSITS IN THE ALPINE GEOSYNCLINAL ZONE: p. 77-78, L'vov.

17. Yenokyan, V.S., Zenchenko, M.S., Vodolazskiy, V.I., and Yatsuk, V.I., 1970, NEW INFORMATION ON STRUCTURE OF UST'-KARA DEPRESSION IN PAY-KHOY. In DATA ON GEOLOGY AND MINERAL RESOURCES OF THE NORTHEAST OF EUROPEAN USSR: vyp. 6, p. 238-242, Syktyvkar.

18. Petrov, V.G., 1970. RING VOLCANIC-TECTONIC STRUCTURES ON THE RUSSIAN PLAIN. In DATA ON GEOLOGY AND MINERAL RESOURCES OF CENTRAL REGIONS OF EUROPEAN USSR: vyp. 6, p. 320-329, Kaluga.

19. Pinchuk, L.Ya., 1971, MORPHOLOGY AND ORIGIN OF THE BEYENCHIME-SALATA BASIN. In KIMBERLITE VOLCANISM IN NORTHEAST OF THE SIBERIAN PLATFORM: Leningrad.

20. Fesenkov, V.G., and Krinov, Ye.L., (Eds.), 1959, 1963, THE SIKHOTE-ALIN IRON METEORITE SHOWER, Vs. 1, 2: Izd-vo AN SSSR, Moscow, 372 pp.

21. Stanyukovich, K.P., and Bronshten, V.A., 1960, THE ROLE OF EXTERNAL COSMIC FACTORS IN THE EVOLUTION OF THE MOON. In THE MOON: p. 299-329, Fizmatgiz, Moscow.

22. Stanyukovich, K.P., and Fedynskiy, V.V., 1947, DESTRUCTIVE EFFECTS OF METEORITE IMPACTS: AN SSSR Doklady, v. 57, no. 2, p. 129-132.

23. Fesenkov, V.G., PRINCIPAL SUCCESSES IN METEORITICS IN RECENT YEARS: Meteoritika, vyp. 30, p. 1-17.

24. Firsov, L.V., 1965, METEORITE ORIGIN OF THE PUCHEZH-KATUNKI CRATER: Geotektonika, no. 2, p. 106-118.

25. Chao, E.C.T., 1973, GEOLOGIC IMPLICATIONS OF THE APOLLO 14 FRA MAURO BRECCIAS AND COMPARISON WITH EJECTA FROM THE RIES CRATER, GERMANY: U.S. Geol. Survey Jour. Res., v. 1, no. 1, p. 1-18.

26. Dence, M.R., 1971, IMPACT MELTS: Jour. Geophys. Res., v. 76, no. 23, p. 5552-5565.

27. ______ 1972, THE NATURE AND SIGNIFICANCE OF TERRESTRIAL IMPACT STRUCTURES: 24th Internat. Geol. Congr., sect. 15, Planetology, Montreal, p. 77-89.

28. French, B.M., and Short, N.M., 1968, SHOCK METAMORPHISM OF NATURAL MATERIALS; Mono Book Corp., Baltimore, 644 pp.

29. Shoemaker, E.M., 1960, PENETRATION MECHANICS OF HIGH VELOCITY METEORITES ILLUSTRATED BY METEOR CRATER, ARIZONA: 21st Internat. Geol. Congr., rept. 18, Copenhagen, p. 418-434.

6

Reprinted from *Internat. Geology Rev.* **14**:327–331 (1971)

The Popigay meteorite crater

V.L. Masaytis, M.V. Mikhaylov, and T.V. Selivanovskaya

In the basin of the Popigay river (a right tributary of the Khatanga river in the north of the Central Siberian Platform) lies a circular depression in the relief, with a diameter of about 70-80 km, that is known as the Popigay basin (3, 5, 8, 9). The floor of this basin is 250-400 m lower than the surrounding watershed areas; it is characterized in plan by a concentric-radial drainage network.

D. V. Kozhevin (1946) interpreted the structure of this depression as a graben, filled with Triassic tuffaceous rocks. The geologists who studied the depression in 1949-1951 shared this view and even believed that the graben was filled with Permian and Triassic tuffaceous rocks and lavas of basic composition.

M. T. Kiryushina was later able to show that some fragments within the tuffs contain remains of a Cretaceous fauna. The vulcanism accompanying the basin's formation was dated from the Cretaceous to the beginning of the Cenozoic, and the volcanogenic rocks were identified as andesites and their tuffs (3).

L. P. Smirnov dated the volcanogenic-sedimentary rocks developed in the depression as Hauterivian to Turonian, and proposed a stratigraphic subdivision of these deposits. The basin itself he considered to be an erosional depression developed on an uplift (9).

P. S. Voronov suggested that the depression might be a collapse crater on the crest of a Tertiary uplift (1).

The unusual occurrence of lavas and tuffs of intermediate composition under platform conditions and the superimposed character of the structure gave rise to attempts to explain the depression's formation from the standpoint of the formation of the Mesozoic arched-block structures that are so widely developed in the eastern part of Soviet Asia (2).

These ideas of the depression's volcano-tectonic and partly erosional origin have been the underlying basis of the geologic survey operations carried out here in recent years. But these views are mutually contradictory in many respects and fail to explain all the observed distinctive features of the depression's structure, the composition and structure of the rocks filling it, and particularly the nature of the contacts between the volcanogenic-sedimentary rocks and the underlying Precambrian and Paleozoic rocks; these contacts had been regarded sometimes as tectonic and at other times as erosional. From the above standpoints it is also difficult to explain the numerous small outcrops of old rocks, including Archean, that can be seen among the younger formations along the basin's periphery and are only a few tens of meters across. Other unusual features are the saturation of the lavas and tuffs with various but predominantly Archean rocks and fragments of their minerals, the absence of layering and gaps in the body of volcanogenic rocks, the lack of separate, individual lava flows and others.

Our field observations in the basin of the middle course of the Popigay river during the summer of 1970, study of the results of previous investigations, interpretation of aerial photos, analysis of the available geologic and geophysical materials and comparison of the structure of this object with that of others of the same kind have shown that the rocks developed within the Popigay depression cannot be classed as volcanogenic or normal sedimentary formations. The depression's distinctive structure, and the composition and structures of the rocks within it and on its periphery can, in our opinion, be explained only if they are interpreted as resulting from the impact of a meteorite. In this hypothesis the Popigay basin is a meteorite crater, and the rocks filling it are impactites and breccias formed in the shattering, shock-metamorphism, fusion and ejection of the various initial rocks. The theory of the formation of such craters, their distinctive morphological features, problems of their identification and of the classification and nomenclature of the rocks in them have been discussed by a number of Soviet and foreign authors (6, 7, 10, 11, 12, 16, 17).

The Popigay meteorite crater is situated on the north slope of the Anabar shield. Here the Archean crystalline rocks are covered by a northward gently sloping body of quartzites, dolomites and limestones of Late Proterozoic and Cambrian age, by a Permian continental coal-bearing body penetrated by sills and dikes of Triassic dolerites. Cretaceous marine and continental deposits are preserved in places around the crater's periphery. Within the depression and along its sides are extensive Quaternary deposits — lacustrine, glacial, glacioaquatic, alluvial and others. The total thickness of the

---

Translated from Popigayskiy meteoritnyy krater, Sovetskaya Geologiya, 1971, no. 6, p. 143-147. The authors are with the All-Union Scientific-Research Geological Institute.

platform mantle here reaches 1.0-1.2 km.

The morphology and structure of the Popigay crater, in our view, are as follows (fig. 1). The crater appears to consist of two funnels of different diameter superimposed on each other — an inner and an outer; the diameter of the former is about 75 km; the latter, about 100 km. The boundary of the depression's floor corresponds approximately to the projection of the inner funnel's contour at the surface; the outer funnel encompasses the depression's slopes all the way to the adjacent watershed areas.

The inner funnel of the crater, which is formed in the basement rocks, is filled with a displaced explosive breccia (allogenic breccia) and impactites lying on an authigenic breccia (not displaced) consisting of biotite-garnet and hypersthene gneisses, plagioclase gneisses and other rocks that emerge in the crater within the semi-annular uplift that was formed after the explosion. The crater's outer funnel, formed of Proterozoic and Cambrian rocks, forms an incomplete ring framing the inner funnel and serves as its outer zone. Proterozoic, Cambrian and some Permian deposits are developed here at the surface. This zone is distinguished by the existence of numerous centrifugal thrusts whose planes dip toward the crater's center at various angles, of anticlines and synclines with strikes parallel to the crater's boundaries and also by the extensive development of brecciated rocks. All these deformations die out with increasing distance from the crater. The allogenic breccia and impactites form lobes extending outward into the outer funnel, where they overlie the deformed sedimentary rocks.

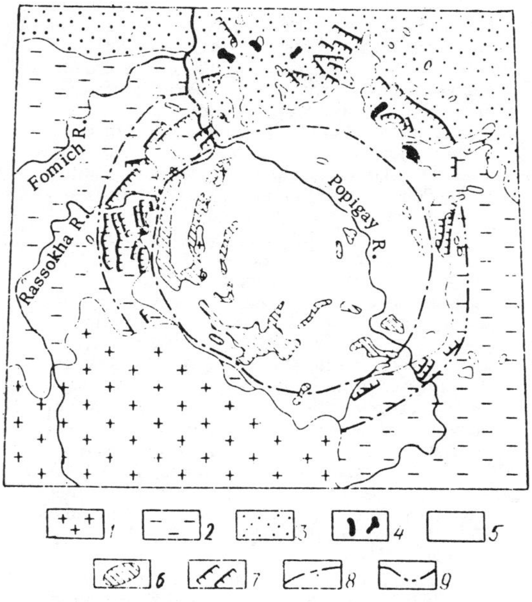

The authigenic breccia consists of shattered gneisses that have been shock-metamorphosed and in places partly fused, resulting in the formation of pseudotachylytes and lenses, veins and dikes of glassy rocks saturated with fused fragments of the gneisses. Separate small fragments and large blocks, sometimes up to many tens and probably even a few hundreds of meters across, of such shock-metamorphosed gneisses occur in the allogenic breccia. These gneisses show specific structures — shatter cones (fig. 2), typical of rocks in areas that have suffered meteoritic impact (12, 13, 15). The minerals in the gneisses, besides shattering, also show signs of shock metamorphism (planar structures, isotropization of the quartz and plagioclase, crumpled biotite and so forth).

FIGURE 1. Structure of the Popigay meteorite crater.

1 - Archean rocks
2 - Proterozoic and Cambrian rocks
3 - Upper Paleozoic rocks
4 - Cretaceous rocks
5 - allogenic breccia and suevites
6 - tagamites
7 - centrifugal thrusts
8 - crater's boundaries within Archean rocks
9 - crater's boundaries within Proterozoic and Cambrian rocks.

The allogenic breccia occurs both within the inner funnel and outside it (in the form of relicts preserved from erosion). At the surface the allogenic breccia crops out predominantly from beneath the impactites covering it along the edges of the inner funnel. The breccia consists of variously sized fragments of partially shattered Archean, Proterozoic, Paleozoic and Mesozoic rocks. The sizes of the fragments and blocks, which are up to several tens of meters across, diminish somewhat with increasing distance from the inner funnel. The allogenic breccia and the separate blocks of rock ejected from the crater have been traced for distances up to 70 km from the center and 40 km from the edge of the inner funnel. In the latter case the fragments and blocks of gneisses lying on the Permian and Cretaceous rocks are still as much as 10-15 m across. The cement of the allogenic breccia usually consists of friable, unlayered sandy or clayey material, and in places is mixed with pieces of glass so that the rock takes on a consolidated structure.

It has been established that the many small outcrops of Archean and partly other rocks, which were previously shown on geologic maps north of the depression as tectonic blocks, are in fact the remnants of allogenic breccia, which in places forms a solid cover (fig. 1). The Cretaceous rocks, which contain fossil fauna and spore-pollen complexes and were thought earlier to be interlayered with "volcanogenic" formations, actually constitute blocks of various sizes included in the allogenic breccia (on the Popigay river, upstream

FIGURE 2. Shatter cones in shock-metamorphosed gneisses occurring as blocks within allogenic breccia (Popigay river, below trading station. Natural size).

of the trading station, on the Namsik-Daldyn river and elsewhere).

The impactites are of two varieties that differ in their mode of occurrence and in the manner in which the material was transported. They are partly crystallized glass, saturated with fragments of gneisses of the basement complex that have been preserved from fusion, and fragments of their component minerals. Pieces of other rocks are occasionally also present. The impactites were previously thought to be andesite porphyrites and their tuffs. Judging from their composition, structure and conditions of formation, the latter should be called suevites (14, 15). They consist of pieces of glass with an admixture of fragments of various rocks (among which shock-metamorphosed Archean gneisses are predominant) and minerals, primarily quartz and plagioclase. Lechatelierite and maskelynite also occur. The suevites were formed as a result of dispersion of the fused basement rocks in the explosion. These rocks form an almost continuous sheet in the crater's interior, and along its periphery they usually lie above the allogenic breccia or sometimes even directly upon the rocks of the Archean, Proterozoic and Paleozoic.

The second variety of impactites consists of massive, sometimes porous, glassy or slightly crystallized rocks. These are also partly saturated with the fused fragments of gneisses and their minerals, and occur as dikes, necks and irregular injection stratal bodies within the suevites and allogenic breccia, but mainly in the form of thick (up to 60 m) discontinuous sheets on the surface of the allogenic breccia and the suevites, predominantly in the northwestern part of the crater. These rocks were previously described as andesite porphyrites, whereas in fact they show only a remote similarity to the latter. They have no analogs among the igneous and metamorphic rocks. We have decided to name them tagamites (after the Tagamy ridge in the eastern part of the crater). The bodies of tagamites were formed as a result of the penetration to the surface of the melt that was at the crater's base and was not scattered in the explosion. In their overall chemical composition, the tagamites correspond to the average composition of the shield's crystalline rocks, whose unfused remnants they contain.

The visible thickness of the allogenic breccia and impactites exposed in the crater is 200-300 m, but erosion has exposed only the upper part of the bodies of which it is composed.

We have dated the meteorite crater's formation from the fact that the rocks deformed by the explosive impact and the allogenic breccia include rocks up to Late Cretaceous inclusive in age; another significant fact is that the breccias are overlain by Lower and Middle Quaternary clays. The radiological age of the hyaline rocks, according to the potassium-argon method, is 28.8 million years (M. N. Vasil'yeva et al.), but this is most likely exaggerated owing to the presence of a marked quantity of feldspar fragments from the gneisses in the glass. Thus it can be said tentatively that the crater was formed during the Neogene.

According to the results of geophysical investigations, the area of the depression contains negative gravitational and magnetic anomalies (5) caused by the lower density of the shattered and re-fused rocks and their weak magnetization.

Among the indications sufficient for the identification of an old meteorite crater (4, 12, 16), and established in the case of the object described here, are the circular depression, the centrifugal thrusts and annular zone of deformation, the semicircular (annular?) uplift and also the shock-metamorphosed rocks with shatter cones, planar structures, isotropized plagioclase and quartz, the glasses and impactites, and the negative magnetic and gravitational anomalies.

It must be stressed that according to the results of studies of the rocks altered in the course of underground nuclear explosions, the pressures required for the formation of planar structures, isotropization of the minerals and so forth considerably exceed the pressures that may be developed in the upper levels of the earth's crust, particularly in volcanic processes. The abovementioned distinctive features of shock-metamorphosed rocks arise at pressures exceeding 30-50 kbar. Judging from the energy of the shock waves, at the points of meteorite impacts the pressure reached 600 kbar and the temperature 2500° (13).

The Popigay crater is the largest object of this kind thus far identified. Of the other known meteorite craters, the Manicouagan crater in Quebec has a diameter of 65 km, the Richat crater in Mauritania 50 km, the Clearwater crater in Canada 32 km, the Nördlinger Ries in Germany 24 km, and so forth (11, 12, 14, 16). The superimposed inset funnels of the Popigay crater resemble the Nördlinger Ries in their distinctive structure. Such funnels are probably characteristic of large craters formed in a multilayered target area. The Nördlinger Ries and Popigay craters are also similar in the presence of breccias containing the products of fusion of the rocks of their base, but in the Popigay crater, because of the great violence of the explosive process, the fusion of the rocks was far more intensive and led to the formation of a series of pseudovolcanic structures and formations.

The energy of the explosion, as calculated from the crater's diameter by the formulas proposed by a number of authors (11), may have been on the order of $n \times 10^{30}$ erg; the thickness of the shattered rocks (authigenic breccia) according to the same estimates is 15-18 km, while that of the allogenic breccia that dropped down into the crater after the explosion is 2.5-3 km. The lens of fused material may reach a thickness of 200 m. The meteorite's impact drove it into the rocks of the sedimentary mantle and basement to a depth of about 1.5 km. The explosion took place in Archean rocks, thus explaining the saturation of the impactites with fragments of rocks of this age. All this indicates the catastrophic character of the event and its considerable geologic importance. The Popigay meteorite crater and its associated geologic formations can be reflected on maps of various scales, including small-scale geologic survey maps.

The materials obtained show that a number of traditional geologic ideas, including those concerning the history of the northern part of the Siberian Platform during the Mesozoic and Cenozoic, must be reexamined. These facts are also of interest to students of planetology, meteoritics, the physics of explosive processes and others. We propose to set forth more detailed results of our investigations of the Popigay crater in later publications.

REFERENCES

1. Voronov, P. S., 1958, RELATION OF CERTAIN REGULAR FEATURES OF THE RELIEF OF NORTHERN SIBERIA TO NEOTECTONIC PROCESSES: NIIGA Trudy, v. 67, no. 7.

2. GEOLOGY OF THE SIBERIAN PLATFORM, 1966: Izd-vo Nedra, Moscow.

3. Kiryushina, M. T., 1959, MANIFESTATIONS OF MESOZOIC AND CENOZOIC VULCANISM ON NORTHERN MARGIN OF SIBERIAN PLATFORM: AN SSSR Izvestiya, ser. geol., no. 1.

4. Krinov, Ye. L., 1962, METEORITE CRATERS ON THE EARTH'S SURFACE: Meteoritika, no. XXII.

5. Levin, D. V., Krykov, S. M., and Karasik, A. M., 1963, AN AEROMAGNETIC SURVEY OF THE CENTRAL SECTOR OF THE SOVIET FAR NORTH. In AEROMAGNETIC SURVEY IN GEOLOGY: Gosgeoltekhizdat, Moscow.

6. Mukhamedzhanov, A. K., 1967, STRUCTURE

OF THE LUNAR CRATERS AND CRATER FIELDS: Astronomicheskiy Vestnik, v. 1, no. 3.

7. Mukhamedzhanov, A. K., and Stanyukovich, K. P., 1966, TOWARD A THEORY OF LUNAR CRATERS: Kosmicheskiye Issledovaniya, v. IV, no. 3.

8. Podkopayev, Ye. I., 1958, UPPER JURASSIC TO LOWER CRETACEOUS ANDESITES IN THE NORTH OF THE SIBERIAN PLATFORM. In GEOLOGIC STRUCTURE OF THE USSR, v. 2: Gosgeoltekhizdat, Moscow.

9. Smirnov, L. P., 1962, STRATIGRAPHY OF CRETACEOUS CONTINENTAL DEPOSITS OF THE POPIGAY DEPRESSION: NIIGA Trudy, v. 121.

10. Stanyukovich, K. P., and Fedynskiy, V. V., 1947, DESTRUCTIVE EFFECT OF METEORITE IMPACTS: AN SSSR Doklady, v. 57, no. 2.

11. Baldwin, R. B., 1963, THE MEASURE OF THE MOON: Univ. Chicago Press, Chicago.

12. Beals, C. S., 1965, THE IDENTIFICATION OF ANCIENT CRATERS: New York Acad. Sci. Annals, v. 123, art. 2.

13. French, B., Editor, 1968, SHOCK METAMORPHISM – SYMPOSIUM VOLUME: Mono Press, Baltimore.

14. Preuss, E., 1963, DAS RIES UND DIE METEORITEN-THEORIE: Fortschr. Mineral., Bd. 41, H. 2.

15. Sauer, A., 1920, ERLAUT. GEOLOG. KARTE WURTTEMBERG 1:50,000: 20, Bopfingen, 2 Aufl., 15.

16. Short, N. M., 1967, ASTROBLEMES AND METEORITE CRATERS. In THE ENCYCL. OF ATMOSPH. SCI. AND ASTROGEOL.: R. W. Fairbridge, Ed. Reinhold Publ. Corp.

17. Shoemaker, E. M., 1963, IMPACT MECHANICS AT METEOR CRATER, ARIZONA. In THE MOON, METEORITES AND COMETS: Univ. Chicago Press, Chicago.

IGR/jr

Part III
STRUCTURES OF CLASS IV IN CANADA AND GREENLAND

No less than 20 structures of class IVa have been recognized within rocks of the Precambrian shield and their Paleozoic cover in Canada and Greenland. All but one of these are situated in Canada. These structures are listed in Table 7 and located in Figure 1.

LAKE ST. MARTIN

The Lake St. Martin structure is situated 240 km north of Winnipeg, Manitoba, in the Interlake area amid terrain covered by glacial deposits, within which there is 1% outcrop. This structure is not covered among the papers included in this volume, and the following discussion will be drawn from the report by McCabe and Bannatyne (1970).

The Silurian *Interlake Group* forms part of a total thickness of 230 m of Lower Paleozoic strata which dip gently (at 2 m/km) toward the west-southwest. In the vicinity of the structure there is a sharp structural uplift, amounting to at least 240 m, bringing basal Paleozoic and Precambrian granite to the surface. Dips off the flank of the uplift are in the order of 55–75 m/km. The elements of the Lake St. Martin structure are:

1. Arcuate rim of uplifted granite and basal Paleozoic, circular, with a diameter of 22.5 km. Structural disturbance extends up to 11 km beyond this rim, the total area of disturbance thus having a diameter of about 45 km. Immediately inside the rim is a complex of brecci-

Table 7 Structures of class IV in Canada and Greenland

	Name	Diameter (km)	Class	Age
(2)	Brent, Ontario	2.9	IVa/1	Middle Ordovician?
(5)	Holleford, Ontario	1.9	IVa/1	Precambrian or Cambrian
(6)	Deep Bay, Saskatchewan	13.6	IVa/1	Jurassic or older
(8)	Lac Couture, Quebec	16	IVa/1?	Mesozoic to Precambrian
(9)	West Hawk Lake, Manitoba	3.5	IVa/1	Jurassic?
(7)	Carswell Lake, Saskatchewan	32	IVa/3	Paleozoic
(4)	Clearwater Lakes, Quebec	32 26	IVa/3	Early Permian
(3)	Manicouagan Lake, Quebec	65	IVa/3	Triassic
(14)	Charlevoix, Quebec	35	IVa/3	Middle Ordovician–late Devonian
(15)	Mistastin Lake, Labrador	20	IVa/2	Triassic
(16)	Lake St. Martin, Manitoba	45	IVa/3	Permian-Triassic
(12)	Steen River, Alberta	24	IVa/3	Pre-Cretaceous
(11)	Nicholson Lake, N.W.T.	12.5	IVa/3	Post-middle Ordovician
(10)	Pilot Lake, N.W.T.	5.7	IVa/1	Ordovician to late Paleozoic
(18)	Gow Lake, Sasketchewan	4	IVa/3	Cretaceous or older
(20)	Haughton Dome, Arctic Archipelago	35	IVa/3	Ordovician to Pliocene
(21)	Slate Islands, Lake Superior	30	IVa/3	Paleozoic?
(13)	Sudbury, Ontario	80	IVa/4	Early Proterozoic
(17)	Lake Wanipitei, Ontario	8.6	IVa/1	Mesozoic or Paleozoic
	Agtö Island, Greenland	10	IVa/5	Archean

Note: Numbers on the left refer to the accompanying map. Number 1 on that map is the New Quebec Crater (McCall, 1977, Papers 23 and 24). Numbers 19 and 22 are Lac La Moinerie and Ile Rouleau, (a) is Hebron and (b–g) are other structures of class V.

ated, downfaulted limestone[1] of Devonian age (at least in part), overlain by aphanitic igneous rocks. Some of the fragments in the carbonate breccia cannot be correlated with any post-Silurian rocks of the outcrop sequence.

2. Together these rocks form part of the Lake St. Martin Series, a complex of faulted and brecciated carbonate rocks, granite breccias, and

[1] It appears that there is a rim and outer slope, an annular depressed zone filled with rocks of the Lake St. Martin Series and a central granite core about 3.5 km in diameter.

aphanitic rocks of trachyandesite composition, overlyinging shocked basement gneiss. All the rocks of this Series occur within the confines of the rim. They comprise pre-Mesozoic crater fill deposited at the time of the crater formation. K/Ar radiometric determinations carried out on the aphanitic igneous rocks (trachyandesites) yielded ages of 200 ± 25 and 250 ± 25 m yr, indicating a Permian age.

3. Overlying this series is a flat-lying sequence of red beds and evaporites of Triassic or Jurassic age, not known to outcrop elsewhere in the immediate vicinity.

The structure was formed during a period of uplift and erosion of the Paleozoic strata. There is evidence that at least the upper 100 m of the Upper Silurian and Devonian strata were present when the structure was formed and that these have been removed by later, post-crater and pre-red-bed erosion. The Precambrian rocks of the basement have been radiometrically dated at 2387 m yr. They are the normal rocks of the Superior Province. There are two outcrops of these rocks in the uplifted core of the structure; they also outcrop along the eastern rim. No other Precambrian outcrops are known for 48 km, at which distance a second, similar disturbance may exist at High Rock Lake. The basement rocks are oligoclase granites, pink biotite granites, microline granites, and pegmatites. Quartz grains in the rocks of the uplifted core show up to six sets of lamellations, and increasing shock intensity produced annealing characteristics and transition to a clear recrystallized mosaic devoid of planar structures. Such recrystallization is especially characteristic of inclusions in the trachyandesite. Quartz grains in mosaics within some granite inclusions in trachyandesite display clear quartz rims of mosaic or spheroidal form, actual melting or recrystallization having occurred along grain boundaries (possible simply due to the heating effect of the hot host). Isotropic quartz, which is ascribed to solid-state disorder, not to melting, is not commonly represented in this specific progressive shock-intensity sequence. Feldspars show comparable shock effects and also development of porous vesiculation structures ($\leqslant$.5 mm diameter), though true maskelynite is rare. Pseudotachylite veins of red to brownish glass or devitrified glass, commonly containing fine mineral and rock fragments, cut the gneiss in the core of the structure. Such fragments display all stages of shock metamorphism. The matrix may be red opaque material; orange to red, translucent, iron-oxide rich glass; patches of devitrified glass; or brecciated fragments of dominantly mafic minerals associated with iron oxide and grading down to the limits of microscopic resolution.

There are many occurrences of carbonate breccia in the Lake St. Martin Series, restricted to a narrow belt close to the rim. Drill core from this series also revealed spectacular polymict breccias. The trachyandesites are purplish to reddish aphanitic igneous rocks, locally vesicu-

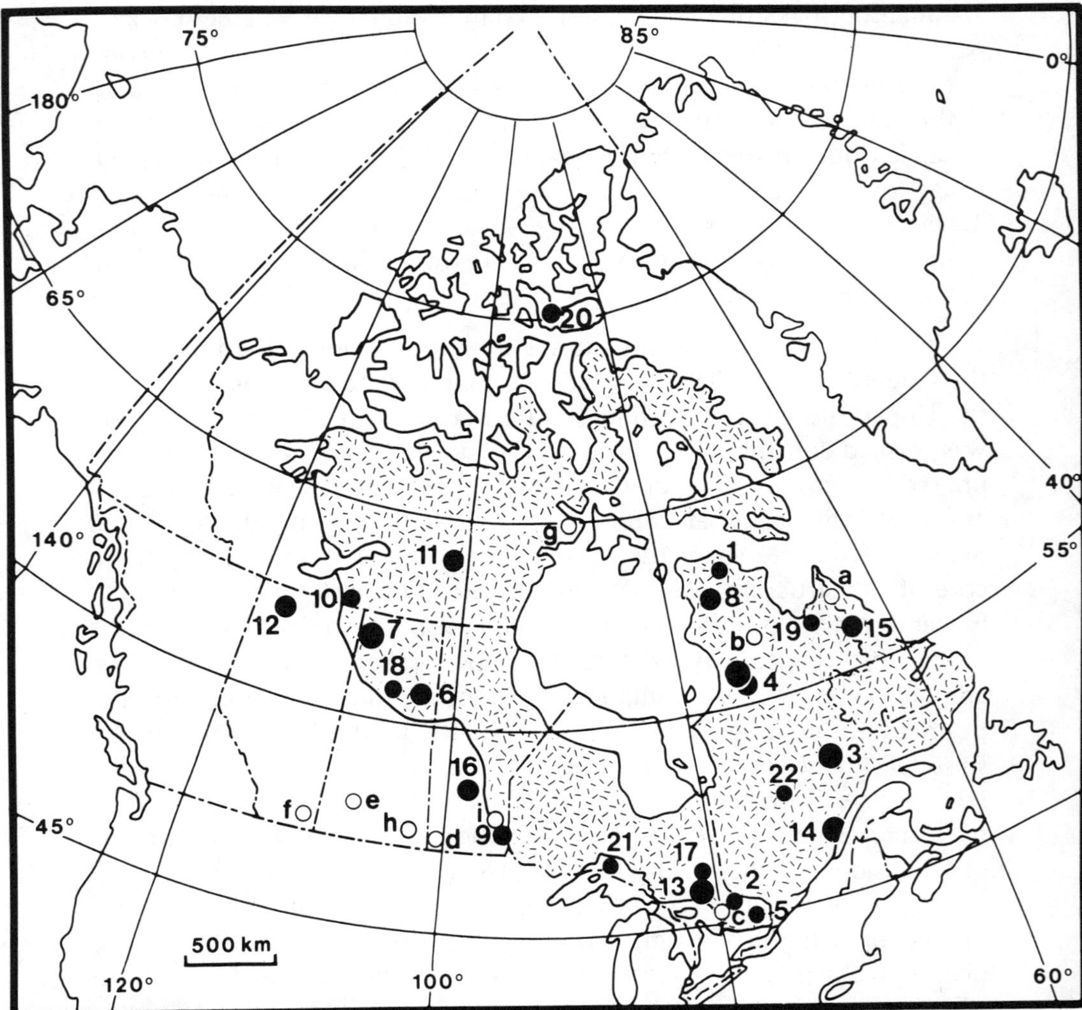

Figure 1. A (above): Distribution of cryptoexplosion structures in Canada (numbers refer to Table 5). B (facing): Diagrammatic sketches of cryptoexplosion structures in Canada. (Reproduced by permission of M. R. Dence, Earth Physics Branch, Dominion Observatory, Ottawa)

lar or amygdaloidal, and containing up to 40% of included mineral or rock fragments. A 70-m section of relatively uniform lava was revealed in one hole, above granitic and carbonate breccias. In another, lava underlay polymict breccia beds, which include what appears to be a fine phase of the trachyandesite. East of Gypsum Lake, an outcrop (of the top of the trachyandesite sequence?) is high vesicular. The thick uniform sequence shows veinlets of calcite at its base. Native copper has been reported from these rocks, but its presence has not been confirmed. Geochemical anomalies of 58 and 490 ppm Cu have also been recorded (igneous rocks of 60–65% SiO_2 of normal magmatic suites nor-

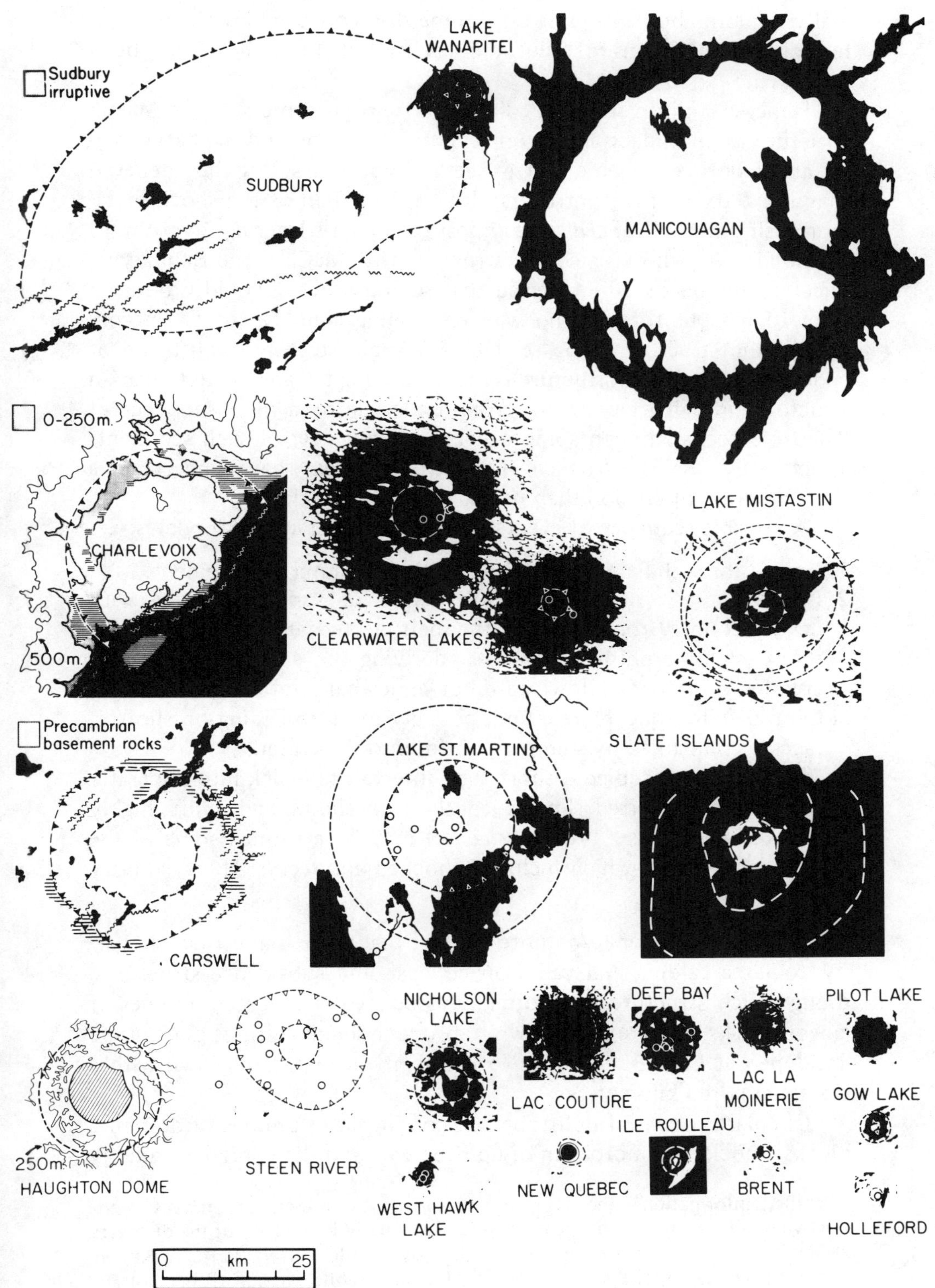
Sudbury irruptive
LAKE WANAPITEI
SUDBURY
MANICOUAGAN
0-250m.
CHARLEVOIX
500m.
CLEARWATER LAKES
LAKE MISTASTIN
Precambrian basement rocks
CARSWELL
LAKE ST. MARTIN
SLATE ISLANDS
NICHOLSON LAKE
DEEP BAY
PILOT LAKE
LAC COUTURE
LAC LA MOINERIE
GOW LAKE
ILE ROULEAU
250m.
HAUGHTON DOME
STEEN RIVER
NEW QUEBEC
BRENT
WEST HAWK LAKE
HOLLEFORD
0 km 25

mally contain about 40 ppm Cu). Some drill core samples are reported to contain inclusions of plutonic rocks, but these have not been described.[2]

Dence examined the rock collection. He regards the Lake St. Martin structure as an impact explosion structure. McCabe and Bannatyne entertained both endogenic and exogenic hypotheses; but they believed that the crux of the controversy lies in the origin of the shock metamorphism and advanced a composite impact-triggering hypothesis. Currie (1971), who visited the structure with McCabe and Bannatyne, rejected this because he believed that such a process would produce a suite of igneous rocks in no way resembling country rocks. He considered the structure to be related to a major tectonic uplift and of endogenic origin. The structure is unique among the analogous Canadian structures in being located in terrain underlain by a thick sequence of Paleozoic rocks, though some others have a veneer of such sediments affected by the disturbance. This structure may have undergone a shorter period of erosion than its analogues in Canada.

A possible sequence of events, according to the impact model is:

1. Impact through Paleozoic cover into the Precambrian basement.
2. Formation of crater: meteorite melted or vaporized along with nearby country rock; most of the melt ejected; uplift of rim; strong shock metamorphism of rocks underlying the crater; intensity decreasing laterally and downward but somewhat erratically.
3. Central uplift due to rebound, possibly by deep sliding on slip surfaces accompanied by slumping of rim into the crater.
4. These events occupied a short time interval; fall-back into the crater would have occurred simultaneously with sliding and uplift; highly shocked and melted breccia (*suevite*) may have formed some of the melt-rock sheets, which included shocked and recrystallized country rock fragments.

McCabe and Bannatyne note the fact that there is a distinct similarity to large caldera and cauldron collapse and subsidence structures, though such structures are normally associated with larger volumes of igneous rocks, and are not known to produce shock features of the type developed at Lake St. Martin. No Permian volcanics are known within several hundred kilometers.

Circularity; central uplift; brecciation; inward slumping of dolomite blocks; shock metamorphism of uplifted gneiss in the core; petrography

[2]Detailed information concerning any plutonic rocks present as enclaves in the trachyandesites would be of great interest, particularly in the light of the discovery at Manicouagan of highly anomalous, apparently mantle-derived picritic rocks containing "cognate xenoliths" of the kimberlite-meimechite kindred (Paper 16) and also in view of the manifest close connection in time between the eruption of the host trachyandesite and the shock brecciation episode.

of melt rocks and inclusions in them; shocking of inclusions; lack of phenocrysts or flow structures; decrease of shock intensity with depth; and lack of nearby volcanic manifestations all support impact origin for Lake St. Martin in the view of McCabe and Bannatyne, who also list some anomalous features:

1. The nature of the trachyandesite layers is more consistent with a series of eruptions from different vents; and the configuration does not entirely conform to the requirements of meteoritic theory—namely, an uppermost layer to the crater fill and a thin skin directly over the basement of the crater floor.
2. There is evidence of intrusion, partial solidification, and subsequent fragmentation in one of the breccias. Though the nature of this breccia could be accommodated to the impact theory, it remains difficult to visualize the complicated series of events required to achieve such accommodation.
3. The composition of the trachyandesite makes it difficult to explain as an impact melt. It appears to be too homogeneous, despite similarity to the country rock composition (no detailed chemical study has been undertaken).
4. The distribution of the trachyandesites accords closely with the requirements of Currie's ring-dike concept.

OTHER STRUCTURES

Space limitations prevent reproduction of the papers in which five other cryptoexplosion structures of class IVa from Canada and one from Greenland are described; these are discussed briefly below.

Pilot Lake, in the Northwest Territories, is a simple structure 6 km in diameter, which has affinities with Brent, Holleford, and West Hawk Lake. It has an island-free, square shape, contrasting with the island-dotted glacial lakes nearby. It lies within Precambrian granite and granodiorite of the Churchill Province. Though its depth is only a fraction of the theoretical expectation for an impact explosion crater of this size (Baldwin, 1963), Dence et al. (1968) believed it to be of meteoritic origin, but severely modified by erosional loss or sedimentary infill. Shock deformation is recorded from boulders on the lake shore, believed to have been dredged from the lake bottom by ice movement. Lamellation is present in quartz grains. Some gneiss fragments are coated with a clastic breccia. Fine-grained breccias with a partly recrystallized quartz matrix and mylonites were also recorded. Microscopic deformations include planar features of types A to D in quartz (Robertson et al., 1968). The age of the structure is unknown, but physiographic arguments favor a late Paleozoic age.

Nicholson Lake, in the Northwest Territories (Dence et al., 1968), is also small, its longest diameter being 12.5 km. However, the structure has affinities with larger, complex structures. Elliptical in shape, it contains an outlier of fossiliferous Middle Ordovician Trenton Limestone on one island. The Precambrian basement rocks are granite and granodiorite gneisses, with bands of amphibolite and massive granite. The gneiss exposed near the lake center displays shatter cones and lamellations in quartz grains (types A and B of Robertson et al., 1968), indicative of weaker shock intensity than at Clearwater Lakes or Manicouagan. Mixed welded breccias overlie the gneisses of the western promontory and two islands closest to the lake center. These breccias were deposited on an irregular surface of more than 50 m relief, and attained a maximum thickness of 330 m. They form a discontinuous annulus, the top of which has been everywhere removed by erosion, so the original thickness is not known. Glassy material is present in these polymict breccias, which include a variety of rock fragments displaying a variety of shock effects including lamellations of type A-D in quartz grains. A negative-gravity anomaly is centered on the lake. Paleozoic cover outcrops in a discontinuous ring 5 km out from the lake center in what may be a ring graben. A maximum Middle Ordovician age thus appears likely. These rocks are believed to have been lowered by 660 m and the center to have been uplifted by 2000 m. The structure was interpreted by Dence et al. as an impact-explosion crater, revealed by erosion compatible with its probable Middle to Late Ordovician age.

Gow Lake, in northern Saskatchewan, is even smaller, 4 km in diameter, but possesses a central uplift (Thomas and Innes, 1977). It is situated on Reindeer Lake, southwest of Deep Bay. The country rocks are Precambrian granite and gneisses. The central island, Calder Island, contains outcrops of clastic breccias, grey-green and brick-red melt breccias. The latter contain shock metemorphosed quartz, felspar, and biotite in a shock-melted glass matrix. Quartz lamellations are of types A, B, C, and D of Robertson et al. (1968), believed to indicate low to moderate shock levels (equivalent to stage 1 of von Engelhardt and Stöffler, 1968). Quartz also shows fracture patterns suggesting inversion from cristobalite. Plagioclase is recrystallized to a finely granular mosaic. Biotite shows deformed, wavy cleavage. The brick-red breccia displays textures characteristic of devitrified glass in the matrix. There is a circular negative-gravity anomaly coincident with the lake and centered on Calder Island. No sediments infill the lake, and this anomaly is attributed to fractured rock beneath the lake. There is no magnetic anomaly to suggest an underlying intrusion. This is the smallest cryptoexplosion structure known in Canada with a central uplift. The structure is believed by Thomas and Innes to be an impact explosion structure, and arguments based on its erosional state compared with Deep Bay suggest an age in excess of 100 m yr.

Haughton Dome, situated in the Canadian Archipelago, is a large, complex structure (Dence, 1972; Robertson and Mason, 1975). It is a structural dome occupied by a topographic depression, incised into a plateau of Cambrian to Devonian rocks. Ordovician evaporites are brought up in the center and the structure has previously been regarded as a salt diapir. The basin has a diameter of 17 km and is enclosed by an annular ridge. Shatter cones are recorded in limestone.

Slate Islands, situated in Lake Superior, is 30 km in diameter (Grieve and Robertson, 1976) and is associated with breccias, shatter cones, and shock metamorphism, including lamellations in quartz.

Agtö Island is the sole cryptoexplosion structure recorded in Greenland (Jensen, 1971). It is the only known example of Archean shock metamorphism (more than 2500 m yr). It is highly anomalous in terms of impact theory and deserves more detailed study. The shock metamorphism is of the usual "unique" type, yet the only possible meteoritic interpretation would involve overprinting of impact on an existing linear fracture system and their reworking. Hyalomylonites yielded Katarchean radiometric ages and, if they are related to impact, it must have occurred then. Shock metamorphism apparently produced veins of hyalomylonite throughout the structure, which is situated in rocks more than 300 m yr old. No physiographic features suggestive of impact are visible on maps. Enderbitic gneisses are more or less tectonically destroyed, and small black veins of pseudotachylite enclosing gneiss fragments have been formed. Decorated and nondecorated planar elements are present in quartz and other minerals in the gneisses and fragments in the pseudotachylite. Diaplectic and fusion glasses are recorded, together with partially fused mineral grains. Strong shock effects are restricted to an area of a few square kilometers and weak effects to an area of 30 km^2. Crush rocks occur along thrust zones up to 130 km long in this region. An age of 3070 ± 70 m yr was obtained by radiometric (K/Ar) dating on weakly devitrified glass from this structure. The lack of devitrification is due to the very "dry" metmorphic environment, in which only thermal reconstitution was possible.

REFERENCES

Baldwin, R. B. 1963. *The Measure of the Moon.* Chicago, Ill.: University of Chicago Press.

Currie, K. L. 1971. Origin of Igneous Rocks Associated with Shock Metamorphism as Suggested by Geochemical Investigations of Canadian Craters. *Jour. Geophys. Research* **76**:5575–5585.

Dence, M. R. 1972. The Nature and Significance of Terrestrial Impact Structures. *Internat. Geol. Cong., 24th, Proc.* **15**:77–89.

Dence, M. R.; M. J. S. Innes; and P. B. Robertson. 1968. Recent Geological and Geophysical Studies of Canadian Craters. In French and Short, 1968, pp. 339–362.

Engelhardt, W. von, and D. Stöffler. 1968. Stages in Shock Metamorphism of Crystalline Rocks in the Ries Basin, Germany. In French and Short, 1968, pp. 159–168.

French, B. M., and N. M. Short, eds. 1968. *Shock Metamorphism of Natural Materials.* Baltimore, Md.: Mono Book Corp.

Grieve, R. A. F., and P. B. Robertson. 1976. Variations in Shock Deformation at the Slate Islands Impact Structure, Lake Superior, Canada. *Contr. Mineralogy and Petrology* **58**:37–49.

Jensen, V. 1971. An Early Precambrian Impact Structure and Associated Hyalomylonites at Agtö, West Greenland. *Nature* **233**:188–190.

McCabe, H. R., and B. B. Bannatyne. 1970. Lake St. Martin Cryptoexplosion Structure and the Geology of the Surrounding Area. *Manitoba Geol. Survey Geol. Paper 3/70,* 79pp.

Robertson, P. B., and G. D. Mason. 1975. Shatter Cones from Haughton Dome, Devon Island, Canada. *Nature* **255**:393–394.

Robertson, P. B.; M. R. Dence; and M. A. Vos. 1968. Deformation in Rock Forming Minerals from Canadian Craters. In French and Short, 1968, pp. 433–452.

Thomas, M. D., and M. J. S. Innes. 1977. The Gow Lake Impact Structure, Northern Saskatchewan. *Canadian Jour. Earth Sci.* **14**:1788–1795.

Editor's Comments on Papers 7 Through 11

7 DENCE and GUY-BRAY
The Brent Crater

8 BEALS
Abstract from and Review of *A Probable Meteorite Crater of Precambrian Age at Holleford, Ontario*

9 INNES
Excerpt from *Recent Advances in Meteorite Crater Research at the Dominion Observatory (The Deep Bay Crater)*

10 BEALS, DENCE, and COHEN
Excerpts from *Evidence for the Impact Origin of Lac Couture*

11 SHORT
Abstract from and Review of *Anatomy of a Meteorite Impact Crater: West Hawk Lake, Manitoba, Canada*

BRENT

Paper 7, by Dence and Guy-Bray, is a concise and up-to-date discussion of the Brent Crater. This structure was described by Millman et al. (1960), and was first observed on air photographs in 1951. The 3 km diameter circular structure lies amid deep forest of the Algonquin National Park. Beals et al. (1963) noted that the rim is almost obliterated, and thus the original crater diameter is uncertain. Paleozoic sediments were found to fill the crater, hollowed out within Grenville (Precambrian) granites and gneisses. Drilling later revealed 200 m thickness of Ordovician sediments at the lake center and a thick breccia section near the shore. Cambrian sediments could be present in the lower part of the fill, which was not penetrated by the drill. The likely age is about 500 m yr, close to the time of filling in, rather than the maximum age of 950 m yr.

Dence (1968) has used Brent as a type example of shock zonation in impact craters within granite/gneiss terrain—there is a distinct zonation of the shock effects, including fracturing, monomict brecciation, shock lamellation of mineral grains, and shock melting (producing possible impactites akin to suevite; Paper 1). The breccias are nickel- and potassium-rich relative to the country rock. The nickel anomaly is attributed to the impacting meteorite, the potassium anomaly to solutions that circulated after impact and caused later metasomatism. Residual heat left in the rock after passage of the shock wave is supposed to have rendered the country rock susceptible to the activity of groundwater and other heated solutions. The shock zones and melt zones represent increased shock pressures (Dence, 1968). This crater displays the most advanced degree of shock modification of rock fragments in the Canadian cryptoexplosion craters.

The Brent Crater is the subject of continuing controversy (Currie and Shafiqullah, 1967). Many of the chemical discrepancies between breccia and host rocks have been attributed alternately to the presence of rare alkaline igneous rock alnoite, and the deep-seated fenitization zones that are supposed to surround a hidden alkaline igneous complex focused beneath the crater. The alkali metasomatism of the country rock is supposed to have been caused by fenitization, not by circulating heated groundwater (Currie, 1971). The dikes are supposed to form part of the igneous complex. Certainly, they resemble those of the Callander Bay Complex near North Bay, to the west, within the same graben zone. The Callander Bay structure, of the same size as Brent, is unquestionably endogenic and of circular shape. There is yet another such complex to the west on the islands within Lake Nipissing. Alnoite does not characteristically occur in extensive regional dike swarms, but is a feature of localized central eruptive bodies associated with carbonatite diatremes. Such igneous rocks carry abundant magnesia and thus could well be associated with a nickel anomaly. The age of the alnoite dikes in relation to the crater formation is not critical, for alnoite commonly occurs in multiphase developments. This controversy is critical to the problem of cryptoexplosion and the validity of the supposed shock metamorphism of unique types as indicators of impact.

HOLLEFORD AND DEEP BAY

Beals, in Paper 8, described the nearby infilled crater at Holleford, Ontario. Because of its length, Beals' work is covered here by the abstract and a brief review.

The Deep Bay structure is discussed, Paper 9, by Innes, who first briefly described the structure in 1957. He reported that no meteoritic material or fragmentary ejecta had been found (Innes, 1957), though the removal of both by glacial erosion appeared reasonable. The lake basin, approximately circular, is much deeper than the rest of Reindeer Lake. No volcanic rocks are known in this region, other than those contained within the Precambrian of the shield. The wall zone has suffered uplift, shattering and jointing. These effects diminish away from the rim. There are numerous radial and oblique faults associated with this deformation. Beals et al. (1963) presented an amplified account, the existence of an outer ring syncline being recognized. Shale fragments recovered from the lake shore were dated as Mesozoic on their fossil content. These shales are unlike any rocks known in the region and are assumed to form the crater fill, later dredged and dumped by ice action. Innes has confirmed this assumption by drilling (see Paper 9). According to the theory of impact explosion, there should be a fill of 427 m in a crater of this size, beneath the shale surface.

LAC COUTURE

Lac Couture (Paper 10), situated in Quebec, and West Hawk Lake (Paper 11), situated in Manitoba, appear to be simple crater structures, though the first could possess a central uplift, hidden beneath the lake. In Paper 10, Beals et al. have presented an account of the complex brecciations and shock deformations at Lac Couture. The lake has no sedimentary infill, and the depth is much less than the theoretical prediction for an impact explosion crater of this size. The lack of uparching of the rim contrasts with Deep Bay, though there is evidence of uparching of the rock formations, and an uparched rim appears to have been eroded to a bevelled surface. This erosion suggests that the structure may be older than Deep Bay, which is supposed to be not less than 150 m yr old; and Lac Couture could even be a Precambrian structure, though Dence (1972) refers it to the Paleozoic or Mesozoic. The destruction of the raised rim could be due to a contrasting potential for glacial erosion in Northern Quebec to that in Saskatchewan, rather than to greater age. If a central uplift did exist at Lac Couture, it would obviate the anomaly of the shallow depth.

WEST HAWK LAKE

West Hawk Lake, covered in Paper 11 by Short, was described by Halliday and Griffin (1963) as an 11,700 ft (3.6 km) diameter crater of simple form and a water depth of up to 365 ft (111 m). Short and Halliday and Griffin (1966) record a residual gravity anomaly. Situated amid Archean clastic sediments, granites, gneisses, and greenstones, the structure was drilled at the center. A thickness of 300 ft (91 m) of sediments was found to overlie 1000 ft (305 m) of brecciated rock, which in turn overlies 700 ft (213 m) of cracked rock. Deformations indicating extreme shock pressure were encountered in the core in some sections. The interface between sedimentary infill and breccia was encountered 300 ft (91 m) higher than theoretically predicted, suggesting that erosional stripping had been more intense than previously supposed, or that the crater has been enlarged by erosion. The layers of slightly shocked rock encountered above 1275 ft (388 m) in the drill core are supposed to have been slumped into position or to have fallen back after explosive ejection. There is a vestige of a raised rim despite prolonged erosion. The age is believed to be Jurassic. Short (Paper 11) described the shock metamorphism in detail. The crater is similar to Brent (Paper 7) in size and simplicity of form, and both are set in crystalline rocks of Precambrian age. Short believes that the crater has been enlarged by erosion, the original diameter being 8000 ft (2.4 km).

REFERENCES

Beals, C. S.; M. J. S. Innes; and J. A. Rottenburg. 1963. Fossil Meteorite Craters. In *The Solar System, vol. IV, Moon, Meteorites, and Comets,* B. Middlehurst and G. P. Kuiper, eds. Chicago: University of Chicago Press, pp. 235-284.

Currie, K. L. 1971. Origin of Igneous Rocks Associated with Shock Metamorphism as Suggested by Geochemical Investigations of Canadian Craters. *Jour. Geophys. Research* **76**:5575-5585.

Currie, K. L., and M. Shafiqullah. 1967. Carbonatite and Alkaline Igneous Rocks in the Brent Crater. *Nature* **215**:725-726.

Dence, M. R. 1968. Shock Zoning at Canadian Craters: Petrography and Structural Implications. In *Shock Metamorphism of Natural Materials,* B. M. French and N. M. Short, eds. Baltimore, Md.: Mono Book Corp., pp. 169-184.

Dence, M. R. 1972. The Nature and Significance of Terrestrial Impact Structures. *Internat. Geol. Cong., 24th, Proc.* **15**:77-89.

Halliday, I., and A. A. Griffin. 1963. Evidence in Support of a Meteoritic Origin for West Hawk Lake, Manitoba, Canada. *Jour. Geophys. Research* **68**:5297-5306.

Halliday, I., and A. A. Griffin. 1966. Preliminary Results from Drilling at West Hawk Lake Crater. *Roy. Astron. Soc. Canada Jour.* **60**:59-68.

Innes, M. J. S. 1957. A Possible Meteorite Crater at Deep Bay, Saskatchewan. *Roy. Astron. Soc. Canada Jour.* **51**:235-240.

McCall, G. J. H., ed. 1977. *Meteorite Craters.* Benchmark Papers in Geology, vol. 36. Stroudsburg, Pa.: Dowden, Hutchinson & Ross, pp. 290-315.

Millman, P. M.; B. A. Liberty; C. F. Clark; P. L. Wilmore; and M. J. S. Innes. 1960. The Brent Crater. *Dominion Observatory Ottawa Pubs.* **24**:1-43.

7

Reprinted from pages 11-18 of *Some Astroblemes, Craters, and Cryptovolcanic Structures in Ontario and Quebec,* Guide to Excursion A65, 24th Internat. Geol. Cong., Montreal, Quebec, 1972, 61pp.

THE BRENT CRATER

M. R. Dence and J. V. Guy-Bray

The Brent Crater

The discovery of the Brent crater in 1951 from air photographs taken by Spartan Air Services Ltd., was in essence the beginning of the systematic investigation of Canadian impact structures by members of the Dominion Observatory. The crater is outlined topographically by an extraordinary circular depression 1.8 miles (2.9 km) in diameter. Extensive topographical, geophysical and geological investigations and diamond drilling programs in 1955, 1959, 1960 and 1967 have shown that the depression contains more than 800 feet (250 m) of sedimentary rocks, underlain in turn by a breccia lens 2,000 feet (610 m) thick in the centre. Altogether twelve holes have been drilled and 16,520 feet (5,035 m) of drill core recovered. From these investigations the geometry of the crater has been established, the distribution of shock effects mapped, and the history of the crater reconstructed in some detail (Dence, 1964, 1968; Hartung et al., 1971).

The crater was excavated in Grenville gneisses on the eastern side of an antiform (Figure 2). In the core of the antiform are quartz - plagioclase - mesoperthite - hornblende gneisses containing some biotite and garnet. This unit underlies most of the crater and is the main rock type found as clasts in the breccias. It is overlain by gneisses in which biotite is more abundant than hornblende, garnet is common and the alkali feldspar is a weakly perthitic microcline. This unit contains a few lenses of garnet amphibolite. It underlies most of the east side of the crater. To the east, the third unit of the antiform is a pink to grey, quartz-plagioclase-microcline-biotite gneiss in which hornblende and garnet are rare. This unit contains a number of basic garnet-amphibolite and garnet-pyroxene coronites as massive, boudin-like structures, cut by large pegmatite dikes. At the eastern side of the Brent region is the western margin of a major synform comprising granitic and basic garnetiferous gneisses interlayered on a scale of centimetres to metres.

In general, lineations in all the Precambrian rocks plunge to the east-southeast, except northwest of the crater, where there is a local reversal of plunge to the northwest. The gneisses are cut by east-west trending faults of the Ottawa Valley fault system, some of which may have been active before the crater was formed. Two faults are well defined, one passing about midway between the hamlet of Brent and the southern margin of the crater, the other being exposed along Cedar Lake, southwest of Brent. Ordovician sedimentary rocks are downthrown on the south side of the second fault.

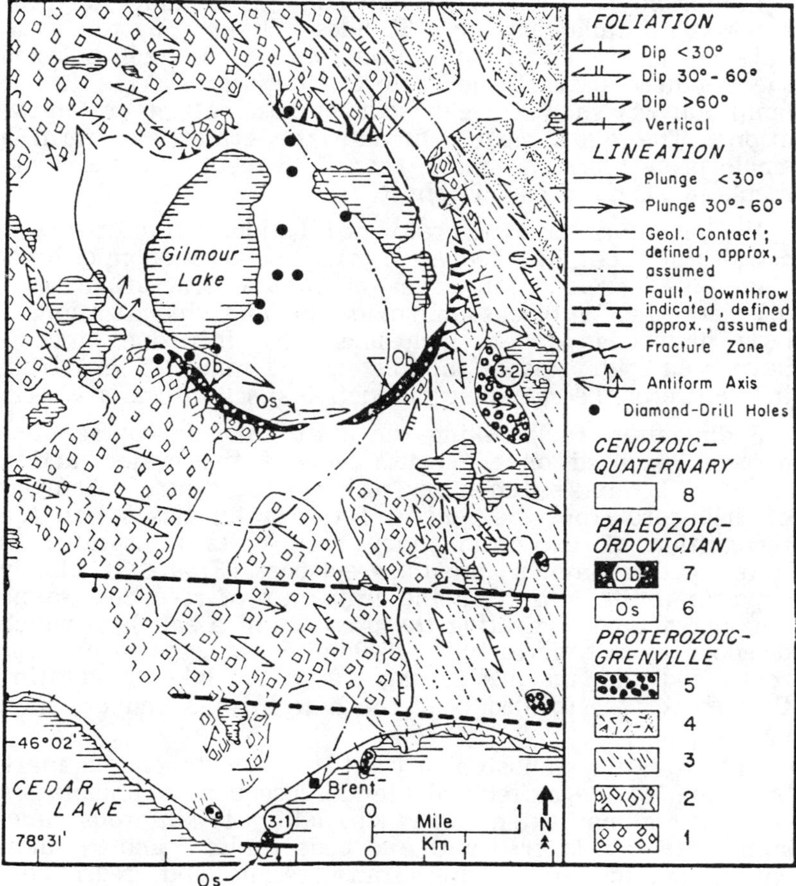

Fig. 2. Simplified geological map of the Brent region. The crater is outlined by unit 7 and the arcuate lakes to the north. The rock units are:

1. Quartz-plagioclase-mesoperthite-hornblende (-garnet-biotite-gneiss).
2. Quartz-plagioclase-microcline-biotite-garnet (-hornblende)-gneiss.
3. Quartz-plagioclase-microcline-biotite-gneiss.
4. Interlayered quartz-feldspar and garnet-amphibolite gneiss.
5. Coronite and garnet-amphibolite gneiss.
6. Dolostones, limestones and crater fill sedimentary rocks of Middle Ordovician age.
7. Talus breccias — crater rim deposits.
8. Glacial drift.

Lamprophyric dikes of Cambrian age, similar to those exposed at Callander Bay 37 miles (60 km) to the west, have not been seen in outcrop but were cut by drill holes in the northern part of the crater. Gneisses adjacent to the lamprophyres show a weak sodic metasomatism or fenitization. The dikes clearly predate the crater and form a small proportion (no more than 5 percent) of the fragments in some breccias within it.

Wall rocks around the crater are distinctly fractured over a width of about 1325 feet (400 m). Locally monomict breccias have developed in zones of intense fracturing, with disorientation of blocks and oxidation and chloritization of both matrix and some inclusions. No intrusive breccias have been found and at this distance (1 mile, 1.5 km) from the centre of the crater no distinctive shock effects are seen.

Sedimentary rocks filling the crater include coarse talus breccias exposed on the south side of the crater, and a series of layered sedimentary rocks in the centre. Analysis of drill core from the centre of the crater has led to the recognition of ten sedimentary rock units interpreted in terms of two major sedimentary episodes (Figure 3). Early deposits (units 1 to 4) in the crater are of restricted saline lake-playa type, indicating the presence of a salt lake which dried up from time to time. Dolomite and gypsum in these rocks indicate marine seepage into the lake, suggesting that the crater formed near the margin of a shallow shelf lagoon sea.

Transgressive sandstones (unit 5) are taken to mark breaching of the crater walls, and subsequent sedimentation was of regional extent. Predominantly fossiliferous limestones are interlayered with red beds (units 7 and 9), indicating periods of marine transgression and regression. There is no indication of any major hiatus in this record, so the Middle Ordovician age of fossils in the sediments is taken as the age of the crater, in reasonable agreement with the most probable age determined by K/Ar methods (Hartung et al., 1971).

Underlying the sedimentary rocks in the crater is a lens of allochthonous breccia, which rests on fractured and locally brecciated country rocks (Figure 4). Shocked fragments are concentrated in suevite-type breccias in the upper 500 feet (150 m) of the breccia lens and at the base in the centre where there is a section 115 feet (35 m) thick referred to the melt-zone. Within this zone fragments of shocked country rocks are held in a matrix of once-molten

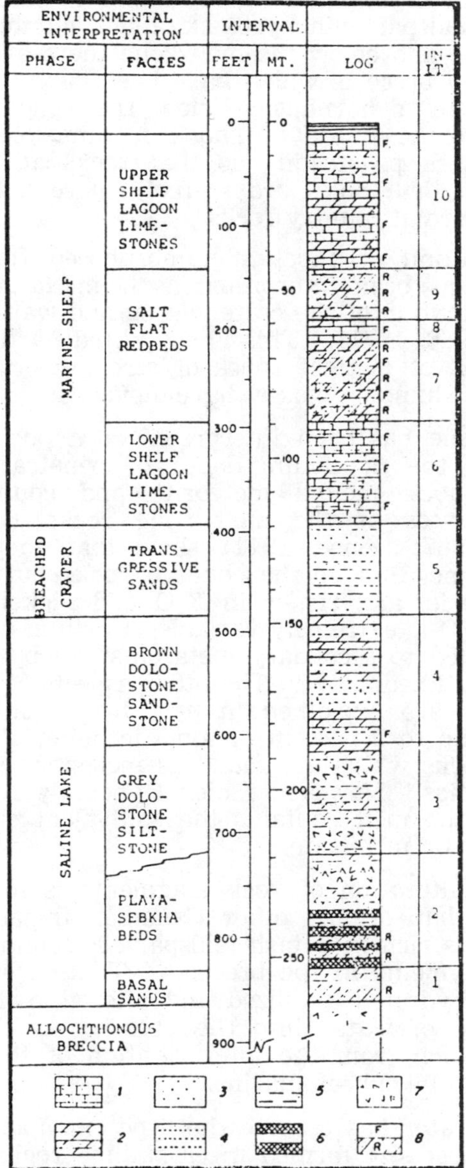

Fig. 3. Sedimentary rocks in the Brent crater. The rock types are:

1. Limestone
2. Dolostone
3. Sandstone
4. Siltstone
5. Shale
6. Gypsum
7. Breccia
8. F = fossiliferous, R = red beds

rock which has crystallized into a distinctive assemblage of zoned alkali feldspar, clinopyroxene and a number of minor phases. Above and below the melt zone there are thermally metamorphosed breccias which have been partly altered and recrystallized to a hornfels. Below the zone of thermal metamorphism the grade of shock metamorphism in the country rocks drops steadily so that rocks at the bottom of the deepest drill holes are scarcely more fractured and altered than normal country rocks.

The distribution of shock metamorphosed fragments in the breccias has been interpreted as being largely the result of redistribution of the crater lining and walls by slumping (Dence, 1968, 1971). This is supported by the presence in the breccias of distinct zones of strongly crushed rocks on which large slumping masses have moved.

The melt zone has been interpreted as a pool of impact melt marking the maximum depth of penetration of the mixture of shock melted meteorite and country rocks. Chemical analyses of the melt zone rocks (Currie and Shafiqullah, 1967; Dence, 1971) show that they resemble the mean composition of the country rocks but are somewhat more basic, and richer in K_2O and nickel. The enrichment in K_2O is extremely variable (Hartung et al., 1971) and is attributed to secondary metasomatism by circulating solutions, as outlined above. The other aspects of the chemical differences are explained in part by including in the melt mixture up to 5 percent of alnöitic material similar to the alnöite blocks which are found as inclusions in the melt zone and breccias. However a nickel anomaly remains and suggests that the melt contains the equivalent of about 1.5 percent chondritic meteorite.

In general, alteration of rock fragments is more advanced at Brent than at any other Canadian impact site and many fragments rich in potash feldspar are found in strongly shocked fragments in the breccias. This exceptional development of hydrated, oxidized and altered rocks may be due to seepage of water into the breccias as they cooled and settled, which from the interpretation of the sedimentary record may have been marine.

The Brent crater has become the type example of impact craters of modest size formed in crystalline rocks. Craters of similar size elsewhere in the Canadian Shield have a similar structure (Beals, 1960; Short, 1970). Craters of this class, which resemble the Barringer Meteor Crater

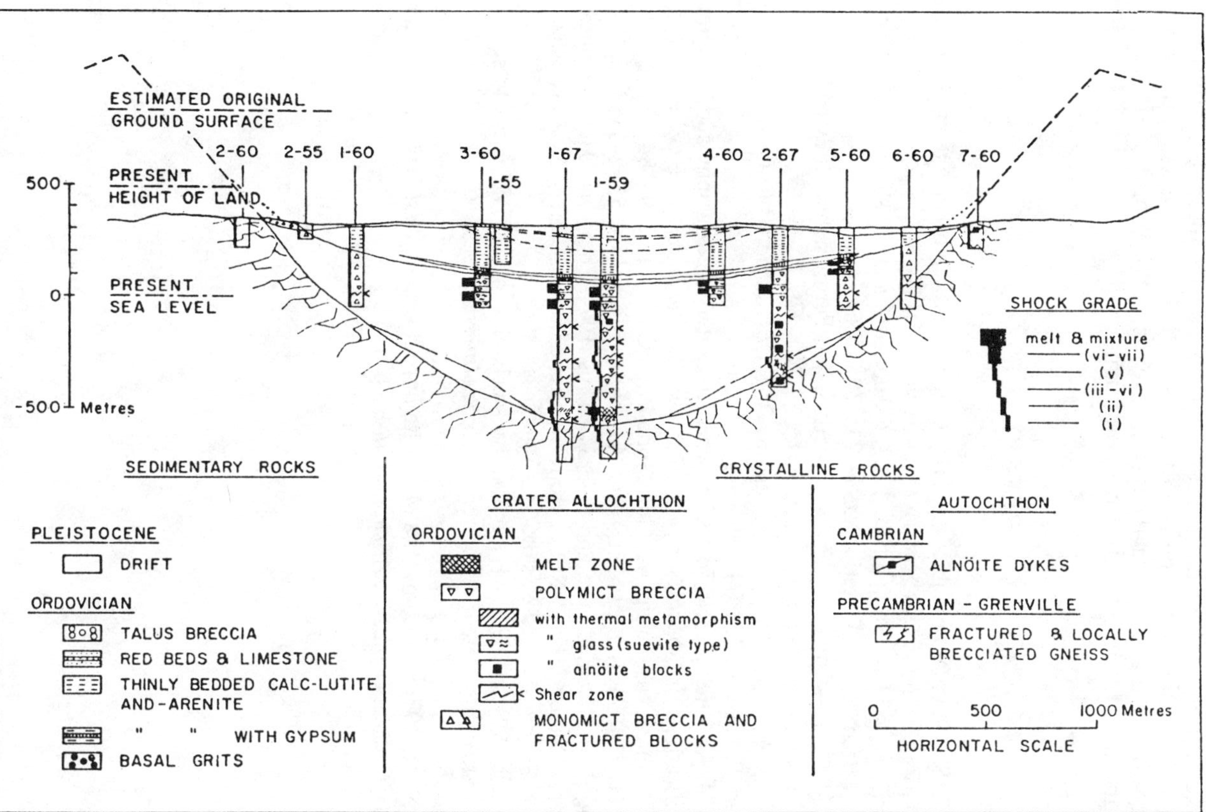

Fig. 4. Cross-section of the Brent crater showing subdivisions of the breccias and autochthonous rocks based on grade of shock metamorphism.

(Shoemaker, 1960) have been termed *simple* impact craters. Larger craters have generally proved to be a *complex* structure involving a central uplift (Dence, 1965) and it is to examples of this type that we now turn.

REFERENCES

Beals, C. S. 1960. A Probable Meteorite Crater of Precambrian Age at Holleford, Ontario. *Dominion Observatory Ottawa Pubs.* **24**:117-142.

Currie, K. L., and M. Shafiqullah. 1967. Carbonatite and Alkaline Igneous Rocks in the Brent Crater, Ontario. *Nature* **215**:725-726.

Dence, M. R. 1964. A Comparative Structural and Petrographic Study of Probable Canadian Meteorite Craters. *Meteoritics* **22**:249-270.

Dence, M. R. 1965. The Extraterrestrial Origin of Canadian Craters. *New York Acad. Sci. Annals* **123**:941-969.

Dence, M. R. 1968. Shock Zoning at Canadian Craters: Petrography and Structural Implications. In *Shock Metamorphism and Natural Materials,* B. M. French and N. M. Short, eds. Baltimore, Md.: Mono Book Corp., pp. 169-184.

Dence, M. R. 1971. Impact Melts. *Jour. Geophys. Research* **76**:5552-5565.

Hartung, J. B., M. R. Dence, and J. A. S. Adams. 1971. Potassium-Argon Dating of Shock Metaporphosed Rocks from the Brent Impact Crater, Ontario, Canada. *Jour. Geophys. Research* **76**:5437-5448.

Shoemaker, E. M. 1969 [sic]. Penetration Mechanics of High Velocity Meteorites, Illustrated by Meteorite Crater, Arizona. *Internat. Geol. Congr., 21st., Rep. Session 18,* pp. 418-434.

Short, N. M. 1970. Anatomy of a Meteorite Impact Crater: West Hawk Lake, Manitoba, Canada. *Geol. Soc. America Bull.* **81**:609-648.

8

Reprinted from page 117 of *Dominion Observatory Ottawa Pubs.*
24:117-142 (1960)

A Probable Meteorite Crater of Precambrian Age at Holleford, Ontario

C. S. Beals

ABSTRACT:—A circular feature 1.46 miles in diameter and 100 feet deep located at longitude 76° 38′ W and latitude 44° 27′ N in southeastern Ontario has been investigated as a possible meteorite crater. Stereoscopic studies of aerial photographs in conjunction with geological and geophysical investigations indicated the presence of a circular depression in Precambrian rock filled in and covered over by Palaeozoic sediments. Three diamond drill holes at distances of 1,400 feet, 2,500 feet and 3,750 feet from the centre were sunk to see whether the depth below the sediments and the sub-surface structure were consistent with a meteoritic origin. The results showed a depth and profile close to those predicted for a meteorite crater of the observed diameter. Below the sediments a thickness of several hundred feet of shattered and pulverized rock was found for which no adequate explanation has yet been found except that of meteorite impact and explosion.

REVIEW*

The Holleford Structure, like Brent [Paper 7] is infilled by ancient sediments and was formed by cryptoexplosion. It also was discovered after a search of air photographs. The crater is 2.4 km in diameter and 30 m deep, to the top of the infill of Paleozoic sediments, which are tentatively ascribed to the Cambrian *Potsdam Formation.* These dip centripetally (cf. Bosumtwi crater, Maclaren, 1931, and Rohleder, 1936). There was massive erosion prior to their deposition, a feature that suggests a possible Precambrian age for the cratering. However, Frarey (1955) suggested an age of 500–600 m yr. Diamond drilling has revealed the physiography of a crater compatible with impact explosion. A thickness of hundreds of meters of sediments overlies an equal thickness of shattered and pulverized rock beneath the crater focus. Undisturbed gneiss and limestone bedrock underlie the breccia zone. The breccias display intense shock deformation of quartz and felspar. This deformation is confined to the central drill cores. Such effects are absent from the surrounding Grenville gneisses (Robertson et al., 1968). The polymict breccias tend to be interbedded with undeformed sediments, and to consist of fragments of at least four rock types, set in a paste of unaltered grains of K-felspar, clinopyroxene, quartz, hornblende, biotite, and rare chlorite. The conclusion is that the breccias were formed more or less in situ, but that the circumstances of the explosion caused a thorough mixing of rock types observed in some drill core sections, and also, elsewhere, uniform, monomict breccias. Short and Bunch (1968) recorded coesite in shocked breccias. Geophysical anomalies over the crater are consistent with a low-density infill, as established by drilling. No meteorite material has ever been recovered from the Holleford crater.

*This review was prepared by the volume editor.

REFERENCES

Frarey, M. J. 1955. Unpublished report on a field study at Holleford, filed at the Geological Survey of Canada.

Maclaren, M. 1931. Lake Bosumtwi, Ashanti. *Geog. Jour.* **78**:270–276.

Robertson, P. B.; M. R. Dence; and M. A. Vos. 1968. Deformation in Rock Forming Minerals from Canadian Craters. In *Shock Metamorphism of Natural Materials,* B. M. French and N. M. Short, eds. Baltimore, Md.: Mono Book Corp., pp. 433–452.

Rohleder, H. P. T. 1936. Lake Bosumtwi, Ashanti, *Geog. Jour.* **87**:51–65.

Short, N. M., and T. E. Bunch. 1968. A World-Wide Inventory of Features Characteristic of Rocks Associated with Presumed Meteorite Impact Craters. In *Shock Metamorphism of Natural Materials,* B. M. French and N. M. Short, eds. Baltimore, Md.: Mono Book Corp., pp. 255–266.

9

Reprinted from pages 224-230 of *Meteoritics* **2**:219-241 (1964)

RECENT ADVANCES IN METEORITE CRATER RESEARCH AT THE DOMINION OBSERVATORY

M. J. S. Innes

[*Editor's Note:* In the original, material precedes and follows this excerpt.]

III. THE DEEP BAY CRATER

Topographical, geological and geophysical studies, have led us to believe that Deep Bay (Innes 1957, 1963), although considerably larger than the Brent crater, has a similar underground structure and is also an impact crater. The unique topographical features of Deep Bay are well brought out in the map (Fig. 4) of the southern end of Reindeer Lake that shows the almost perfect circularity of Deep Bay and the

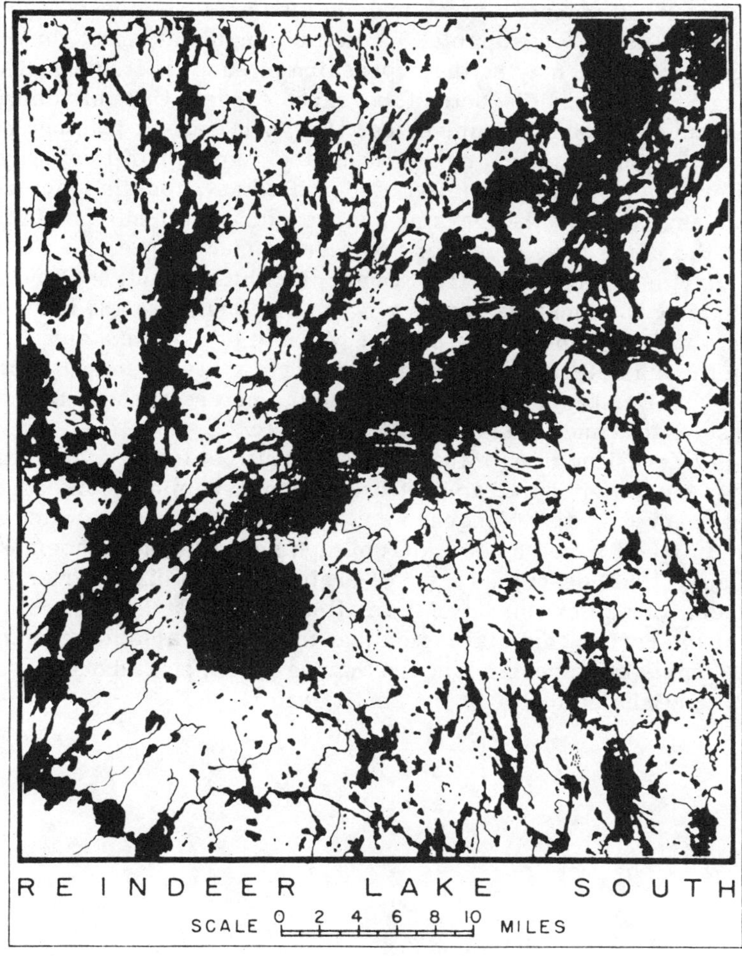

Fig. 4. Map of Reindeer Lake south; water areas are in black.

absence of islands within it. In marked contrast, Reindeer Lake has a highly irregular shoreline and is dotted by numerous rocky islands, both large and small. Careful examination of the structural relationships of the rocks surrounding Deep Bay in no way suggests a geological origin. Three main units of granitic gneisses are recognized; their trend is north-easterly and approximately the same on both sides of the crater. Although deeply eroded, much of the bedrock portion of the crater's original rim remains and stands on the average about 270 feet above the waters of the bay. The present rim diameter as marked by the height of land surrounding the bay, is about 8-1/2 miles. The map shows clearly that the drainage into the bay, with the exception of the three broad channels into Reindeer Lake, is restricted to short rivers no greater than two miles in length.

That Deep Bay is the result of impact is clearly indicated from the intense fracturing and shattering of the granitic rocks, which is most pronounced in the vicinity of the shoreline. Deformation is present on all scales, from a close-set pattern of jointing near the bay to large fault zones extending for great distances. Fig. 5 is a map showing prominent lineaments due for the most part to faults and fractures. The lineaments have been identified as fault zones from ground observation.

A system of concentric fractures is also well developed; the narrow arcuate lake, for example, 3 miles in length and about three miles east of the crater, may be the expression of such fracturing. The drainage pattern and dissected topography suggest that this depressed zone is much longer, circumscribes the whole crater, and has a diameter of about 12 miles, nearly twice the diameter of Deep Bay. This circular zone, which is tentatively identified as a tensile fracture, may be analogous to the markings surrounding the lunar maria, although the latter have diameters on a scale much larger than Deep Bay. Hartman and Kuiper (1963) interpret the concentric structures surrounding the lunar basins as ring faults, formed at the time of the impacts which formed these basins. A table by Hartman and Kuiper giving the diameters of the lunar concentric systems indicates that for the smaller basins the ratios of the diameters of the outer rings to those of the inner basins vary from 1.9 to 2.2.

It is interesting to note that the same ratio is found if the dimensions of extremely small impact craters are examined. In experiments to investigate the mechanism and conditions under which shatter cones may be formed by shock from high speed impact, Shoemaker and Gault (1962) described a small crater 5 cm in diameter and 1.3 cm deep, formed by impact of a small aluminum pellet into dolomite. It is of interest to this discussion that the impact also formed an almost completely circular fracture having a diameter about 11 cm, or about twice the diameter of the small crater.

At Deep Bay the circular fracture zone marks approximately the outer limit of the fractured zone and may be interpreted as a hinge line about which the granitic rocks forming the rim have been uplifted. Indeed the rocks within this circle, have the general appearance of having been uplifted and shattered into huge blocks without having undergone much horizontal movement.

It seems appropriate to emphasize here, that the raised rim of an impact crater lies within the fracture zone, and is produced by upward and outward compression of the surface rocks, rather than simply a piling up of fragmental material ejected from the crater. Uplift of the granite gneiss surrounding Deep Bay can be seen in the photograph (Fig. 6), which shows easterly facing rock exposure on the western margin of the bay. This exposure is located well within the fracture zone, at least one mile beyond the rim of the original crater. Uplift of the gneiss here is remarkably similar to that exhibited by the

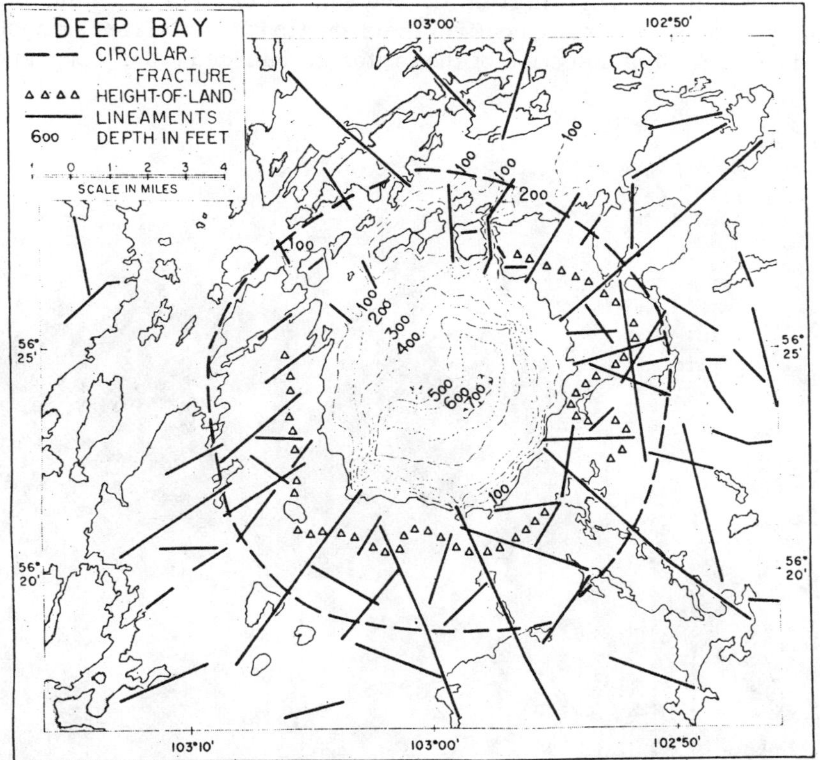

Fig. 5. Large fracture zones in the vicinity of Deep Bay as inferred from lineaments appearing in aerial photographs.

deformed sedimentary strata forming the rim of the Barringer Crater in Arizona. A photograph from D. M. Barringer's (1910) original paper is reproduced in Fig. 7 for comparison.

In the many arguments and discussions that have proceeded for years, in which the outward slopes of lunar and terrestrial craters have been compared, it has not usually been recognized that a considerable portion of the rim results from uplift of the surrounding rocks due to up-thrusting, and to an expanded crust within the fracture zone. Indeed, this matter appears not to have been considered by Baldwin (1963) in his recent book "The Measure of the Moon." As a result, he finds the sub-surface crater volume to be always greater than the volume of the rim. This is at variance with the result of a previous analysis (Innes 1961) in which it was found that expansion of the crustal rocks due to fragmentation and uplift is appreciable and that the volume of a meteorite crater's rim is usually greater than that of the volume below plain.

The most recent investigation of the Deep Bay crater has been a diamond drilling program from the ice surface of the Bay during the winter months, to test or confirm the deductions drawn from previous studies. Fig. 8 is a section of the crater as inferred from topographic

Fig. 6. An easterly facing rock exposure on the western margin of Deep Bay showing evidence of uplift.

Fig. 7. A view of the inner wall of the Barringer Crater (from D. M. Barringer, 1910).

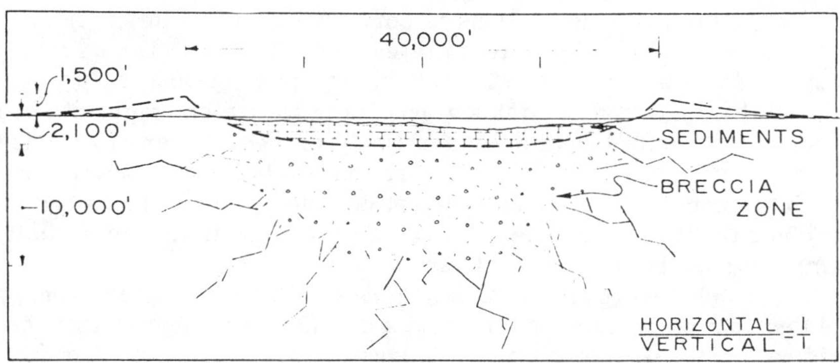

Fig. 8. Section of the Deep Bay Crater as inferred from topographical and gravity observations.

and gravity observations. The original crater profile is shown by the broken line, and is based upon Baldwin's (1949) depth-diameter relationship for impact craters. The rim diameter is estimated to be about 40,000 feet on the assumption that the present shoreline of the bay, having a mean diameter of 6-1/2 miles, represents the sectioning of the crater at its original ground level. The site for the first drilling was located about three-quarters of a mile from the center of the crater where, over a small area, the water depth is slightly less than 400 feet. Here one would predict that the drill would penetrate at least 1600 feet of sedimentary rock before striking the breccia zone. Unfortunately at a depth of 591 feet the drill rods jammed, and it became necessary to abandon this site before reaching our full objective. However, the drilling here did reveal the existence of at least 100 feet of undisturbed shale, confirming our prediction that sedimentary rock now fills the crater. On the basis of fossil evidence the age of this rock is considered to be Mesozoic, probably late Jurassic or Cretaceous. This age, about 140 million years, may be taken also as the minimum age of the crater.

The second drill site was located about one mile from the eastern shoreline of the bay, where the water depth is only 200 feet. The hole was finally collared into bedrock at a depth of 240 feet and put down to a total depth of 1420 feet. On examination of the drill core it was found that no sedimentary rock was encountered in the upper part of this hole, although about 500 feet of sediments would be predicted for a crater of this size. The rock types encountered lie entirely within the biotite-garnet gneiss unit recognized in the general field geology of this region. The drill core shows evidence of all grades of deformation; there are long sections of core up to 400 feet in length in which only occasional fractures are evident. These sections grade above and below into zones of more intense deformation. At its most intense, the deformation yields a breccia of wall rock fragments up to several centimeters, in a matrix of finely crushed material also derived from the wall rock. The general nature of the deformation in which there is little or no evidence of complete mixing, leads one to conclude that this drill hole is situated very close to the outer limit of complete rupture, but does not lie within it.

The conclusion then, that this hole lies within the fracture zone, and the fact that no sedimentary rock was encountered suggests that the original crater diameter was considerably less than 40,000 feet as first estimated. The revised crater model, consistent with this new information and based upon Baldwin's (1963) depth-diameter relationships, is illustrated in Fig. 9. The revised crater dimensions are as follows:

Original Rim Diameter	31,000 feet
Apparent Crater Depth	3,200 feet
Rim Height	890 feet
True Depth	2,310 feet

The most important result from the drilling operation has been confirmation of the interpretation given the geophysical data: that the negative anomaly field reflects deformation of the rock underlying the waters of Deep Bay in the form of intense fracturing and fragmentation. Although it may now be predicted more confidently that crater filling breccia underlies the sedimentary rocks within the bay, it is extremely important to follow up this work by successfully putting down a drill hole at the center of the crater.

REFERENCES

Baldwin, R. B. 1949. *The Face of the Moon.* Chicago, Ill.: Univ. Chicago Press.

Baldwin, R. B. 1963. *The Measure of the Moon.* Chicago, Ill.: Univ. Chicago Press.

Barringer, D. M. 1910. Meteor Crater (formerly called Coon Mountain or Coon Butte) in Northern Central Arizona, private pub. 24.

Hartmann, W. K., and G. P. Kuiper. 1963. Concentric Structures Surrounding Lunar Basins. *Comm. Lunar and Planetary Lab.* **1**:51.

Innes, M. J. S. 1961. The Use of Gravity Methods to Study the Underground Structure and Impact Energy of Meteorite Craters. *Jour. Geophys. Research* **66**:2235.

Shoemaker, E. M., D. R. Gault, and R. V. Lugn. 1961. Shatter Cones Formed by High Speed Impact in Dolomite. *Short Papers, Geol. Hydrol. Ser.,* Articles 293, 438. D. 865.

10

Reprinted from pages 410–415 and 418–426 of *Dominion Observatory Ottawa Pubs.* **31**:409–426 (1967)

EVIDENCE FOR THE IMPACT ORIGIN OF LAC COUTURE

C. S. Beals, M. R. Dence and Alvin J. Cohen

Abstract:—Lac Couture is a circular lake of diameter 16 km with a central island-free area of 10 km diameter and depth 150 metres. The central deep area is surrounded by a shallower zone characterized by numerous islands and peninsulas all of which show marked indications of glacial action, the direction of ice movement being from east to west.

An examination of the bedrock of the islands showed many locations of heavy shattering and other indications of disturbance with sheeting planes and joints departing from the usual horizontal-vertical pattern common to areas surrounding the lake. Most of the islands and peninsulas examined showed numerous boulders and other types of glacial debris and on a number of islands on the westward side of the lake this debris included large quantities of rock breccia of the type normally associated with meteorite impact. Since the direction of glacial motion was from east to west and since no other explanation could be found for the breccia it was concluded that it had been dredged from the lake bottom by the moving ice.

Mineralogical examination of breccia fragments indicated the presence of numerous crystals deformed and altered by heavy shock. The evidence suggests an ancient and eroded impact crater of 12 km diameter of which the circular fringe of islands and peninsulas represents the remains of the rim, formerly several hundred metres in height.

[*Editor's Note:* Figures 4, 5, and 8 have been omitted, as has the 8-mile map sheet 35 SE that is included in a pocket at the back of the original publication.]

Introduction

The possibility that Lac Couture, Quebec (long. 75°20′W, lat. 60°08′N) might have an origin entirely different from most of the lakes in the Canadian Shield was first suggested by an examination of the 8-mile map sheet 35 SE published in 1955*. A more recent edition of this map is contained in the pocket in the back of this publication. On this map it will be seen that the lake is of roughly circular form with a diameter of the order of 8½ miles (13 km) with a shoreline characterized by numerous inlets which extend its dimensions in certain directions from 16 to 18 kilometres. The most striking feature of the lake is a central island-free area of about 10-km diameter surrounded by a fringe of islands and peninsulas which give it a unique and artificial aspect suggestive of some unusual mode of origin.

An aerial mosaic of Lac Couture is shown in Figure 1. This mosaic emphasizes its circular outline but unfortunately the lake was full of floating ice at the time the photographs were taken so that the details of islands and promontories are obscured. Figure 2 shows the same mosaic with the ice blocks painted out while Figure 3 is a map of the lake originally drawn to a scale of one mile per inch. This map shows in considerable detail the islands and the irregularities of the shoreline as well as the numerous lakes in the surrounding country.

*The New Quebec Crater (long. 73°40′W, lat. 61°17′N) Meen 1950, 1957, and Millman, 1956, also appears on this map.

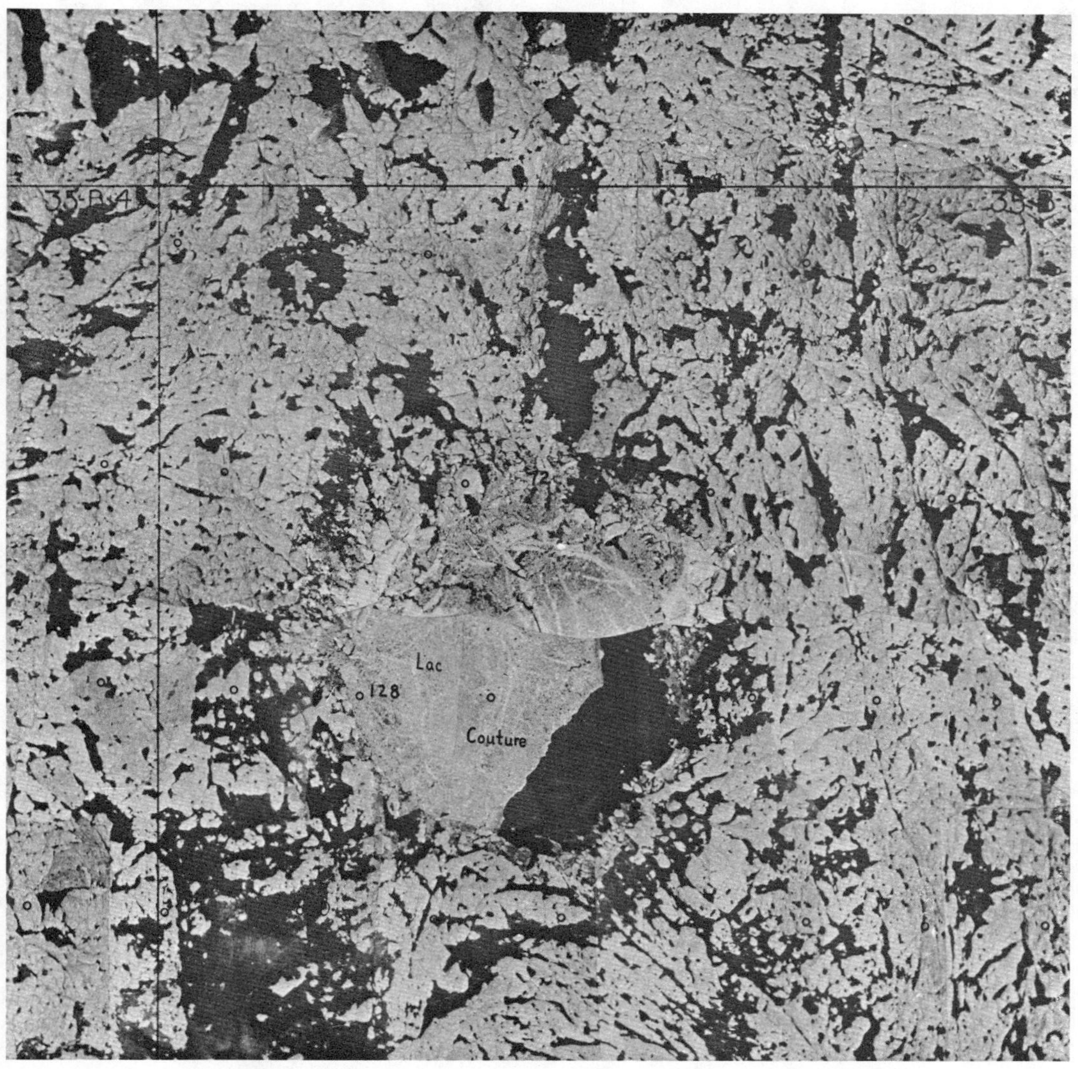

FIGURE 1. Aerial mosaic of Lac Couture; islands partly obscured by floating ice.

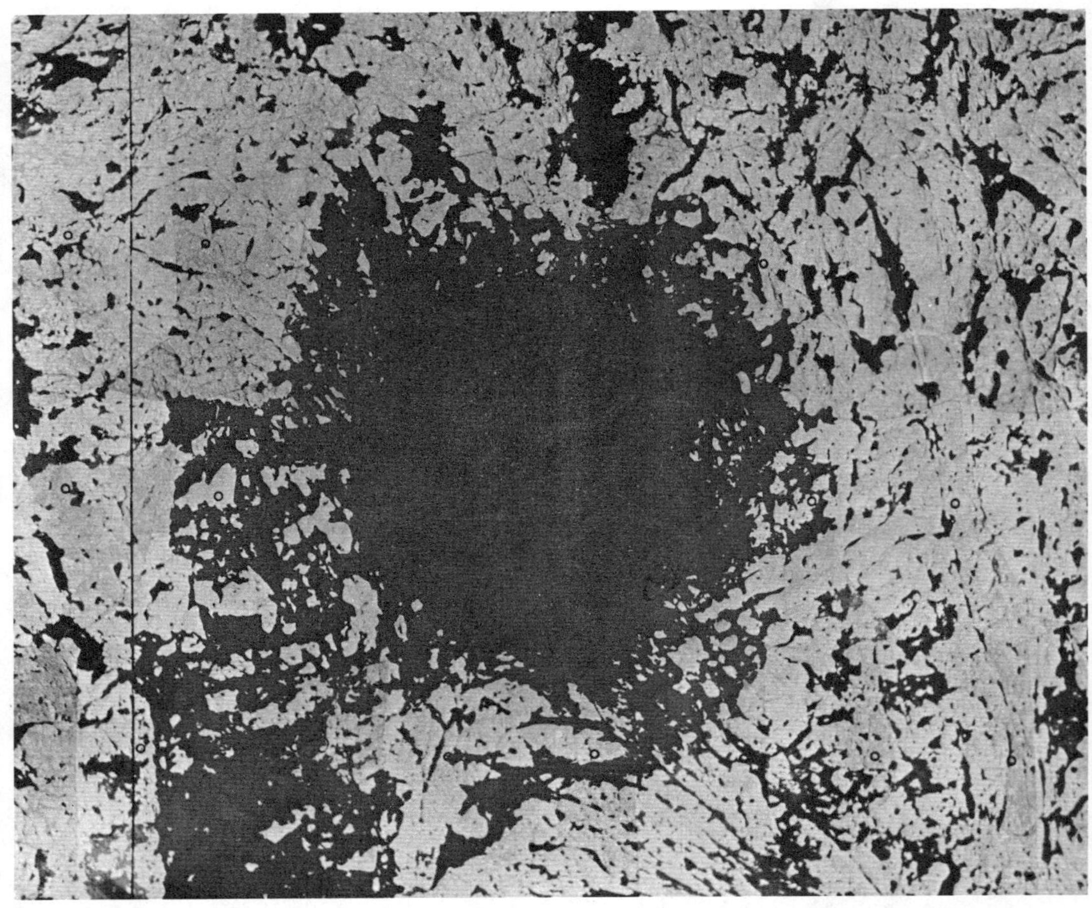

FIGURE 2. Same as Figure 1 with ice painted out to show distribution of islands.

Superimposed upon the map is a depth traverse in an approximately north-south direction indicating maximum depths of 120 metres. The depth traverse was carried out by R. K. McConnell and J. G. Tanner in connection with a gravity survey of this general area in 1959 and 1960. The measurements were made from a Beaver aircraft which was taxied across the lake using a simple mechanical depth sounder and dead reckoning positions. The accuracy was sufficient to show clearly the considerable variations in depth along the traverse and the fact that the greatest depths were not in the centre of the lake. This recalls a similar situation in the Deep Bay Crater (Innes *et al.*, 1964) where the deepest area consisted of a kidney-shaped depression southeast of the centre, possibly gouged out to that shape by glacial action. As in the case of Deep Bay the measured depths of Lac Couture indicate a lake much deeper than is ordinarily found in the Canadian Shield but shallow relative to that of an impact crater of this size. Indications of a rim remnant are found in the islands and peninsulas surrounding the lake, a part of a crater more likely to be subject to erosion than the unshattered mainland area farther from the centre.

The sum total of the preliminary information about Lac Couture outlined above indicates a definite possibility that this lake may be the remains of a very ancient impact crater nearly filled with sediments and/or glacial debris with its rim almost entirely removed by erosion. It was accordingly decided to carry on

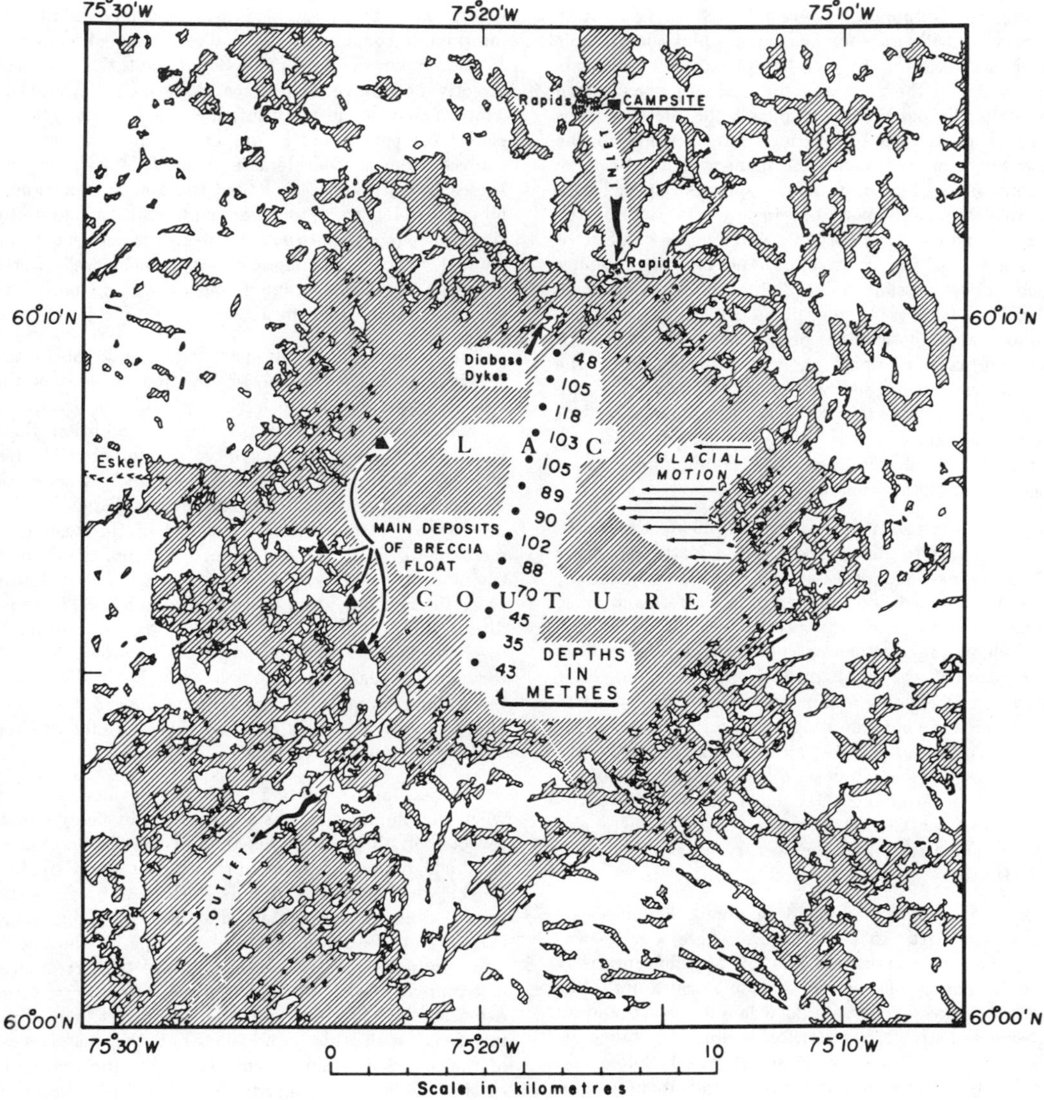

FIGURE 3. Map of Lac Couture showing depths, glacial motion and breccia locations.

further investigations of the crater and to search for additional evidence which might confirm or deny these tentative conclusions.

Field Expedition to Quebec Crater Sites

The first opportunity to do this was provided by an expedition to northern Quebec organized by the Dominion Observatory in August 1963 for the purpose of studying a number of crater sites of which Lac Couture was one. The party consisted of C. S. Beals and M. R. Dence of the Dominion Observatory, Alvin J. Cohen of the Department of Earth and Planetary Sciences, University of Pittsburgh, Geoffrey Charlewood of Heath and Sherwood Drilling Co., and the pilot of the aircraft,

Paul Saunders. Transportation was provided by a two-engined Grumman Goose amphibian of Laurentian Air Services with seats for five persons plus considerable additional space for carrying equipment. Arrangements were made for the pilot and the plane to remain with the party throughout the trip and the mobility thus provided proved to be of the greatest value to the whole operation. Although the plane was a little large for some of the numerous island landings required, this was more than compensated for by the skill of the pilot in manœuvering his craft and by the comfort of the seats and the relative quietness of the engines which made possible long flights without excessive fatigue. A feature of the plane which proved very valuable was the window beside the seat of the co-pilot, normally occupied by a passenger. When the window was open there was no draft from the slip stream and while the aeronautical implications of this were not entirely clear to us, we took full advantage of it in taking photographs and making direct visual observations of the ground below.

Leaving Ottawa on August 1, 1963, the party travelled north to James Bay via the Nottaway River and up the coast of James Bay and Hudson Bay to Great Whale River which was reached in late afternoon. Since one major objective of the expedition was to observe the island chain hugging the east coast of Hudson Bay from Long Island in the south to McCormack Island, Elsie Island and Portland Promontory in the north, the trip was resumed the next morning, following the coast as far north as Povungnituk. Most members had not flown this route before and were impressed by the magnitude of the rivers flowing into James Bay and Hudson Bay (particularly the Fort George River or Grand Rivière and Great Whale River) and the scenery on the east coast of Hudson Bay.

From Povungnituk, a flight of about 90 miles east brought the party to Lac Couture where a camp was set up. The site chosen was on a lake, or more precisely the enlargement of a stream leading south into Lac Couture, where there was a good landing beach and an unobstructed stretch of water for landings and take-offs. This location, indicated on Figure 3, was 11 kilometres north of the centre of Lac Couture and about 6 kilometres from the northern border of the island-free area of the lake.

Lac Couture from the Air

The aerial mosaic (Figure 1), its modification (Figure 2), and the large- and small-scale maps already mentioned give a general idea of what Lac Couture should look like from the air. Indeed experience in the study of aerial photographs has proved so valuable and the amount of information which can be gained from them is so great (particularly when suitable sequences of stereoscopic pairs are available) that it might well be questioned whether any additional information could be gained by direct visual observation. It should however be pointed out that the eye used directly does have some advantages over the camera. These include a wider instantaneous field of view, better colour perception and in our case ability to view the subject from a considerable variety of heights and angles. When to these is added the ability of a human observer to integrate, to judge and to make substantially simultaneous comparisons between the subject being studied and its surroundings, it seemed well worth while to spend some time flying over Lac Couture to study it directly from the air.

Most members of our party had had substantial previous experience in the observation of lakes of the Precambrian Shield which was valuable for purposes of comparison with Lac Couture. In addition there were, in its immediate vicinity, a number of bodies of water which might be termed generally characteristic of the Shield area which could be compared with it directly by no more than a turning of the head or a mere re-directing of the glance occupying a small fraction of a second. We flew around Lac Couture twice the first day at an altitude of about 2,500 feet and during the next four days numerous flights were made over the lake at different altitudes in order to reach the various islands and shorelines to be studied from the ground. Some photography was done but in view of the coverage already existing, this was subordinated to direct visual observations.

The first impression of Lac Couture, gained on a day when the illumination was good with light fleecy clouds reflected in the water, was that of a very fine visual spectacle of a kind often met with in the Canadian Shield with numerous islands dotting a broad water area having an irregular shoreline. Looking more carefully, the essentially circular shape of the shoreline gradually separated itself from the irregularly distributed inlets and indentations. One very large extension of the lake to the southwest somewhat interfered with this impression but the massiveness of the peninsulas and islands separating the two units carried the line of circularity around and left it reasonably intact. The next and most vivid impression was that of the broad circular island-free area in the centre of the lake outlined by some hundreds of islands of sizes ranging from a mile or more across to sizeable rocks projecting above the surface of the water. The islands, like the surrounding inlets were rather irregularly distributed, being to some extent bunched together in groups separated from one another by considerable stretches of water. When the distribution was looked at as a whole however, the impression of circularity reasserted itself and it became clear that we were looking at a distribution of land and water that was unique in our experience and probably not duplicated anywhere else in the Canadian Shield.

Once established, the visual pattern struck the observer with a force far greater than any photograph. Any doubts in our minds about the exotic origin of Lac Couture were almost completely eliminated by the impact of the visual observations. Of course we were aware that no general impression, however powerful, could take the place of detailed ground observations of the kind associated with circular objects of confirmed asteroidal impact origin. As a stimulus to further observations however, the direct studies from the air had a profound effect which was of the greatest value in overcoming initial setbacks in the ground studies and encouraging persistence beyond the time which had been initially assigned to this part of the expedition.

Topography of the Islands and Surroundings of Lac Couture

Lac Couture lies in the northern part of the Superior Tectonic Province of the Canadian Shield (Lowdon *et al.*, 1962), and some 60 miles (100 km) south of the southern margin of the Wakeham Bay-Cape Smith fold mountain belt. The regional terrain is generally subdued with local relief rarely in excess of 100 feet (35 metres). Bedrock is well exposed and glacial debris is commonly restricted to a thin veneer of erratics. The main controls on the topography are the strike of the regional gneissosity; linears and steeply dipping joint sets in the gneisses; strong horizontal sheet jointing in typical development; and the last direction of glacial scour. The many small lakes and the general drainage pattern are controlled essentially by the strike of the gneissosity and the jointing, producing elongated to highly irregular, shallow bodies of water in which N-S to NW-SE trends dominate. The last glaciation moved from east to west and was mainly instrumental in moulding outcrops into typical *roche moutonnée* form. Weathering controlled by the horizontal sheeting has later modified some of these forms into step-like terraces, giving blocky profiles to many of the hillocks of the area.

Petrography of the Area

The gneisses which crop out around the lakeshore and peripheral islands are of granodioritic to quartz dioritic composition, the content of mafic material varying from less than 5 per cent to about 20 per cent. Pink, coarse-grained pegmatites of simple mineralogy occur as bands, veins and small, irregular masses cutting the gneisses. A few small diabase dykes also cut the gneisses and represent the latest event in the orogenic history of the area.

At the north end of the lake the gneisses are almost all massive, medium- to coarse-grained leucocratic rocks with less than 5 per cent of biotite and other mafic minerals. They are cut by a small swarm of northwest-trending diabase dykes up to 50 cm wide. To east, south and west the gneisses have a more emphatic foliation due to the distribution of mafic bodies, making up 10 to 20 per cent of the rock, as bands and inclusions. The mafic inclusions are predominantly medium-grained aggregates of biotite, hornblende and plagioclase. Where the banding is most pronounced many minor faults and small shears are defined. The average strike of the gneisses is north-northwest with dips in excess of 60 degrees. The southern margin of the lake corresponds to the northern closure of a major fold which is clearly defined in aerial photographs. The banding is particularly regular in the gneisses forming the fold, the alternations of light and dark material taking place over thicknesses of 2 to 10 cm.

During the late stages of consolidation of the gneisses, pegmatites of quartz, alkali feldspar, biotite and magnetite were injected to form irregular pods and veins up to 1 metre across which crosscut and locally deform the foliation. They themselves are little deformed and orogenic movement must have ceased at the time of their emplacement, save for the late injection of the diabase dykes previously mentioned.

Joints and Fractures

Observations to the north, west and east of the lake show that vertical sets of joints and horizontal sheet joints spaced a metre or more apart are common in the gneissic terrain (Figure 4a). However, on the islands and promontories of the lake margin these regular systems are not found. The joints that are observed dip at a variety of angles. At some localities (e.g. along the northern margin of the lake) regular surfaces which may correspond to the regional horizontal sheeting dip away from the centre of the lake at angles of 25 degrees or more. More severe disturbances are suggested by outcrops along the western side of the lake where joints of similar character dip at angles which approach the vertical. Figure 4b illustrates one location where jointing departs from the usual pattern.

Besides such signs of disturbance of pre-existing structures, more intense fracture zones are common on the outcrops of the peripheral islands. These fractures are less regular than the regional joints, are oriented at all attitudes and in many cases are concentrated in zones 10 to 30 cm across in which the individual fractures are 1 to 5 cm apart. Zones of this character are particularly common in the leucocratic members of the assemblage, notably in the pods and veins of pegmatites. Examples of such fractures are shown in Figures 5 and 6.

[*Editor's Note:* Figures 4, 5, and 6 have been omitted.]

More detailed and extensive surveys are needed to establish the degree of derangement of primary structures and introduction of later fracturing, but the observations recorded above compare closely with marginal structures from other circular structures for which a meteoritic origin has been advanced, e.g. Deep Bay and Clearwater Lake (Innes, Pearson and Geuer, 1964; and Dence, Innes and Beals, 1965).

Discovery of Breccia on the Western Islands

By analogy with other ancient Canadian craters suspected to be of meteorite origin it was expected that Lac Couture would be at least partially filled with consolidated sediments which would conceal the original crater surface. The presence of such sediments is known both from studies of surface geology and diamond drilling at the Holleford (Beals, 1960) and Brent (Millman *et al.*, 1960) craters. At the Deep Bay Crater (Innes, Pearson and Geuer, 1964) which is filled with water, the presence of sediments was suspected from the presence of Mesozoic shale pebbles and boulders located on the south shore of the lake and presumably dredged from the lake bottom by glacial action. This was later confirmed by diamond drilling near the centre of the lake which turned up shale cores similar to the specimens mentioned above as well as indicating the presence of rock breccia at the expected depths under the sediments (Innes, private communication). A similar situation existed at Clearwater Lake where numerous specimens of Palaeozoic sedimentary rocks were found on the western shores of the lake, presumably transported there by an east to west motion of glacial ice. Again the presence of sediments in this lake was confirmed by diamond drilling which succeeded in penetrating the sediments to the crater surface beneath (Dence, Innes and Beals, 1965). The same general pattern was repeated at West Hawk Lake (Halliday and Griffin, private communication 1965).

At Lac Couture, evidence for fragments of sedimentary rocks on the shorelines was lacking although, as already mentioned, heavy rock shattering and anomalously dipping gneisses were observed. What may well be the reason for the lack of sedimentary fragments was brought to our attention when numerous pebbles and boulders of rock breccia began to appear. The first finds were made at a location on an island southwest of the lake centre and consisted of a few small but very interesting breccia fragments. Later as our party moved to other locations, numerous boulders of all sizes (Figures 7 and 8f) many smaller fragments and a number of "beds" of breccia, presumably due to the disintegration of large boulders by frost action, were found. The breccia was very similar to that found in other ancient meteorite craters. A majority of the specimens showed evidence of crushing and fracturing only but some showed evidence of heating as well, offering a very complete analogy with the Holleford and Brent breccia which has been uncovered by diamond-drilling methods.

Unfortunately the time available for study at Lac Couture was limited by commitments for similar studies at Clearwater Lakes and Manicouagan so that it was not possible to make a detailed study of the distribution of breccia locations relative to their surroundings. Such studies will certainly be made in the future in connection with geophysical and other investigations of this area. However, the very unusual nature of the material, and the difficulty of accounting for it by ordinary geological processes combined with the unusual character of the lake itself already mentioned, make an association between the two highly probable. All the breccia fragments were found west of a north-south line passing through the lake centre and it seems a reasonable conclusion that they had been dredged from the lake bottom by the east-to-west movement of glacial ice. If this is the true explanation it further seems probable that Lac Couture was never filled with consolidated sediments and that this fortunate circumstance has made possible the shoreward movement of the breccia discovered by our expedition.

FIGURE 7. Characteristic breccia boulder on island west of lake centre.

The large size (up to 1 metre across) and the generally weak, friable nature of the boulders as shown by their tendency to disintegrate in place, makes it unlikely that they could have been transported more than a few miles and reinforces the probability that they come from the lake bottom. Again, no other possible source is known. Some of the breccia boulders may be derived from crater breccias slightly reworked to form the base of the sedimentary sequence as at Brent and Clearwater East. There is still a possibility that a veneer of basal coarse sediments may be preserved in the centre of the crater even if no representatives of finer-grained sediments remain, if they were ever present.

Petrography of the Breccia Boulders

The boulders of brecciated and altered gneiss observed at Lac Couture ranged from large, subrounded, massive blocks up to 1 metre across to small, soft platy cobbles averaging 5 cm across. It was soon apparent that the large blocks consisted of relatively unaltered fragments of gneiss up to 10 cm across in a grey-green to red-stained matrix, while the smaller erratics, more extensively altered and metamorphosed, were of medium to fine grain-sizes and varied in colour from pale grey-green to yellow and buff-pink shades. In some of the latter the groundmass was of igneous texture with small vesicles lined with quartz clearly visible. Some inclusions were chalky-white, distorted and in some cases vesicular. It was clear that while the material in general was well-consolidated it would not stand up long to continuous mechanical attrition. The boulders with fragmental matrices were generally poorly cemented and hence friable and easily broken apart, while those with partly or completely recrystallized matrices parted readily along curved joints. In many cases large boulders were found to have broken up on standing, presumably through frost action penetrating and parting the rock on planes of weakness. The exceptions were large boulders with a strong framework of well-packed cobbles of gneiss and relatively little fine-grained matrix. These proved tough and resistant to mechanical breakdown.

An effort was made to collect as broad a selection of boulder types as space and time would allow. A rough agreement was maintained between the proportion of the different varieties collected and those observed in the field, although the collection tended to be weighted in favour of the rarer types. The general field classification into massive, coarse, relatively unaltered breccias and gneiss fragments and those of fine grain-size and clearly altered or recrystallized was found on preliminary laboratory examination to be capable of elaboration into two categories of gneisses showing either weak or advanced alteration and metamorphism and five categories of breccias. The latter form a progressive series reflecting ever more severe alteration, deformation and ultimately recrystallization. The sequence is gradational and to that extent divisions are arbitrary. Nevertheless, all specimens examined to date can be assigned to one or another category with little difficulty or ambiguity, so the classification appears adequate for present purposes.

Whereas samples of gneiss collected from outcrop show only minor fracturing and alteration with the introduction of small amounts of epidote, chlorite and sericite, similar but stronger changes are seen in some gneiss fragments. One change consists of limited brecciation with development of fine-grained epidote in abundance, with chlorite and sericite also common; another of a more typically cataclastic texture. Shearing, recrystallization of quartz and deformation of primary feldspar is accompanied by development of epidote, sericite and hematite, as before.

Far-reaching metamorphism is apparent in other specimens of single gneissic fragments, which probably were large inclusions in the more highly metamorphosed breccias described below. The original distribution and relatively coarse grain-size of primary quartz, feldspar and mica is still recognizable in these specimens but original crystals are recognizable only as vague relicts. Primary crystals have been fractured and deformed and boundaries smeared. Original quartz is entirely replaced by fine-grained mosaics of secondary quartz. In a few cases primary twin lamellae in feldspars can be recognized but in general the crystals are now partly vesicular, nearly isotropic mixtures including relicts of primary domains, secondary feldspar of extremely fine grain-size and possibly some glass. Minute inclusions of epidote and iron ore are common. Biotite and to a lesser extent apatite has also been altered.

Breccias of Type I, which include the largest, coarsest and most abundant members of the breccia population, have fragments up to 10 cm across held in a matrix of the angular quartz, feldspar, mica, epidote and iron ore. They are distinguished by the extent to which their fragments bear secondary minerals (epidote, sericite) in similar development to that of the weakly altered gneissic fragments, while showing little evidence of important structural deformation. In some cases the matrix is also extensively altered and may be largely replaced by iron oxide, chlorite and epidote.

In breccias of Type II, secondary minerals, particularly epidote, are not so important as in Type I breccias, though iron oxide is again prominent. A more significant distinction lies in the more extensive deformation of individual fragments. Some granitic fragments have fracture surfaces of shatter cone type. Under the microscope, deformation of feldspar is seen to involve fracturing along irregular planes, bending of twin lamellae and the development of patchy, undulous extinction. In mica, kinking and

(a)
Thin section in plane light of breccia of Type II. The dark minerals are biotite mica showing well-developed kinking. Bottom left and far right are quartz grains with strong planar structures.

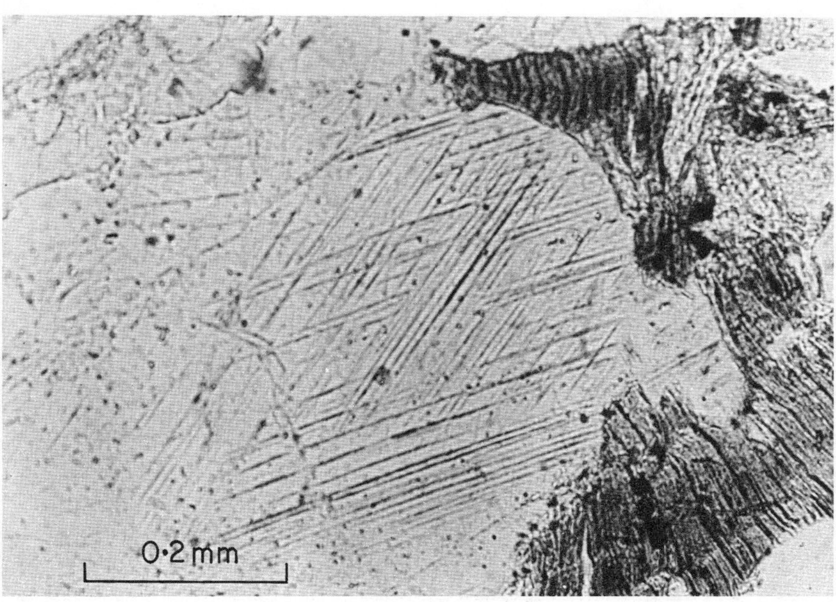

(b)
Detail of (a). A quartz grain in plane light showing two well-developed sets of planar structures. Associated mica is kinked.

FIGURE 9. Photomicrographs of thin sections of Lac Couture breccia.

(c)
The same as (b) under crossed nicols.

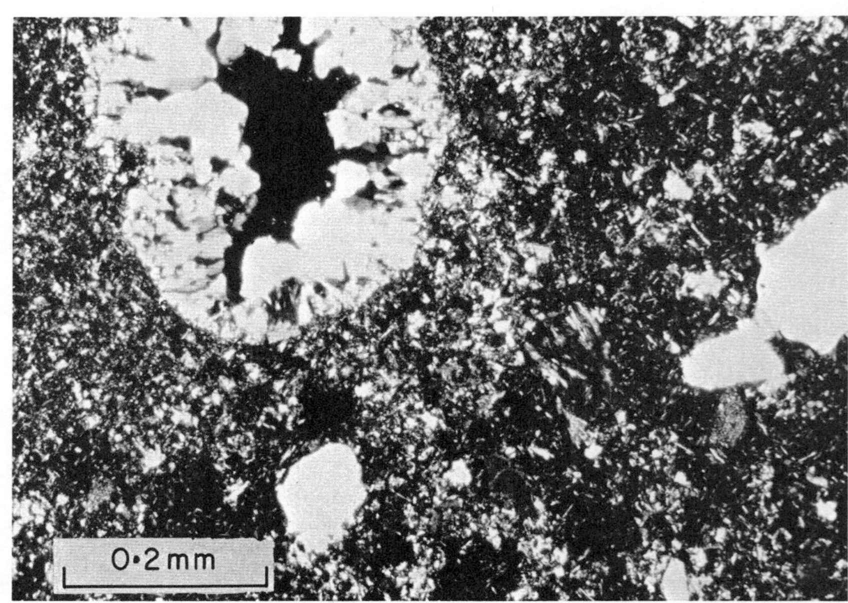

(d)
Breccia of Type V with a few fragments of quartz in a completely recrystallized matrix of fine quartz and feldspar. The large circular structure is a vesicle partly filled with secondary quartz.

morphology of the Brent, Deep Bay, New Quebec and Clearwater Lake craters.

A further comment concerns the composition of the gneisses. Of all the Canadian craters Lac Couture shows the most restricted range in the composition of its country rocks. With few exceptions these are rocks rich in quartz and feldspar with minor biotite. They therefore have a bulk composition close to that of normal granodiorite or granite and far removed from meteoritic compositions whether stony or metallic. This encourages a particular hope that intensive search may lead to positive identification of meteoritic material even if well mixed, metamorphosed and altered. However, the most favourable site for such a search would be the centre of the crater, which can probably only be investigated by diamond-drilling methods.

General Conclusions

Reviewing the evidence collected in an attempt to ascertain the origin of Lac Couture, it may be pointed out that the initial observations of its circular shape, unusual for so large a lake and the unique distribution of islands and peninsulas around the deep, island-free central area, were suggestive of an unusual origin rather than diagnostic of the actual physical process which formed the lake. This suggestion of an unusual origin was however reinforced by the observation of extensive rock shattering on the western islands and the indications of disturbance presented by the attitudes of sheeting planes and joints on the islands and around the shorelines.

The final conclusion that the lake and its surroundings probably represent the eroded remnant of an ancient impact crater depends mainly on the observation of many tons of rock breccia on the western islands (presumed to have been removed from the lake bottom by glacial action) for which no other reasonable explanation has been found. The validity of the conclusion that the breccia was due to asteroidal impact rather than some other cause depends first on the similarity of its general character to that associated with reasonably well authenticated craters such as Holleford, Brent, West Hawk, and Deep Bay; and secondly, on the microscopic indications of crystals, particularly quartz, deformed and altered by shock in a manner so far encountered in nature only at sites suspected to be of meteorite impact origin. In this connection it is interesting to recall suggestions made by Beals, Innes and Rottenberg (1963) about the comparative pressures to be expected in meteorite impact and volcanic explosive sites. They point out that a meteorite striking the earth at a velocity of some tens of kilometres per second, generates a peak pressure of the order of 10^7 atmospheres, a pressure which is only attained lithostatically at a depth of 2,500 km below the surface of the earth. Even allowing for the attenuation of such pressures as the shock wave moves outward from the point of impact, it would still be expected to be many times greater than that of volcanic gas explosions venting near the surface of the earth. (Even at the depth of the Mohorovicic layer the lithostatic pressure is only a few per cent of a megabar.) It is scarcely surprising therefore to find crystallographic evidence of intense shock at meteorite impact sites and it is becoming increasingly probable that shock-fractured crystals represent a definite criterion for distinguishing between impact and volcanic features.

It is unfortunate that the relatively inaccessible location of Lac Couture has so far not permitted the sampling of the lake bottom itself by diamond drilling or other methods and this is something for the future. Some scattered gravity observations have been made but they are not sufficiently numerous or well distributed for a determination of the subsurface distribution of breccia by gravity methods.

In spite of the obvious lacunae in the observations which can only be filled in by extensive geophysical observations combined with diamond-drilling techniques, we consider it justifiable at this time to add Lac Couture to the growing list of probable meteorite impact sites. The number of such locations for craters over a mile in diameter is now approaching fifty (Beals and Halliday, 1965) and only recently we have received information about new discoveries in Sweden (W. von Engelhardt, private communication) which may bring the total above this figure.

It must of course be acknowledged that many scientists still regard the evidence for the impact origin of some or all of these craters to be insufficient. We would have to agree that in practically all cases a great deal of vital information is still lacking; short of the complete excavation of a crater some miles in diameter, it may never be possible to speak of final proof. In this as in many fields of astronomical and geophysical science it is necessary to rely for our conclusions on considerations of general probability combined with the gradually accumulating weight of evidence for or against a given hypothesis. With regard to the latter, the additions to the evidence for asteroidal impact as an important process on the surface of the earth and moon have been impressive over the past decade. As to the former, it would appear that the general credibility of the asteroidal impact process throughout the solar system has received a considerable impetus from the Mariner IV observations of the surface of Mars (Sky and Telescope, 1965). The remarkable photographs resulting from these observations give clear evidence of a surface dominated by circular features of a kind which is characteristic of high velocity impact. The fact that such features have now been observed on three members of the solar system, Earth,

Moon and Mars, is important evidence suggesting that similar processes are probably causing interesting markings of this kind on bodies throughout the solar system at least within the orbit of Jupiter. While it can be argued that the common factor is volcanism rather than impact, the similarity of the lunar and Martian features and their resemblance to terrestrial impact craters is difficult to reconcile with such a view. The appearance of closely similar features on three planets with such widely different characteristics as Earth, Mars and the Moon would at least suggest that the crater forms are characteristic of the mechanism of formation rather than the nature of the surface or the value of gravity on the planet where they are observed. As space exploration proceeds, it will be interesting to see what further evidence turns up on the surface of other planetary bodies unobscured by heavy atmospheres, e.g. Mercury and the satellites of Mars, Jupiter and Saturn.

Age of the Crater

While the absence of any stratigraphic evidence in the data collected so far presents a rather discouraging picture for age determination, it is perhaps worthwhile to compare this crater with Deep Bay, a feature of similar size and with a reasonably reliable minimum age. Deep Bay has a rather precipitous rim rising in places to heights of the order of 100 metres above its surroundings. This at least suggests that Lac Couture, whose rim gives little evidence of even minor elevation above the surrounding plain, is the older of the two. Mesozoic strata penetrated by drilling near the centre of Deep Bay indicate a minimum age of 150 million years and Innes (1964) has suggested, on the basis of surrounding topography, that this may be close to the true age. From this it appears that Lac Couture could be considerably older and might even be Precambrian in age.

While the observations made at Lac Couture suggest that this crater may never have been filled with sediments, this is by no means certain. Until this possibility has been more thoroughly investigated by diamond drilling or other methods, it seems best to reserve judgment on the age of Lac Couture.

Reconstruction of the Original Crater Form

In conclusion it is interesting to make use of the data on terrestrial and lunar impact craters to attempt a reconstruction of the original form of Lac Couture. Following Baldwin (1963) we made use of his empirical equations (7–2A) p. 137 and (7–6A) p. 143 to calculate the apparent depth of the crater and the height of the rim above the surrounding plain.

$$D = 0.0256d^2 + 1.0264d - 2.3461 \qquad (7\text{–}2A)$$

$$Rh = 0.004366D^3 - 0.008506D^2 + 0.9098D + 1.5987 \qquad (7\text{-}6A)$$

Here D is the logarithm of the crater diameter in kilometres measured at the rim summit, and

d is the logarithm of the crater depth in metres measured from the rim summit of the apparent crater to the crater floor.

Rh is the logarithm of the rim height above the surrounding plain. The diameter of the original crater is estimated at 12 km—the diameter of the median line of the circular complex of islands and peninsulas surrounding the central island-free area of Lac Couture.

The resulting value of 1,250 m for the depth of the crater and 378 m for the height of the rim are necessarily approximate but should give a reasonably clear picture of what a crater of these general dimensions should be like. A profile of the crater before erosion compared with present topography is shown in Figure 10. If this model is even approximately correct it is evident either that the crater has been largely filled up by erosional products from the rim and glacial debris or that the Precambrian plain surrounding the crater has itself been eroded by an amount of the order of several hundred metres since the crater was formed. No doubt both processes have been going on but it seems likely that the former would be predominant.

While current ideas about explosion craters suggest that the greatest depth is in the centre, there is increasing evidence that some craters have central mounds or uplifts due to some sort of rebound mechanism associated with the impact process (Beals and Halliday, 1965). The dotted line of Figure 10 indicates this possibility which receives some credence from the depth soundings. If the original crater had a massive central uplift, then the above suggestions, relative to the amount of erosion would have to be modified.

The size of the meteorite which could produce a crater of this diameter has been calculated from Baldwin's equation (Baldwin, 1963, p. 176)

$$D = 0.3284E - 7.9240 \qquad (8-12A)$$

where D is the logarithm of the crater diameter in kilometres and E is the logarithm of the number of ergs of kinetic energy of the impacting meteorite. Assuming a rock meteorite of density 3.4 and a velocity of 20 km/sec, the calculated energy is 2.60×10^{27} ergs and the diameter 901 metres. If the crater were formed by an iron meteorite then the calculated diameter would be 683 metres.

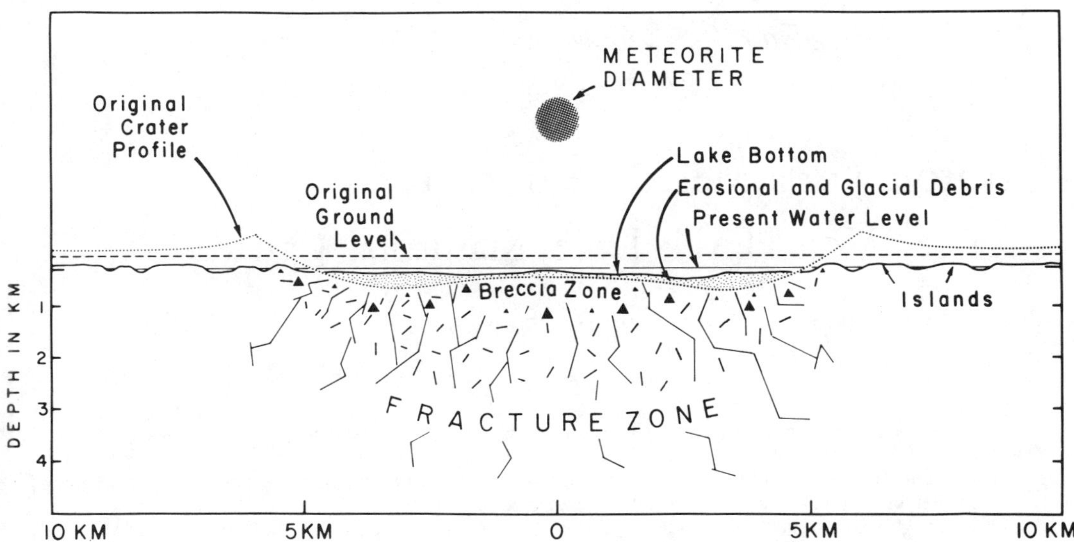

FIGURE 10. Suggested relationship between original crater profile and present topography. Order of diameter of stone meteorite capable of producing crater at velocity of 20 km/sec, 900 m.

References

Baldwin, R. B., 1963. The Measure of the Moon, *Univ. Chicago Press.*

Beals, C. S., 1960. A Probable Meteorite Crater of Precambrian Age at Holleford, Ont. *Pub. Dom. Obs.*, **24**, 117.

Beals, C. S., Innes, M. J. S. and Rottenberg, J. A., 1963. "Fossil Meteorite Craters" *in* The Solar System, ed. by Barbara Middlehurst and G. P. Kuiper. *Univ. Chicago Press. Contr. Dom. Obs.*, **5**, No. 20.

Beals, C. S. and Halliday, Ian, 1965. Impact Craters of the Earth and Moon. *J.R.A.S.C.*, **59**, 199 *Contr. Dom. Obs.*, **4**, No. 19.

Bunch, T. E. and Cohen, A. J., 1963. Coesite and Shocked Quartz from Holleford Crater, Ontario, Canada. *Science*, **142**, 379.

Currie, K. L., 1964. Analogues of Lunar Craters on the Canadian Shield, *Ann. New York Acad. Sci.*, **123**, 915.

Currie, K. L., 1965. The Geology of the New Quebec Crater. *Can. J. Earth Sci.*, **2**, 141.

Dence, M. R., 1964. A Comparative Structural and Petrographic Study of Probable Canadian Meteorite Craters. *Meteorites*, **2**, 249. *Contr. Dom. Obs.*, **6**, No. 3.

Dence, M. R., 1965. The Extraterrestrial Origin of Canadian Craters. *Ann. New York Acad. Sci.*, **123**, 941.

Dence, M. R., Innes, M. J. S. and Beals, C. S., 1965. On the Probable Meteorite Origin of the Clearwater Lakes, Que. *J.R.A.S.C.*, **59**, 13. *Contr. Dom. Obs.*, **6**, No. 7, 1965.

Engelhardt, W. von and Stöffler, D., 1965. Spaltflächen im Quarz als Anzeichen für Einschläge grosser Meteoriten. *Die Naturwissenschaften*, *52*, 489-490.

Halliday, I. and Griffin, A. A., 1963. Evidence in Support of a Meteorite Origin for West Hawk Lake. *J. Geoph. Res.*, **68**, 5297.

Innes, M. J. S., Pearson, W. J. and Geuer, J. W., 1964. The Deep Bay Crater. *Pub. Dom. Obs.*, **31**, 19.

Lowdon, J. A., Stockwell, C. H., Tipper, H. W. and Wanless, R. K., 1962. Age Determinations and Geological Studies. *Geol. Surv. Can.*, Paper 62-17, Fig. 2.

McIntyre, D. B., 1961. Impact Metamorphism at Clearwater Lake, Quebec. (Abstract) First Western National Meeting, *Am. Geoph. Union*, Dec. 27-29.

Meen, V. B., 1950. The Chubb Crater, Ungava, Que. *J.R.A.S.C.*, **44**, 169.

Meen, V. B., 1957. Chubb Crater—A Meteor Crater. *J.R.A.S.C.*, **51**, 137.

Millman, P. M., 1956. A Profile Study of the New Quebec Crater. *Pub. Dom. Obs.*, **18**, 61.

Millman, P. M., Liberty, B.A., Clark, J.F., Willmore, P. L. and Innes, M. J. S., 1960. The Brent Crater. *Pub. Dom. Obs.*, **24**, No. 1.

Milton, D. J. and De Carli, P. S., 1963. Maskelynite: Formation by Explosive Shock. *Science*, **140**, 670.

Sky and Telescope, 1965. Special supplement on Mars photos, **30**, 155.

11

Reprinted from page 609 of *Geol. Soc. America Bull.* **81**:609–648 (1970)

Anatomy of a Meteorite Impact Crater: West Hawk Lake, Manitoba, Canada

NICHOLAS M. SHORT *University of Houston, Houston, Texas*

ABSTRACT

West Hawk Lake is a near-circular, 11,700-ft-wide, deep lake located 110 mi east of Winnipeg, Canada. The lake now completely submerges an ancient meteorite impact crater. A central drillhole passes through nearly 1080 ft of slump and fallback breccia into 655 ft of rupture zone bedrock below a well-defined crater base. Petrographic analysis of core from this hole, a hole near the original crater rim, and an off-center hole reveals numerous shock features which, supplementing earlier geophysical studies, demonstrate an impact origin for this structure. These results are significant:

(1) The original crater was approximately 8000 ft wide; the rim has recessed by erosion; the target rock consists of steeply dipping units of metasediments and metavolcanics.

(2) Microscope criteria for shock include: abundant planar features in quartz; diagnostic parallel extension fractures; extensive isotropization of tectosilicates in quartz-mica schists (analcite is a common product of recrystallization of the isotropized phases); variable decomposition of biotite and actinolite; abundant glassy fragments and matrix-binding "melt," now generally devitrified; shock-induced twinning and granulation in pre-impact carbonates.

(3) Distribution of shock features is strongly dependent on mineral composition and particle size. Unless intensely shocked, amphibole-rich rocks are usually devoid of shock evidence, even where fragments are intermixed with shocked schistose metasediments.

(4) The vertical distribution within the central hole core of intensity of shock in the breccias, as plotted on a *shock log*, is notably erratic.

(5) Autobrecciation, cataclastic fragmentation, and injection of shocked breccia into open fractures account for observed effects in the rupture zone.

(6) Zoisite and secondary calcite develop extensively within the breccia lens by solution and neomineralization; serpentine var. antigorite forms with the breccia matrix, and in a network of veinlets in fragments and in the rupture zone; sodium-calcium zeolites form in possibly replaced melt material at the crater base.

(7) Analysis of strongly shocked rock fragments from the breccias by infrared absorption, X-ray fluorescence, electron microprobe, and argon loss methods demonstrates significant changes in the crystal structure and chemistry of constituent minerals in these materials as compared with unshocked equivalents collected off-structure.

West Hawk Lake is compared with the Brent structure, the best-studied model of a *simple* impact crater in crystalline rocks.

REVIEW*

The most obvious shock effects at West Hawk Lake site are planar elements in quartz and isotropism in other minerals, also the development of glass. Autobrecciation and cataclastic fragmentation are other shock effects. The shock intensity shows an erratic vertical distribution. Secondary zeolite, of both sodium and calcium varieties; zoisite; calcite; chlorite; montmorillonite; biotite and antigorite serpentine indicate that there was complex superimposed hydrothermal alteration.

*This review was prepared by the volume editor.

Polymict breccias are regarded as fall-back breccias (suevites). Crystalline rock fragments in them display diverse shock effects and are associated with glassy bodies. Dominant lithotypes may number from one to four, intermingled. Derivation is in many cases indicated from a small volume of target material without intermixing or by-passing in transport. Shock-induced melting is described in some detail along with the more familiar lamellation effects in quartz. Melting involves loss of crystal outlines; extreme contortion in regions of glass-like material; increased uniformity of color due to spread of iron oxide; schlieren, bubble trains, and trichyte alignments due to fluidal flow; and varying degrees of vesiculation. Within 10 ft (3 m) of the presumed crater base in the core, a dark blackish disseminated phase is evident in the matrix of the melt rock, and specks of magnetite and pyrite are megascopically evident. This appears to be a hyaline melt phase. Most glass bodies are partly or entirely replaced by secondary calcite or the zeolites gmelinite and gismondine. Shocked calcite is a feature of the cores, but is difficult to utilize diagnostically because of the difficulty in deciding which are the shock-induced modifications. Alteration is most marked within the breccia lens. Age dating has met with little success and a latitude from 100 to 990 m yr remains, though preservation of a breccia lens would be unlikely in a very ancient impact explosion structure and Dence et al. (1968) regard the structure as Mesozoic.

REFERENCE

Dence, M. R.; M. J. S. Innes; and P. B. Robertson. 1968. Recent Geological and Geophysical Studies of Canadian Craters. In *Shock Metamorphism of Natural Materials*, B. M. French and N. M. Short, eds. Baltimore, Md.: Mono Book Corp., 339–362.

Editor's Comments on Papers 12 Through 16

12 INNES

Excerpt from *Recent Advances in Meteorite Crater Research at the Dominion Observatory (The Carswell Lake Structure)*

13 CURRIE

Abstract from and Review of *Geological Notes on the Carswell Circular Structure, Saskatchewan (74K)*

14 DENCE, INNES, and BEALS

On the Probable Meteorite Origin of Clearwater Lakes, Quebec

15 MURTAUGH

Shock Metamorphism in the Manicouagan Cryptoexplosion Structure, Quebec

16 CURRIE

Abstract from and Review of *Geology and Petrology of the Manicouagan Resurgent Caldera, Quebec*

CARSWELL LAKE

The first of the more complex cryptoexplosion structures in Canada to be described was the Carswell Lake structure. This structure is covered in Paper 12 by Innes and Paper 13 by Currie. The structure is situated amid flat-lying *Athabasca Sandstone* in north Saskatchewan. Blake (1956) noted arcuate ridges of dolomite in featureless Proterozoic sandstone terrain, and Gussow (1959) remarked on the circular ridge outcrop with an inside diameter of 29 km and an apparent ridge width of 4.8 km (Fahrig, 1961). The dolomite overlies the sandstone and is intensely folded and brecciated, with some overturning toward the center of the ring. Concentric rings of sandstone within the dolomite ring are also steeply inclined. Erratics of brecciated *Athabasca Sandstone* occur within the ring. Innes (Paper 12) regarded the structure as a deeply eroded meteorite crater modified by isostatic uplift, the crater area corresponding roughly to that of the now-exposed gneiss. Fahrig precluded meteorite impact on account of the basement gneiss outcrop at the center of the structure. Dence (1968, 1971) considered it to be a completely readjusted structure, the lower levels of an astrobleme in which erosion has removed most of the breccia etc. of the superstructure and only weak shock effects are preserved.

Currie favors Fahrig's endogenic explanation and makes some important observations in his "Notes on the Carswell Structure." Because of its considerable length, Currie's account is represented in Paper 13 by the abstract and a summarizing review by this editor.

CLEARWATER LAKES

The structure formed by the two merging Clearwater Lakes in Quebec, covered in Paper 14 by Dence et al., has been briefly described by Beals et al. (1960, 1963). These lakes are 32 and 25 km in diameter. Their circularity suggests meteoritic origin. A ring of lava islands is situated within the larger western lake. Twin impacts have been postulated and comparisons have been drawn with crater pairs on the moon. The igneous rocks were supposed to be a secondary eruptive product. Kranck and Sinclair (1963) inferred volcano-tectonic origin and collapse of basins on circular fractures. They described dacite lava and volcanic breccias with quite normal hemicrystalline lava microtextures. Dence (1965) described the western structure as having a central uplift probably matched by an annular depression; the eastern structure may, as drilling results show (Dence et al., 1965), also possess an uplift. Dence et al. (Paper 14) consider the structure to be the lower structural section, revealed by erosion, of a site of impact explosion (Brent, the Rieskessel, and Popigay would be higher structural sections). Though the country rocks are Precambrian gneisses, gritty sandstone and limestone/shale were revealed by drilling beneath 87 m of water and 46 m of unconsolidated bouldery materail under the eastern lake. These sediments extend to 245 m (Dence, 1965). Spheroids produced by liquid imiscibility have been described from glasses of the western lake (Dence et al., 1974).

Bostock (1969) described diabase intruding both Precambrian gneiss and Ordovician marble in the island ring of the western lake. The volcanic sequence consists of argillized breccia, contact breccia, and quartz-latite. Serpentine-chlorite veins and maskelynite were both recorded. The argillization and formation of friable breccias is believed by Bostock to have occurred early and to have been followed by partial fusion, recrystallization, and the formation of spurrite (a double lime carbonate characteristic of contact-metamorphosed limestone). The quartz-latite yielded a radiometric age (K/Ar) of 300 m yr. The volcanic rocks have 1% more potassium than the plutonic rocks, but nickel, iron, and magnesium contents are similar. There is an apparent low-to-high-temperature progression in the volcanic rocks, not, Bostock believed, consistent with a meteoritic origin. He favored collapse after doming of a magma chamber, with maskelynitization occurring in successive explosive escapes of volatiles along fractures. The process Bostock favored was resurgence of a collapsed block in a volcanic complex. The geochemistry of the volcanic rocks from Clearwater—and from Carswell, and Manicouagan (which is considered in Papers 15 and 16)—has been summarized by Currie and Shafiqullah (1968). Impoverishment in soda and increase in potassium in the case of the eastern Clearwater lake was attributed to the actual originating process. Millerite-tetrahedrite-galena mineralization was noted in cores from the western lake. Heavy metals are significantly enriched in the igneous rocks. The fractionation of alkalies or their depletion could not, these authors believed, be caused by later circulation of groundwater heated by the impact explosion, as has been suggested in the case of Brent (see the editor's comments on Paper 7). The igneous rocks show no evidence of alteration and there is close agreement between igneous and country rock compositions which favors neither a pure volcanic hypothesis nor impact triggering of primary mantle-derived volcanism. Currie and Shafiqullah suggested a connection between the alteration effects and fenitization, which involves generation of alkaline solutions at the margins of igneous complexes of the ultrabasic-alkaline and carbonatite kindred (see editor's comments on Paper 7). From the eastern lake, Currie (1971) obtained an age of 265 m yr on green dacite. Zinc, nickel, and chromium anomalies were noted in igneous rocks compared with values for country rocks. Red dacite, of homogeneous character, from the western lake yielded a radiometric age of 292 m yr, but there the igneous rocks show impoverishment in nickel, magnesium, and chromium relative to those from the eastern lake. This would, according to accepted magmatic theory, indicate a higher degree of differentiation in the case of the western suite. Like Carswell Lake, the Clearwater Lakes structures remain controversial.

MANICOUAGAN LAKE

The Manicouagan structure, the largest Phanerozoic cryptoexplosion structure in Canada, is covered by Murtaugh in Paper 15 and Currie in Paper 16. Here, Dence (1965) has described an eccentrically placed, irregular central uplift, Mont de Babel, 15 km long and standing more than 300 m above the surrounding plateau of "allochthonous breccia/

lava series." This plateau dips gently inward; and an inner lowland of no great extent intervenes between it and the uplift, which is largely composed of anorthositic gneiss—mostly crushed and with plagioclase converted to maskelynite. Pods and veins of pseudotachylite are spaced at intervals of 50 m through it. The sequence of breccias and lavas surrounding the uplift resembles that at Clearwater Lakes (Paper 14) in setting, sequence, and petrography; also in the relative dimensions of the individual layers. Between the uplift and the circular depression occupied by the lakes there is a subdued ring uplift, analogous to the island ring in the western Clearwater Lake. The lakes themselves occupy an annular graben. The breccias have been raised in a ring uplift enclosed by this graben. Downfaulted, folded, and fractured Paleozoic limestone outcrops on their inner shores. Similar rocks do not outcrop within 200 km of the structure, and they are only preserved here because on the downthrow of at least 1 km in the graben. The veneer of carbonates has analogies with Carswell and Clearwater lakes. Never very thick, the limestone mantled the Precambrian basement at the time of formation of these structures. Dence suggested that erosion was less advanced at Manicouagan than at Clearwater, and that the greater variety of country rocks there produced greater structural complexity. The cover series is less oxidized and the pseudotachylite veins are larger than at Clearwater Lakes, and the pseudotachylite contains bodies of partly altered glass. Energy considerations based on the proportion of fused glass to crushed rock, and evidence derived from nuclear tests, suggested to Dence that a large-scale impact explosion would liberate sufficient energy to produce such an immense structure. He classified the structure as a complex crater and suggested that the floor of the crater has been eroded by at least 50 m. He preferred shock wave rebound to isostatic adjustment to account for the central uplift.

In Paper 15, Murtaugh covers the shock effects, grain deformations, fusion effects, shatter cones, and maskelynitization of the Manicouagan structure. No coesite or stishovite has been recognized. Impact explosion is favored by Murtaugh, even though he acknowledges the controversial nature of attribution of the volcanic rocks to secondary, impacted-generated processes. These rocks give radiometric ages of 208 ± 30 and 208 ± 25 m yr (K/Ar), and paleomagnetic evidence also suggests a Triassic age. Murtaugh suggested cometary impact to account for the central uplift and shallowness of the crater. Currie has produced a very full geological appraisal, something of a tour-de-force. More than 100,000 words long and well illustrated, Currie's report can only be represented here (Paper 16) by its abstract and a summarizing review by this editor. Currie regarded the structure as a resurgent caldera, of volcanic origin. His paper is of critical significance to any discussion of cryptoexplosion.

Dence (1977) has presented a new interpretation of Skylab imagery, in which he recognized a possible outer ring to this structure 150 km in diameter. If valid, this would make Manicouagan the largest of all cryptoexplosion structures. However, the definition of this outer ring is not very satisfactory.

REFERENCES

Beals, C. S.; M. J. S. Innes; and J. A. Rottenburg. 1960. The Search for Fossil Meteorite Craters. *Current Sci.* **29**:205-218, 249-262.

Beals, C. S.; M. J. S. Innes; and J. A. Rottenburg. 1963. Fossil Meteorite Craters. In *The Solar System,* vol. IV, *Moon, Meteorites, and Comets,* B. Middlehurst and G. P. Kuiper, eds. Chicago, Ill.: University of Chicago Press, pp. 235-284.

Blake, D. A. A. 1956. Geological Notes on the Region South of Lake Athabasca and Black Lake, Saskatchewan and Alberta. *Canada Geol. Survey Paper 55-13,* p. 12.

Bostock, H. H. 1969. The Clearwater Complex, New Quebec. *Canada Geol. Survey Bull. 178,* 63pp.

Currie, K. L. 1971. Origin of Igneous Rocks Associated with Shock Metamorphism as Suggested by Geochemical Investigations of Canadian Craters. *Jour. Geophys. Research* **76**:5575-5585.

Currie, K. L., and M. Shafiqullah. 1968. Geochemistry of Some Large Canadian Craters. *Nature* **218**:457-459.

Dence, M. R. 1965. The Extraterrestrial Origin of Canadian Craters. *New York Acad. Sci. Annals* **123**(article 2):941-969.

Dence, M. R. 1968. Shock Zoning at Canadian Craters: Petrography and Structural Implications. In *Shock Metamorphism of Natural Materials,* B. M. French and N. M. Short, eds. Baltimore, Md.: Mono Book Corp., pp. 169-184.

Dence, M. R. 1971. The Nature and Significance of Terrestrial Impact Structures. *Internat. Geol. Cong., 24th Proc.* **15**:77-89.

Dence, M. R. 1977. Manicouagan Impact Structure Observed from Skylab. In *Skylab Explores the Earth,* Earth Phys. Branch, Ottawa, Contr. 544, pp. 175-189.

Dence, M. R.; M. J. S. Innes; and C. S. Beals. 1965. The Nature and Significance of Terrestrial Impact Structures. *Royal Astron. Soc. Canada Jour.* **59**:13-22.

Dence, M. R.; P. B. Robertson; and R. L. Wirthlin. 1974. Coesite from the Lake Wanipitei Crater, Ontario. *Earth and Planetary Sci. Letters* **22**:118-122.

Fahrig, W. F. 1961. The Geology of the Athabasca Formation. *Canada Geol. Survey Bull. 68,* 41pp.

Gussow, W. C. 1959. The Athabasca Formation of Western Canada. *Geol. Soc. America Bull.* **70**:1-18.

Kranck, S. H., and G. W. Sinclair. 1963. Clearwater Lake, New Quebec. *Canada Geol. Survey Bull. 100,* 20pp.

12

Reprinted from pages 234–239 of *Meteoritics* **2**:219–241 (1964)

RECENT ADVANCES IN METEORITE CRATER RESEARCH AT THE DOMINION OBSERVATORY

M. J. S. Innes

[*Editor's Note:* In the original, material precedes and follows this excerpt.]

V. THE CARSWELL LAKE STRUCTURE

Another of nature's interesting ring structures, to which the Observatory recently has given attention, is the Carswell Lake structure (Figs. 1 and 13). It is located a few miles south of Lake Athabasca in northern Saskatchewan, in a large area of the Canadian Shield underlain by flat-lying and relatively undisturbed sandstone, known as the Athabasca formation. This interesting feature was first described by Blake (1956) who was surprised to find in this area of featureless terrain, high ridges of dolomitic limestone forming an arcuate belt which extends some twenty miles both east and west from Carswell Lake. Gussow (1959) made a photogeologic map and interpretive study of the same area and pointed out that the carbonate rocks outcrop in a complete circle. According to Fahrig (1961) who since examined the area in considerable detail the dolomite ring has an inside diameter of about 18 miles and apparent width of three miles. The dolomite sequence, which he believes to lie stratigraphically above the sandstone, is intensely folded, contorted, and brecciated, with much overturning of the beds, particularly those closest to the center of the ring. There is evidence also that the Athabasca sandstone suffered similar deformation. Concentric outcrops of the sandstone lying within the ring of dolomite, to the west of Carswell Lake, are steeply dipping; also in this area, large erratics of brecciated Athabasca strata have been found. Dr. Fahrig produces evidence to show that basement gneiss underlies the center of the feature at about the same topographical level as the surrounding folded rocks. He feels that the presence of gneiss within this ring eliminates meteoritic explosion as a possible origin for the structure. He postulates that the deformation is caused by upward and outward pressure from below, due to the intrusion of an igneous mass that failed to reach the surface, and has tentatively classified the feature as crypto-volcanic.

There are many similar circular features that have been so-named, and the question of their origin, whether meteoritic or volcanic, has been the subject of controversy for many years. However, many scientists, geologists and astronomers alike, now conclude that most of the so-called crypto-volcanic structures find their most logical explanation in meteoritic impact. Wilson (1953) has called attention to the presence of large quantities of rock breccia and massive deformation around the margins of these features but finds no evidence of volcanism. Dietz (1963) has examined most of the known structures in North America and abroad and found shatter cones associated with them. He also finds that the orientations of the shatter cones are such as to indicate that the forces come from above rather than below. High pressure forms of silica have been discovered at several of these circular structures, indicating that shock pressures of great magnitude accompanied their formation. For these reasons further investigations of the Carswell Lake feature seemed desirable.

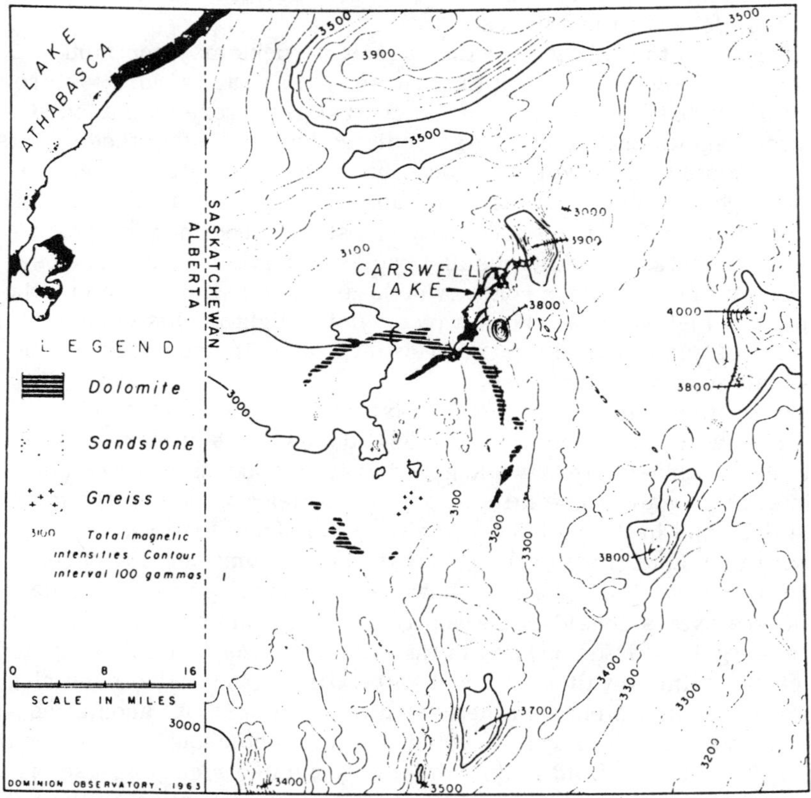

Fig. 13. Map of the Carswell Lake area showing anomalies of the total magnetic intensity contoured at intervals of 100 gammas.

In what follows only the preliminary results of magnetic and gravity investigations of the Carswell feature are given; a more complete analysis and report will be presented elsewhere (Innes and Dence 1964). Fig. 13 is a compilation of the aeromagnetic data showing anomalies of total magnetic intensities contoured at intervals of 100 gammas. The information is based upon one mile to one inch magnetic anomaly maps prepared by the Geological Survey of Canada. The most significant feature is the marked contrast between the intensities within and without the dolomite ring. While the anomalies are highly variable without, the maximum variation over the feature itself is only about 200 gammas. As mentioned previously, magnetic fields of low intensity are characteristic of all the impact craters so far studied, and are believed due to the blanketing effect of brecciated and highly shocked material underlying the crater floors. The greatest magnetic disturbance within the circle is in the northeast quadrant and this may be interpreted as an indication that uplift of the central area has been more effective here, whereby less disturbed basement gneiss now lies closer to the surface.

Changes in the lithology of the basement rocks may contribute to the higher variation. During the course of the gravity work, outcroppings of the basement gneiss in the central area were examined at the locations shown on the map. As predicted by Dr. Fahrig these rocks are brecciated. However, the rocks have not been deformed as one would expect from gradual tectonic uplift, but exhibit all degrees of intense deformation and shock. Fig. 14 shows specimens of gneiss from these outcrops which display radiating and striated surfaces, characteristic of shatter cones, which, as some believe, can only be formed by high velocity shock waves. In the light of this evidence, the absence of highly variable magnetic fields over the central area of this feature is well explained.

During the course of regional gravity surveys in northern Saskatchewan (Innes and Tanner 1964) during the summers of 1960 and 1963, approximately two hundred gravity stations were established at intervals of about one mile to provide a clear picture of the gravity field over and in the vicinity of the Carswell Lake feature. From the gravity map (Fig. 15) in which total Bouguer anomalies contoured at intervals of two milligals are shown, it is obvious that the circular structure exerts considerable control over the anomalies. Generally the gravity field is depressed in the vicinity of the dolomite ring and reaches minimum values of about -60 milligals, near the inner ring, underlain by sandstone. In the central area the gravity anomalies are higher and reach peak values which appear to be about normal for the area. The gravity field is not completely symmetrical because of regional effects. A large negative anomaly which dominates the pattern to the east is due to a large basement structure that extends for miles across the Churchill province.

The variation in gravity along a northeast-southwest diameter, which parallels the axis of the large regional disturbance, may reasonably be taken as representative of the gravity variation due almost entirely to the Carswell structure. This is shown in Fig. 16 along with a structural interpretation. As noted on the gravity map the field is depressed about 10 milligals in the vicinity of the dolomite ring. The dolomite, with a density of 2.85 gms/cc, is heavier than any rock in the area. It is obvious, then, that the dolomite does not exert major control over the anomalies and, therefore, does not persist with depth. Major control of the negative anomalies most likely stems from local thickening of the sandstone and also from a doughnut-shaped lens of impact breccia that has escaped erosion following the central uplift. The thickness of low density material required to explain the negative anomaly is estimated to be from 2500 to 5000 feet, depending upon the density assumed for the basement rocks.

It appears most unlikely that any process of nature other than meteoritic impact is capable of producing the Carswell Lake feature. However, in reconstruction of its history the role of isostatic adjustment must be considered. Gravity values over the central area are

Fig. 14. Specimen of shocked gneiss from an outcrop near center of the Carswell Lake structure—note radiating and striated surface, characteristic of shatter cones.

about normal for the region indicating that uplift has been effective here and isostatic equilibrium has been reached. It has already been pointed out that the depth of deformation due to impact increases with increasing crater size very nearly as the cube of the diameter. For smaller craters, perhaps up to about 10 miles in diameter, the strength of the crust is adequate to support indefinitely the negative load produced by such deformation. On the other hand, for an impact crater the size of the Carswell Lake feature, the mass deficiency remaining in the crust, long after the superstructure of the original crater has been eroded, would be of sufficient magnitude to promote slow upward isostatic adjustment. Assuming that the cube law holds, the minimum uplift required to restore equilibrium following the deformation resulting from impact is estimated to be from 3500 to 5000 feet depending upon the depth at which the compensation takes place.

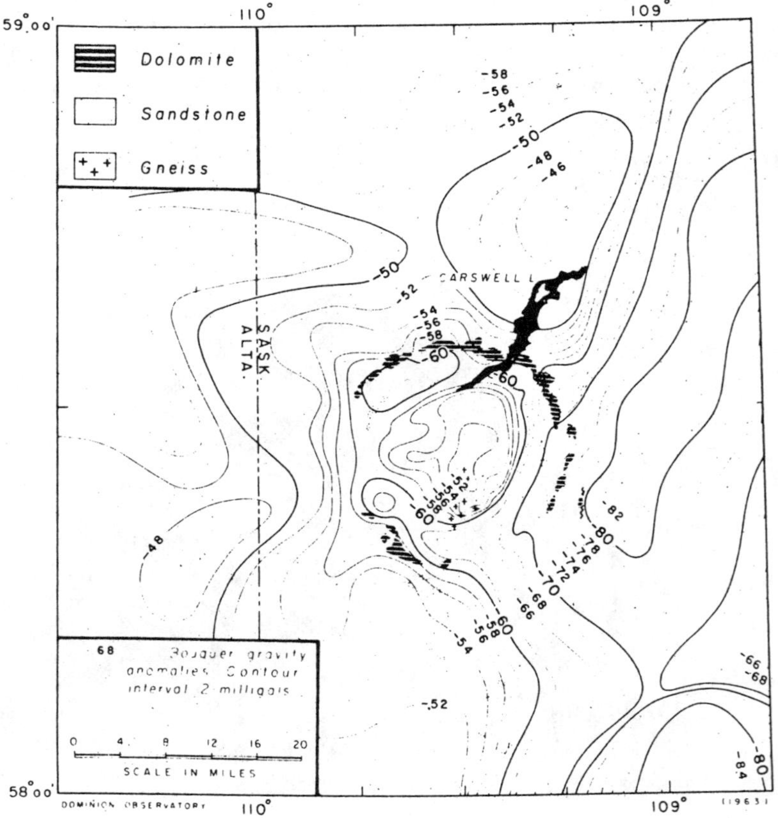

Fig. 15. Map of the Carswell Lake feature showing Bouguer gravity anomalies contoured at intervals of two milligals.

It should also be mentioned that the Carswell Lake feature produces a gravity field that is remarkably similar to that produced by the Vredefort ring structure in South Africa (Maree 1944), a structure long considered by many scientists to have been formed by meteoritic impact. Recently shatter cones have been discovered (Hargraves 1961) in abundance in the deformed rocks of the Vredefort ring giving added support to this hypothesis.

The interpretation offered here, then, is that the Carswell Lake feature is the expression of a deeply eroded meteorite crater. Following its formation the crater floor has been uplifted in response to isostatic forces so that the greater part of the impact breccia has been eliminated by erosion. The outer limit of brecciation defining the true floor of the original crater may be identified as the highly fractured basement rocks, now exposed in the area of central uplift.

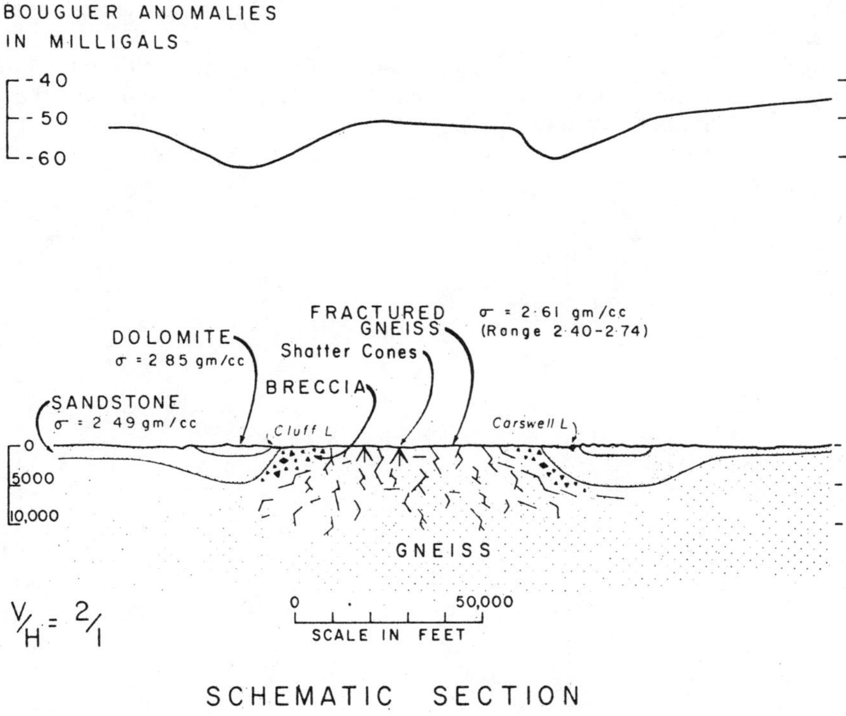

Fig. 16. The variation in gravity across the Carswell Lake feature and structural interpretation.

REFERENCES

Blake, D. A. A. 1956. Geological Notes on the Region South of Lake Athabasca and Black Lake, Saskatchewan and Alberta. *Geol. Survey Canada Paper 55–33,* p. 12.

Dietz, R. S. 1963. Ancient Meteorite-Impact Structures on the Earth. In *The Moon, Meteorites and Comets,* G. P. Kuiper and B. Middlehurst, eds. Chicago, Ill.: Univ. Chicago Press, p. 285.

Fahrig, W. F. 1961. The Geology of the Athabasca Formation. *Geol. Survey Canada Bull.* **68**:41.

Gussow, W. C. 1959. Athabaska Formation of Western Canada. *Geol. Soc. America Bull.* **70**:1.

Hargraves, R. B. 1961. Shatter Cones in the Rocks of the Vredefort Ring, *Geol. Soc. S. Africa Trans.* **64**(for 1964):147.

Innes, M. J. S., and M. R. Dence. 1964. The Carswell Lake Ring Structure (in preparation).

Innes, M. J. S., and J. G. Tanner. 1964. Gravity Anomalies in Northern Saskatchewan and Alberta, Gravity Map Series (in preparation).

Maree, B. D. 1944. The Vredefort Structure as Revealed by Gravimetric Survey. *Geol. Soc. S. Africa Trans.* **47**:183.

Wilson, C. W. 1953. Wilcox Deposits in Explosion Craters, Stewart County, Tennessee and Their Relations to Origin and Age of Wells Creek Basin Structure. *Geol. Soc. America Bull.* **64**:753.

13

Reprinted from p. v of *Canada Geol. Survey Paper 67-32,* 1969, 60pp.

GEOLOGICAL NOTES ON THE CARSWELL CIRCULAR STRUCTURE, SASKATCHEWAN (74 K)

K. L. Currie

ABSTRACT

The Carswell circular structure comprises a nearly circular downdropped block about 25 miles (39 kilometres) in diameter, set in undisturbed clastic sedimentary rocks of the Athabasca Formation. Inside the circular block, the Carswell dolomite, 610 feet (185 metres) thick, is exposed in a marginal ring syncline intricately folded and deformed. The Athabasca Formation, relatively flat lying and undeformed is exposed in a ring between the Carswell Formation and a central core of basement complex about 11 miles (18 kilometres) in diameter. The basement complex is characterized by deformation lamellae on quartz, ascribed to high shearing stress under high confining pressure, and by local potash metasomatism and iron depletion. Both of these features are associated with fault zones in which lenticular masses of Cluff breccia are found. Breccias of similar style are also found in fault zones in the Athabasca and Carswell Formations.

An hypothesis of origin by extraterrestrial impact is considered and rejected. The cause of the structure is considered to be diapirism of the basement caused either by igneous intrusion, or more probably, by concentration of volatiles at depth. The rising diapir dragged up the Athabasca Formation along its flanks, reducing it to a series of concentric fault slivers. The Carswell Formation at first domed over the rising diapir, then slid down the flanks of the dome accumulating in folds concentric to the dome. The last stage of activity was marked by escape of volatiles, producing metasomatized breccia zones along the edges of the diapir, and causing eventual collapse of the structure. This stage culminated about 475 million years before present.

The Carswell circular structure is structurally similar to other circular structures displaying central uplifts and shock metamorphism, but differs from them in the lack of igneous phenomena.

REVIEW*

The Carswell structure has a marginal depression, inner slope and central highland (11 km wide). The depression, immediately inside the rim, is a ditch-like zone 1-3 km wide. The ground rises from its edge to the central core, which stands 30-60 m above the rim. The area is swampy and poorly exposed. The *Carswell Dolo-*

*This review was prepared by the volume editor.

mite (Fahrig, 1961) is believed to be of continuous deposition with the underlying sandstone. The central core of gneiss, granite, and pegmatite displays southeasterly strike and steep dips; also strong alteration and a rubbly, fractured character (possibly not related to the event which formed the structure). The age of the *Athabasca Sandstone* is 1500 m yr. The Carswell formation displays brecciation related to folding, which affects its lowest member; all outcrops display complex, small-scale folding. The *Cluff Breccia* forms a series of dikes which cut all other formations and follow linear zones. Multihued, they contain deformed fragments of sandstone and basement set in a fine base of crush or vitric material. Dikes that cut the Carswell formation contain fragments of both it and the basement. Currie differentiates between disturbed basement; fragmental breccias with particulate matrix; and largely monomict breccias of Athabasca and Carswell rocks. Quartz in fragments within glass-bearing breccias contains shock lamellations, but neither coesite nor stishovite could be found. A specimen of glass and quartz fragments was separated into two color elements which yielded K/Ar ages of 469 ∓ 28 and 486 ∓ 55 m yr. The breccias are manifestly fault controlled. The polymict breccias never contain fragments of upper units incorporated within lower units, the movement of fragments being consistently upward. One breccia vein of sandstone composition has apparently moved up 90 m through the enclosing *Carswell Dolomite.* Seismic investigations show that the structure lies astride an east-west arch in the basement.

Shatter cones are present but are atypical: there are no radial striae or conical surfaces when broken across the cone axis (Dietz, 1959). Currie doubted whether they are *shatter cones* within the limits of the definition. The quartz lamellae follow crystallographic planes and do not extend from one grain to another: multiple sets occur within single grains; and the maximum plot of the poles of the lamellations against *c*-axes lies at a greater angle than is normal for tectonites, a condition believed to be diagnostic of shock metamorphism (Short, 1966). Currie believed that such lamellations could be produced in tectonic deformation, under quasistatic load and confining pressures up to 30 kb and temperatures above 200°C. Short obtained such lamellations, under implosive loading conditions, on quartzose rocks. Carter (1965) obtained basal and inclined lamellae under confining pressures as low as 8 km; and at 16 kb, under critical resolved shear stress. Currie believed the fact that the *Athabasca Sandstone* overlies basement rocks that display quartz lamellations but does not itself display them cannot be simply due to a different response to shock, for the two rocks have approximately equivalent response. He concluded that the shock could not have come from above (as is required in the impact model), and that the distribution of the lamellations around the edge of the central core precludes impact prior to Athabasca and Carswell deposition. Impact, if it occurred at all, must have occurred after the deposition of the rocks of the deformed and brecciated dolomite ring. In view of this, he recognized a major anomaly in the lack of deformation in *Athabasca Sandstone* sandwiched between basement and the Carswell formation. Outliers of the dolomite inside the ring are undeformed and flat-lying. He believed that there was much decollement in the *Carswell Dolomite,* and this must be allowed for in any interpretation.

The Carswell structure differs from simpler structures in the presence of an uplift and downdropped zone. There is evidence of geochemical anomalies associated with it, especially K-enrichment. There is also evidence of highly volatile, reactive solutions being active during the formation of the structure, evidence which led Currie to favor nonphreatic, endogenic chemical explosion. He noted that high pressures can also develop endogenically simply due to certain subsurface configurations of confined rocks (Jamieson, 1963). He pointed out that experimental data suitable for assessing the possibility of high shock pressure in geologic processes are virtually nonexistent. The Carswell structure cannot, he believed, be reasonably explained by impact: endogenic explosion was more easily reconciled with the evidence, but the exact nature of the process remained obscure, though buoyant forces within the basement appear to be the fundamental cause for the structure.

REFERENCES

Carter, N. L. 1965. Basal Quartz Deformation Lamellae—A Criterion for Recognition of Impactites. *Am. Jour. Sci.* **263**:786–806.

Dietz, R. S. 1959. Shatter Cones in Cryptoexplosion Structures (Meteorite Impact?). *Jour. Geol.* **67**:496–507.

Fahrig, W. F. 1961. The Geology of the Athabasca Formation. *Canada Geol. Survey Bull.* **68**:41 pp.

Jamieson, J. C. 1963. Possible Occurrence of Exceedingly High Pressures in Geological Processes. *Geol. Soc. America Bull.* **74**:1067–1070.

Short, N. M. 1966. Shock Processes in Geology. *Jour. Geol. Education* **14**:149–166.

14

Reprinted from *Royal Astron. Soc. Canada Jour.* 59:13–22 (1965)

ON THE PROBABLE METEORITE ORIGIN OF THE CLEARWATER LAKES, QUEBEC

By M. R. Dence, M. J. S. Innes and C. S. Beals

Abstract

Gravity observations over the Clearwater Lakes have indicated the presence of negative anomalies of the kind associated with meteorite craters whose sub-surface structure is characterized by large volumes of low density breccia. Surface geological observations as well as diamond drilling have revealed the presence of large quantities of rock breccia and shattered basement rocks in accord with the gravity observations, adding weight to the probability that the craters were formed by the impact of twin meteorites.

Much of the breccia is characterized by a recrystallized matrix evidently subjected to great heat. While this has been interpreted by Kranck and Sinclair (1963) as due to normal volcanism, preliminary calculations indicate that it can be accounted for equally well by heat energy supplied by the impact process.

The craters are shallower than predicted and a number of factors may have influenced their appearance and structure including excessive erosion of above surface crater structure and surrounding terrain, the operation of isostatic compensation, uplift by a rebound mechanism and possibly the occurrence of impacts in an area covered by water.

Introduction. The sequence of events following the collision of a large meteorite with the earth has been worked out by a number of investigators and may briefly be outlined as follows:

The mass of air displaced by a meteorite of thousands or millions of tons will be insufficient to produce appreciable retardation during its passage through the atmosphere and it will reach the surface of the earth with little loss of velocity. Upon striking the solid earth it will suffer a very rapid exponential deceleration, coming to rest within a few meteorite diameters below the earth's surface. If the velocity of impact is say, 20 km./sec., the peak pressure generated will be of the order of 10 megabars, comparable to that at the centre of the earth. Such a high instantaneous pressure, which will be shared by the meteorite itself as well as considerable volumes of rock into which it imbeds itself, will have two principal effects.

First, a pressure of this magnitude will give rise to a shock wave which will vaporize a certain proportion of the material involved, melt a somewhat larger proportion and shatter large volumes of rock extending to distances of the order of 50 times the diameter of the impacting meteorite.

Secondly, as the meteorite comes to rest, the mixture of highly compressed rock and meteorite fragments will receive an upward and outward acceleration mainly from the decompression of shocked material.

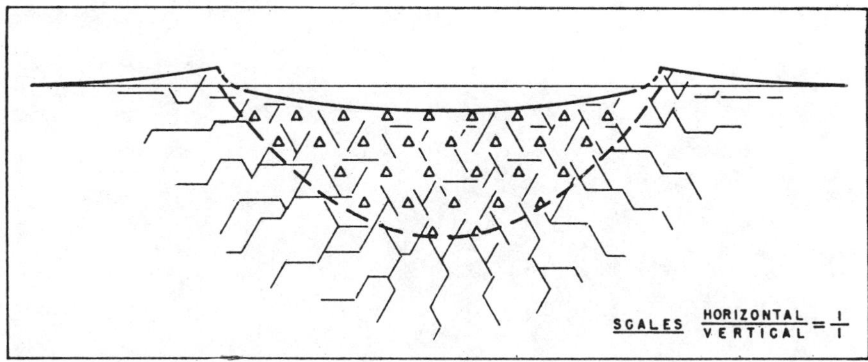

FIG. 1—Section of typical meteorite crater illustrating relative insignificance of surface features when compared with volume of sub-surface fractured rock.

The effect will be comparable to that of a powerful explosive, producing the external aspect of the crater by ejecting large quantities of debris and uplifting the surrounding strata to form the familiar depression surrounded by the raised rim typical of a meteorite crater. Without going further into a process which has already been described elsewhere (Beals, Innes and Rottenberg 1963, Shoemaker 1963) the end result is illustrated by figure 1 which shows a section of a typical meteorite crater.

It will be seen from this illustration that the typical depression surrounded by a raised rim which we are accustomed to think of as the important parts of a meteorite crater are in fact relatively insignificant compared with the large hemispherical volume of shattered rock produced by the shock wave. It will also be obvious that the raised rim and the relatively shallow depression will be particularly vulnerable to the geological processes of erosion and deposition, so that any crater more than a few million years old will, *apart from its circularity*, bear little external resemblance to a recent crater such as Barringer or New Quebec (Chubb). For the study of such ancient craters we are therefore compelled to rely on observations of underground structure and, to date, the more fruitful methods have been the somewhat indirect methods of geophysics (mainly gravity and seismology) and the use of diamond drilling which can reveal the presence of shattered rock produced by the meteorite impact.

Observations at Clearwater Lakes. In figure 2 is shown an aerial mosaic of the two Clearwater Lakes (lat. 56°N., long. 74° 30′W.) with diameters

Fig. 2—2 Mosaic of Clearwater Lakes showing location of drill holes and circular ring of islands in West Lake. Diameters: West Lake 32 km., East Lake 21 km.

of 32 km. and 21 km. respectively. The circular form of these lakes is in marked contrast to the elongated bodies of water caused by glacial erosion which are characteristic of this part of the Canadian Shield. Both lakes are relatively shallow, the large westerly lake with its numerous islands having an average depth of the order of 30 m. The central depth of the eastern lake which has no island far from the shore line is 160 m. This is rather deep for a lake in this region but shallow compared with that of an unaltered meteorite crater of this diameter. The circular ring of islands in the west lake has been shown from observations by McIntyre (1961) and by Dence (1965) to be rock breccia, the bulk having a crystalline matrix, overlying strongly shattered and altered basement gneisses.

The possibility of an impact origin for these lakes was first suggested in 1956 (Beals, Ferguson and Landau). Asteroid pairs of the size required to form craters of the dimensions of the Clearwater Lakes are not known from direct observation and are probably rare (Rabe 1960, Ruskol and Safronov 1961), but are suggested by the presence of several crater pairs of similar dimensions on the moon. Tanner (1963) has shown that such a pair of meteorites once it has formed would possess sufficient stability to exist without much change in the separation of the two meteorites under the action of planetary perturbations in the solar system and during capture.

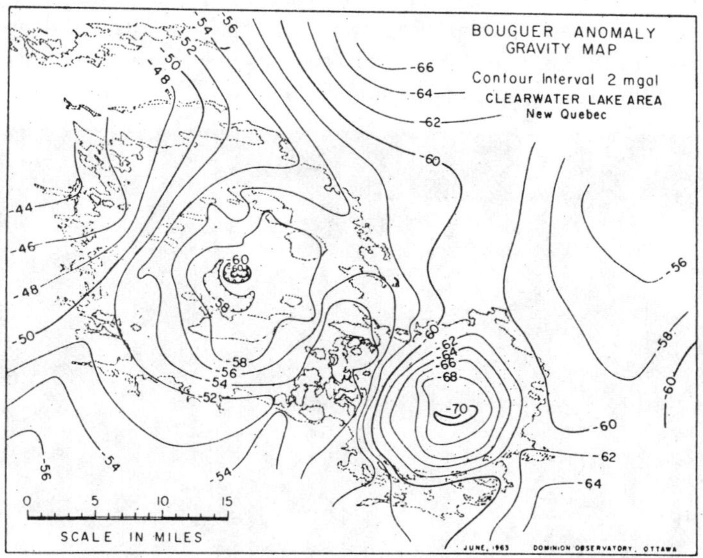

FIG. 3—Gravity contours over Clearwater Lakes illustrating negative anomalies caused by fractured sub-surface rock.

Gravity observations of the two lakes are illustrated in figure 3. After making the necessary corrections for depth of water it will be seen that there are massive negative anomalies of the kind expected for a crater underlain by large volumes of breccia and shattered rock. The gravity observations do not suggest the presence of high density volcanic materials under the surveyed parts of the lakes. The similarity of the Clearwater gravity contours to those of the more definitely identified craters of Brent and Deep Bay is very marked and may be verified by reference to previous publications (Beals, Innes and Rottenberg 1963).

Drilling operations from the ice, in winter, were carried out on both the lakes. The locations of the holes may be seen from the aerial mosaic of figure 2 with the holes numbered from 1 to 5. Hole 1 was drilled in shallow water near the centre of the west lake, Hole 3 was drilled in shallow water midway between the centre of the west lake and the ring of large islands. Hole 4 was located in shallow water close to the inner rim of the ring of islands. Hole 4a was drilled near hole 4. Hole 5 was drilled in shallow water 245 m. nearer the lake centre than hole 4. Hole 2 was drilled in 87 m. of water near the centre of the east lake.

The drilling results may be summarized as follows. Hole 1, after passing through 20 m. of water and unconsolidated material was drilled to a depth of 396 m. in shattered basement rock lithologically similar to the Precambrian gneisses surrounding the lake. Hole 3, 392 m. deep, gave results essentially similar except that the basement was encountered at 7 m. below the water-level of the lake.

FIG. 4—Varieties of breccia encountered in Hole 4. Diameter of core 2.85 cm.

Holes 4, 4a and 5 in the west lake, 119, 64 and 222 m. deep respectively, encountered breccia similar to that found by studies of surface geology, followed at greater depths by shattered basement. Figure 4 illustrates the breccia found in Hole 4 while the shattered basement of Holes 1, 4 and 5 is shown in figure 5.

Hole 2 in the east lake passed through 87 m. of water followed by 46 m. of unconsolidated material, including boulders which made drilling difficult. Sediments consisting of interbedded limestone, shale and gritty sandstone were then encountered and these extended to a depth of 245 m. where rock breccia made its first appearance. The breccia contains small

FIG. 5—Shattered basement in Holes 1, 4 and 5. Diameter of core 2.85 cm.

fragments of Precambrian rocks typical of the region. The fragments are partly fused and vesiculated in a manner consistent with shock-induced heating. The matrix is completely crystalline. At the top of the section it is dark green, soft and extremely fine-grained. This material grades down into a medium to light grey, highly vesicular and slightly coarser phase which in turn gives place to a noticeably darker non-vesicular, still coarser phase over the bottom 15 m. of the section. The composition and texture of the matrix at this level closely resembles the matrix of similar grain size in the breccias of the western island ring. This sequence suggests that the breccia consolidated as a pool of molten material enclosing partly melted fragments, the whole cooling as a single unit. The relatively coarse grain size of the matrix at the bottom of the hole suggests that drilling, which extended to 340 m., stopped near the middle of the breccia lens. Specimens of the rock cores are shown in figure 6.

Conclusions and Discussion. Summarizing the results of the geophysical and drilling operations we reach three main conclusions.

1. The gravity field is consistent with the presence of the shattered

rock pattern of an ancient meteorite crater and is difficult to explain on other grounds.

2. The presence of rock breccia and shattered rock basement is in general agreement with the idea of impact mechanics such as normally produces an impact crater. The general character of the rock breccia forming the greater part of the ring of islands in the west lake and penetrated by the drilling in the east lake indicate that a great quantity of rock has been melted and consolidated in a similar manner to lava. Melting has been previously observed at smaller craters such as Barringer and Brent, but the quantities of fused rock generated in those cases has not been large enough to form extensive melt layers.

While there is no quantitative theory of the partition of the energy of an impact between heating and various kinds of mechanical action, general considerations would suggest that a substantial proportion of the total energy would be transformed into heat. If we make use of Baldwin's formula (1963, p. 176)

$$D = 0.3284\,E - 1.9041$$

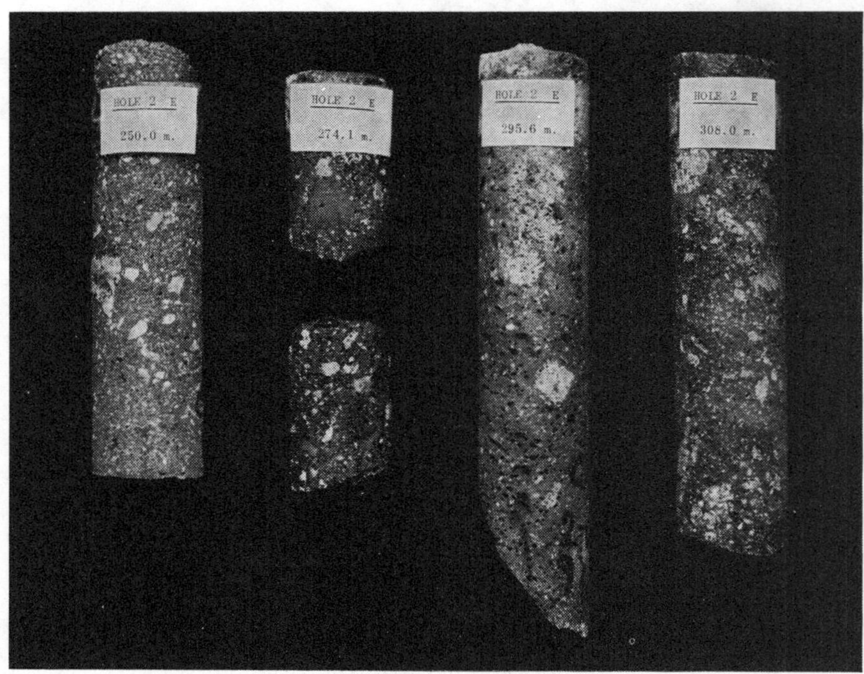

FIG. 6—Specimens of breccias from Hole 2 in East Lake. Note evidence of heating at 295.6 m. Diameter of core 2.85 cm.

relating the logarithm of the diameter of a crater D (in feet) to the log total energy E (in calories) we obtain a value for the energy of west Clearwater of 1.2×10^{21} calories.

A rough calculation of the volume of melted rock in the island ring of the west lake allowing for erosion, places it at about 50 cubic kilometres. Adopting values of 2.70 gm./cc. and 0.25 calories/gm. for the density and specific heat respectively, 1,500° C. for the melting point and 70 cal./gm. for the heat of fusion (Birch *et al.* 1942), leads to a figure of 6×10^{19} calories required to produce the estimated 50 cubic kilometres of recrystallized matrix material, being only 5 per cent of the calculated total energy. While the relation between crater diameter and impact energy is still too uncertain to make this result definitive it does at least suggest that impact alone may be adequate to explain the so-called volcanism of the Clearwater Lakes.

An alternative interpretation of the lava-like material on the islands in terms of a volcanic extrusion has been put forward by Kranck (Kranck and Sinclair 1963) who has suggested a ring-dike fed from a reservoir of lava under the west lake. This suggestion finds no clear support on the ground and is difficult to reconcile with the gravity observations presented earlier in this paper. As has already been mentioned, the gravity contours of figure 3 indicate the presence of a body of low density material under the lake, a condition precluding the presence of a significant amount of magma of normal density. Nor can the comparatively thin (<150 m.) blanket of crystalline breccia on the island ring of the western basin be compared directly with the great thicknesses of light tuffs and ignimbrites which contribute to the negative gravity anomalies of large volcanic calderas (Yokoyama 1963).

3. The profiles of smaller craters such as the Barringer crater and the profiles revealed by drilling at the Brent and Holleford craters are in agreement with that of the crater model of figure 1. However, simple extrapolation of these parameters to terrestrial craters of the size of the Clearwater Lakes shows substantial differences between the theoretical model and the dimensions indicated by the drilling programme. Several explanations for these discrepancies may be advanced.

The floor of the eastern crater, interpreted as the top of the breccia layer, was penetrated at a depth of 245 m. below the surface of the lake. If the crater originally had a profile agreeing with theoretical prediction, then a general erosion of about 1 km. of the upper parts of the crater structure and of the surrounding bedrock is indicated. It is difficult to decide what weight to assign to this suggestion in the absence of reliable estimates of the rates of erosion in Precambrian Shield areas.

Even if such gives a satisfactory explanation of the profile of the eastern crater, it will not suffice alone for the western crater. The breccias on the ring of islands correlate with the breccia layer in the eastern crater, but stand about 350 m. above their eastern counterparts. In addition the central islands consist of basement gneisses which are fractured and deformed but are not severely brecciated indicating that the centre of the western crater has been uplifted.

Mechanisms which may be suggested to account for this uplift include isostatic adjustment and a form of rebound at the time of crater formation. Isostasy operating over a period of some thousands to millions of years could gradually elevate the floors of the craters. The areas involved are rather small for such a process to operate, and clearly much of the anomaly remains. However, it is possible that the disturbed conditions caused by the impact would facilitate action of this kind. Rebound on the other hand would occur within a matter of seconds to minutes after impact, but an adequate theory has yet to be advanced. The fact that it has clearly been much more important in the western crater suggests that the mechanism may require the reflection of the shock wave from an important discontinuity such as the Conrad or Mohorovicic discontinuities. It thus seems probable that for craters formed in the Precambrian basement, only for those of diameter comparable to the thickness of the crust would this mechanism be important.

A third factor contributing to the apparent shallowness of the craters might come into play if the craters had been formed in an area covered by a considerable depth of water. If so, a substantial proportion of the kinetic energy of the meteorites would be used up in displacing the water, and the resulting craters would have smaller depth to diameter ratios than normal (Gilvarry 1962). However, there is little direct evidence that such has been the case at Clearwater.

References

Baldwin, R. B. 1963. "The Measure of the Moon", Univ. of Chicago Press, Chicago, Ill.

Beals, C. S., Ferguson, G. M., and Landau, A. 1956. *R.A.S.C. Jour.*, vol .50, pp. 203–250.

Beals, C. S., Innes, M. J. S., and Rottenberg, J. A. 1963. The Moon, Meteorites and Comets, "The Solar System", vol. IV, Ed. B. M. Middlehurst and G. P. Kuiper, pp. 235–284.

Birch, F., *et al.* 1942 (reprinted 1961). "Handbook of Physical Constants", Geol. Soc. of America.

Dence, M. R. 1965. *Meteoritics* (in press).

Gilvarry, J. J. 1961. *Nature*, vol. 190, pp. 1048–1053; 1962. *Nature*, vol. 196, p. 975.

Johnson, G. W., Higgines, G. H. and Violet, C. E. 1959. *J. Geophys. Res.*, vol. 64, no. 10, p. 1457.

Kranck, S. H. and Sinclair, G. W. 1963. *Geol. Surv. Can. Bull.* 100, 25 pp.

McIntyre, Donald B. 1961. (abstract) Am. Geophys. Un., First National Western Meeting, 1961, p. 55.

Rabe, E. 1960. *Ap. J.*, vol. 131, p. 231.

Ruskol, E. L. and Safronov, V. S. 1961. *Soviet Astr. A. Jour.*, vol. 5, no. 2, p. 203.

Shoemaker, E. M. 1963. The Moon, Meteorities and Comets. "The Solar System", vol. IV, Ed. B. M. Middlehurst and G. P. Kuiper, Chpt. II, pp. 301–336.

Tanner, R. W. 1963. *R.A.S.C. Jour.*, vol. 57, p. 109.

Yokoyama, I. 1963. *J. of the Fac. of Sci.*, Hokkaido Univ., Japan. Sec. VII, vol. II, no. 1.

15

Reprinted from *Internat. Geol. Cong., 24th, Proc.* 15:133–139 (1972)

Shock Metamorphism in the Manicouagan Cryptoexplosion Structure, Quebec

JOHN G. MURTAUGH,
U.S.A.

ABSTRACT

The outer limit of shock metamorphism at Manicouagan is marked by kink-bands in biotite in Precambrian country rocks 33 km from the center of the structure. At 27 km quartz is cleaved, and at 20 km deformation bands in hornblende appear along with weakly developed shock lamellae in quartz. Within 10 km of the center shock metamorphic features formed include: kink-bands and deformation bands in mafic minerals; shock lamellae in felsic minerals; and solid-state shock vitrification of felsic minerals, indicating shock pressures of 250-350 kb. Radially oriented shatter cones have apices pointing up and out, probably because of overturning during uplift of the central massif.

Impact breccias (suevite) overlie and intrude shocked Precambrian rock and Ordovician limestone. The breccias, containing shocked and unshocked fragments mixed with blebs and schlieren of glass of diverse composition, were formed from rocks and minerals that were rapidly mixed, melted and quenched. The formation of such rocks is best explained by hypervelocity impact, generating pressures in excess of 600 kb.

The radial distribution of intensity of shock metamorphism, along with the high pressures indicated, suggest formation of the structure by a sudden central source of energy; that is, hypervelocity impact of a cosmic body.

Permian-Triassic monzonitic igneous rocks, which overlie and intrude the shocked country rocks and impact breccias, may either be impact melt or impact-triggered volcanic rocks. They are interpreted here as impact melt, and their age is assumed to date the impact.

INTRODUCTION

MANICOUAGAN is one of the largest of a group of controversial structures that may have been formed by the hypervelocity impact of a cosmic body (Freeberg, 1966; Short and Bunch, 1968). Such structures characteristically show the effects of shock metamorphism, a process defined by French (1968) as including all changes in rocks and minerals which result from the passage of a transient, high-pressure shock wave. Aspects of the shock metamorphism at Manicouagan have previously been described (Bunch, *et al.*, 1967; Robertson, *et al.*, 1968; Murtaugh and Currie, 1969; Dworak, 1969; Dressler, 1970; Wolfe and Hörz, 1970). This paper describes the evidence for an impact origin of Manicouagan based upon intensity and distribution of shock metamorphic features. The data on which the observations are based were collected during four seasons of mapping by geological field parties of the Geological Exploration Service, Department of Natural Resources, Quebec.

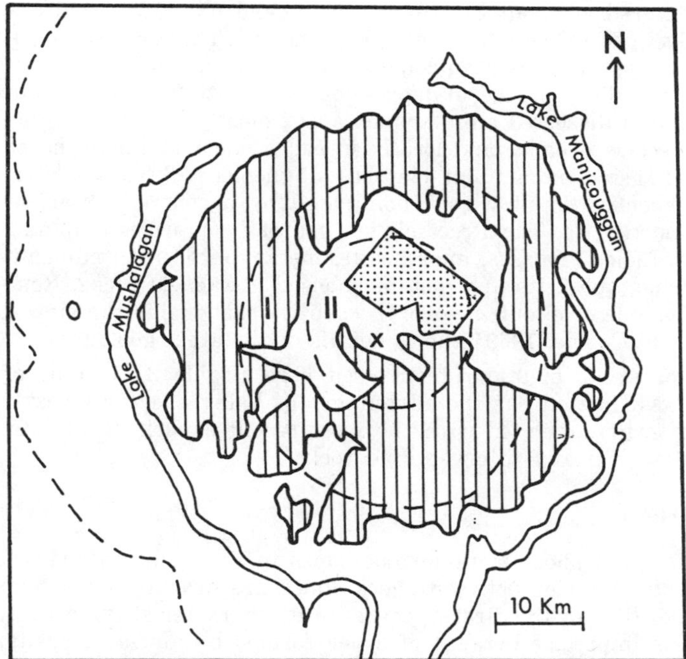

FIGURE 1 — Map of some elements of the geology of the Manicouagan structure. Shaded area = annular plateau of igneous rocks and impact breccias (Table 1). Dotted area = central uplift. X = approximate geometric center of structure. Numerals = shock stages in country rocks (Table 2). Dashed line = approximate boundaries of transitions between shock stages. (Outer limit of shock stage 0 is not well-defined, except where it is roughly coincident with a drainage divide on the west, as described in the text.)

TABLE 1 — Stratigraphic Sections of Manicouagan Units*

Section at outer edge of annular plateau of igneous rock	Thickness (meters)	Section at inner edge of annular plateau of igneous rock	Thickness (meters)
Monzonite	0-20	Monzonite	0-100
contact not observed		*contact abruptly gradational*	
Latite	0-200	Latite	0-100
contact not observed		*contact gradational*	
Basalt	0-15		
contact locally intrusive, locally gradational		Red impact breccia	0-5
Suevite	0-7		
contact locally intrusive, locally gradational		*contact locally intrusive, locally gradational*	
Autochthonous breccia, weakly shocked country rock	unknown	Weakly to moderately shocked country rock	unknown

*Sections are composites of several localities, and thicknesses are estimates. Complete sections are not exposed anywhere.

The structure is centered at about 51° 20′N latitude and 68° 40′W longitude in the Precambrian rocks of the Grenville province of Quebec. The apparent periphery, marked by the valleys of lakes Manicouagan and Mouchalagane, is about 65 km in diameter (Fig. 1). The peripheral depression is essentially circular, but octagonal in detail. Inside, a dissected annular plateau of mostly horizontal igneous rocks and impact breccias overlies and locally intrudes shock-metamorphosed country rocks (Table 1). Inside the annular plateau of igneous rocks is a central region of more highly shocked country rock dominated by a central massif of shock-metamorphosed anorthosite. The Precambrian bedrock comprises amphibolite facies and granulite facies gneisses, anorthosite, metagabbro, metasedimentary iron formation, marble, quartzite and metamorphosed ultrabasic rocks. Remnants of Ordovician limestone crop out in several localities just inside the rim. A recent map (Murtaugh and Currie, 1969) shows details of the areal geology.

The classification of shock metamorphism used here (Table 2) is that of Stöffler (1971). This classification has the advantages of being based on clearly defined petrographic criteria correlated with the most recent experimental data, and of being applicable specifically to crystalline rocks.

SHOCK METAMORPHISM

The approximate limits of shock-metamorphic zones in Precambrian bedrock are shown in Figure 1. Boundaries between shock zones are drawn on the map and described below according to the first appearance of characteristic features. Location of boundaries is imprecise because of cover formed by glacial deposits, vegetation and the annular plateau of igneous rocks. In addition, megablocks up to 30 m in diameter are included in the basal igneous units, and it is not everywhere possible to be sure that outcrops of shocked rocks are really bedrock.

Stage O. Deformation, in the form of an increase in the variety and intensity of jointing, first appears about 45-50 km from the structure center (Roy, 1969). The increase in jointing is roughly coincident with a drainage divide (shown as the outer limit of Stage 0 in Fig. 1) that may represent the original extent of the crater rim. Deformation bands (kink-bands) appear in biotite about 33 km from the center. Artificially produced kink-bands in mica form at shock pressures as low as 9 kb (Hörz and Ahrens, 1969). Autochthonous breccia, consisting of rotated local fragments in a comminuted matrix, appears at about 29 km. Quartz becomes fractured and cleaved at 27 km.

TABLE 2 — Classification of Shock-Metamorphosed Crystalline Rocks and Breccias (after Stöffler, 1971)

Shock Stages	Shock Effects	Peak Pressure (approx., in kb)
0	Fractured quartz and feldspar	<100
I	Diaplectic* quartz and feldspar	100-350
II	Diaplectic quartz and feldspar glasses	350-450
III	Fused feldspar (vesiculated glass)	450-550 or 600
IV	Heterogeneous rock glasses	600-800
V	Silicate vapor	>800

*Diaplectic refers to products formed by shock without melting (von Engelhardt and Stöffler, 1968).

Stage I. Deformation bands in hornblende (Fig. 2) and planar elements (shock lamellae) in quartz appear at about 20 km from the center. Farther in, multiple sets of planar elements in quartz are abundant, and planar elements appear in feldspar (Fig. 3). Deformation bands are formed in garnet. Near the approximate boundary of Stage II, at about 10 km, reduced birefringence becomes apparent in tectosilicates.

Shatter cones (Fig. 4) are most common near the border between Stages I and II, but are present in a few outcrops farther from the center. Plots of shatter-cone striae at two localities 7 and 10 km from the center (after the method of Manton, 1965) indicate that they are approximately radially oriented, and point up and out from the center. At other structures such orientation is caused by overturning of rocks around the central uplift (Vredefort Ring, Manton, 1965; Sierra Madera, Howard and Offield, 1968).

Stage II. Diaplectic glass (formed by solid-state vitrification) appears in isolated patches, beginning at about 10 km from the center (Fig. 5). This glass was first described at Manicouagan by Bunch *et al.* (1967) and later by Dworak (1970). In a few localities, virtually all the tectosilicates have been converted to diaplectic glass (Fig. 6). Experimentally, such glass is formed by shock pressures of 250-350 kb. Scapolite (Wolfe and Hörz, 1970) and apatite (Dworak, 1969) have planar

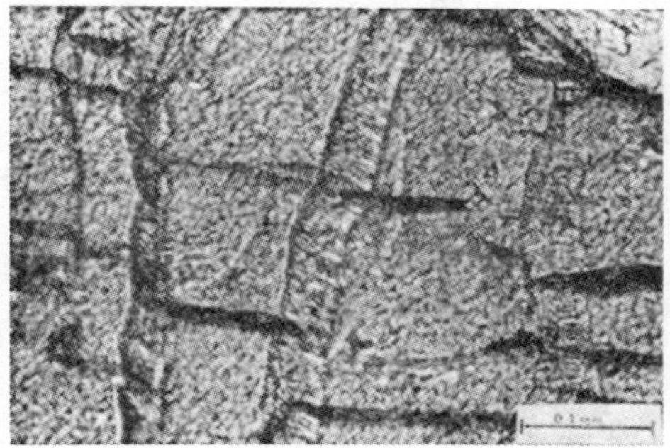

FIGURE 2 — Photomicrograph of deformation bands (vertical) crossing and bending cleavage (horizontal) in hornblende. (plane light)

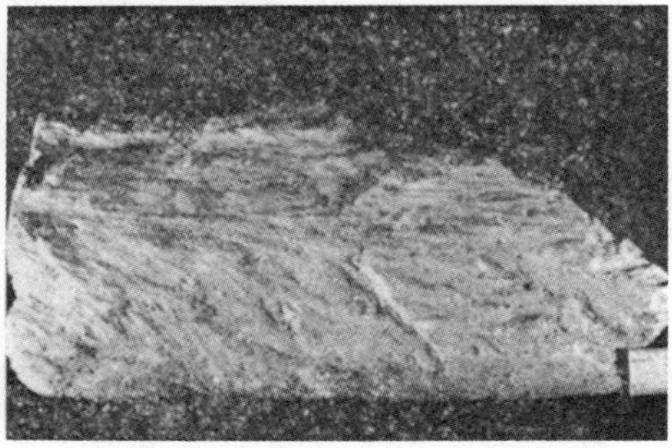

FIGURE 4 — Shatter-cone segment in gneiss. (centimeter scale)

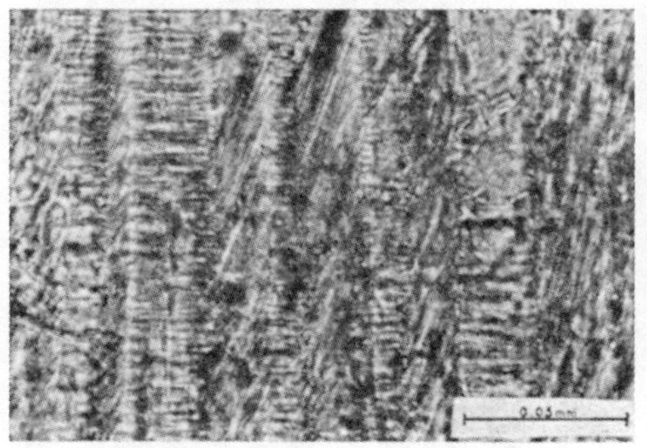

FIGURE 3 — Photomicrograph of crystallographically controlled development of planar elements in vertical twin lamellae in plagioclase. (plane light)

FIGURE 5 — Photomicrograph of partial development of maskelynite (diaplectic plagioclase glass) in anorthosite from the central uplift. Twin lamellae (some in extinction position) are vertical. Wavy black zones trending horizontally are maskelynite. (crossed polarizers)

elements. In this zone, disruption of the rocks by the uplift of the central massif gives rise to a very erratic distribution of shock features, so that rocks nearer the center are commonly less shocked than those farther out. Zeolitization, hornfelsing, local vesiculation and melting by the igneous rocks commonly obscure shock effects in this zone.

Stages III and IV. Suevite is an impact breccia containing diverse fragments of shocked and unshocked rocks and minerals, along with glass blebs of different composition, in a glassy or clastic matrix (Fig. 7 and 8). This rock differs from the type at the Ries in having less well developed bombs (Fladen), in being the basal unit, rather than the upper unit, of allochthonous impact material (Table 1) and in being intrusive locally (Fig. 9). The red impact breccia at the base of the inner edge of the annular plateau (Table 1) of igneous rocks differs from the suevite in being monomict, and in containing virtually all shocked fragments.

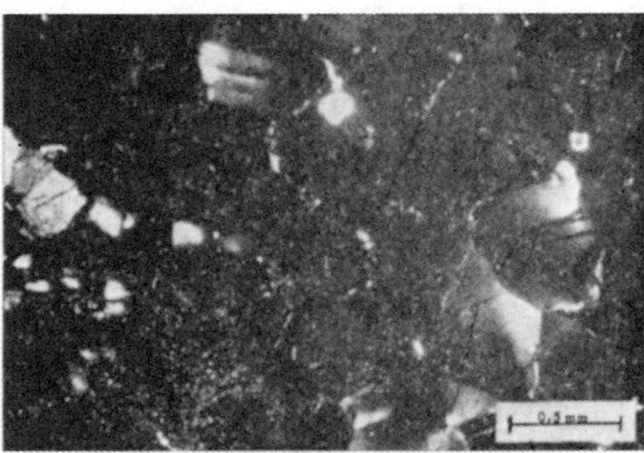

FIGURE 6 — Photomicrograph of extensive development of maskelynite in anorthosite of the central massif. Plagioclase is almost totally converted to maskelynite, but grain boundaries (partly recrystallized) are preserved. A few grains of plagioclase are still partly anisotropic. Dark euhedral grains are garnet. Bright grains are pyroxene or scapolite. (crossed polarizers)

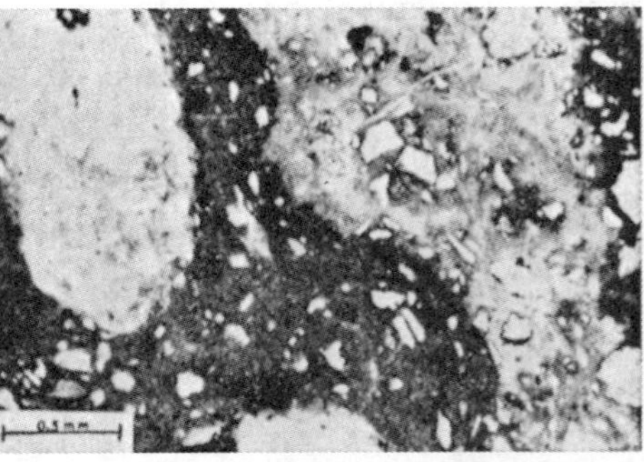

FIGURE 8 — Photomicrograph of thin-section of suevite specimen shown in Figure 7 with white, grey, brown and black schlieren and blebs of glasses of different composition. Angular grains are mineral fragments. (plane light)

FIGURE 7 — A variety of suevite, showing flow structure. (centimeter scale)

FIGURE 9 — Dike of suevite (vertical) in gneiss 27 km from the center of the structure.

Pseudotachylite is similar to the red impact breccia, but forms small dikes and irregular veins in the shock-metamorphosed rocks of the interior. The nature of these rocks indicates that they were formed by rapid melting, mixing and quenching. Similar glasses in the Reis suevite are interpreted as having formed after undergoing shock pressures of about 600 kb (von Engelhardt and Stöffler, 1968).

ORIGIN OF THE IGNEOUS ROCKS AND AGE OF THE STRUCTURE

The igneous rocks (monzonite, latite, basalt) forming the annular plateau (Fig. 1) have been interpreted both as impact melt rocks (Dence, 1971) and as normal volcanic rocks (Currie, 1971; Currie and Shafiqullah, 1968). In a paper in preparation for the Quebec Department of Natural Resources I interpret these rocks as impact melt products based on their gradational relationships with each other and with the shock-metamorphosed rocks, and on their apparent contemporaneity with the impact event. Discussion of details is beyond the scope of this paper, but the reader should be aware of the unresolved controversy.

The igneous rocks have been dated at 225 ± 30 m.y. by the K-Ar whole-rock method (Wanless, *et al.*, 1966), 208 ± 25 m.y. by the fission track method (Fleischer, *et al.*, 1969) and have Lower Triassic paleomagnetic poles (Larochelle and Currie, 1967; Robertson, 1967). If the igneous rocks are crystallized from an impact melt, then the time of the impact is determined as Early Triassic or Late Permian.

Wolfe (1971), using K-Ar techniques, dates the igneous rocks at 210 ± 4 m.y. He also obtained a tentative date of 280-300 m.y. for the shock outgassing of a suite of shock-metamorphosed rocks, thus suggesting that the igneous rocks are much younger, and therefore of endogenous origin. However, the individual dates of the shocked rocks are variable, and one specimen gives the same age as the igneous rocks, a fact that he admits casts doubt on his conclusion.

CONCLUSIONS

The inward increase of intensity of shock metamorphism in country rocks strongly suggests a single central energy source as the cause of formation of the structure. The high-pressure high-temperature quench phenomena in the impact breccias seem inexplicable except by rapid mixing of shocked and unshocked material along the floor of the crater. The most likely source of energy for such an event is the hypervelocity impact of a cosmic body. A recent study (Roddy, 1968) suggests that shallow craters with central uplifts may be the result of cometary impact. Manicouagan may be the result of such an event.

The Manicouagan structure was formed in late Paleozoic or Early Mesozoic time, but determination of the precise age depends on interpretation of the origin of the rocks within the crater. If these rocks are endogenous volcanic rocks, then the crater may be 280-300 m.y. olds, as Wolfe (1971) suggests. If, however, these rocks are formed from impact melt, then their age of 225-208 m.y. is the time of crater formation. I interpret the igneous rocks as impact melt.

ACKNOWLEDGMENTS

The Geological Exploration Service of the Department of Natural Resources, Quebec, supported the field and laboratory investigations. Part of the laboratory investigation was supported by the Bownocker fellowship at The Ohio State University.

REFERENCES

Bunch, T. E., Cohen, A. J., and Dence, M. R., 1967. Natural terrestrial maskelynite. Am. Mineral., 52, p. 244-253.

Currie, K. L., 1971. Origin of igneous rocks associated with shock metamorphism as suggested by geochemical investigations of Canadian craters. J. Geophys. Res., 76, p. 5575-5585.

Currie, K. L., and Shafiqullah, M., 1968. Geochemistry of some large Canadian craters. Nature, 218, p. 457-459.

Dence, M. R., 1971. Impact melts. J. Geophys. Res., 76, p. 5552-5565.

Dressler, B., 1970. Die Beanspruchung der präkambrischen Gestine in der Kryptoexplosionstruktur von Manicouagan in der Provinz Quebec, Canada. Doct. Dissert., Univ. Munich, 108 p.

Dworak, U., 1969. Stosswellenmetamorphose des Anorthosites vom Manicouagan Krater, Québec, Canada. Contrib. Mineral. Petrol., 24, p. 306-347.

Fleischer, R. L., Viertle, J. R. M., and Price, P. B., 1969. Age of the Manicouagan and Clearwater Lakes craters. Geochim. Cosmochim. Acta, 33, p. 523-527.

Freeberg, J. G., 1966. Terrestrial impact structures—a bibliography. U.S. Geol. Surv. Bull. 1220, 91 p.

French, B. M., 1968. Shock metamorphism as a geological process. *In* French, B. M., and Short, N. M. (*Editors*). Shock Metamorphism of Natural Materials. Mono Book Corp., Baltimore, p. 1-17.

Hörz, F., and Ahrens, T. J., 1969. Deformation of experimentally shocked biotite. Am. J. Sci., 267, p. 1213-1229.

Howard, K. A., and Offield, T. W., 1968. Shatter cones at Sierra Madera, Texas. Science, 162, p. 261-265.

Larochelle, A., and Currie, K. L., 1967. Paleomagnetic study of igneous rocks from the Manicouagan structure, Quebec. J. Geophys. Res., 72, p. 4163-4169.

Manton, W. I., 1965. The orientation and origin of shatter cones in the Vredefort ring. Annals N.Y. Acad. Sci., 123, art. 2, p. 1017-1049.

Milton, D. J., and De Carli, P. S., 1963. Maskelynite: formation by explosive shock. Science, 140, p. 670-671.

Murtaugh, J. G., and Currie, K. L., 1969. Preliminary study of Manicouagan structure. Quebec Dept. Nat. Resour. Prelim. Rep. No. 536, 9 p.

Robertson, P. B., Dence, M. R., and Vos, M. A., 1968. Deformation in rock-forming minerals from Canadian craters. *In* French, B. M., and Short, N. M. (*Editors*). Shock Metamorphism of Natural Materials. Mono Book Corp., Baltimore, p. 435-452.

Robertson, W. A., 1967. Manicouagan, Quebec, paleomagnetic results. Can. J. Earth Sci., 4, p. 641-649.

Roddy, D. J., 1968. The Flynn Creek Crater, Tennessee. *In* French, B. M., and Short, N. M. (*Editors*), Shock Metamorphism of Natural Materials. Mono Book Corp., Baltimore, p. 291-322.

Roy, D. W., 1969. Etude de la fracturation dans la partie ouest de la structure circulaire de Manicouagan. M.S. thesis, Univ. Montreal, 98 p.

Short, N. M., and Bunch, T. E., 1968. A worldwide inventory of petrologic features characteristic of rocks associated with presumed meteorite impact craters. *In* French, B. M., and Short, N. M., (*Editors*). Shock Metamorphism of Natural Materials. Mono Book Corp., Baltimore, p. 255-266.

Stöffler, D., 1971. Progressive metamorphism and classification of shocked and brecciated crystalline rocks at impact craters. J. Geophys. Res., 76, p. 5541-5551.

von Engelhardt, W., and Stöffler, D., 1968. Stages of shock metamorphism in the crystalline rocks of the Ries Basin, Germany. *In* French, B. M., and Short, N. M., (*Editors*). Shock Metamorphism of Natural Materials. Mono Book Corp., Baltimore, p. 159-168.

Wanless, R. K., Stevens, R. D., Lachance, G. R., and Rimsaite, J. Y. H., 1966. Age determinations and geological studies: K-Ar isotope studies, Rept. 6. Geol. Surv. Can. Pap. 65-17, p. 77-78.

Wolfe, S. H., 1971. Potassium-argon ages of the Manicouagan-Mushalagan Lakes structure. J. Geophys. Res., 76, p. 5424-5436.

Wolfe, S. H., and Hörz, F., 1970. Shock effects in scapolite. Am. Mineral., 55, p. 1313-1328.

16

Reprinted from p. xiii of *Canada Geol. Survey Bull. 198,* 1972, 153 pp.

GEOLOGY AND PETROLOGY OF THE MANICOUAGAN RESURGENT CALDERA, QUEBEC

By

K. L. Currie

Abstract

The Manicouagan resurgent caldera is an octagonal depression 38 miles across, with a mountain of Precambrian rocks about 4 miles north of the centre. The moat between the mountain and the walls of the depression is partly filled with volcanic rocks of the Triassic Manicouagan complex. These rocks are composed of varying fractions of alkali olivine basalt mixed with country rocks. The Precambrian rocks beneath the Manicouagan complex, comprising Archean gabbro, anorthosite, mangerite, and charnockite, overlain by Proterozoic quartz-feldspar gneisses containing quartzite–marble–iron-formation pods, have been erratically shock metamorphosed to assemblages containing one or more of the following features: vesicular glass, deformation lamellae, maskelynite, optically anomalous minerals, hornfelsic texture, and metasomatic alteration. Field and experimental evidence suggest that metamorphism occurred under conditions of very high temperature, moderate pressure, and extremely rapid oscillation of both temperature and pressure. These conditions are deduced to be connected with a hydrogen-rich character of the parent magma.

Structural analysis of the area reveals that a long history of Precambrian deformation preceded formation of the Manicouagan structure. Development of this structure began with strong upwarping in the areas now occupied by Manicouagan and Mouchalagane lakes. Development continued with radial volcanic blasts followed by subsidence along pre-existing zones of weakness. Doreitic magma was poured out into the depression, and the central mountain rose on the back of intrusions at depth. The structural pattern of the igneous rock sheets agrees with geophysical evidence by indicating that Precambrian rocks beneath the Triassic igneous rocks are essentially unbrecciated.

Hypotheses supposing the Manicouagan structure to result from or to be initiated by extraterrestrial impact, are considered and rejected as contrary to field evidence.

The geometric form and history of development of the Manicouagan structure are similar to those of resurgent calderas. The structure is concluded to be an unusual variety of resurgent caldera.

REVIEW*

Four physiographic zones are distinguished within the arcuate lakes of the Manicouagan structure: an *outer lowland,* which Currie does not regard as a ring uplift—it is composed of brecciated basement rocks and pipes of ultrabasic material (unit 14 in Table 8); an *annular plateau;* an *inner lowland;* and a *central mountain.*

The Ordovician limestone-dolostone and sandstone cover contains an arctic fauna of the shallow temperate seas which covered most of northeast Canada at this time. These are not deposits of restricted, preexisting fjord-like basins (Hammond, 1946; Ambrose, 1964). Fragments of these rocks in Triassic volcanics have been carried inward and upward, and this suggests a source in the ring of narrow lakes for these eruptives.

Metamorphic effects include decomposition of minerals with local production of vesicular glass; lamellar structure in grains; maskelynitization; optically anomalous plagioclase or pyroxene; development of zeolites, scapolite, and potash felspar. Remarkable specimens of vesicular gneiss are illustrated. The sequential nature of the mineral breakdown is noted. The decomposition and vesiculation are attributed to rapid heating and quenching, effects not inconsistent with magmatic processes. Currie accepts that lamellation in quartz is due to shock, but questions whether it can be used as an indicator of varying degrees of shock (Hörz, 1968; Robertson et al., 1968). He believes that other factors besides shock must contribute to maskelynitization. Anomalous plagioclase, pyroxene, and garnet; hornfelsing associated with fine-grained igneous rocks of units 15 and 16 (Table 8); and hydrothermal metamorphism involving the zeolite thomsonite, scapolite, and iron oxides are described. Zeolites occur as massive rocks, and their distribution suggests that they formed prior to the volcanic activity. They are only present in the igneous rocks as xenolithic fragments. Pseudotachylite is the only rock believed to contain a zeolitic igneous component. Zeolites cannot be easily referred to the volcanism for they apparently developed between the time of formation of the pseudotachylite with accompanying shock metamorphism and the considerably later onset of andesitic volcanism. Currie concluded that the nature of the shock metamorphism does not prove the passage of a shock wave with a peak pressure exceeding 100 kb. He favours one or more pulses at lower pressures. The erratic distribution of shock effects suggests to him that there were several localized pressure peaks, not a single central pulse. On the basis of theory and observation, Currie proposes that endogenic processes such as gas explosion could produce pressures of the required magnitude. Shock metamorphism under pressures of less than 10 kb would require temperatures of 800–1200°C according to experimental evidence cited. Such high temperatures must have been abruptly quenched. He favors shock metamorphism at high to extreme temperature, but low to moderate pressure, with rapid PT variations: conditions which, he believes, could arise in volcanic and subvolcanic vent environments through the intervention of a reactive, combustible gas phase.

Currie concluded that:

1. The igneous rocks are all olivine normative and alkaline.
2. The ultrabasic rocks crystallized at considerable depth under moderate to high water pressures.

*This review was prepared by the volume editor.

Table 8 (after K. L. Currie)

CENOZOIC (Pleistocene and Recent)		Glaciofluvial deposits, minor drift
- - - - - - - - - -		unconformity - - - - - - - - - -
	UNIT 16	Grey to brown, medium- to coarse-graned larvikite with phenocrysts of augite and hypersthene
		- - - - - - abrupt gradational contact - - - - - -
MESOZOIC (Triassic)	UNIT 15	Red to brownish fine-grained to aphanitic rocks ("trachy-andesites"): (a) grain size .5 mm; (b) fine-grained spherulitic rocks; (c) red monomict breccia with shocked fragments, igneous matrix
		- - - - - - - relationships uncertain - - - - - - -
Manicouagan complex 700 feet thick	UNIT 14	Heterogeneous basic and ultrabasic rocks: peridotite, picrite and basalt: red-stained breccia with ultrabasic matrix
		- - - - - - - - - intrusive contact - - - - - - - - -
		Pseudotachylite dikes
		- - - - - - - - - intrusive contact - - - - - - - - -
	UNIT 13	Breccias with clastic matrix: (a) autochthonous; (b) polymict with glassy fragments (suevite); (c) "ball fracture"
		- - - - - - - - unconformity, cross-cutting contact, part intrustive - - - - - - - -
PALEOZOIC (ORDOVICIAN)		Thinly bedded limestone and fossiliferous dolostone, basal sandstone, and green shale (*Trenton* and *Wilderness* stages)
		- - - - - - - - - - unconformity - - - - - - - - - -
Hadrynian		Dikes (and sills?) of andesite and basalt; ultrabasic plutonites; dunite, peridotite, garnet peridotite, troctolite
Elsonian		
Aphebian ?		Metasediments: quartzite, marble, iron formation
		Quartz-felspar-biotite gneiss complex
PRECAMBRIAN		Hybrid gneisses (charnockitic, biotite gneisses): granitoid gneisses: migmatite grading to hybrid charnockitic gneisses
Aphebian		Granular, massive to faintly foliated granitoid rocks: granite gneiss, migmatite grading to hybrid charnockitic gneiss
		Enderbitic and mangeritic rocks
		Leucocratic garnet anorthosite
		Melanocratic hornblende-garnet-plagioclase gneiss
		Granulitic gabbro, green gnèiss, minor charnockite
		Sillimanite-hypersthene-garnet gneisses: calcsilicate rocks and graphitic gneiss

3. This material is cognate, derived from an alkali basalt stem.
4. Pseudotachylites are not simply modified country rocks, but are mixtures of finely crushed or melted country rock with 10–70% of picritic or alkali basalt component.
5. The igneous breccias and the suevite contain an intermediate, igneous component of the unusual *doreite* composition.
6. The igneous rocks are not of equivalent composition to the country rock average, but the more siliceous of them could have resulted from mixture of 15% alkali basalt component with country rock.

Shatter cones are found up to 10 km from the central peak, northwest and southeast from it; they occur as 10 cm long segments. They are attributed to tension fracturing (Manton, 1965). Such cones have been produced in cratering experiments at pressures of only 5 kb, according to Currie, and he believes they could represent extremely rapid release of shock pressure pulses and quite low peak pressures. Evidence from the older structural patterns precludes any subsequent rotation at Manicouagan (Manton, 1965). The only vector obtained by Currie indicated a shock wave emanating from the focus and moving outward, not downward.

The fault patterns are complex. Many faults reactivate preexisting lineaments. Mont de Babel was uplifted on faults and the limestone-filled graben dropped by at least 800 m at some time later than 210 m yr ago. The mountain is in no way circular and is cut by oblique faults. The entire area was initially updomed by at least 200–250 m, the dome affecting an area that extends beyond the outer margins of the arcuate lakes. Flexuring was particularly intense in the zone of the lakes. Explosions later breached this great, gentle dome on radial fractures, with wholesale shattering and intensification of preexisting fracture zones.

Currie lists three possible origins of the Manicouagan structure:

1. *Volcanic:* The igneous rocks are products of deep-seated magmatism, and the shock effects due to unusual volcanic processes; the shocked rocks include a magmatic component; the chemistry of the country rocks only exerts a control on contaminated magmatic rocks.
2. *Meteoritic:* The igneous rocks are all products of mixing and melting of finely brecciated and pulverized country rocks, the average composition of which they should display.
3. *Impact-volcanic:* The pure magmatic rocks are due to impact-triggered, lower-crust or mantle-derived eruptivity; compositions may be unrelated to those of country rocks; the shock-metamorphic and melt rocks should display no chemical affinity to the igneous rocks, and should precede them by a long interval of time; they should themselves have the average composition of the country rocks.

The *meteoritic* hypothesis finds its main support in the nature of the shock metamorphism—in the form of the structure and in its isolation from contemporaneous volcanic or tectonic events. Despite the asymmetry of the fault-bounded central peak, it has been attributed to either *prompt rebound*—simultaneous with the excavation of the crater (Roddy, 1968a,b)—or *delayed rebound*—avalanching of debris from crater walls or isostatic adjustment on curved planes of movement. The first concept leans heavily on the "Snowball Experiment" whereby low-density cometary impact, in which energy is expended in surficial explosion, is invoked. Currie considered that there is insufficient evidence of deformation adequate to allow the required mass flowage from side to center. Likewise, he discounted the avalanching process. Experiments by Danes (1962) have shown that a crater 100 km in diameter could assume the form of Manicouagan by isostatic adjustment in an elapsed time of 250 m yr after crater formation, but these experiments concerned a highly viscous, fluid medium. Faulting in a brittle medium like that at Manicouagan could possibly achieve the same result—and uplift at Manicouagan manifestly occurred on faults, not plastically—but the nature of the uplift and the relationships of the various igneous units at Manicouagan are difficult to reconcile with such a hypothesis. Compensation by marginal down-sinking is involved in the process studied by Danes, and the plastic flowage that would be required to effect this appears to be absent. Compensation by movement on a set of conjugate fractures is not supported by the evidence. Voids up to 50 km in diameter have remained isostatically uncompensated for periods greater than that for which this structure has existed. Jeffreys 1959) considered compensation unlikely for a structure of this size during the period available since the Mesozoic. And it must be remembered that Manicouagan has many much smaller analogs of equivalent age, possessed of central peaks (for example Nicholson Lake, 12.5 km diameter, p. **112**; Gow Lake, 4 km diameter, p. **112**). Theoretical considerations show that in such cases isostatic compensation could not have produced a central peak in the time elapsed. There are severe objections, therefore, to explaining central peaks by either version of the delayed-rebound hypothesis.

The *impact-volcanic* hypothesis requires a chain of tectonic and eruptive events to be set off by impact, thermally or mechanically triggered. However, maskelynitization, pseudotachylite formation, suevite and picrite/ultrabasic eruptivity are all demonstrably early events at Manicouagan, penecontemporaneous with crater formation, a fact that appears to be fatal to the thermal alternative. Melting of alkali basalts within the country rocks can be discounted, for there are none. Impact-triggered tapping of a hearth of differentiating magma of such composition is open to a number of objections (French, 1972), but cannot be discounted completely, even though such a process requires a coincidence of a cosmic body striking the surface immediately above the only known differentiating magmatic hearth for hundreds of kilometres.

Currie favors the volcanic hypothesis, drawing comparisons with resurgent calderas. Manicouagan differs from other known resurgent calderas mainly in the nature of the shock metamorphism, which he attributes to the explosive effects of hydrogen liberated in an explosive gas phase. Such a gas phase in unusual but not of unique character, he believes.

The evaluation of the energy involvement in the Bezymyanny volcanic explosion (pressures of up to 3.5 kb; Gorshkov, 1969) cannot be taken as a rigid limit to pressure involvement in endogenic explosion, and the time sequence—doming, collapse, resurgence—is familiar in cauldron/caldera development. Resurgent calderas are larger than normal volcanic craters, show integration with regional fault systems, possess central uplifts within dropped areas, and occur in horst and graben structural settings. They are commonly associated with intermediate eruptives like those at Manicouagan, which tend to be explosive because of contained volatile phases, as well as with fluidization, assimilation of country rock, and pseudotachylites. Deep-seated alkaline rocks in such associations contrast with calc-alkaline intermediate suites in the same way as at Manicouagan. Breccia and tuffisite developments are common.

REFERENCES

Ambrose, J. W. 1964. Exhumed Peneplains of the Canadian Shield. *Am. Jour. Sci.* **262**:817–857.

Danes, Z. F. 1962. Isostatic Compensation of Lunar Craters. *Research Inst. Puget Sound, Document RIR-GP-62-1.*

French, B. M. 1972. Production of Deep Melting by Large Scale Meteorite Impacts: The Sudbury Structure, Canada. *Internat. Geol. Cong. 24th, Proc.* **15**:125–132.

Gorshkov, G. S. 1969. Gigantic Eruption of the Volcano Bezymyanny. *Bull. Volcanol.* **20**:77–109.

Hammond, W. P. 1946. Geology of Mouchalagane Lake Area, Saguenay County, Quebec. M.A. thesis, University of Toronto.

Hörz, F. 1968. Statistical Measurements of Deformation Structures and Refractive Indices in Experimentally Shock-loaded Quartz. In *Shock Metamorphism of Natural Materials,* B. M. French and N. M. Short, eds. Baltimore, Md.: Mono Book Corp., 243–254.

Jeffreys, H. 1959. *The Earth: Its Origin, History and Physical Constitution,* 4th ed. Cambridge: Cambridge Univ. Press.

Manton, W. I. 1965. The Orientation and Origin of Shatter Cones in the Vredefort Ring. *New York Acad. Sci. Annals* **123**(article 2):1017–1049.

Robertson, P. B.; M. R. Dence; and M. A. Vos. 1968. Deformation in Rock Forming Minerals from Canadian Craters. In *Shock Metamorphism of Natural Materials,* B. M. French and N. M. Short, eds. Baltimore, Md.: Mono Book Corp., 433–452.

Roddy, D. J. 1968a. Comet Impact and Formation of Flynn Creek and Other Craters with Central Peaks. *Am. Geophys. Union Trans.* **49**:272 (abstr.).

Roddy, D. J. 1968b. The Flynn Creek Crater, Tennessee. In *Shock Metamorphism of Natural Materials.* B. M. French and N. M. Short, eds. Baltimore, Md.: Mono Book Corp., 291–322.

Editor's Comments on Papers 17, 18, and 19

CHARLEVOIX

The Charlevoix (La Malbaie) structure was attributed to meteorite impact by Robertson (1968). It is considered here by Rondot in Paper 17. Set in Grenville (Precambrian) charnockite and granite gneiss terrain, and 35 km in diameter, the grain of the ancient terrain seems to pass more or less undisturbed through the structure, indicating little lateral displacement. Limestone of Ordovician age is preserved in an annular graben, though absent due to removal by erosion inland; it has been displaced downward by 780 m. The southern part of the structure has been removed by a major fault. Ordovician sediments of quite different facies to the limestones forming the Isle aux Coudres. Shatter cones are best developed in a zone 7 km out from the focus in the central uplift (Mont des Eboulements). They show increasingly poor development inward and outward from this zone and die out 12 km out. Shock metamorphism, including lamellation in quartz, is concentrated around a peak zone in the Mont des Eboulements. This metamorphism was weak to moderate, a fact that Robertson attributed to removal of blankets of impact melt and fall-back breccia, together with the upper levels of the shocked basement by erosion. The eccentricity of the structure, including the shock zones, was attributed by him to oblique impact, rock type variation, post- and pre-crater deformation patterns, and subsequent erosive effects. Rondot (Paper 17) suggested that the crater was originally 28 km in diameter, and that lithostatic adjustment followed its formation, being reflected in the relative position of the Paleozoic sediments, which he believed are now displaced by as much as 1300 m in the annular ring. The uplift is believed by Rondot to have been produced by the relative displacement of the fault blocks. It displays only minor fractures, up to 100 m or so in length. This block is supposed to have been derived from the foundation beneath the crater and to have been vertically raised by 5 km. Pseudotachylite fills minor fractures in the central peak, up to 3 km from the focus, and as far out as the enclosing scarp the blocks are separated by mylonite composed of breccia fragments enclosed in a thread-like, varicolored matrix. Rondot suggested that 1000 km^3 of rocks were excavated at the moment of impact. Some of the blocks so produced are, he believed, now to be found at the base of the glacial deposits. They include "impactites," products of extreme heat and pressure. Rondot mentions "impact ignimbrites," but does not describe them. The allochthonous breccias are black, of finely crystalline, aphanitic, cryptocrystalline, or fluidal texture, and generally of pale tines and prehnitized. Hydrothermal alteration has strongly affected the center of the structure, the dikes of breccia and their immediate surroundings, and possibly also some "fall back breccia" layers. The minerals formed are zeolites, including chabazite, laumontite, stilbite, and prehnite. The curved planes of sliding on which Rondot envisaged adjustment as having been effected are believed by him to be typical of a special type B crater, formed by relatively slow-impacting bodies. Such readjustment is, he believed, on a scale proportional to the width of the structure, and characteristic of large astroblemes. Gault and Heitowit (1963) and Braslau (1970) have stated that repartition of energy should be about half toward excavation of the crater, a quarter to compaction and fragmentation of the substratum, and the remainder to thermal effects. Gault (pers. comm. to Rondot) later suggested that increased speed produces more thermal effects and less fragmentation. Type B craters are supposed to involve considerable fragmentation and fracturing, which allow readjustment.

Currie (1972, p. 139) noted a radiometric age of 335 m yr for this structure. Robertson believed the age of the structure to be no greater than 460 m yr (based on the youngest rocks affected) and no less than the Devonian Acadian orogeny (365 m yr) to which the thrusting of the Isle aux Coudres that complicates the structure to the south is attributed. Currie (1972) reported the presence of carbonatite dikes, comparing the structure with Hicks Dome, Illinois in its symmetry and in its association with fluorite development. He regarded the structure as of endogenic origin, a resurgent caldera, much eroded.

MISTASTIN LAKE

Mistastin Lake, is covered in this volume by Taylor and Dence, in Paper 18. It is a roughly circular lake in northern Labrador, with a large horse-shoe shaped island. The structure attracted attention because of the association of flat-lying igneous rocks with a subcircular lake and arcuate island. The lake is rimmed by aphanitic to medium-grained and porphyritic igneous rocks containing fragments of country rock. Strong and irregular fracturing, shock lamellation of quartz, and slight hydrothermal alteration of the rocks on the island characterize the basement rocks. Near to the outcrop of igneous rocks, the adamellite is thermally metamorphosed and injected by glassy veinlets. Weak to moderate shock metamorphism has affected the rocks of the island, and weak shock affected the adamellite prior to the thermal metamorphism. Inclusions within the igneous rocks have suffered varying degrees of shock prior to incorporation in the melt, but are not recrystallized. Fusion of grains is observed in shocked adamellite. Taylor and Dence put the original crater diameter at 20 km, and believed that the structure is more deeply eroded than Clearwater Lakes. Shock metamorphism, together with the fracturing of anorthosite and adamellite on the island are supposed to indicate meteorite impact origin (Paper 18).

Currie (1971) advanced an alternative explanation in a detailed account, believing that the structure is of endogenic origin, a resurgent cryptoexplosion crater—we consider Currie's report at some length below. He published a brief initial account of Mistastin Lake, in which he stressed ellipticity (17 X 11 km) and the fact that only two of the eight similar structures in Canada were, in fact, truly circular (Currie, 1968). He also stressed the complete preservation of the volcanic apparatus—ring dikes feed a 40 m thick volcanic sequence and there is preservation of poor shatter cones pointing down to a central focus in anorthosite. Currie recorded augen granites and granodiorites, with amphibolite schlieren and concordant strips of anorthosite and mangerite as forming the country rocks. Indirect evidence of K/Ar dating indicates an age for these of 1600 ± 100 m yr. Brecciation is evident on the islands and also following an elliptical belt of igneous rock outcrops, which shows widenings like a string of beads and follows the lake shore. Between these occurrences, fracturing alone is evident. The brecciated rocks may display glass; a dull white color accompanying distortion of grain lattices; glassy or red aphanitic veinlets; and intense cleavage or powdering without microbrecciation. One or all of such modifications may be present, and they increase in intensity and number toward both the focus and the igneous rock belt around the lake shore. Monomict breccias consisting of angular, rotated country rock fragments with little or no matrix occur near the lake shore, and grade either into fractured rock or stronger crush, showing optically abnormal mineral grains. Optically anomalous plagioclase, passing into true maskelynite; heterogeneous and homogeneous vesicular glasses; and pseudotachylite veinlets characterize the breccia zone.

Currie described the igneous rocks as of four distinct types: early tuffisites; tuff breccia, including shocked fragments (suevite); massive, black or highly vesicular andesites; and younger tuff breccia containing fragments of the older volcanics, which it caps. The tuffisite and tuff breccia form a thin unconformable overlay on the basement rocks and dikes, passing gradationally up into andesite and down into basement. The breccia includes both basement and vesicular andesite fragments. Scapolite is a surprising constituent. The matrix is mainly of fine mineral or rock fragments. Shock effects are lacking. Locally the tuff-breccia is intruded by andesite.

Currie regards suevite as an ephemeral precursor to normal volcanic activity, related to high-temperature gas explosions, and derived from wall rocks or shock processes affecting fluidized systems. The volume of suevite at Mistasin Lake is negligible compared with other melt rocks. It shows anomalously high Mg, Fe content, and low alumina, and varies from sheets of almost monomict breccia of weakly shocked material at the base of the andesite sheets along the lake shore, to complex, swirling veins and lenticles of highly shocked rocks that display turbulent flow structures; more than half of it occurs as dikes that show flow structures and are commonly blind upward. It cuts across or unconformably overlies moderately to severely shocked rocks, and is unconformably overlain by andesite, which intrudes it as dikes. The matrix consists of fine, angular opaque fragments or opaque dust. The shocked fragments show isotropism of plagioclase (maskelynite). Currie finds the simple shock-wave explanation for maskelynite (Bunch et al., 1967) to be inadequate to explain the chemical fractionation that is evident in its formation. Chao and Bell (1969) found that maskelynite could be formed experimentally at only 10 kb pressure by melting, given rapid temperature rise followed by rapid quenching. Chemical explosion, Currie suggests, could produce just such an effect:

> During combustion, temperatures and pressures in a confined space rise very rapidly. If the confining walls fail as a result of explosion, the temperature drops very rapidly due to adiabatic expansion of the gaseous explosion products. Temperatures of 1500°C and pressures of 10 kb could easily be reached during a partially confined explosive combustion of hydrogen, for example. The time during which this would be maintained depends on the rapidity of burning of the gas and the nature of the walls . . . periods ranging from a fraction of a second to several tens of seconds seem reasonable. Chemical explosions due to combustion of hydrogen are a well-known feature of many volcanic eruptions (Jaggar, 1947). The close association of . . . maskelynite with rocks of volcanic appearance suggests a close association between the two. The occurrence . . . in the central core of plutonic rocks could be explained by supposing this core to have become saturated with explosive gases along fractures, and a general explosion or a series of explosions to have occurred. Formation of maskelynite by purely mechanical shock seems inadequate to explain the composition of the Mistastin

> Lake maskelynite. While temperature rises during impact, Milton and Di Carli (1963) estimated the maximum temperature in their maskelynite producing experiments not to exceed 350°C. If the temperatures reached at peak shock melting were sufficiently high to produce melting, the residual temperature would produce annealing of the glass to atmospheric pressure properties. Maskelynite appears to be a product of shock metamorphism, that is a very rapid rising of temperature and pressure. The shock producing of maskelynite at Mistastin Lake was a high temperature, moderate pressure event. Such shock is compatible with chemical explosion, but not with purely mechanical shock, such as that caused by impact (Currie, 1971)

Currie notes that whereas homogeneous glasses are common in volcanic edifices, heterogeneous or mixed glasses are rare.* Thetomorphic glasses such as maskelynite are not known from volcanic edifices, yet they are apparently represent decompression effects that could occur at quite low static pressures; and their occurrence in volcanic edifices is, therefore, theoretically possible. If the thetomorphic glass at Mistastin is due to simple shock loading, two sources of shock would have to have been involved, one at the focus and the other along the lake shore; or else there has been a complex displacement of the shock zones, of tectonic character. Both alternatives appeared unlikely to Currie, and he invoked multiplication of pressure of the type used in high-pressure devices (Jamieson, 1963), developed locally in a volcanic edifice at unusually high temperatures (1100–1200°C at 10 kb). Heterogeneous glasses could be produced this way. The generation of high temperature was attributed to circulation of hot reactive gas, undergoing exothermic chemical reactions to maintain heat supply. The cooling of this gas by adiabatic expansion connected with explosive failure of the high pressure cell provides the required quenching mechanism.

STEEN RIVER

Drilling in the Lower Cretaceous Loon River Shale in 1963 revealed an elliptical, buried structure, the Steen River structure in Alberta. This structure, discussed in Paper 19 by Winzer, measures 24 X 19.5 km and affects basement rocks that stand 762 m above their regional level. The flat muskeg terrain is devoid of any surface indication of the presence of a circular structure. There are three structural elements: a rim with a steep inward fall and gentle outer slope shown by geophysical evidence to be cut by radial faults; a rim syncline in the form of an uneven annular depression; and a central uplift. Winzer reports that a hole drilled in

*Mixed glasses are, however, quite common in banded lava and ash-flow suites in volcanic edifices.

the flank of the uplift revealed in its lower 795-ft (242-m) section abundant well-foliated gneiss, with injection veins of finely comminuted rock powder and dark greenish-grey suevite breccia with an aphanitic matrix, commonly vuggy or vesicular with pink zeolite and green chlorite filling cavities. Rounded to subrounded gneiss fragments are evident or frothy, pumice like fragments of vitrified gneiss. Comminution is more common toward the top of the section and melt toward the bottom. Both matrix and vitrified fragments show plastic deformation. The melt phase has enhanced K_2O and decreased Na_2O and SiO_2 content compared with the other rocks. The borehole terminates in dense black partly devitrified glass, which shows a chilled margin. An anomalous temperature of 55°C was encountered at this depth. Other holes drilled within the rim syncline encountered sediments of unknown age. It is not known whether these display shock effects indicative of involvement in the disturbance.

Carrigy and Short (1968) described the shock metamorphism in detail. It includes cataclastic deformations (microbreccias, rock dust), planar elements in quartz and felspar grains, and isotropization of felspar (maskelynite). Coesite has not been detected. The melt rocks resemble in their chemistry the biotite and hornblende granites of the basement. The "pitchstone" is an unusually water-rich trachyandesite, the water being locked up in the glass and in zeolites that fill amygdules. One drillhole described by Carrigy and Short penetrated Cretaceous, Mississippian, and Devonian strata and appeared to encounter a repetition of the stratigraphy, terminating in Middle Devonian beds. The section shown by these authors appears to represent a volcanic interpretation, for it includes a volcanic neck. It is different from Winzer's section. Both are probably highly interpretative, considering the obscured nature of the structures. Drilling in the ring syncline showed the Slave Point Formation to be 2000 ft (610 m) lower than predicted from adjacent well data, indicating that the syncline is a zone of subsidence. Carrigy and Short concluded that there is insufficient evidence to determine the origin of the shock.

REFERENCES

Braslau, D. 1970. Partitioning in Hypervelocity Impact Against Loose Sand Targets. *Jour. Geophys. Research* **75**:39–87.

Bunch, T. E.; M. R. Dence; and A. J. Cohen. 1967. Natural Terrestrial Maskelynite. *Am. Mineralogist* **52**:721–733.

Carrigy, M., and N. M. Short. 1968. Evidence of Shock Metamorphism in Rocks from the Steen River Structure, Alberta. In *Shock Metamorphism of Natural Materials,* B. M. French and N. M. Short, eds. Baltimore, Md.: Mono Book Corp., pp. 367–378.

Chao, E. C. T., and P. M. Bell. 1969. Annealing Characteristics of Dense Felspar Glass. *Carnegie Inst. Washington Year Book* **67**:126–130.

Currie, K. L. 1968. Mistastin Lake, Labrador, A New Canadian Crater. *Nature* **222**: 776–777.

Currie, K. L. 1971. Geology of the Resurgent Cryptoexplosion Crater at Mistastin Lake, Labrador. *Canada Geol. Survey Bull. 207,* 62pp.

Currie, K. L. 1972. Geology and Petrology of the Manicouagan Resurgent Caldera. *Canada Geol. Survey Bull. 198,* 153pp.

Gault, D. E., and E. D. Heitowit. 1963. The Partition of Energy for Hypervelocity Impact Craters Formed in Rock. *Hypervelocity Impact Symp., 6th, Cleveland, Ohio, Proc.* **2**:419–456.

Jaggar, T. A. 1947. Origin and Development of Craters. *Geol. Soc. America Mem. 21,* 508pp.

Jamieson, J. C. 1963. Possible Occurrence of Exceedingly High Pressures in Geological Processes. *Geol. Soc. America Bull.* **74**:1067–1070.

Milton, D. J., and P. S. Di Carli. 1963. Maskelynite Formation by Explosive Shock. *Science* **140**:670–671.

Robertson, P. B. 1968. The La Malbaie Structure, Quebec, A Paleozoic Meteoritic Impact Site. *Meteoritics* **4**:89–112.

17

Reprinted from *Internat. Geol. Cong., 24th, Proc.* 15:140–147 (1972)

Géologie de la Structure de Charlevoix

JEHAN RONDOT,
Canada.

RÉSUMÉ

La Structure de Charlevoix, située au bord de la fosse d'effondrement du Saint-Laurent, comprend les faciès représentatifs de la Province de Grenville, les roches de sa couverture sédimentaire et l'extrémité d'une nappe de charriage appalachienne.

Parmi les roches précambriennes, l'anorthosite de Saint-Urbain présente des caractères magmatiques évidents, schlieren de roches basiques, dyke d'anorthosite plus acide. Le bord NW du massif est rubané, et disloqué au contact en blocs à pendages divers. La série charnockitique, postérieure à la mise en place de l'anorthosite (faciès de bordure à grain fin, dyke dans l'anorthosite) présente les caractéristiques d'une suite palingénétique de haute température et pauvre en eau (fusion partielle de roches déjà métamorphiques). Elle comprend des migmatites charnockitiques avec quelques lambeaux de paragneiss et des intrusions successives, d'abord acides et syncinématiques, puis granodioritiques et dioritiques et nettement intrusives.

La transgression marine ordovicienne s'est heurtée à l'escarpement que représente le bord SE du Bouclier. Les premiers dépôts (détritiques) sont limités à la bordure. Ce n'est qu'à la fin de l'Ordovicien moyen que la mer envahit le Bouclier et y dépose des calcaires. A la même époque, il se produit des effondrements sur cette même ligne et les calcaires fraîchement déposés glissent vers le SE. Les derniers dépôts, toujours au SE de la faille d'effondrement, sont silteux et argileux. C'est à ce moment que s'avance la nappe de charriage des roches appalachiennes, comblant en partie le fossé d'effondrement.

Seul, un événement tectonique particulier permet d'observer les dépôts de couverture sur le bouclier et les dépôts de base au S de la faille d'effondrement. Cet événement est l'impact au Dévonien d'une grosse météorite (de l'ordre de 1 km de diamètre). L'impact a provoqué un métamorphisme de choc, des shatter cones et, par la force de pénétration, des surfaces spiraloïdes de glissement qui ont permis, après l'évidement d'un cratère de 28 km de diamètre, un réajustement lithostatique tendant à niveler le cratère au détriment des roches adjacentes. Il en résulte une dépression de 56 km de diamètre, un synclinal en graben annulaire, protégeant de l'érosion, les calcaires de couverture et une remontée centrale laissant apparaître la faille d'effondrement et les dépôts de base.

LA STRUCTURE DE CHARLEVOIX est un des accidents topographiques et géologiques circulaires considérés comme astroblèmes dont le nombre va croissant depuis le début du siècle. Elle est située à quelque 100 km au NE de Québec dans une zone habitée en bordure du fleuve Saint-Laurent. Son origine météoritique, suggérée par la découverte de shatter cones (Rondot, 1968a) a été confirmée par un groupe de membres de la "Meteorotical Society" à l'occasion d'une excursion de cette société (Rondot, 1968b), par l'observation de zone de métamorphisme de choc du quartz (Robertson, 1968) et par l'observation de la réduction et la transformation des courbes de thermoluminiscence du calcaire de l'extérieur vers le centre de la structure (McDougall, 1970).

Par sa taille (enceinte de 56 km de siamètre), la Structure de Charlevoix fait partie des astroblèmes les plus importants reconnus sur terre. Elle peut être comparée par exemple à celle du Ries ou à celle de Manicouagan. Cependant, sa situation privilégiée en bordure du fleuve St-Laurent a permis une élimination presque complète des retombées, laissant le socle à nu et découvrant une cuvette de 1,000 m de relief.

L'étude géologique de cette structure qui affecte à la fois les roches précambriennes du Bouclier Canadien, sa couverture paléozoïque, la faille d'effondrement le long du Saint-Laurent et les roches appalachiennes charriées de l'Ile aux Coudres permet de reconnaître les modifications survenues au cratère primitif après son excavation.

ASPECT GÉNÉRAL

La cuvette semi-circulaire d'une altitude moyenne de 400 m entaille le bouclier précambrien représenté dans cette partie du Québec par un plateau arasé d'environ mille m d'altitude, déchiqueté le long du fleuve Saint-Laurent (Fig. 1).

Dans le détail, cette cuvette se compose de plusieurs éléments topographiques concentriques:

1° l'enceinte qui en forme le bord est bien délimitée par un escarpement d'une hauteur de 200 à 500 m, formant un demi-cercle interrompu seulement au N par la vallée de la rivière Malbaie qui a creusé son lit dans une granodiorite charnockitique plus tendre que les autres roches précambriennes.

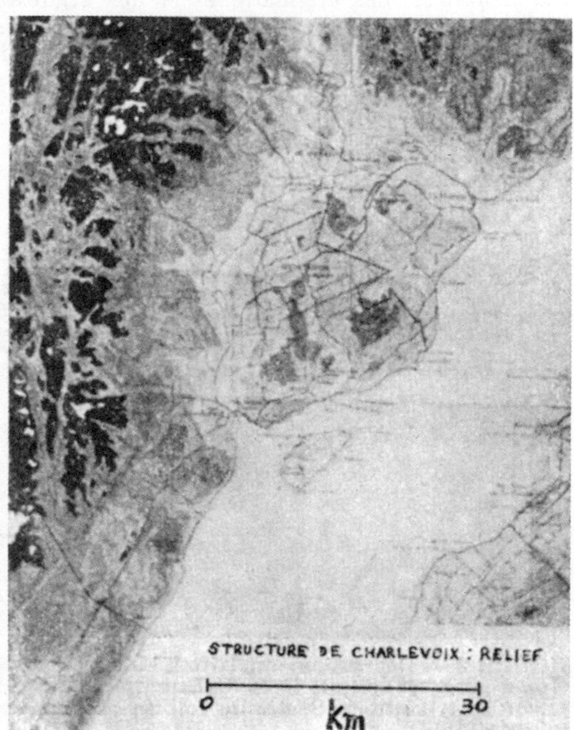

FIGURE 1 — Photographie d'une carte hypsométrique dont les différentes teintes s'étagent de 200 m en 200 m; les parties les plus claires à l'W et au NW représentent des sommets dépassant 1,000 m d'altitude.

2ᵉ une dépression semi-circulaire d'un diamètre de 36 km est occupée par les rivières du Gouffre (Fig. 2) et Malbaie et par une suite de lacs. On y observe quelques affleurements de calcaire ordovicien généralement très faillé et localement plissé. En examinant plus attentivement cette dépression, on remarque qu'elle est plutôt polygonale que circulaire.

3ᵉ la partie interne de la structure (en deça de la dépression) est un plateau peu élevé d'où émerge une couronne de collines séparées les unes des autres par des vallées radiales. Plus près du centre, une deuxième dépression, subcirculaire, d'une diamètre moyen de 7 km et dans laquelle les shatter cones sont les mieux développés, fait ressortir le pointement central (Fig. 3) dont le sommet au centre exact de la structure atteint 768 m d'altitude et où l'on observe des figures de métamorphisme de choc dans les minéraux.

PRÉCAMBRIEN

Le Précambrien du Bouclier Canadien est particulièrement intéressant à la hauteur de la Structure de Charlevoix parce qu'il fait partie d'une zone (Province de Grenville) où des séries métasédimentaires et des intrusions d'âges précambriens très divers (Wynne-Edwards, 1969) ont été repris par un phénomène tectonique attribué par Schenk (1971) au télescopage de l'Afrique et de l'Amérique du Nord. Aux transformations métamorphiques des minéraux, en particulier des micas dont la datation au K/Ar est de 955 M.A., s'ajoute un système de fracturation NNE de la limite S de cette zone datée de 900 M.A. (Philpotts et Miller, 1963).

La Structure de Charlevoix affecte deux ensembles métamorphiques différents dont la limite est sensiblement NNE; l'un à l'E, groupe des roches métasédimentaires, des migmatiques et des granite à amphibole et mica; l'autre, à l'W, groupe des lambeaux de métasédiments disloqués par des intrusions anorthositiques et repris par une série palingénétique charnockitique.

FIGURE 2 — Dépression annulaire à la hauteur de St-Urbain. Cette vue, prise des anciennes mines d'illménite de St-Urbain, montre les divers aspects de la morphologie du massif anorthositique. Seules les collines de l'arrière plan à droite ne font pas partie du massif.

FIGURE 3 — Vue du massif du Mont des Eboulements prise de la route qui passe au S du massif. La colline en pain de sucre à droite possède à son sommet et à sa base, des lambeaux d'anorthosite arrachés au socle.

Deux petites masses anorthositiques (de 5 km de long et 1 km de large au maximum) affleurent l'une au centre l'autre à l'extrémité SW de la zone interne de la structure. Une troisième masse beaucoup plus importante (30 km x 10 km) est coupée par la dépression semi-circulaire de la structure près de Saint-Urbain (Fig. 2, 4). D'origine magmatique profonde, ce massif, composé de schlieren basiques et d'intrusions successives de plus en plus acides jusqu'à des dykes d'anorthosite pegmatitique, a déformé les structures métasédimentaires préexistantes. Cette montée par gravité comme Martignole et Schryver (1970) l'ont suggéré pour le massif de Morin n'a pas été sans déformation des bords de l'anorthosite et l'on observe par endroits près du contact avec les autres roches une brèche à blocs et ciment anorthositique.

Les roches du complexe charnockitique se composent d'une série gneissique (migmatite charnockitique) avec lambeaux de métasédiments transformés, et d'une suite palingénétique composée d'intrusions syncinématiques acides (charnockite, monzonite quartzitique, syénite et monzonite à pyroxène) puis d'intrusions discordantes et plus basiques (granodiorite) et enfin de petites intrusions et d'un dyke dioritique à grain fin. Cette succession est à l'inverse d'une succession magmatique normale mais correspond bien à une suite palingénétique où l'élévation de température permet la fusion partielle d'un composé de plus en plus basique (Winkler, 1965).

COUVERTURE ORDOVICIENNE

La pénéplanation du bouclier précambrien était achevée à l'Ordovicien ce qui a permis une transgression très rapide sur presque la totalité de la plateforme laurentienne à l'Ordovicien moyen. Dans la région de Charlevoix cependant la limite SE du Bouclier représentait déjà une discontinuité topographique et les formations les plus anciennes (début de l'Ordovicien moyen) de même que les dépôts détritiques de la base de la série n'apparaissent qu'au SE de cette ligne.

En de nombreux endroits du Bouclier les dépôts de calcaire préservés par des accidents particuliers (graben, astroblèmes) témoignent de l'extension de la transgression marine. Dans Charlevoix, cette couverture devait être de l'ordre de 200 m.

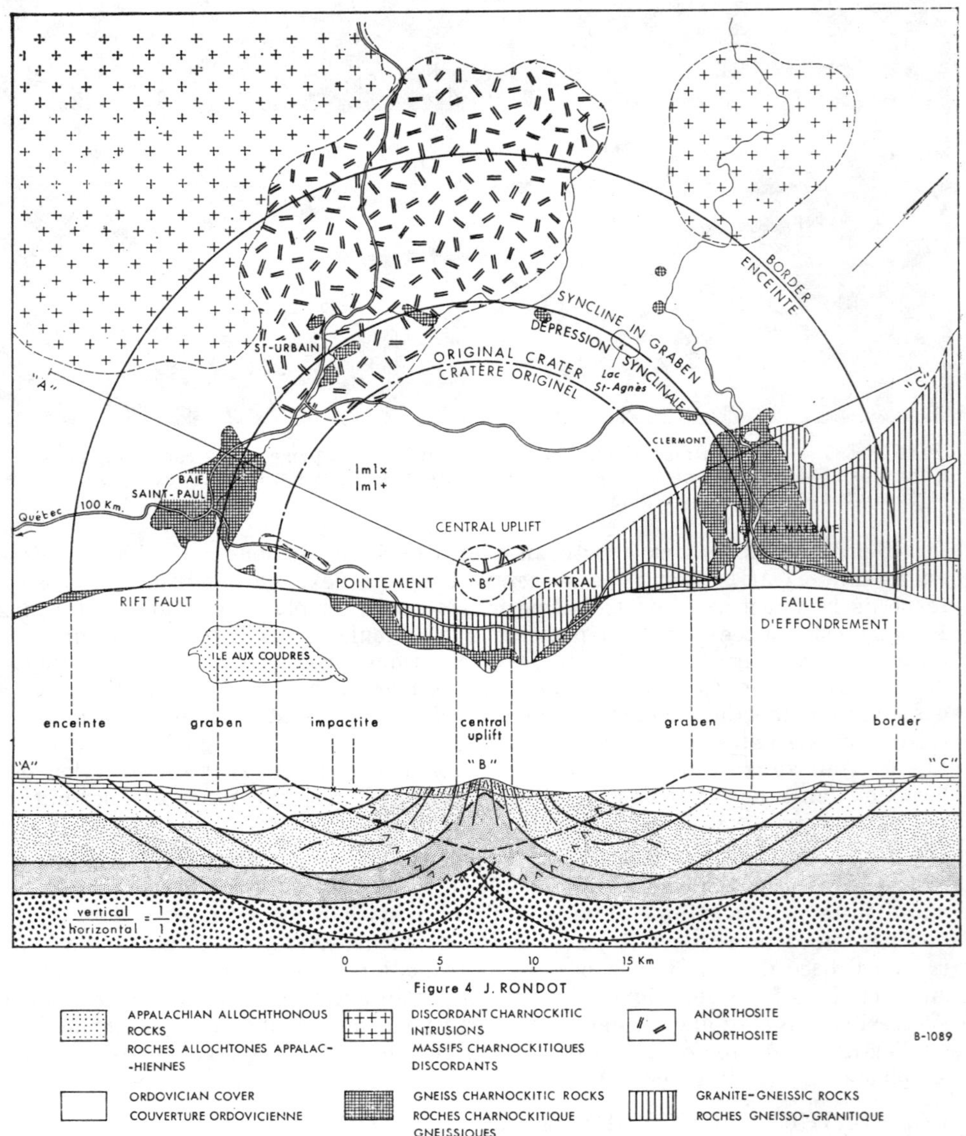

FIGURE 4 — Carte géologique simplifiée de la structure de Charlevoix et coupe avant érosion basée sur la disposition et l'attitude du contact Précambrien-Paléozoïque. Dans la coupe la ligne discontinue représente le cratère originel sans les retombées. Les lignes épaisses séparent des niveaux hypothétiques primitivement horizontaux du Précambrien pour faire ressortir les glissements et l'épaississement de ces niveaux vers le centre de la structure. Les surfaces de glissements sont schématisées par des lignes continues. Les V représentent la zone de développement maximum des "shatter cones" et les traits verticaux indiquent la zone de métamorphisme de choc maximum.

FAILLE D'EFFONDREMENT ET NAPPES DE CHARRIAGE APPALACHIENNES

Vers la fin de l'Ordovicien moyen, les premières manifestations tectoniques de l'orogenèse taconique se traduisirent par le glissement des calcaires et schistes accumulés sur la pente que constituait à ce moment la bordure du Bouclier (Rondot, 1972). A partir de ce moment, la partie SE s'effondre et la sédimentation se dissocie de celle qui va se poursuivre sur la plate-forme laurentienne à l'Ordovicien supérieur (Riva, 1969).

La faille d'effondrement de direction NE qui limite actuellement le Bouclier a un rejet vertical qui peut atteindre 2 km. Elle est visible le long de la côte en dehors de la structure et sur 20 km dans la partie interne de la structure.

Les dépôts qui se sont accumulés au SE de cette faille: schiste, microgrès, grès et conglomérats, de type turbidite contiennent de très nombreux fragments qui ne peuvent provenir que de la destruction de roches appalachiennes. L'entraînement des dépôts de remplissage de ce bassin se faisaient vers le SW comme en témoignent les nombreuses flûtes à la base des bancs de grès (Rondot, 1972).

Le charriage des roches appalachiennes de l'Ile-aux-Coudres qui viennent buter sur l'escarpement de faille termine l'époque orogénique taconique. Il ne semble pas qu'il y ait eu dans la région d'autres dépôts ou mouvements tectoniques importants avant l'impact au Devonien (350 M.A.) d'une météorite de l'ordre d'un km de diamètre.

LES EFFETS DE L'IMPACT SUR LES ROCHES

Du point de vue géologique, les effets de l'impact sont le métamorphisme de choc, le réajustement lithostatique, la sédimentation des roches fondues et des retombées et l'altération hydrothermale.

Les effets du métamorphisme de choc sur les roches du massif du Mont des Eboulements se reflètent dans leur aspect général: le quartz en particulier est devenu laiteux tandis que le feldspath potassique a pris une teinte rouge-brique. Les figures planaires des quartz et des feldspaths et leur isotropisation partielle sont les indices les plus évidents. La fracturation particulière due au passage de l'onde de choc diminue rapidement du centre vers l'extérieur mais ce n'est qu'entre 3 et 10 km du centre de la structure que les shatter cones sont bien formés.

L'étude du réajustement lithostatique qui a suivi l'évidemment du cratère est particulièrement intéressante dans les roches cristallines grâce à la présence, lors de l'impact, d'une couverture paléozoïque horizontale. La présence de cette couverture nous permet d'estimer le diamètre du cratère originel (28 km) et d'apprécier le déplacement vertical dû à l'enfouissement des formations dans la dépression semi-circulaire (environ 1,300 m) (Rondot, 1970). De plus, grâce à l'effondrement de la vallée du Saint-Laurent, le contact précambrien-paleozoïque est reporté à un niveau plus bas. La disposition actuelle de ce contact primitivement subhorizontal, sauf à quelques centaines de m de part et d'autre de la faille d'effondrement, nous indique que la remontée centrale s'est effectuée par le déplacement de blocs les uns par rapport aux autres. Il est à noter que le bloc central au Mont des Eboulements ne montre que des cassures mineures sur une centaine de m d'affleurement. On peut estimer que ce bloc provenant du fond du cratère originel a été remonté verticalement de près de 5 km.

Les pseudotachylites remplissent les cassures mineures du massif du Mont des Eboulements, tandis qu'à 3 km du centre et jusqu'au niveau de l'enceinte les blocs sont séparés par une mylonite particulière nommée provisoirement dyke de brèche à cause de sa disposition filonienne et de sa constitution bréchique particulière. Ces dykes de brèches, qui ont servi de lubrifiant aux déplacements relatifs des blocs, ont des épaisseurs variables. Certains atteignent 1 m et plus. Ils sont caractérisés par la présence de fragments de brèche et de mylonite de teintes diverses.

Bien qu'environ 1,000 km^3 de roches aient été excavées au moment de l'impact, le tapis actuel des retombées est peu épais et consiste en blocs remaniés à la base des dépôts glaciaires et en deux petits affleurements à 9 et 10 km à l'W du centre de la structure (Rondot, 1971).

Un de ces affleurements d'impactite (roche allochtone fondue par la pression et la chaleur dues à l'impact) montre sur 2 m de bas en haut à partir du plancher précambrien la succession suivante: 1° brèche avec nodules de roches cryptocristallines; 2° niveau cryptocristallin massif puis nodulaire et enfin à petites alvéoles circulaires de minéraux secondaires; 3° brèche soudée (impact-ignimbrite) surmontant les dépôts des derniers nodules.

Les brèches allochtones, remaniées, sont soit noires et finement cristallines, cryptocristallines ou aphanitiques et fluidales, soit de teintes pâles et généralement préhnitisées.

L'altération hydrothermale a en effet fortement affecté la partie centrale de la structure, les dykes de brèche et leur voisinage immédiat et probablement aussi certaines couches des retombées. Les minéraux formés sont principalement les zéolites, chabazite, laumonite, stilbite et surtout la préhnite.

CONCLUSION

A cause de sa situation particulière à cheval sur deux domaines géologiques différents, l'astroblème à pointement central de Charlevoix nous révèle que le cratère originel dû à l'impact n'a que la moitié du diamètre maximum de la structure et que le réajustement s'est effectué par le déplacement relatif de blocs plutôt que par plissement.

Comment un tel réajustement a-t-il pu s'effectuer? Lorsqu'une force est appliquée en un point du sol comme sous les barrages par exemple il se produit des plans de glissements de forme spiraloïde (Fig. 4, coupe) la formation de ces plans qui ont pu servir pour le réajustement ont nécessité une force d'une durée relativement longue ce qui peut être le cas lors de l'impact d'une grosse météorite à vitesse lente.

Cependant, si la plupart des grands astroblèmes sont de ce type (Rondot, 1970) et montrent un réajustement proportionnel à la taille de la structure, d'autres ont par contre gardé sensiblement la forme originelle en cuvette (Ries, Rochechouart, Sudbury, Wanapitei, etc.) ce type de cratère Type A présente des phénomènes superficiels importants (déplacements d'énormes blocs dans le Ries) et une forte proportion de roches fondues, mais la fracturation du socle diminue rapidement avec la distance. Les astroblèmes à réajustement, Type B, présentent au contraire de grands bouleversements dans le socle.

Gault et Heitowit (1963) et Braslau (1970) ont constaté lors de leurs expériences que la répartition de l'énergie cinétique lors d'un impact est distribuée pour moitié environ en énergie mécanique d'excavation, pour ¼ en compaction ou fragmentation du substratum et le reste en chaleur, mais que, d'après Gault (communication personnelle), la quantité de chaleur produite et donc la proportion de matériel fondu augmentent rapidement avec la vitesse de la météorite à l'impact, au détriment de la fragmentation. C'est bien ce que l'on constate entre les deux types de cratères. L'énergie de l'impact de l'un, Type A, est distribuée superficiellement, fusion et éjection, et n'affecte que peu le socle tandis que, dans le Type B, la fragmentation et la fracturation du sol sont suffisamment importantes pour provoquer un réajustement tendant à niveler le cratère, tandis que le matériel fondu est relativement moins important.

Je remercie Michael Dence et Peter Millmann qui m'ont incité à écrire cet article, publié avec la permission du Ministre des Richesses naturelles, Québec.

OUVRAGES CITÉS

Braslau, David, 1970. Partitioning in hypervelocity impact against loose sand targets. J. Geophys. Res., 75, p. 39-87.

Gault, D. E., et Heitowit, E. D., 1963. The partition of energy for hypervelocity impact craters formed in rock. Proc. 6th Hypervelocity Impact Symp., Cleveland, Ohio 2, p. 419-456.

Martignole, J., et Schrijver, K., 1970. Tectonic setting and evolution of the Morin Anorthosite, Grenville province, Quebec. Bull. Geol. Soc. Finland 42, p. 165-209.

McDougall, D. J., 1970. Natural thermoluminescence of calcareous rocks from the Charlevoix (Malbaie) structure. Meteoritics, 5, p. 75-83.

Philpotts, A. R., et Miller, J. A., 1963. A Precambrian glass from St. Alexis-des-Monts, Quebec. Geol. Mag., 100, No. 4, p. 337-344.

Riva, John, 1969. Middle and Upper Ordovician graptolite faunas of St. Lawrence Lowlands of Quebec, and of Anticosti Island, North Atlantic. Geology and Continental Drift. Mem. 12, Am. Assoc. Pet. Geol.

Robertson, P. B., 1968. La Malbaie structure, Québec, a Paleozoic meteorite impact site. Meteoritics, 4, p. 89-112.

Rondot, Jehan, 1968a. Nouvel impact météoritique fossile? La structure semi-circulaire de Charlevoix. Can. J. Earth Sci., 5, p. 1305-1317.

———, 1968b. Excursion géologique: Structure de Charlevoix. Contr. Congr. Meteorit. Soc. Minist. Rich. nat. Qué., Québec, Canada.

———, 1970. La structure de Charlevoix comparée à d'autres impacts météoritiques. Can. J. Earth Sci. 7, p. 1194-1202.

———, 1971. Impactite of the Charlevoix Structure, Quebec, Canada. J. Geophys. Res., 76, p. 5414.

———, 1972. (En Préparation). La transgression ordovicienne dans Charlevoix, Québec.

Schenk, Paul E., 1971. Southeastern Atlantic Canada, Northwestern Africa, and continental drift. Can. J. Earth Sci., 8, p. 1218.

Winkler, H. G. F., 1965. Petrogenesis of metamorphic rocks, Springer-Verlag.

Wynne-Edwards, H. R., 1969. Tectonic overprinting in the Grenville province, Southwestern Quebec. Geol. Assoc. Can. S.P. 5, p. 163-182.

18

Reprinted from *Canadian Jour. Earth Sci.* **6**:39–45 (1969)

A probable meteorite origin for Mistastin Lake, Labrador

F. C. TAYLOR
Geological Survey of Canada, Ottawa, Canada

AND

M. R. DENCE
Dominion Observatory, Ottawa, Canada

Received October 18, 1968
Accepted for publication November 8, 1968

Mistastin Lake, Labrador, latitude 55°53′ N, longitude 63°18′ W, is a roughly circular lake with a large central island lying in a marked depression. A meteorite impact origin is indicated by shock metamorphism features and fracturing of anorthosite and adamellite on the central island and of inclusions in flat-lying igneous rocks on the western shore. The crater had an original diameter of about 20 km and an age, by whole-rock K–Ar determination, of 202 ± 25 m.y.

Introduction

An exposure of flat-lying igneous rocks, at latitude 55°51′ N, longitude 63°26′ W in northern Labrador was brought to the attention of the Geological Survey of Canada by geologists of the British Newfoundland Exploration Limited (Brinex) as a volcanic occurrence. An examination of a topographic map of the area showed the igneous rocks lie near the shore of Mistastin Lake (Fig. 1), a roughly circular lake 13 to 16 km (8 to 10 miles) across, containing a large arcuate island and a few smaller islands near its center (latitude 55°53′ N, longitude 63 °18′ W). The similarity between the physiographic character of Mistastin Lake coupled with the nearby presence of igneous rocks, with lakes considered to be meteorite craters, such as Clearwater Lake, suggested that Lake Mistastin may also be of impact origin. Consequently a short visit was made by Taylor to the igneous rock outcrop and the islands in the center of the lake and a few representative samples were collected. A brief air examination of the immediate area surrounding the lake failed to show any other outcrops similar to the flat-lying igneous rock exposure.

Although a raised rim, a characteristic of young impact craters, is not clearly evident, Mistastin Lake partly fills a distinct topographic basin enclosed by a discontinuous ring of low hills with a diameter of 26 km (16 miles) and an

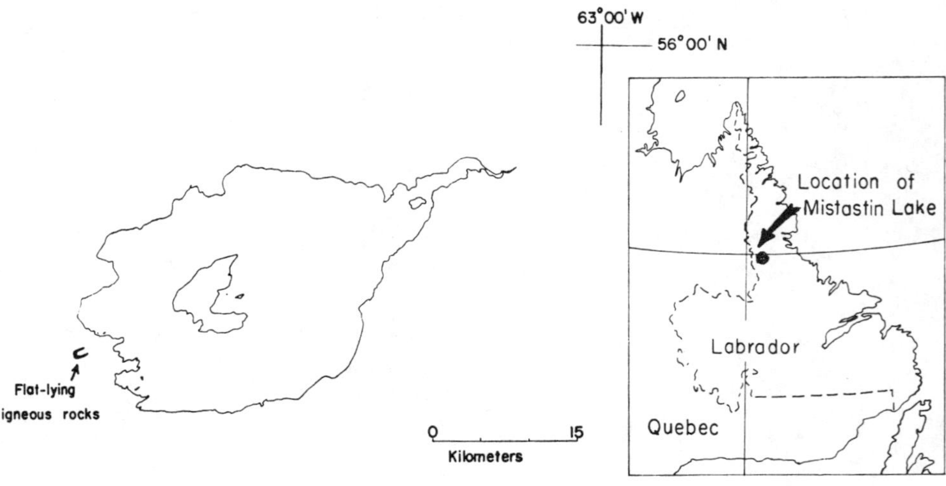

FIG. 1. Sketch map of Mistastin Lake, Labrador, showing the location of the outcrop of igneous rocks on the western shore.

elevation of up to 670 m (2200 ft) above sea level. The lake surface is at 338 m (1110 ft) above sea level, about 150 m lower than the mean regional elevation. The only prominences within the basin are the hill of flat-lying igneous rocks and the central island, which rise 250 m and 130 m respectively above the lake level.

Mistastin Lake lies in a geologically little known part of the Canadian Shield. Brinex company geological maps show the general bedrock in the area to consist dominantly of granite (adamellite) with small bodies of anorthosite erratically distributed throughout. A K–Ar age of 1340 ± 50 m.y. was obtained on biotite from adamellite collected 15 km (9 miles) south of Mistastin Lake (C.S. Stockwell, *in* Wanless *et al.* 1965); this date has been interpreted as the time of intrusion of the adamellite. The islands in the lake are composed primarily of adamellite with small amounts of anorthosite. Both are massive, coarse-grained rocks similar to those mapped by Brinex geologists, and are referred to here as basement or plutonic rocks to distinguish them from the flat-lying igneous rocks exposed on the lake shore. Contact relationships between the plutonic rocks on the islands are not known, but both rock types are in part closely fractured, although large-scale brecciation is not present.

Flat-lying igneous rocks are exposed over an area about 500 m^2 on a butte-like hill about 1¼ km southwest of Mistastin Lake (Fig. 1). Outcrop is present on all sides of the hill except the north-northeast, which coincides with the downstream direction of glacial advance in the region. A few small scattered outcrops are also present on the top of the hill. The top of the outcrop area is about 180 m above the surface of Lake Mistastin. On the western side of the outcrop the rock is columnar-jointed from about 30 m above the base of the outcrop. Although from a distance the dividing line between the columnar rock and that lying beneath appears to be a sharp break, when examined in detail no well-defined break exists. On the southern side of the hill about 60 m of columnar rock is present and forms the entire outcrop.

Adamellite forms the country rock in the immediate vicinity of the igneous rock outcrop, and is exposed on the north and west sides and on top of the hill. On the north side large adamellite blocks are nearly surrounded by igneous rock and may be detached from the underlying country rock. A similar relationship on the west side shows a block of adamellite about 5 m wide and 12 m long surrounded by the younger igneous rock. At the top of the hill dikelets of aphanitic igneous rock apparently fill fractures in the adamellite. Whether the adamellite at the top of the hill is a loose block or part of the country rock is not known.

The flat-lying igneous rock is chiefly medium gray, with local medium dark gray and light olive-gray shades. Texturally it is quite diverse and ranges from a rarely aphanitic to a fine- to medium-grained massive rock, characteristic of the columnar parts of the outcrop, to a breccia with fragments up to 10 cm in diameter but chiefly less than 1 cm. The fragments are angular to subrounded. Some parts are porphyritic with plagioclase laths up to 5 mm in a fine-grained matrix. Locally the rock is vesicular and in some places the vesicles, which measure up to 3 cm, are wholly or partly filled with stilbite.

Petrography

Twelve samples, including plutonic adamellite and anorthosite from the central island, adamellite from the hill on the western shore of the lake, and several phases of the flat-lying igneous rocks have been examined in thin section. The rocks are generally fresh and textures are clearly preserved. The adamellites and anorthosite have compositions and fabrics distinguished by: (a) strong, irregular fracturing affecting all minerals; (b) development of several sets of fine planar features in quartz and feldspar; (c) in the case of the central island samples, a slight hydrothermal alteration; (d) in the case of the adamellite near the igneous rocks, the presence of injected glassy veinlets and general thermal metamorphism of the original minerals. The igneous rocks comprise extremely fine grained to glassy varieties, with inclusions of the plutonic rocks and mineral fragments, many of which are strongly deformed or recrystallized, and coarser-grained rocks in which few inclusions can be recognized, and evidence of recrystallization is general. Universal-stage measurements have been made on samples of each of the main rock types, using procedures outlined by Robertson *et al.* (1968) for the measurement of planar features and the Turner-Slemmons method (Slemmons 1962) for the determination of feldspar optical parameters.

The results recorded below show that the rocks of the central island have undergone weak to moderate shock metamorphism, that the adamellites from the shoreline have been weakly shocked before being thermally metamorphosed by the igneous rocks, and that unrecrystallized inclusions in the latter have experienced various grades of shock metamorphism before incorporation into the melt.

Basement Rocks

The anorthosites of the central island are coarse-grained, granular rocks with andesine ($An_{43\pm2}$) showing albite, pericline, and carlsbad–albite twins as the predominant mineral. The dark minerals (2 to 10% of the mode) include distinctly pleochroic hypersthene, clinopyroxene, iron ore, apatite, and in one sample biotite intergrown with a little muscovite. Small crystals of quartz are present in amounts up to 2% of the mode. All the anorthosite samples show intense fracturing. The fractures have curved or irregular traces, cut grain boundaries in many instances, and are spaced from a few microns to rarely a few hundreds of microns apart. Lateral displacements of up to 20 microns are seen along some of the fractures, but in most cases the fractures are tensional and have acted as channelways for later alteration.

Planar features are more strongly developed in the anorthosites than in the other basement rocks. All quartz grains have normal optics, but show strong development of several planar sets, an example being shown in Fig. 2. The planes are sharp and straight, with strict parallelism between members of the same set. The more prominent planes, many of which are accentuated by fine decorations or secondary mineralization are 5 to 20 microns apart, but finer, more closely spaced planes are visible down to the resolving power of the microscope. Three sets are visible in Fig. 2. Universal stage measurements show, with the usual ambiguity as to polarity, that the two conspicuous sets correspond to two members of the form ω $\{10\bar{1}3\}$. Two other ω sets of opposite polarity are accessible at high tilt angles, while the weak set visible in Fig. 2 is of the form ω $\{10\bar{1}2\}$. The appearance and orientation of the planar sets is in complete agreement with deformation features recorded from other Canadian impact craters and identified as due to shock metamorphism (Robertson *et al.* 1968). In the scheme adopted for the other craters, the presence of weak features of the π form identifies the deformation of Fig. 2 as a low-grade member of Type D. Of a total of six grains measured in the thin section, four are of Type D and two of Type C, with an average of 4.5 planar sets per grain, of which 80% are of the ω form.

The plagioclase ($An_{43\pm2}$) grains are less conspicuously deformed. Their optics are normal ($2V_z = 80° \pm 2$) but they also show local development of one to three sets of planar features, of which an example is given in Fig. 3. The plagioclase grain shows normal polysynthetic albite twin lamellae 10 to 30 microns wide. Within one twin set are developed fine, planar deformation features up to a few microns wide and of much lower refractive index than the rest of the grain. The deformation set seen in the photomicrograph is oriented close to the basal plane (001), while a second weaker set is close to the form (012). In other grains planar sets near (150) or (010) are seen. Where planar features are strongly developed lower birefringence and disturbances of extinction position of up to 10° are also observed. These results are in accord with measurements on shocked plagioclase from other Canadian craters (Robertson *et al.* 1968) and the detailed study by Stöffler (1967) of shocked feldspars from the Ries crater, Germany.

Except for some kinking of biotite grains and the general fracturing mentioned earlier, the dark minerals are unaffected by shock. Altogether the deformation recorded from the anorthosite samples corresponds to low zone (v) shock metamorphism in the nomenclature of Dence (1968). Subsequent to the deformation a slight to moderate hydrothermal alteration has affected feldspars and dark minerals, mainly at grain boundaries, fractures and regions of strong deformation, precipitating iron oxides, chlorite, white mica, and other fine-grained secondary minerals.

The adamellites of the central island are coarse-grained, allotriomorphic granular rocks in which the quartz and feldspar grains measure as much as 1 cm long. An approximate mode is quartz 20%, alkali feldspar (a finely perthitic microcline with $2V_x = 68° \pm 2$) 20%, plagioclase ($An_{33\pm2}$) 35%, clinopyroxene intergrown with bronzite in lamellae averaging 50 microns wide 15%, green hornblende mantling the pyroxenes 5%, and iron ore, apatite, biotite, and

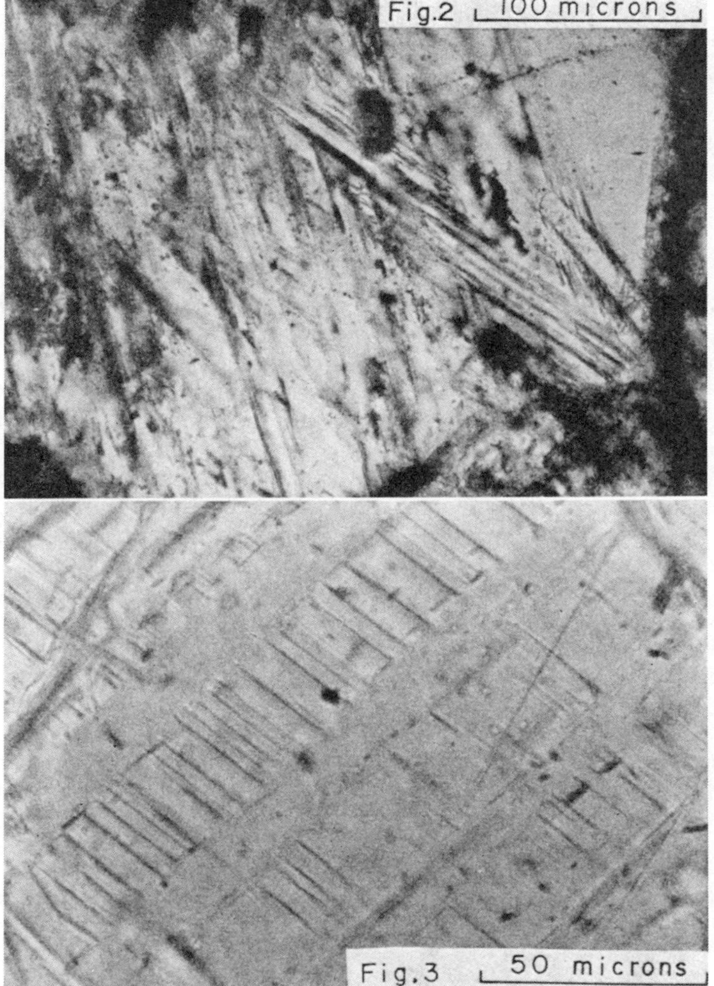

FIG. 2. Photomicrograph of part of a quartz grain in anorthosite from the central island. Two sets of planar features of the ω form and a weak set of the π form are visible.

FIG. 3. Photomicrograph of andesine from the same sample as Fig. 2, showing planar features of low refractive index approximately parallel to (001) confined to one set of original albite twin lamellae.

sphene 5%. The rock is fractured in similar degree to the anorthosite but the shock deformation is considerably weaker. Of ten quartz grains examined, only half have planar features, most of which are confined to the margins of the large grains. The planar sets developed are parallel to the base, *c*(0001), or to the ω form, the deformation corresponding to Types A or B from other craters. As plagioclase is devoid of planar sets and the mafic minerals are merely fractured, the shock grade corresponds to high zone (ii). The difference between this grade and the higher grade of shock metamorphism seen in the anorthosites suggests faulting between the two rock types late in the structural history of the crater. Further investigation is required.

The adamellites of the western shore closely resemble those of the central island in composition and fabric, in degree of fracturing, and in grade of shock metamorphism, but are cut by veinlets of glass. Shock deformation has resulted in the formation of one or two sets of planar features in at least parts of most quartz grains, which may thus be assigned to Types A or B shock deformation. No planar features are seen in the feldspars. By comparison with the central island adamellites, as well as many examples from other craters, it would be expected that the feldspars would show normal optics. However, this is not the case. Both plagioclase and alkali feldspar have strongly dispersed extinction and show changes in optic orientation and optic axial angle indicating a disordered, "high-temperature" state. In particular the alkali feldspars are sensibly uniaxial ($2V_x < 10°$) and the plagioclase ($An_{33\pm2}$) has $2V_z$ as high as 104°. This disordering is clearly the result of thermal metamorphism due to the adjacent igneous rocks. The mafic minerals are even more drastically affected. Original iron ore, apatite, sphene, and to some extent clinopyroxene are relatively unaffected except where in direct contact with the glassy veinlets, but primary orthopyroxene and amphibole have completely recrystallized to pseudomorphous aggregates of fine-grained pyroxene, iron ore, and glass. The veinlets of glass occur in cracks and along grain boundaries. Several varieties of glass are seen, a brown glass where in contact with feldspar, and colorless or yellow glass where quartz or mafic minerals are adjacent. Much of the glass is clear, but some recrystallization to small tabular feldspars and trichites of indeterminate composition has taken place. For the most part the glass is probably the product of local melting at grain boundaries, but some injection of melt from the nearby igneous rocks may also have occurred.

Flat-Lying Igneous Rocks

The medium to dark gray, aphanitic igneous rocks have numerous inclusions of country rock and mineral fragments ranging in size from tens of microns across to boulders held in a matrix averaging a few microns in grain size. Two textural, and probably compositional varieties are observed. One is illustrated in Figs. 4 and 5. Inclusions of quartz, feldspar, mafic minerals, and glass are enclosed in an equigranular matrix of feldspar microlites, pyroxene, and iron ore. A striking feature is the development of reaction coronas around quartz inclusions, such as the triangular inclusion near the center of Fig. 4 and the large inclusion shown in more detail in Fig. 5. The coronas consist of pale green clinopyroxene adjacent to the quartz grain, mantled in turn by clear brown glass. The glass is isotropic and fresh. Curved trichites and a few tabular grains of feldspar have crystallized from the glass.

The quartz inclusions generally have a cloudy appearance and an irregular, mosaic extinction pattern. Under high-power magnification, closely spaced planar features in several sets are visible, an example being shown in Fig. 5. Altogether, nine sets are accessible to measurement in this grain, four of the sets being of the ω and three of the π forms, with two sets inclined at high angles to the base. Similar results, averaging five or more sets per grain, have been obtained from other quartz inclusions, indicating that most have been shocked to Type D grade deformation. The few mafic inclusions and the abundant feldspar inclusions lack distinct reaction coronas with the matrix, but commonly have a mottled or sieve texture due to partial melting, or have recrystallized to fine-grained aggregates. Distinct evidence of shock is lacking in these grains, but many of the feldspar fragments have highly contorted outlines indicating vitrification and distortion before recrystallization to an aggregate of fine, felty microlites.

Also seen in Fig. 4 are several round patches of brown glass up to a millimeter across, containing numerous trichites and some tabular crystallites of feldspar. Like the glass reaction rims around quartz they are isotropic, but lack cores of granular pyroxene and remnants of shocked quartz. Other glassy patches are of irregular shape and in some cases are associated with spherical vesicles, which are partly or completely filled with green chloritic material. Further investigation is needed to determine the composition of the glasses and whether they were formed by shock fusion or by reaction of highly shocked fragments with the melt.

The second variety of the aphanitic igneous rocks is light gray, and has numerous small fragments of quartz and feldspar held in a subtrachytic matrix of feldspar microlites, pyroxene, iron ore, and a little brown mica. Except for a few quartz inclusions with planar features, most of which are Type B, evidence of shock deformation is lacking. Feldspar inclusions are slightly corroded around the margin and, in some cases, have high-temperature optics. Quartz fragments are overgrown by quartz in optical continuity and small vesicles are lined with quartz and clay minerals. Thus the rock is distinctly more siliceous than the first variety and presumably has drawn

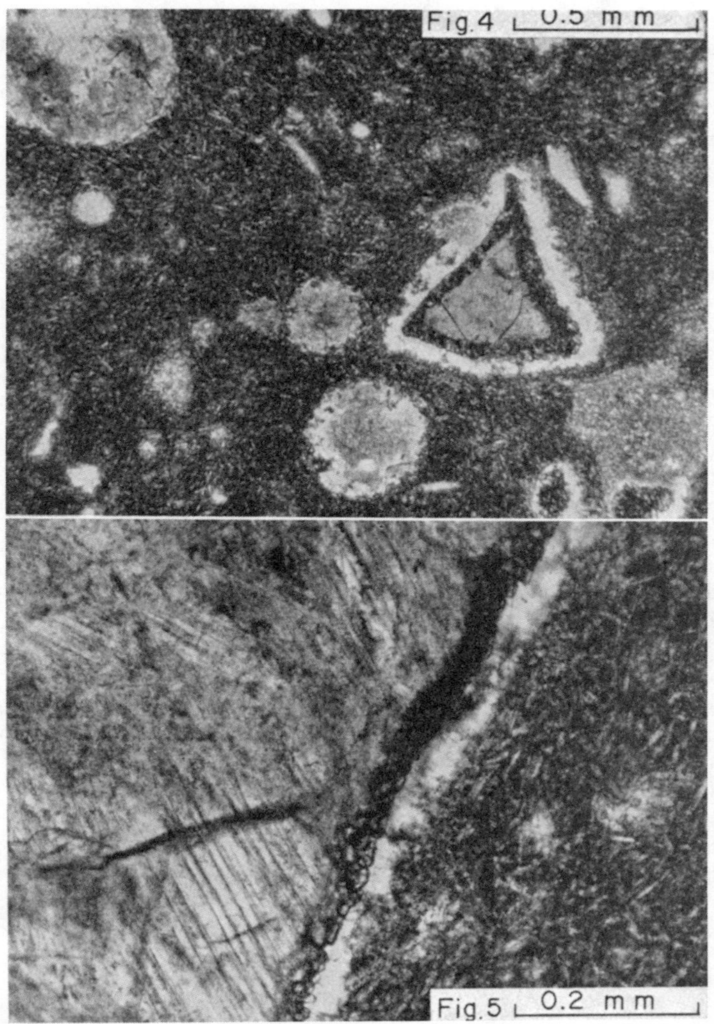

FIG. 4. Photomicrograph of aphanitic igneous rock with numerous inclusions of shocked quartz and feldspar and globules of glass (center and top left). Note reaction rim of pyroxene and clear glass around the triangular quartz grain in the center.

FIG. 5. Photomicrograph of part of a shocked quartz grain of Type D with a similar reaction rim to that shown in Fig. 4. The prominent planar sets are of the ω form.

most of its inclusions from a less strongly shocked region.

The coarser-grained igneous rocks are more uniform in appearance. Tabular feldspar grains as microphenocrysts 0.5 to 5 mm across, and some larger inclusions are held in a fine-grained, granular matrix of zoned plagioclase, iron ore, clinopyroxene, acicular apatite, and orange-brown mesostasis. The small phenocrysts of plagioclase are of subidiomorphic habit. Their mottled interiors are due to irregular small inclusions of andesine, iron ore, and pyroxene. The enclosing plagioclase is normally zoned from calcic labradorite centers to margins of sodic andesine or oligoclase. In a few specimens thin discontinuous rims of alkali feldspar are present. In addition there are a few partly corroded inclusions of feldspar, and larger inclusions of vesicular, recrystallized feldspar, with vesicles filled with stilbite. The presence of undigested fragments of country rock and the overall irregularity of grainsize and mineral distribution is compatible with these rocks being more slowly cooled equivalents of the aphanatic rocks, starting with similar inhomogeneities. The feldspar phenocrysts and some pyroxenes have grown poikilitically and do not represent an earlier crystallized fraction.

Discussion

An origin by meteorite impact for the Mistastin Lake crater, suggested by its topographic and structural similarity to other Canadian craters of meteorite origin, is confirmed by the presence of shock metamorphism features. The closest comparison is with the Clearwater Lake craters (Dence 1965, 1968), which have a similar distribution of igneous rocks with shocked inclusions around central mounds of moderately shocked basement rocks. Of the three craters Mistastin Lake is the most deeply eroded. Its original diameter is estimated to have been about 20 km, compared with about 15 km for the East Clearwater and 32 km for the West Clearwater craters. The igneous rocks at the three craters have many textural features in common and are considered to be the product of shock melting of the country rocks with admixtures of fragments of less strongly shocked materials. The variations in texture and composition in the Mistastin Lake igneous rocks are probably the result of mixing in different proportions of shocked fragments and melts of the adamellite and anorthosite host rocks, together with variations in cooling history.

Age

A sample of the fine-grained olive-gray igneous rock with small feldspar phenocrysts was analyzed for K and Ar in the laboratories of the Geological Survey of Canada. The determination gave an age of 202 ± 25 m.y. ($K = 1.90\%$; $^{40}Ar = 0.0125$; radiogenic $Ar = 72\%$) and is accepted by the authors as indicating a Triassic age for the crater.

DENCE, M. R. 1965. The extraterrestrial origin of Canadian craters. Ann. N.Y. Acad. Sci., **123**, pp. 941–969.

——— 1968. Shock zoning at Canadian craters: petrography and structural implications. *In* Shock metamorphism of natural materials (B. M. French and N. M. Short, *Ed.*). Mono Book Corp., Baltimore, pp. 169–184.

ROBERTSON, P. B., DENCE, M. R., and VOS, M. A. 1968. Deformation in rock-forming minerals from Canadian craters. *In* Shock metamorphism of natural materials (B. M. French and N. M. Short, *Ed.*). Mono Book Corp., Baltimore, pp. 433–452.

SLEMMONS, D. B. 1962. Determination of volcanic and plutonic plagioclases using a three- or four-axis universal stage. Geol. Soc. Amer., Special Paper 69, 64 pp.

STÖFFLER, D. 1967. Deformation and Umwandlung von Plagioklas durch Stosswellen in den Gesteinen des Nördlinger Ries. Contr. Mineral. Petrol., **16**, pp. 51–83.

WANLESS, R. K., STEVENS, R. D., LACHANCE, G. R., and RIMSAITE, R. Y. H. 1965. Age determinations and geological studies, Part 1 — Isotopic ages, Report 5. Geol. Surv. Can., Paper 64–17 (Part 1), p. 120.

19

Reprinted from *Internat. Geol. Cong., 24th, Proc.* 15:148-156 (1972)

The Steen River Astrobleme, Alberta, Canada

STEPHEN R. WINZER,
Canada

ABSTRACT

The Steen River structure occurs in the Interior Plains of northwestern Alberta as an anomalous feature affecting the crystalline basement. From boreholes a roughly elliptical anomaly approximately 24 by 19.5 km (15 by 13 mi.) is defined. The outer boundary is defined by a rise in crystalline basement, followed by a sharp dropoff into a "rim syncline". The central portion is uplifted an average of 1280 m (4200 ft.) above the base of the rim syncline, and lies 762 m (2500 ft.) above the normal basement elevation in this region. The rim syncline is filled with sediments of unknown age. The entire structure is covered with undisturbed Cretaceous strata, and there are no surface indications of the structure.

Thirty-one rock specimens recovered from two boreholes have been examined. Suevite breccias and melt phases similar to those of the Ries Crater can be identified in the Steen River core. From quartz grains, the orientations of 6,472 planar deformation structures were obtained. The most common orientations were parallel to $\{10\bar{1}2\}$, $\{10\bar{1}3\}$, $\{10\bar{1}1\}$, $\{10\bar{2}2\}$ and $\{10\bar{3}3\}$. X-ray "single crystal" diffraction patterns show varying degrees of asterism, indicative of lattice disorientation resulting from extreme shock pressures. Maskelynite, thetomorphic quartz glass and flowed glass have all been identified, indicating that some of the rocks have reached middle or upper Stage II of shock metamorphism corresponding to maximum shock pressures of 400-500 kb and temperatures of 500-1000° C.

INTRODUCTION

DURING December, 1963, Imperial Oil Enterprises, Ltd., drilled a wildcat exploration hole into the Steen River geophysical anomaly (LSD 12, Sec. 19, Twp. 121, Range 21 W 5; see Fig. 1). After passing through 184 m (603 ft.) of lower Cretaceous Loon River shale, the drill penetrated crystalline rocks lying approximately 762 m (2500 ft.) above the normal regional level of Precambrian basement. The hole was extended down to 516.6 m (1695 ft.) and then abandoned. Cores were taken at irregular intervals by Imperial Oil geologists (Fig. 2).

The region in which the holes were drilled is one of extensive flat muskeg lying 194 km (121 mi.) SSW of Hay River and 668 km (430 mi.) WNW of Edmonton. No surface indications of any circular feature are present, either in drainage or topography (Fig. 3). Basement elevations, determined from several drill holes in the area, range from 1280 to 1432 m (4200 to 4700 ft.) below ground level. The regional dip of the basement is gently southwest. The center of the feature lies about 63 km (38 mi.) NW of a large NE-trending Precambrian lineament, the Hay River fault (?), which extends from the south shore of Great Slave Lake through northwestern Alberta.

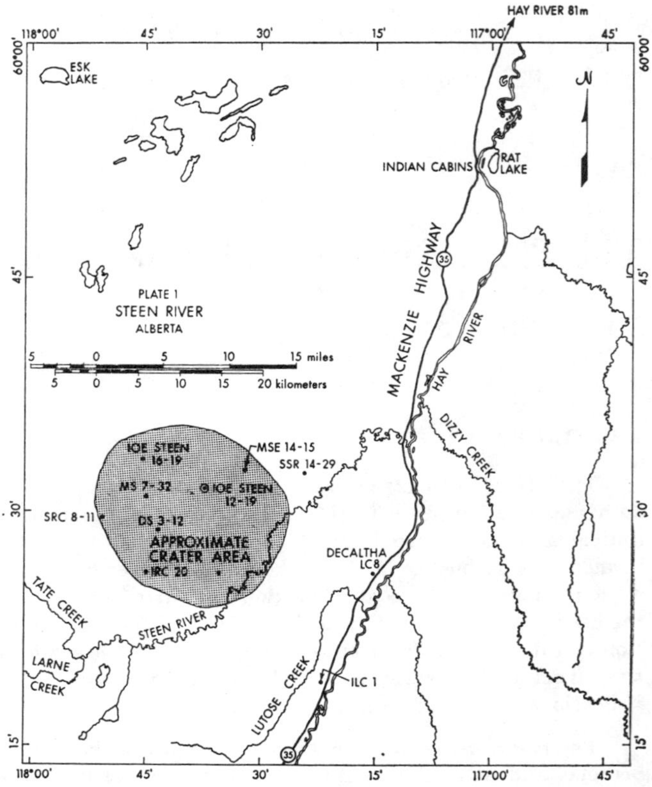

FIGURE 1 — Location, Steen River, Alberta.

A rough idea of the physiography of the structure has been ascertained from borehole data, basement contouring and discussion with geologists from Imperial Oil, Ltd. In plan, the structure is elliptical, 24 km (15 mi.) long and 19.5 km (13 mi.) across (Fig. 1). The basement rises gently to the rim from outside the structure, then drops 180 m (600 ft.) on the southeast and 520 m (1700 ft.) on the northwest. About two-thirds of the total area of the structure is occupied by this uneven annular depression or "rim syncline". The central part rises an average of 1280 m (4200 ft.) from the bottom of the rim syncline and lies 760 m (2500 ft.) above the normal basement elevation for this area (Fig. 3). Geophysical data indicate that the rim is cut by numerous radial faults. Other boreholes (IOE Steen 16-19, MSE 14-15, MS 7-32, DS 3-12, IRC 20) drilled within the rim syncline indicate the presence of sediments, but their age is unknown.

MEGASCOPIC FEATURES OF ROCKS FROM THE STEEN RIVER STRUCTURE

The well log (Fig. 2) shows the lithology of crystalline rocks put together from several cored intervals taken from the lower 242 m (795 ft.) of the borehole. The top 184 m (603 ft.) is composed of non-calcareous Loon River shale, followed by 121 m (397 ft.) of rock described in the Imperial log as "volcanic effusia". Cuttings of this rock resemble the suevite breccias recovered from the 427-m (1400-ft.) level.

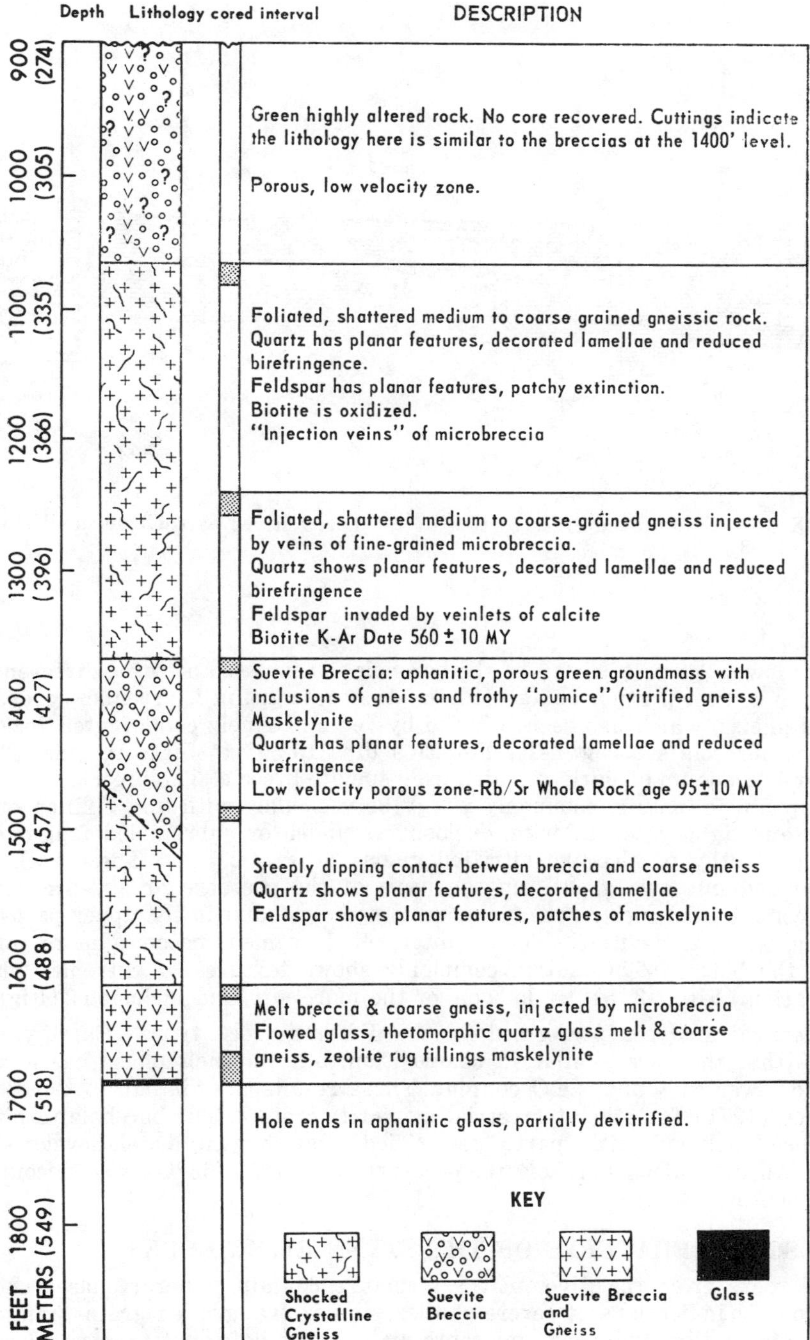

FIGURE 2 — Lithologic log from 900 to 1695 ft. (IOE Steen 12-19).

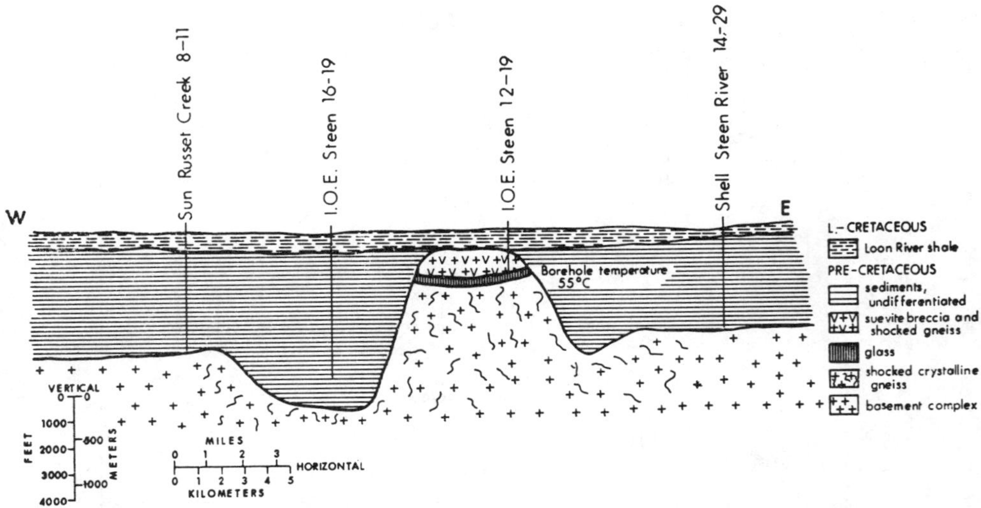

FIGURE 3 — Generalized cross section of the Steen River astrobleme modified after Carrigy, 1968).

The lower 242 m (795 ft.) of the borehole is composed of three different rock types, the largest portion being medium- to coarse-grained, greenish- grey, well-foliated gneiss, which has been injected by veins of finely comminuted rock powder (microbreccia see Fig. 4-1). Foliation dips 15-50° to the core axis. The remainder is made up of dark greenish-grey suevite breccia. The breccia has a dark green aphanitic matrix, commonly vuggy or vesicular, with vug fillings of pink zeolites and light green chlorite, enclosing rounded or subrounded fragments of gneiss or frothy pumice-like vitrified gneiss (Fig. 4-2, 3, 4). These rocks have been divided into two groups on the basis of the presence or absence of melt. Finely comminuted rock powder is more commonly found in the upper part of the borehole, whereas devitrified glass, interpreted as melt, becomes more common toward the bottom. The matrix commonly shows textures of extreme "plastic" deformation (Fig. 4-3, 5), as do some of the more vitrified inclusions (Fig. 4-2).

Chemical analyses performed on the different rock types (Carrigy, 1968) indicate that they are similar in composition, but the melt phase has a higher K_2O and lower SiO_2 and Na_2O content. These results are similar to those found by Dence (1971) for other terrestrial meteorite craters. The borehole terminates in a dense black, aphanitic, partly devitrified glass showing a well-developed chill border. An anomalous borehole temperature of 131°F (55°C) was encountered at this depth.

MICROSCOPIC FEATURES OF THE STEEN RIVER ROCKS

All rocks from the Steen River structure contain mineral phases showing the following indications of severe shock: angular fragmented grains, microfractures, deformation twinning, microbreccia, planar deformation structures and thetomorphic minerals. Devitrified glass is most common in the suevite breccias; two samples of undevitrified glass are found in the shocked gneiss. Thetomorphic quartz glass was found only in the suevite breccia.

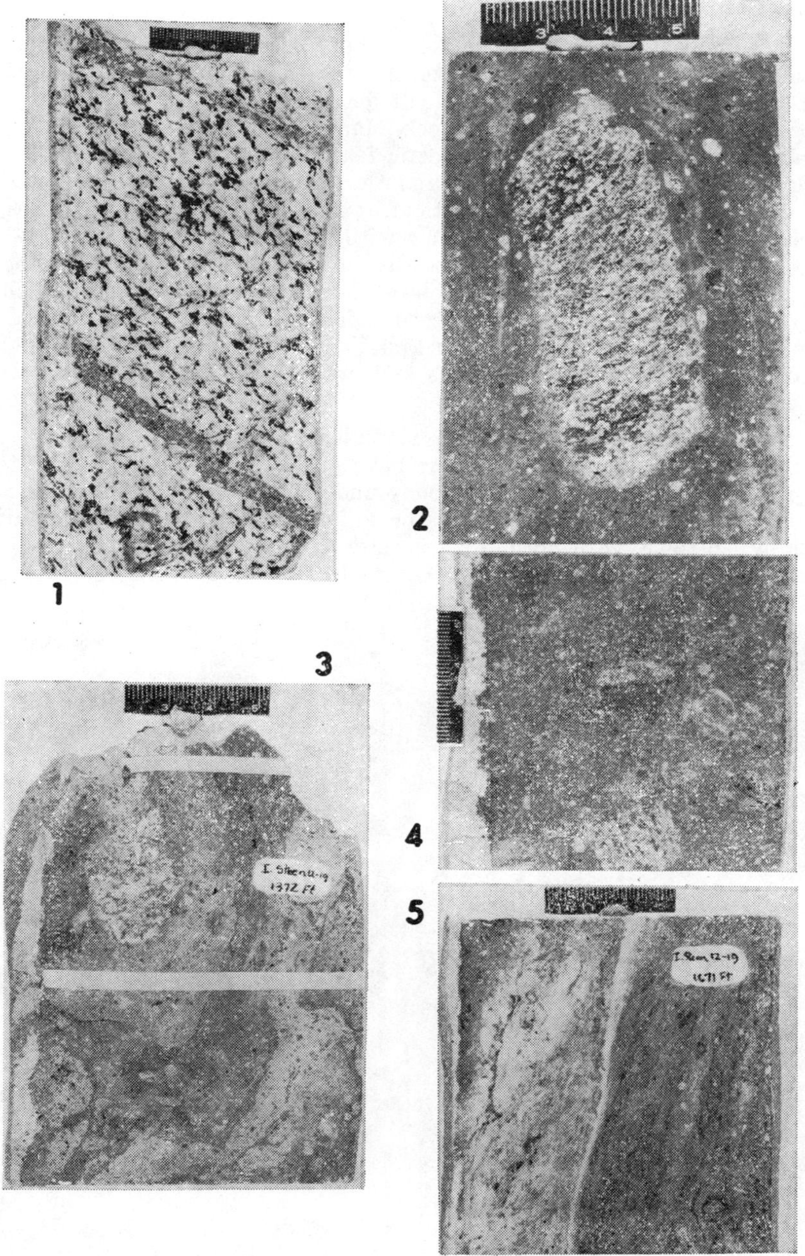

FIGURE 4 — (Scale in centimeters) — (1) 1247′ Gneiss injected by microbreccia. (2) 1369.5′ Partly vitrified gneiss fragment. (3) 1372′ Suevite breccia. (4) 1380′ Suevite breccia. (5) 1671′ Flow texture in suevite breccia.

Feldspar

Potassium feldspar and plagioclase are both present in the Steen River rocks, with the greater proportion of the feldspar porphyroblasts being potassic. Differentiating between the two is difficult optically, as many diagnostic features have been destroyed apparently by shock. Many grains show mechanical twinning in a pattern which masks such diagnostic features as albite twinning. Decorated lamellae, commonly wavy and of irregular thickness, appear on many of the larger grains and many show one or more sets of lamellae, or thin planar features. These features commonly are decorated either with mineral matter or by tiny bubbles or mineral inclusions. Grains of both plagioclase and K-feldspar show deformation bands and patchy extinction. The deformation bands tend to extinguish either slightly before or slightly after the crystal during rotation under crossed nicols, and are similar to the thick lamellae or kink bands in biotite. In some cases numerous planar features cross these bands and, where they do, take on an undulose appearance (Fig. 5-1, 2).

A number of specimens contain partially or completely isotropic feldspar (maskelynite). Totally isotropic grains have outlines similar to normal feldspar grains in the rock, and are clear to cloudy under plane light. No interference figures can be obtained from these grains (Fig. 5-3). Another common type of grain is partially isotropic, with normal plagioclase or potassium feldspar grading into patches of isotropic feldspar. In some grains this appears as a core (Fig. 5-4), or

FIGURE 5 — (1) 1257′ Deformation bands and mechanical twinning, feldspar. (2) 1621′ Planar features and lamellae, feldspar. (3) 1492′ Maskelynite. (4) 1689′ Maskelynite core, feldspar. (5) 1080.5′ Quartz. Granulated boundaries and planar deformation structures.

FIGURE 6 — (1) 1689′ Planar deformation structures in quartz. (2) 1484.5′ Decorated lamellae and planar elements in quartz. (3) 1682′ Thetomorphic quartz glass (unconfirmed by refractive index measurements. (4) Kink band in biotite. (5) 1689′ Flowed glass.

as a core altered to zeolite or chlorite. The interference figure is either poorly defined or absent. In plane light the isotropic portion is somewhat cloudy and has a lower refractive index than the surrounding feldspar grain. This cloudy nature is probably due to alteration, possibly to zeolite. X-ray photographs of single crystals of feldspar show varying degrees of asterism indicative of lattice disordering from strong shock pressures (Dachille *et al.*, 1968).

Quartz

The quartz from this suite of rocks shows a variety of textures. Most quartz grains are somewhat cloudy, and in rocks of the suevite phase and lower gneisses the quartz is cloudy almost to the point of opacity. This cloudiness is caused by tiny bubbles or glass contained within the quartz grain. All quartz, with the exception of later "mineralized" veins, has signs of shock metamorphism. Larger grains have been shattered, giving patchy extinction due to differences in C-axis orientation. Boundaries between grains are granulated (Fig. 5-5).

All quartz grains show planar deformation structures, most of which are decorated with mineral matter or inclusions, either along the trace of the intersection of the plane with the grain surface or extending along the plane into the host grain. The latter type especially is noticeable in cloudy grains, and commonly provides the only means of measuring these features. All grains have at least two sets of planar deformation structures, and some as many as four (Fig. 6 — 1 and 2). Measurement of 6,472 of these structures indicates that they are parallel to the forms $\{10\bar{1}2\}$, $\{10\bar{1}3\}$, $\{10\bar{1}1\}$, $\{10\bar{2}2\}$ and $\{10\bar{3}3\}$. Two occurrences of thetomorphic quartz glass (not confirmed by refractive index measurements) have been found in these rocks, both occurring in the suevite breccia (Fig. 6–3).

Biotite

The biotite shows two features which indicate severe strain; kinking on a very small scale (Fig. 6-4) and partial decomposition and oxidation. In most specimens, the biotite has been fragmented such that only very fine, almost opaque grains are found scattered throughout the rock. Biotite is virtually absent from the suevite breccia.

Glass and Other Minerals

Devitrified glass constitutes over 50% of the suevite matrix and is filled with minute crystallites and larger vesicles or vugs containing chlorite or zeolite. The alignment in the crystallites is interpreted as flowage planes in the glass.

Two pieces of undevitrified, flowed glass were found in the lower gneisses, adjacent to shocked quartz and feldspar. The glass is clear, with a refractive index lower than that of Canada balsam, and shows no chill border (Fig. 6-5).

X-RAY SINGLE CRYSTAL DIFFRACTION STUDIES

Single crystal X-ray photographs were taken to determine the similarity of diffractograms of quartz and feldspar with those of quartz and feldspar from other impact structures. Crystals from seven rock specimens were examined. A small amount of rock was crushed and crystals from 0.2 to 0.5 mm were picked and examined in oil to insure that they were single crystals. They were then mounted on a fine glass needle in a 114.6-mm Debye-Scherrer (Straumanis) powder camera and irradiated with Ni-filtered Cu radiation for 1.5 hours.

The patterns showed varying degrees of asterism, indicative of crystalline damage. Back reflections were faint or completely absent (Fig. 7). This type of

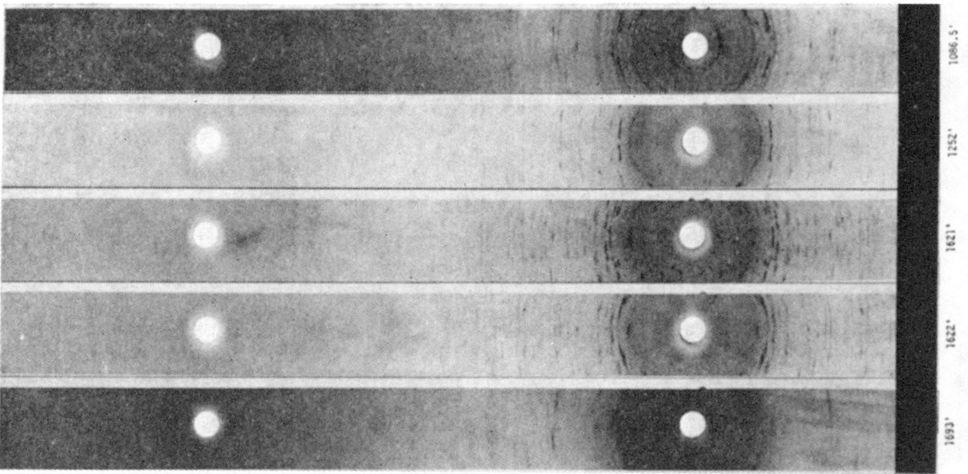

FIGURE 7 — 331.1 m (1086.5'), Quartz; 381.6 m (1252'), Quartz; 494.1 m (1621'), Plagioclase; 494.4 m (1622'), Quartz; 516 m (1693'), Quartz.

pattern correlated well with patterns found by Dachille *et al.* (1968) and Chao (1968).

The powder patterns were indexed to determine the type of quartz and feldspar present. Apparently the lattice of quartz has been distorted by the shock event, as the patterns do not give an exact match with quartz polymorphs. Alpha quartz can be identified with certainty, but patterns show lines which might be accounted for by small quantities of beta quartz and coesite, although neither of these can be positively identified. Possibly, minute crystallites of these minerals exist as submicroscopic inclusions along slip planes, as seen in decorated lamellae.

Coesite

An attempt was made to separate coesite from three samples of Steen River rocks. Although lines attributable to coesite appeared in single crystal photographs, no coesite could be found using Stöffler's method (Dence, personal communication). Further work is being done on quartz separated from Steen River rocks.

Carrigy (1968) concluded from his study of seven samples from the Steen River rocks that there was insufficient evidence to decide the cause of shock metamorphic features found in these rocks. The present author has shown that the Steen River rocks show all the main petrographic features found in terrestrial meteorite impact craters. with the exception of coesite. Studies conducted by Hörz (1968), Chao (1968), Engelhardt and Stöffler (1968) and Stöffler (1971) enable determinations to be made of pressures and temperatures generated in shock-metamorphosed rocks. According to the scale of Stöffler (1971), the presence of glass, maskelynite and thetomorphic quartz glass, as well as planar deformation structures, indicates that middle or upper Stage II was reached, corresponding to maximum pressures of 400-500 kb and temperatures between 500 and 1000°C. The absence of natural terrestrial processes capable of producing pressures of this magnitude provides convincing evidence that these pressures are impact-generated. Thus, the features found in rocks from the Steen River structure can best be explained by the impact of a large meteorite.

ACKNOWLEDGMENTS

I would like to thank Dr. R. A. Burwash of the University of Alberta for providing me with specimens from Steen River and to thank Drs. R. A. Burwash, R. E. Folinsbee, R. D. Morton and D. G. W. Smith for their helpful comments regarding the manuscript. Thanks go to Imperial Oil, Ltd. for providing cuttings and data from the Steen River area. Thanks also go to my wife, Kathy, for helping in the laboratory and with the typing.

REFERENCES

Carrigy, M., 1968. Evidence of shock metamorphism in rocks from the Steen River Structure, Alberta. *In* French, B. M., and Short, N. M., *(Editors)*. Shock Metamorphism of Natural Materials. Mono Book Corp., Baltimore, p. 367-378.

Chao, E. C. T., 1968. Pressure and temperature histories of impact metamorphosed rocks, based on petrographic observations. *In* French, B. M., and Short, N. M., *(Editors)*. Shock Metamorphism of Natural Materials. Mono Book Corp., Baltimore, p. 135-158.

Dachille, F., Gigl, P., and Simons, P. Y., 1968. Experimental and analytical studies of crystalline damage useful for the recognition of impact structures. *In* French, B. M., and Short, N. M., (*Editors*). Shock Metamorphism of Natural Materials. Mono Book Corp., Baltimore, p. 555-569.

Dence, M. R., 1971. Impact melts. J. Geophys. Res. 76, No. 23, p. 5552-5565.

Engelhardt, W. v., and Stöffler, D., 1968. Stages of shock-metamorphism in crystalline rocks of Ries Basin, Germany. *In* French, B. M., and Short, N. M., (*Editors*). Shock Metamorphism of Natural Materials. Mono Book Corp., Baltimore, p. 159-168.

Hörz, F., 1968. Statistical measurements of deformation structures and refractive indices in experimentally shock loaded quartz. *In* French, B. M., and Short, N. M. *(Editors)*. Shock Metamorphism of Natural Materials. Mono Book Corp., Baltimore, p. 243-254.

dices in experimentally shock loaded quartz. In French, B. M., and Short, N. M.eCc

Stöffler, D., 1971. Progressive metamorphism and classification of shocked and brecciated crystalline rocks from impact craters. J. Geophys. Res., 76, No. 23, p. 5541-5551.

ERRATA

Page 148, line 14 should read: "$\{10\bar{1}1\}$, $\{21\bar{3}1\}$ and $\{51\bar{6}1\}$. . . ."
Page 154, lines 20 and 21 should read: ". . . $\{21\bar{3}1\}$ and $\{51\bar{6}1\}$. . . ."

Editor's Comments on Papers 20 Through 24

20 GUY-BRAY
Introduction (to the Geological Association of Canada Special Paper Number 10)

21 DENCE
Meteorite Impact Craters and the Structure of the Sudbury Basin

22 FRENCH
Shock-Metamorphic Features in the Sudbury Structure, Ontario: A Review

23 GUY-BRAY and GEOLOGICAL STAFF (International Nickel Co., Copper Cliff)
Shatter Cones at Sudbury

24 DENCE and POPELAR
Evidence for an Impact Origin for Lake Wanipitei, Ontario

SUDBURY

The Sudbury, Ontario, and Vredefort, South Africa, cryptoexplosion structures are the largest and most complex known. Dietz (1964) suggested meteoritic origin for Sudbury on account of similarities to the Vredefort structure. He predicted that he would find shatter cones at Sudbury, and this proved to be the case. But he may have gone too far in suggesting (Dietz, 1972) that the nickel ores at Sudbury are themselves of meteoritic origin. Such direct derivation of the ores is accepted by few of the later workers on the Sudbury geology, though many of them favor an impact origin for the complex itself. It is true that traces of copper have been reported from both iron and stony meteorites, but direct origin of the ores would require the erection of a new and hitherto unsuspected ad hoc class of cosmic body, with a copper content many orders higher than that of any known meteorite. There is an immense amount of copper and nickel ore in the irruptive, and the copper-nickel ratio of the ores is similar to that of nickel sulphide ores in gabbro-norite associations throughout the world. Also probably unacceptable is the model advanced by Dietz, in which the entire irruptive was attributed to impact-induced melting (see below). Nevertheless, his was the pioneer impact interpretation of Sudbury, and the discovery of shatter cones did establish that the structure was initiated by an immense-scale explosive event, even if it did not establish ipso facto that Sudbury is an astrobleme—there are far too many conflicts of evidence for this to be accepted as proven.

The modern interpretations of Sudbury as an impact structure are covered by four papers reproduced here (Papers 20-23) and by three additional papers by French (1968, 1970, 1972) briefly discussed below. Guy-Bray's introductory paper (Paper 20) is outstanding and speaks for itself. He does not evade the problems that confront this new interpretation of a long-famous geologic occurrence, and leaves no doubt that he favors it. He presents the philosophical background to the controversy and the reasons for seriously entertaining impact origin. Dence (Paper 21) puts the case in greater detail. In particular, he discusses the lack of any central uplift at Sudbury. The limits of the irruptive are considered to correspond to those of an uplift that foundered after an ephemeral existence. The overall structure is believed to be more extensive and to be defined by an uplifted area circumscribed by a downfaulted ring of Huronian sediments (a concept following Speers, 1957). Dence interprets the famous "sublayer" as a basal breccia—a novel idea that is bound to be argued by other geologists, because the sublayer seems to have a genetic relationship to the hidden lower ultrabasic rocks of the irruptive and is not entirely unlike other recorded nickel sulphide breccias in mineral deposits (for example, the breccia at the Thompson Mine, Manitoba). Dence's hypothesis requires the sublayer to be an early development. Its time relationship to other events at Sudbury remains obscure, but it is by no means certain that it is an early emplacement. Dietz (1972) also related the sublayer directly to the impact explosion.

French (Paper 22) described shock deformation in the Onaping Formation of the Whitewater Series, immediately above the irruptive. Though Dietz regarded this formation as part of the irruptive, this is not accepted either by French, who favors impact origin, or by Stevenson (1972), who rejects it. The Onaping Formation appears to have been emplaced before the layered rocks of the irruptive, and to be invaded and metamorphosed by the micropegmatite at the top of the irruptive (Stevenson, 1963). French described and illustrated shocked fragments within the Onaping Formation, a vast and monotonous ignimbrite-like pile of dacites and rhyodacites, which he interpreted as a fall-back breccia related directly to an impact explosion. He could see objections to such a proposal, however, in the abundance of carbonaceous material and the ubiquity of disseminated nickel sulphides in the Onaping Formation. Shock effects of the foot-wall rocks to the irruptive were also described by French. Both these and those within the Onaping Formation are of the familiar "unique" type described from many other sites of cryptoexplosion.

Guy-Bray et al. (Paper 23) provide a lucid account of shatter cones at Sudbury.

Some papers not reproduced here must be mentioned briefly. French (1970) found that the total theoretical requirement of melt for an impact explosion of Sudbury dimensions is accounted for by the shocked fragments within the Onaping Formation. This conflicts with the attribution of the entire irruptive to such melting by Dietz (1964). Dence accepts a much larger overall structure, and this allows him to propose that one-third to one-half of the volume of the irruptive stemmed from melting of crustal rocks. Frence (1972) considered the origin of the irruptive itself to be the fundamental problem facing any impact hypothesis. He considered the possibility of generating the magma in the lower crust or upper mantle by *impact-induced offloading* (French, 1968). He recognized a major discrepancy between the volumes of igneous rocks produced at Sudbury and Vredefort, cryptoexplosion structures of like age and size. Copius generation at the first contrasts with meager generation at the latter. This discrepancy requires variations in the geothermal gradient during the early Proterozoic. Computations of a theoretical nature require that the Sudbury structure would have to be formed by a meteorite that encountered an area of exceptionally high gradient or a locus of preexisting mantle melting at depth. Either the mantle at this time had a lower minimum melting curve than pyrolite or it possessed a gradient equal to or above the present suboceanic gradient (much higher than the continental gradient). French cited the radiometric evidence of Souch et al. (1969), which suggested that the age derived by Faure et al. (1964), widely accepted for the micropegmatite, might be too low. If the true age of the irruptive is greater than 2000 m yr, then it might be close to that derived for the Nipissing diabase (van Schmus, 1965). This suite was derived by widespread mantle melting of endogenic origin. French's idea was that the impact fortuitously hit this preexisting magma pool below or within the lower crust, and it was remobilized by offloading.

He found that it was not possible to produce the vast volumes of silica-rich micropegmatite in the irruptive by assimilation of shock-melted country rock by the magma at its level of generation. The norite alone could conceivably have been so produced, but this leaves the problem of the micropegmatite unsolved. He considered the possibility that it was directly derived by melting of granitic target rocks as suggested by Dence, also the possibility of assimilation occurring higher in the crust than the generation level of the Nipissing diabase. The initial stronium isotope ratios of the micropegmatite do not, however, appear to favor crustal origin for the micropegmatite in any manner, and French's conclusion that the origin of the irruptive is not clear, in the light of the impact theory, is a fair statement of the case. It is a fact that neither direct impact melting of crustal rocks nor impact-induced offloading satisfactorily explains the immense generation of magma at Sudbury.

Stevenson (1961, 1963, 1972) is foremost among those who favor endogenic origin for the entire Sudbury structure, rejecting impact origin. He described the breccia at the base of the Whitewater Series (beneath the Onaping Formation) as of sedimentary origin, though felspathization of the quartzite breccia has produced chessboard albite in a crushed quartzite matrix. He attributed this process to contact metamorphism by the subjacent irruptive. In his second paper, Stevenson added a more detailed description of these contact effects. Guy-Bray and Peredery (1971) have interpreted this same zone as a complex association of impact-melted rocks and shocked explosion breccias. The irruptive would certainly be expected to produce very marked contact effects, and the editor did find this zone, on a field visit, to be not unlike the contact zones of the immense Binneringie Dike in western Australia (McCall and Peers, 1971). The shock effects noted by Guy-Bray and Peredery are real, but the alternative possibility should not be discounted; that the very complex relationships recorded at this contact represent the superimposition of the contact zone of the micropegmatite upon country rocks that were earlier affected by intense shock and fluidization up the fracture that provided the channel of access for the magma, a cone-sheet-like fracture up which the irruptive was emplaced. The shocked rocks of the foot-wall and of the hanging-wall (the zone under discussion) were, of course, in close contiguity prior to being wedged apart to their present separation of many kilometers by the emplacement of the irruptive.

Stevenson (1972) interpreted the Onaping Formation as a fissure-erupted ash-flow or ignimbrite mass. An erractically distributed basal breccia and lapilli tuff unit were recognized passing up into mixed lapilli tuff and then to ashy tuff. Gradation through this sequence without abrupt boundary demarcations extends to the base of the Onwatin Siltstone and Slate Formation. This ash-flow ranks among the world's largest, covering 350 mi^2 (906 km^2) averaging 6000 (1.83 km) in thickness. A volume of 400 mi^3 (1664 km^3) is estimated. It lacks the laminated bedding characteristic of air-fall tuffs and the flow structure typical of lavas. However, several successive flow units are recognized, and, though welding and metamorphism make recognition of primary texture difficult, vitroclastic and devitrification textures entirely typical of volcanic rocks can be made out. A source in dispersed fissure, not a single center, is suggested. Eruption of successive flows was so rapid that a single cooling unit comprises several such flows. Rapid eruption and coalescence into a single cooling unit characterizes ash flows of immense size throughout the world. Individual clast populations are laterally persistent, but there is abrupt variation vertically through the formation, marking boundaries between individual flows. Such boundaries are now difficult to trace. Thin beds of air-fall tuff and slate succeeded the Onaping Formation, forming a passage into the Onwatin Siltstone and Slate. The presence of these tuff layers suggests prolonged eruption continuing during sedimentation, and does not support instantaneous impact origin for the Onaping Formation. Silica decreases upward through the ash-flow. Such a progression is supposed to preserve in reverse the compositional sequence of the magma chamber source (cf. Creede, Colorado; Ratté and Steven, 1976). Similarly, inverse relationship between potassium and sodium vertically through the ash-flow is characteristic of ash-flow units and attributed to alkali exchange during cooling or differentiation-controlled stratification in the magma chamber. Ash flow developments of this type are normally associated with large volcano-tectonic depressions.

Sedimentological papers by Rousell (1972), and Cantin and Walker (1972) are also critical to the Sudbury controversy. The Whitewater Series, though found nowhere else in Canada, is composed in its upper section of sediments of familiar depositional facies, not in themselves unique. Rousell found little evidence to suggest that the Onwatin was derived from the Onaping beneath. Its source was probably the tonalitic terrain of Archean age exposed north of the Sudbury basin. The Chelmsford Formation was deposited in a basin with a rectilinear axis, unlike the present basin, and less elliptical, though it had the same general trend of elongation and southwestward paleoslope. Cantin and Walker believed that the longitudinal basin was already established at the time of Chelmsford deposition, for sedimentary structures indicate derivation of the sediments from the sides of an elongate trough by turbidity current action. Flow along the axis appears to have been dominant. There was regional tilt to the southwest during deposition, controlling movement along the trough. These authors suggested that the uplift necessary to supply the immense volumes of sand may have been due to the forces that produced the elongate form of the basin in pre-Chelmsford time. Guy-Bray (Paper 20), considering this evidence, suggested that impact occurred in water producing a subdued crater that merged with the existing crater. Deposition of the Onaping Formation followed by a long period of mud deposition could have smoothed out the topography, removing the morphological expression of the crater by the time of the Chelmsford deposition. Paleocurrent trend would in this case be unrelated to configuration of the crater, being controlled entirely by the larger basin. However, uplift of the source area would still be required, to explain the turbidites. The lack of turbidite development outside the Sudbury basin is against this explanation. Two models, both involving impact, were entertained by Cantin and Walker, but neither is entirely satisfactory. The sedimentological evidence raises the possibility that the basin may never have been circular. There is an area of doubt whether the structure of the south range could ever have been derived from a sector of a circle (a requirement of all versions of the impact hypotheses). Certainly, it appears unlikely that the south range could have been so modified by the rather meager post-irruptive faulting and without more significant metamorphism than is evident. If the irruptive forms only a part of the larger structure, which includes the cluster of Huronian granites surrounding it, as suggested by Dence (Paper 21), an analogy can well be drawn with Vredefort (Paper 30). Sudbury may be a two-phase structure, formed by updoming of Archean rocks with some early granite emplacement. An immense-scale explosive phase may then have opened up cone-sheet fractures and subsequent foundering been followed by eruption of the Onaping ash flow, rapidly succeeded by sedimentation within the basin. The irruptive and sublayer may have later been emplaced up ring fractures. In such a model, which is endogenic in concept, the structure may never have been circular. The original dome may have been elongated and also the cone-sheet structures. This concept is in line with observation of similar boat-shaped magmatic complexes in the Great Dike of Rhodesia and Jimberlana Dike, Western Australia, where layered sequences of igneous rocks have a *primary* elongation, owing nothing to subsequent deformation (Campbell et al., 1970). In such cases, eruptivity was controlled by a linear fracture zone. Sudbury's closest analogue, Vredefort, is controlled by a tectonic lineament (Bucher, 1963), but linearity is neither primarily introduced or secondarily imposed, and the layered igneous rock and ash-flow eruptive association is absent. Vredefort simply did not develop beyond the initial explosive phase and doming, and associated eruptivity was meager. Yet the pseudotachylite developments at Vredefort and the Levack breccia at Sudbury are so similar that there can be little doubt of one thing: whatever the origin of these two structures, endogenic or exogenic, the same fundamental processes produced both of them.

The Sudbury norite is not unique. Though no such magmatic development is seen at Vredefort, there is an analog in the Archean Dampier Archipelago Complex of Western Australia (see Part VI), where there is another great elongated magmatic complex with micropegmatite dominant, as at Sudbury. Other such complexes are also known to the south, near Mount Magnet, the latter being also of

Archean age. This raises the question whether these immense differentiations of micropegmatite in gabbro and norite complexes of this ancient age are not simply due to evolutionary changes in differentiation patterns throughout the history of this planet. Ultrabasic and basic layered intrusions and lava suites of the Barberton and Pilbara regions of Africa and Australia (McCall, 1977a) show differentiation patterns unknown in later rocks.

LAKE WANIPITEI

Dence and Popelaar (Paper 24) recognized a much younger cryptoexplosion structure at Lake Wanipitei, contiguous to the east end of the Sudbury basin. This is probably of late Paleozoic or Mesozoic age. There is a geophysical anomaly and boulders of fragile glassy and fluidal fragmental rocks occur among glacial boulders at the southwest corner of the lake, which has crude circularity and unusual bathymetry. Drilling has confirmed the source beneath the lake. Coesite has been recognized in one boulder. The familiar shock lamellation patterns are also present (Dence et al., 1974). The juxtaposition of this younger structure with the ancient Sudbury structure is an amazing coincidence if these are both meteorite impact sites (see McCall, 1977b, p. 246 for discussion of the similar coincidence of the Pretoria Salt Pan with the Bushveld Structure).

REFERENCES

Bucher, W. H. 1963. Cryptoexplosion Structures Caused from Within or Without the Earth (Astroblemes or Geoblemes?). *Am. Jour. Sci.* **261**:567-649.

Campbell, I. H.; G. J. H. McCall; and D. S. Tyrwhitt. 1970. The Jimberlana Norite, Western Australia—A Smaller Analogue of the Great Dyke of Rhodesia. *Geol. Mag.* **107**:1-12.

Cantin, R., and R. G. Walker. 1972. Was the Sudbury Basin Circular During the Deposition of the Chelmsford Formation? *Geol. Assoc. Canada Spec. Paper 10*, pp. 93-101.

Dence, M. R.; P. B. Robertson; and R. L. Wirthlin. 1974. Coesite from the Lake Wanipitei Crater, Ontario. *Earth and Planetary Sci. Letters* **22**:118-122.

Dietz, R. S. 1964. Sudbury Structure as an Astrobleme. *Jour. Geol.* **72**:412-434.

Dietz, R. S. 1972. Sudbury Astrobleme, Splash Emplaced Sublayer and Possible Cosmogenic Ores. *Geol. Assoc. Canada Spec. Paper 10*, pp. 29-40.

Faure, G.; H. W. Fairbairn; P. M. Hurley; and W. H. Pinson. 1964. Whole Rock Rb-Sr Age of Norite and Micropegmatite at Sudbury, Ontario. *Jour. Geology* **72**: 848-854.

French, B. M. 1968. Sudbury Structure, Ontario: Some Petrographic Evidence for and Origin by Meteorite Impact. In *Shock Metamorphism of Natural Materials,*

B. M. French and N. M. Short, eds. Baltimore, Md.: Mono Book Corp., pp. 383–412.

French, B. M. 1970. Possible Relations Between Meteorite Impact and Igneous Petrogenesis as Indicated by the Sudbury Structure, Ontario, Canada. *Bull. Volcanol.* **34**:466–517.

French, B. M. 1972. Production of Deep Melting by Large Scale Meteorite Impacts: The Sudbury Structure, Canada. *Internat. Geol. Cong., 24th, Proc.* **15**:125–132.

Guy-Bray, J. V., and W. V. Peredery. 1971. Field Guide for Excursion 1, Sudbury Basin. *Geol. Assoc. Canada/Mineral. Assoc. Canada Joint Meeting, Sudbury, Proc.*, 19pp.

McCall, G. J. H. 1977a. *The Archean.* Stroudsburg, Pa.: Dowden, Hutchinson & Ross.

McCall, G. J. H. 1977b. *Meteorite Craters.* Benchmark Papers in Geology, vol. 36. Stroudsburg, Pa.: Dowden, Hutchinson & Ross.

McCall, G. J. H., and R. Peers. 1971. The Geology of the Binneringie Dyke, Western Australia. *Geol. Rundschau* **60**:1174–1263.

Ratté, J. C., and T. A. Steven. 1976. Ash-Flows and Related Volcanic Rocks Associated with the Creede Caldera, San Juan Mountains, Colorado. *U.S. Geol. Survey Prof. Paper 524H,* 58pp.

Rousell, D. H. 1972. The Chelmsford Formation of the Sudbury Basin. *Geol. Assoc. Canada Spec. Paper 10,* pp. 79–91.

Souch, B. E.; T. Podolsky; and Geological Staff. 1969. The Sulphide Ores of Sudbury: Their Particular Relationship to a Distinctive Inclusion-Bearing Facies of Nickel Irruptive. *Econ. Geology Mon.* **4**:252–261.

Speers, E. C. 1957. The Age Relation and Origin of the Common Sudbury Breccia. *Jour. Geology* **65**:497–514.

Stevenson, J. S. 1961. Recognition of Quartzite Breccia in the Whitewater Series, Sudbury Basin, Ontario. *Trans. Roy. Soc. Canada* **55**:57–66.

Stevenson, J. S. 1963. Upper Contact Phase of the Sudbury Micropegmatite. *Can. Mineralogist* **7**:413–419.

Stevenson, J. S. 1972. The Onaping Ash-Flow Sheet, Sudbury, Ontario. *Geol. Assoc. Canada Spec. Paper 10,* pp. 41–48.

Schmus, R. van. 1965. The Geochronology of the Blind River-Bruce Mines Area, Ontario, Canada. *Jour. Geology* **73**:755–780.

20

Reprinted from pages 1-5 of *Geol. Assoc. Canada Spec. Paper 10,* 1972

INTRODUCTION

J. V. Guy-Bray
The International Nickel Company of Canada, Limited,
P.O. Box 44, Toronto-Dominion Centre, Toronto 111, Ontario

This volume is an outgrowth of the twenty-fourth Annual Meeting of the Geological Association of Canada held at Laurentian University, Sudbury, Ontario, in May, 1971. It was only natural for this meeting to include a session on the seemingly inexhaustible subject of Sudbury geology. However, current studies of the traditional subjects of inquiry, namely the Nickel Irruptive and its ores, proved to be in short supply. This provided a unique opportunity to present, for the first time before a general audience of Canadian geologists, details of the evidence for and against the view that the Sudbury structure is an astrobleme: Dietz's term for an ancient, eroded meteorite impact scar.

All of the principal figures in the controversy took part. All of the papers presented in the Special Session at Sudbury are now published in this volume which also includes a valuable contribution, by Cantin and Walker, from another session of the meeting.

Not all of the studies reported here were directly addressed to the impact theory – hence the title of the volume – but all are relevant to it. Because of the great economic importance of the Sudbury structure it is desirable that new information about it should, as far as possible, be available for interpretation. Workers at Sudbury have always known that the local geology was much more singular than might appear from the frequent references in the literature to "Sudbury-type" ore deposits. However, adoption of the impact hypothesis for Sudbury can lead to the conclusion that its nickel-copper deposits are *unique.* Judging from my own experience at Sudbury, which coincided with the period (1964-1971) during which the astrobleme theory grew from a possibility to a probability, the implications of all this have indeed been most readily assimilated by exploration geologists.

Serious interest in the theory is now widespread, but our geological community's general knowledge of impact studies and shock metamorphism remains low, even in this age of Apollo, and even though Canada leads the world in field investigations of impact craters.

There are valid reasons for this. Impact studies are a relatively new development, comparatively few workers are involved, and until recently little of their work was published or discussed in the more widely circulated journals. Much of the research has been done in government agencies, not in universities, and consequently the subject is not taught. Thus unless one is a participant, as all Sudbury geologists must be, the matter is easy to overlook; particularly as impact geology appears to represent a kind of new Catastrophism, whereas in our formal training most of us were taught to seek more Uniformitarian solutions.

It is therefore hoped that the present volume will not only shed light on Sudbury geology, but may also serve a wider purpose by providing an introduction to the study and literature of impact processes and shock metamorphism for the general reader.

THE PAPERS

The first two papers, by Dence and French, are especially well suited to this latter purpose: they place the Sudbury problem in the context of other crater studies, such as the notable series of investigations pioneered at Ottawa by the Dominion Observatory workers under Beals. French, who discovered shock metamorphic textures in the Onaping Formation, describes in detail these petrographic features which are widely believed to be unique indicators of meteorite impacts.

Dietz, who not only re-interpreted Sudbury geology in terms of an impact origin but also discovered the first hard evidence for it, in the form of shatter cones, here defends his theory's even more controversial corollary, that the ores themselves are cosmogenic.

The nature and origin of the Onaping Formation (hitherto commonly called the Onaping tuff) is central to the Sudbury controversy. Stevenson, who has been involved in International Nickel's detailed mapping of this and other local rocks for two decades, presents the Onaping as a volcanic ash-flow deposit. Peredery, a former student of Stevenson's and a graduate of the same mapping program, presents new chemical evidence for an origin by impact.

Accurate ages for Sudbury rocks are essential for reconstruction of the history of events, yet many of the major formations, including the norite, have not yet been satisfactorily dated. Gibbins, Adams and McNutt, by Rb-Sr dating of a granite dike, raise the possibility that the norite cut by it is older than 2 b.y.

Card and Hutchinson, interpreting regional structure, history, stratigraphy, petrology and metallogeny in a valuable and stimulating review, suggest that Sudbury's setting may be too special to be the location of a fortuitous impact.

Rousell presents the first formal description of the Chelmsford Formation in terms of modern sedimentological concepts. Cantin and Walker use paleocurrent measurements to suggest that the Sudbury Basin was not circular during deposition of the Chelmsford.

Popelar's detailed gravity study places additional constraints on the interpretation of Sudbury geology, and also provides important evidence for Dence and Popelar's identification of Lake Wanapitei as a second, smaller and much more recent Sudbury impact site. This last paper, by describing a fairly "typical" younger crater, provides a helpful comparison with the studies of the older structure, where the evidence is necessarily more obscure owing to a long and eventful geological history.

BACKGROUND

A short descriptive review of local geology by the present writer preceded the other papers in the Special Session at Sudbury, to provide a basis for discussion. This introduction will not be repeated here, but readers unfamiliar with the evolution of conventional geological studies at Sudbury during recent years are urged to consult the papers by Naldrett *et al.* (1970) on the norite-micropegmatite, by Souch, Podolsky *et al.* (1969) on the sulphide ores, and by Fairbairn *et al.* (1968) on age determinations. The Ontario Department of Mines (1969) map of the Sudbury area is the best available. The most recent field guides are those prepared by Bray and Peredery (1971) and Naldrett *et al.* (1971) for the Geological

Association of Canada – Mineralogical Association of Canada 1971 joint meetings. Other guidebooks have been prepared for the 1972 International Geological Congress field excursions at Sudbury.

The impact theory itself was effectively launched by Dietz's classic paper (1964). Its development is marked in several publications by French (e.g., 1967, 1970). The reader seeking background information on shock metamorphism and impact studies generally is referred to the excellent compilation by French and Short (1968) and the more recent one by Hōrz (1971).

Before discussing the meteorite impact controversy, I must declare my own views, which do not necessarily coincide with those of my colleagues or of other Sudbury workers. I consider that the Sudbury Basin structure was initiated by the explosive impact of a large meteorite; I do not believe the ores came from the bolide; but I do think its impact acted as a trigger for endogenic magmatism that produced the Sudbury Nickel Irruptive, of which the ores are a part.

SHOCK METAMORPHISM

The reason for belief in an impact origin is simple: *shock metamorphism*. "The term *shock metamorphism* has been created to describe all changes in rocks and minerals resulting from the passage of transient, high-pressure shock waves. The great importance of such effects in geological studies arises from the fact that the only known natural method of producing such effects is the hypervelocity impact of a large meteorite. The terms *shock metamorphism* and *impact metamorphism* are thus closely related, and the unusual shock-metamorphic effects in rocks become important as criteria for the recognition of meteorite impact in structures so old that meteorite fragments no longer survive". (French, 1968, p.2).

Evidence of shock metamorphism is both widespread and common at Sudbury, and its presence – though not its origin – is no longer in dispute. It comprises not only the shatter cones discovered by Dietz, but the petrographic features discovered by French in rock fragments in the Onaping Formation, and now known to occur widely in inclusions in the Sudbury breccia and in footwall rocks adjacent to the Irruptive. Field and laboratory studies, including examinations of rocks involved in nuclear explosions, permit calibration of such features: it is apparent that shock metamorphism can involve pressures up to the megabar (1000 kb.) level, temperatures in excess of 2,000°C. and extremely high strain-rates.

Such conditions are in striking contrast to those of "normal" geological processes, and indeed the resulting effects are both distinctive and unusual. In spite of this, years of search have so far failed to demonstrate their presence in any unequivocally volcanic rocks, or in any rocks affected only by intense tectonic activity. This is perhaps not surprising: it is difficult to see how conventional processes can match the physical conditions of explosive meteorite impact, in which an enormous kinetic energy is instantaneously dispersed at the Earth's surface. For example, the most violent volcanic explosions are estimated to attain pressures well below ten kilobars – which is an order of magnitude less than the pressure required for the first appearance of unequivocal petrographic shock features.

The writer finds it difficult to disagree with Dence's evaluation (1968, p. 179): "the contention that volcanism can generate strong shocks . . . (has) . . . no supporting evidence either in nature or in theory".

THE IMPACT THEORY

So far, then, the evidence seems to point to an impact origin for the Sudbury structure. But are the other geological data at Sudbury consistent with such an hypothesis? Can, for instance, the Irruptive and its ores be fitted into the framework of an impact crater? Here opinions differ.

The contrary arguments will be reviewed below. However, it is to be noted that no single piece of evidence now at hand makes an impact interpretation untenable. The "Sudbury event" has long been viewed as one involving a large and violent release of energy, as evidenced by the extraordinary scale and variety of brecciation it displays. The mechanisms proposed hitherto have been volcanic and/or tectonic in nature, but a meteorite impact is certainly no less acceptable in this regard, and a major one would provide an excellent trigger for magmatic activity.

The Precambrian crust was surely not thicker than at present. Creation of an impact crater 100 km in diameter (the approximate size indicated by the exposed zone of brecciation and shatter coning, and by the remnant ring of Huronian strata to the north) would fracture the crust profoundly; the effects would extend to the mantle. Shock heating of the rocks, and pressure reduction by excavation of the crater, would both tend to promote partial melting at depth. Migration of the melt would be facilitated by the impact fracturing, and be promoted also by rising temperatures due to the insulating effect of the lens of breccias in the crater above. The Irruptive as we see it today is where it should be in such an interpretation: it occupies the zone of most intense disturbance, between the fractured footwall and the overlying blanket of chilled melt and impact debris – the Onaping Formation.

It should not be forgotten that the Sudbury structure lies in a nickel- and copper-bearing metallogenic province, in which deposits associated with basic rocks extend from Duluth, Minnesota, in the southwest to Cobalt, Ontario, in the northeast. Moreover, given our present uncertainty about the ages of the breccias and the norite, it is not impossible that the Sudbury event took place soon after the magmatism that produced the Nipissing diabases and gabbros and their Fe-Ni-Cu sulphides. In such circumstances it would not be so surprising if a meteorite impacting at Sudbury triggered the evolution of an ore-bearing basic magma. The final composition of the resulting igneous body might well show certain peculiarities (such as an uncommonly high quartz content) reflecting its unusual origin (Naldrett *et al.,* 1970, p. 147).

The foregoing scenario is presented not as an "Authorized Version", but simply to show that a viable working hypothesis can be constructed on an impact basis. Advocates of the impact theory regard shock metamorphism as a central fact of Sudbury geology. In the apparent absence of a satisfactory alternative explanation for the shock, it is difficult to escape the conclusion that an impact must have occurred.

EVIDENCE AGAINST IMPACT

Others see it differently. Their arguments are for the most part directed to a number of aspects of Sudbury geology that appear to relate it uniquely to its regional setting, making it a necessary and integral part of its environment, and thus implying an unacceptably high degree of coincidence in the site of a fortuitous impact. Other arguments relate to differences between the geology of the Sudbury structure and that of a "typical" impact crater.

Responsible critics do not dispute the presence of shock metamorphism at Sudbury, nor do they deny that meteorite impacts cause shock. They imply, rather, that the observed shock effects can also be produced in other places in other ways, for example by unusually violent forms of explosive volcanism, of which the Onaping "event" would be a prime example. The experimental "calibration" of shock processes is questioned on the basis of its completeness and general applicability to natural systems; the failure to recognize shock features in volcanic and other rocks lacking impact association is attributed to their being overlooked because of unfamiliarity (as happened at Sudbury).

These divergent views really reflect different ways of looking at and weighting the evidence, and there is no easy means of resolving them at present. Both sides might concede that most geologists prefer, at least initially, the answer that does the least violence to general geological principles and their personal experience. Let us now briefly consider some of the problems raised by the impact theory.

The major objections to impact at Sudbury are well summarized by Card and Hutchinson in this volume. The view that the basin is uniquely situated near the junction of three structural provinces of the Canadian Shield, and at the intersection of two regional fault systems, is a variant of the "lineament" argument often raised against proposed impact sites to show that their distribution is not random. French (1968, pp. 8-9) discusses this argument and some general objections to it. In the present specific case it may be observed that: at the time of the event Huronian rocks covered the area and the Superior-Southern province boundary was probably far removed from its present location; the Grenville front seems to be a younger phenomenon (although Sudbury is close to a major Huronian depositional axis); the Onaping-Timiskaming and Murray fault systems have many other points of intersection and Sudbury is not their focus.

A more remarkable coincidence is suggested by recent paleocurrent and stratigraphic studies of the Huronian which indicate that the Sudbury area was a positive element influencing Huronian deposition in pre-event times. If further work confirms this interpretation and shows that the Sudbury structure is locally unique in this respect, an endogenic origin would be favoured. Other, post-event, paleocurrent data are discussed by Cantin and Walker in this volume. In the absence of any detailed model for Sudbury, or any general understanding of the post-impact evolution of craters of this exceptional magnitude, the critical implications of their work are difficult to evaluate, but it appears to pose problems for the impact theory.

A possible explanation of the relationship of the Irruptive to other mineralized, ring-shaped, differentiated, gabbroic – and undoubtedly endogenic – intrusions in the Sudbury area has been suggested in the preceding section.

Card and Hutchinson also raise the intriguing and cogent question of metallogenic development at Sudbury. They suggest that Sudbury is like many another centre in this respect, showing the characteristic mineral deposits "generated during successive stages in the prolonged evolution of a major Proterozoic volcanic-sedimentary complex". On this view the nickel-copper sulphides form an integral part of a succession from sulphide-facies iron formation in the lower Huronian to the lead-zinc-copper deposits in the Whitewater Group, which are interpreted as typical volcanogenic sulphides, consanguineous with felsic pyroclastics. Again, the nature of the Onaping Formation is at the heart of the problem: is it volcanic, as originally thought, or the product of impact, as apparently demanded by its shock metamorphism?

Before discussing specific differences between Sudbury and other presumed impact craters, it is well to recall that Sudbury may be not only a unique structure, but also a unique astrobleme. Among possible terrestrial examples, only Vredefort is comparable in size and age, and its geology is quite different. The characteristics of "typical" meteorite impact craters are by now fairly well known, and as Dence points out in this volume Sudbury is non-typical in several important respects. Without a model for comparison, Sudbury – sui generis as ever – will continue to resist interpretation.

Sudbury lacks the central uplift that characterizes large, complex craters. In its place we find the basin structure of the Whitewater rocks, and the Irruptive itself. Dence and French (this volume) present several possible explanations for this, but in the present state of our knowledge we cannot judge them.

Sudbury also lacks circularity. This lack is found to be more apparent than real once it is appreciated that the irregularly elliptical Irruptive lies well within the proposed original crater margin, and may be controlled by fracturing. Observed compression and faulting are usually attributed to Grenville (northwest-southeast) deformation.

The emplacement of the Irruptive, and the various other igneous and metamorphic events in Sudbury's long history, may have combined to obscure many other typical impact features, such as a negative gravity anomaly and the more unstable mineral products of intense shock such as stishovite and coesite. The passage of two billion years is sufficient alone to ensure the disappearance of lechatelierite, maskelynite and other glassy shock features. Note that all these except stishovite have already been found at the Wanapitei structure.

Several miles of erosion would readily account for a lack of preserved meteoritic fragments. Such material is in any case vanishingly rare at all astroblemes, though fragments are found in the surface debris from recent craters. Note also that whereas iron meteorites may be fairly conspicuous, stony ones are easily overlooked and the frozen particles composing comet heads would leave no trace.

Large impacts produce large volumes of melt, by fusion of the target rocks. Sudbury, it appears, is unique among possible terrestrial impact sites in having no recognizable major melt unit but instead an apparently endogenic igneous complex at its centre (though other, unrecognized examples may exist: see French, 1970). Small melt bodies are found at the base of the Onaping (Peredery, this volume). Proponents of impact believe that the abundant glassy fragments throughout the Onaping represent the missing impact melt – perhaps disrupted by quenching in water – but might the Irruptive itself be, at least in part, a direct product of impact? Dence proposes that most of the Irruptive may have arisen in this way; Dietz certainly answers yes for the sub-layer and particularly its constituent sulphides. Much work remains to be done on this topic.

If the shock metamorphism in the Onaping Formation presents a major obstacle for the conventional (volcano-tectonic) theory, the abundance of carbon and iron sulphide in the same rocks is a considerable problem on the impact side. The prominence of these components – which account respectively for the rock's black appearance when fresh and its orange colour when weathered – distinguishes it from many otherwise similar "suevite" breccias at the Ries Kessel (the type astrobleme) and other more or less well authenticated impact sites. It has been suggested that impact took place in a body of water which contained primitive life forms that provided a source of carbon for the Onaping's matrix. Not only the matrix but many inclusions carry sulphides; this may reflect the greater chemical mobility of sulphur or the existence of sulphide rich rocks in the target area, but the mere presence of sulphur in the rock constitutes a notable anomaly. Possible explanations for the carbon are presented in this volume by both Peredery and French. The origin of the sulphur is even less understood. To the extent that sulphide rich rocks are a distinguishing feature of Sudbury geology generally, it might be argued that this favours an endogenic process. The Onaping sulphides are nickeliferous (Desborough and Larson, 1970): this observation can of course be interpreted in more than one way.

CONCLUSION

This brief introduction cannot do justice to the complexities of the Sudbury story. The impact side of the argument has deliberately been emphasized here, because this was after all a main purpose of the Special Session at Sudbury. This volume is intended to answer a demand for information that is not readily provided by the existing literature.

The reader must study the papers and judge the impact controversy for himself. Probably he will feel, like most geologists who have tried to decipher the Sudbury record, that further work needs to be done. It is likely that there will always be pieces missing from the Sudbury puzzle; for the present, it is worth noting that at Sudbury itself many workers have taken the impact theory seriously enough to adopt it as a working hypothesis.

If it can be shown that some as yet undocumented form of volcanism can reproduce the features of shock metamorphism, then a real alternative to impact will be available. In the meantime, it has been argued that there is a certain lack of intellectual satisfaction in a hypothetical cryptovolcanic process that must mimic so exactly the evolution of a meteorite crater.

No doubt the future holds solutions for the problems we debate. How can we shorten the waiting period? Continuing terrestrial and lunar studies will do much to advance our general knowledge of both impact and volcanic processes. Greater awareness of shock metamorphism, throughout the geological community, will probably lead to some discoveries and perhaps some surprises.

At Sudbury, a most pressing need is for additional, high quality radiometric age determinations, to provide a more secure framework for the observed events. Both sides in the impact controversy will presumably agree that the shatter cones, the pseudotachylitic Sudbury breccia and the Onaping melts and glass-rich breccias, most if not all of which are shock products whatever their ultimate origin, are closely related in time and mark the formation of the Sudbury "cryptoexplosion" structure. However, neither the North Range Sudbury breccia matrix nor the melt – the most promising candidates – have yet been successfully dated. Moreover, although Fairbairn and his associates obtained an excellent 1.7 b.y. Rb-Sr isochron for the North Range Irruptive, the norite itself was represented by only one sample and its age cannot be regarded as settled. Especially as subsequent attemps to date it (Souch, Podolsky *et al.*, 1969; Gibbins *et al.*, this volume) have produced evidence that it may be much older, and not necessarily consanguineous. Several radiometric dating programs are currently attacking these problems, and successful results could place useful constraints on our theories. It should not be forgotten, however, that so far Sudbury has proved to be something of a type area for demonstrating the difficulties of obtaining meaningful ages from rocks affected by multiple igneous and metamorphic events.

Chemistry provides another promising approach. Systematic studies of the elemental and isotopic compositions of the Irruptive and the Onaping Formation can shed light on their origins (see Peredery, this volume) and also on their mutual relationships. Major element chemical analyses of the Irruptive rocks are presently scheduled, as an extension of the work reported by Souch, Podolsky *et al.*, (1969) and Naldrett *et al.*, (1970).

More complete data on shock metamorphic zoning at Sudbury (cf. Dence, 1968 and this volume) might aid in reconstructing the immediate post-event configuration of the rocks, now much obscured by folding, faulting and intrusion. In particular, the mapping of shatter cone orientations (Bray *et al.*, 1966) in full detail could yield important dividends, because studies at several other cryptoexplosion structures have demonstrated that cone apexes initially tend to point towards a common origin.

The suggestion is often made that a deep borehole in the centre of the Sudbury Basin would solve many problems. This may be so, but existing information, and the great scale of the structure, alike suggest that many miles of difficult and expensive drilling separate us from direct examination of the roots of this problem. There may be more to gain from indirect geophysical investigations of the deep structure. Popelar (this volume) shows what can be done with gravity techniques; in the future, improved seismic methods may also prove useful.

Last but not necessarily least, the potential of careful, conventional field observation and mapping should not be overlooked. If nothing else, recent developments at Sudbury have clearly shown that after more than three quarters of a century of geological study, there may still be new – even shocking – discoveries to be made.

ACKNOWLEDGEMENTS

The National Research Council of Canada contributed greatly to the quality of the Special Session at Sudbury by providing travel grants for the two speakers from the United States, R. S. Dietz and B. M. French. The Geological Survey of Canada has generously made funds available to assist in the publication of this volume.

The organization and running of the Special Session itself would have been impossible without the assistance of the other Sudbury committee members and of Prof. A. J. Naldrett, the co-chairman. The writer also gratefully acknowledges the generous help and encouragement of his colleagues, and thanks The International Nickel Company of Canada, Limited, for permission to undertake the project and to publish this introduction.

REFERENCES

Bray, J. G. and Geological Staff (1966)–Shatter cones at Sudbury; J. Geol. vol. 74, pp. 243-245.

Bray, J. G. and Peredery W. V. (1971) – Field guide for excursion (1), Sudbury Basin, 19 p., Geol. Assoc. Can. – Mineral Assoc. Can., Joint Annual Meeting, Sudbury.

Dence, M. R. (1968) – Shock zoning at Canadian craters: petrography and structural implications; *In* Shock Metamorphism of Natural Materials, French, B. M. and Short, N. M. *(Editors)* pp. 169-184, Mono Book Corp., Baltimore.

Desborough, G. A. and Larson, R. R. (1970) – Nickel-bearing iron sulfides in the Onaping Formation, Sudbury Basin, Ontario; Econ. Geol. vol. 65, pp. 728-730.

Dietz, R. S. (1964) – Sudbury structure as an astrobleme; J. Geol., vol. 72, pp. 412-434.

Fairbairn, H. W., Hurley, P. M. and Pinson, W. H. (1968) – Rb-Sr whole-rock age of the Sudbury lopolith and basin sediments; Can. J. Earth Sci., vol. 5, pp. 707-714.

French, B. M. (1967) – Sudbury structure, Ontario: some petrographic evidence for origin by meteorite impact; Science, vol. 156, pp. 1094-1098.

......(1968) – Shock metamorphism as a geological process; *In* Shock Metamorphism of Natural Materials, French, B. M. and Short, N. M. *(Editors)* pp. 1-17, Mono Book Corp., Baltimore.

......(1970) – Possible relations between meteorite impact and igneous petrogenesis, as indicated by the Sudbury structure, Ontario, Canada: Bull. Volcanologique. vol. 34, pp. 466-517

French, B. M. and Short, N. M. *(Editors)* (1968) – Shock Metamorphism of Natural Materials; 644 p., Mono Book Corp., Baltimore.

Hörz, F. *(Editor)* (1971) – Meteorite Impact and Volcanism; J. Geophys. Res. vol. 76, 23, pp. 5381-5798.

Naldrett, A. J., Bray, J. G., Gasparrini, E. L., Podolsky, T., and Rucklidge, J. C. (1970) – Cryptic variation and the petrology of the Sudbury Nickel Irruptive; Econ., Geol., vol. 65, pp. 122-155.

Naldrett, A. J., Armstrong, T., Hewins, R. H., Pattison, E. F., and Phipps, D. (1971) – Field Guide to the Sudbury Nickel Irruptive; 27 p., Geol. Assoc. Can. – Mineral Assoc. Can., Joint Annual Meeting, Sudbury.

Ontario Department of Mines (1969) – Sudbury Mining Area, Sudbury District, Map 2170, 1 inch = 1 mile.

Souch, B. E., Podolsky, T. and Geological Staff (1969) – The sulfide ores of Sudbury: their particular relationship to a distinctive inclusion-bearing facies of the Nickel Irruptive; Econ. Geol. Monograph 4, pp. 252-261.

21

Reprinted from pages 7-8 and 10-18 of *Geol. Assoc. Canada Spec. Paper 10,* 1972.

METEORITE IMPACT CRATERS AND THE STRUCTURE OF THE SUDBURY BASIN

M. R. Dence
Gravity Division, Earth Physics Branch, Department of Energy, Mines and Resources, Ottawa, Ontario

[*Editor's Note:* Figure 1 has been omitted.]

INTRODUCTION

The seed sown by V. B. Meen when he began scientific study of the New Quebec Crater in 1950 (Meen, 1950) has grown into a comprehensive investigation of possible impact structures in Canada. Predictions that the Canadian Shield would be a rich hunting ground for such sites (Beals *et al.*, 1956) have proved essentially correct so that in world lists of somewhat more than one hundred sites about twenty are in Canada (Short and Bunch, 1968; Millman, 1971; Dence, 1972). Among them, on the basis of the original discoveries of Dietz (1964), and the later work of Guy-Bray *et al.* (1966) and of French (1967), is numbered the Sudbury Basin.

The addition of Sudbury to these lists is the result of steady progress in the study and understanding of impact structures in many parts of the world which makes it possible to indicate within reasonably well defined limits the effects of differences in size, setting, level of erosion and subsequent geological events. Certainly the Sudbury Basin does not conform to the main criteria – circularity, uplifted and overturned rim rocks, negative gravity anomaly – which guided the initial stages of impact crater studies. Nor does it have a distinct central uplift such as those now recognized at most large impact structures. Rather the case for the impact origin of the Basin rests on the presence within the structure of characteristic rock deformations attributed uniquely to the action of high pressure shock waves. Such deformations include shatter coning (Dietz, 1968) recognized at the mesoscopic scale, and a variety of microscopically recognized mineralogical changes collectively known as shock metamorphism. Such changes have been described by many authors (see papers in French and Short, 1968; Hörz, 1971) and will not be enumerated here. Suffice it to note that where careful work has been done clear distinctions have been drawn between mineral deformations carried out at slow (static) strain rates and those at high (dynamic) strain rates (Borg and Heard, 1970; Carter, 1971), in keeping with the fundamental differences between the pressure-temperature fields of endogenic regional and exogenic shock metamorphism (Stöffler, 1971).

Earlier interpretations of the Sudbury structure as being of volcanic-intrusive origin (Collins, 1934; Thomson, 1956; Williams, 1956; Wilson, 1956) drew many parallels with ignimbritic volcanic complexes and with basic layered intrusions but had difficulty accounting for several unique aspects of the Sudbury structure, including the abundance and variety of breccias (Speers, 1957; Stevenson, 1961; Souch *et al.*, 1969) and the relatively siliceous composition of the Irruptive rocks (Collins, 1934; Naldrett *et al.*, 1970). The impact hypothesis, viewed simply, appears to alleviate the problem of the breccias, but not of the Irruptive. However closer comparison with large impact structures shows that the Sudbury structure has several unique characteristics which have prompted consideration of models combining impact with endogenic igneous events (French, 1971; Thomson, 1969).

In this article some familiarity with the geology of the Sudbury Basin is assumed. The salient structural features of large circular impact structures are reviewed and compared with those of the Sudbury Basin and surrounding region. Some new data on the distribution of shock metamorphism in the country rocks are introduced and comparisons are made between the igneous rocks and breccias of the Sudbury structure and those from well known impact sites. From this basis a scenario is developed in which the evolution of the basin is seen as the outcome of impact into a crust in which the thermal gradient was high relative to standard cratonic areas.

STRUCTURE OF IMPACT CRATERS

The study of young meteorite craters with which meteorites are associated has enabled a sequence of crater types to be recognized (Krinov, 1963). The smallest are simple pits, but craters more than 90 m across have among their ejecta glassy "bombs" or "impactites" formed by the action of strong shock waves. This is in agreement with calculations that such craters are formed by meteorites which are sufficiently large that they land with hypervelocity, typically about 15 km/sec, having been only partly decelerated by atmospheric retardation. Shock waves, high pressure waves travelling at supersonic velocities, are generated at impact in both target rocks and the meteorite and result in vaporization, fusion, phase changes or deformation depending on the pressures attained. The mechanics of cratering by hypervelocity impact have been described by Shoemaker (1960, 1963), Bjork (1961), Gault *et al.* (1968) and others and the resulting shock effects in the ejecta described by Chao (1967 a, b), Keiffer (1971) and others for craters such as Henbury, Wabar and Barringer Meteor Crater.

In older craters erosion and weathering effectively remove all sign of meteoritic material so that identification is based on structural, geophysical and petrographic similarities to young craters. Of the various criteria used, that of shock metamorphism is the most powerful and direct and has enabled the recognition of impact sites which otherwise show little apparent resemblance to craters of the Barringer type. The differences are not merely a matter of depth of erosion or of coverage by younger sediments or water, but result from major differences in crater structure which appear to be mainly functions of crater size and country rock type.

Craters similar to Barringer Meteor Crater have been called *simple* (Dence, 1964) and are characterized by uplifted rims around a central depression underlain by a lens of breccia which is thickest in the middle. Craters in this class are relatively small. Indeed, Barringer (1.2 km) is the largest, well known, simple crater formed in sedimentary rocks and Brent, estimated to have been about 4 km across when fresh, the largest simple crater in crystalline rocks. All larger impact sites for which sufficient information is available have been shown to have a *complex* structure with a central uplift and a subdued or even depressed zone at the margin. Distinct central uplifts are apparent in even relatively small structures in sedimentary rocks (e.g. Steinheim Basin, 3.5 km), but central uplifts are not conspicuous in craters with diameters less than 15 km formed in crystalline rocks.

The transition in crater structure with increasing size is clearly illustrated by the Clearwater Lake craters (Dence *et al.*, 1965), the smaller East crater having a central uplift which was revealed only by drilling (Dence, 1965), the larger West crater having a distinct central uplift and a well-developed ring structure (Figure 1). The latter consists of a central peak of moderately to weakly shocked rocks, an annular ring of breccias and melt rocks, a graben-like depression and a peripheral aureole of mildly disturbed rocks. The axis of the depression is usually taken as the circumference of the crater. It consists of downdropped rim rocks, among which in this case Ordovician carbonate rocks are preserved. Other large impact craters such as Carswell (Innes, 1964), Charlevoix (Rondot, 1968; Robertson, 1968), Siljan (Svennson, 1971) and Manicouagan (Dence, 1965; Murtaugh and Currie, 1969) show the same general features though differing in relative level of erosion. Another example, closer to the Sudbury structure in size and age and in level of erosion, is the Vredefort dome (Manton, 1962, 1964; Hargraves, 1971).

Several proposals have been made to explain the differences between simple and complex craters. Roddy (1968) and others suggest that complex craters result from the impact of large, low density bolides, such as comets, forming relatively shallow craters in which rarefactions of the shock wave lead to an inwards and upwards motion in the centre of the crater. Dence (1968, 1971) also favours rapid formation of the central uplift but at a slightly later stage driven mainly by gravitational collapse of the uplifted rim. This may be illustrated by several examples (Figure 2). In each case the transient cavity excavated by the shock wave and its rarefactions is drawn at its maximum development. In the case of the simple Brent crater the final crater profile changes relatively little, except for slumping and erosion. The other three have central uplifts. The well stratified, originally sub-horizontal country rocks at Gosses Bluff and Vredefort allow independent estimates of the profile of the transient cavity to be made by simply returning the uplifted strata to their original positions. At Charlevoix only the cover of lower Paleozoic sedimentary rocks can be treated in this way, but the crater is in many ways typical of craters of this size formed in crystalline rocks and favourable exposures have allowed many data to be amassed there on the distribution of shock effects within the crater (Robertson, 1968). This distribution is indicated in the diagram by vertical line patterns of varying intensity for crystallographic indications of shock deformation, and by inverted 'V's for shatter coning. Interpreted in this way the relative dimensions of the transient cavity vary little with crater size for similar target rocks, the ratio of depth to diameter being approximately 1 to 4. In terms of the final dimensions of a large complex crater, the depth of the transient cavity is about one-fifth of the diameter of the axis of the ring depression or one-eighth to one-tenth the diameter of the peripheral disturbed zone. At Vredefort, for example, where stratigraphic controls indicate uplift in the centre of 15 to 20 km. the axial trace of the ring synclinal structure has a diameter of 100 km and the outer limits of the structure are at a distance of 80 km from the centre (Manton, 1962).

Applying this analysis to the Sudbury structure, it is apparent that it differs from the typical complex crater in having a central basin rather than a central uplift. However Speers (1957) recognized that the basin lies inside a domal structure which is outlined by a discontinuous ring of downfaulted and tilted blocks of Huronian sedimentary rocks on the north and west side (Figure 4). Altogether about 120 degrees of arc of radius about 50 km are outlined by the Huronian rocks. On the south and east sides relationships are complicated by subsequent faulting and folding but the domal structure is still apparent as the rocks show a general decrease in age away from the centre. Thus the outer parts of the Sudbury structure show a peripheral depression similar to that at Vredefort, and indeed the two structures have similar dimensions, although at Sudbury the outermost limits are not apparent in the gneisses of the Superior province. In the centre there are significant differences between the two, and these are considered in a later section.

SHOCK ZONING

The distribution of zones of shock metamorphism in the basement rocks of complex craters is a further aspect of their structure by which relative dimensions can be compared. Such zones are exposed only in craters in which the breccias and melt rocks have been mostly removed by erosion, the best example being the previously mentioned Charlevoix crater, Quebec (Robertson, 1968). The pattern established by Robertson is based largely on intensity of planar features in quartz, according to the general scheme set up by Robertson *et al.* (1968). At Charlevoix concentric zones of shock metamorphism were established showing a decrease in intensity with distance from the central peak. The weakest indications of planar features in quartz were mapped at 10 km from the centre. This distance corresponds to about 70 per cent of the radius of the inferred transient cavity (Figure 2). Less complete results for other large craters are in general agreement with this figure, including results for samples kindly provided by W. I. Manton and D. P. Gold from the Vredefort ring. Shatter cones, by comparison, extend to somewhat greater distances from the centre of the structure than do planar features in quartz, depending in both cases on the lithology of the rocks involved.

At Sudbury, extensive sample collecting connected with the regional gravity survey by Popelar (1971, 1972) together with thin sections of samples kindly provided by K. Card, has enabled the main features of the distribution of quartz planar elements and other manifestations of shock metamorphism to be mapped. As shock in the Onaping Formation and adjacent quartzite breccias has been described by French (1968, 1969) and Peredery (1972) and the distribution of shatter cones in the country rocks by Dietz and Butler (1964) and Guy-Bray *et al.* (1966), attention was concentrated on petrographic study of the footwall rocks.

Recrystallization by regional metamorphism has wiped out any petrographic evidence for shock in the footwall rocks of the South Range. North and northwest of the Irruptive, however, the rocks have been only gently annealed, except immediately adjacent to igneous intrusions, and have retained

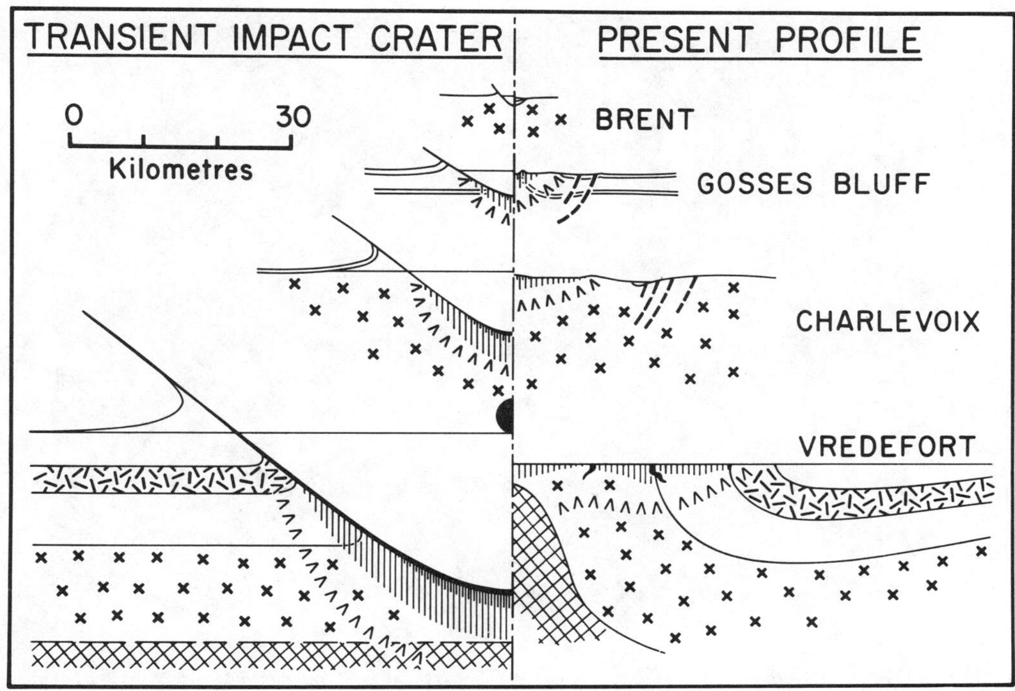

Figure 2: Cross-sections of four representative impact craters showing (right) the present profile and (left) a reconstruction of the transient cavity (primary crater) excavated by the shock wave. Immediately afterwards the crater is modified by slumping (Brent) or central uplift (others). Legend: Crosses: crystalline rocks; horizontal ruling: sedimentary rocks; stipple: volcanic rocks; cross-hatch: deep crustal rocks (mafic?); vertical ruling: zones of shock metamorphism; inverted 'V': outer limit of shatter coning; black: impact melt and bolide (Vredefort, only).

the texture they had prior to the Sudbury event. In these rocks shock deformation structures are weakly developed. The grade of shock metamorphism is low, so it is expressed primarily by kinking in mica and the development of planar elements in quartz. Typical examples of the latter are shown in Figure 3. They possess the sharpness, parallelism, spacing and somewhat irregular development typical of decorated planar elements in quartz (Engelhardt and Bertsch, 1969). The examples shown, and the great majority of others measured in these rocks, are oriented parallel to the basal plane of the crystal containing them. They thus are called Type A planar features in the terminology of Robertson *et al.* (1968). A few are of Type B with planar features parallel to the omega form ($10\bar{1}3$). This association is in good agreement with experience from other craters such as Brent and Charlevoix.

The distribution of planar features in quartz in the footwall rocks north and west of the Irruptive is shown in Figure 4. A simple classification was employed using three categories: (i) rocks with no planar features in quartz, (ii) rocks with Type A basal features weakly developed, (iii) rocks with Type A features distinctly developed as in Figure 3. The rare examples of Type B features were included in the latter category for convenience. The number of data points are relatively few owing to the generally poor access into the area, but nonetheless a distinct pattern is evident. Quartz planar features are present wherever primary (unrecrystallized) quartz is preserved in a zone adjacent to the North Range of the Irruptive and up to 8 km wide. Both the Levack complex migmatites and Cartier granite are so deformed. Within the zone there is some irregularity in the intensity of planar feature development, possibly as a result of later faulting and other disturbances.

Shatter coning is observed within the same zone, but is also found somewhat farther from the Irruptive, having been recorded up to 13 km north of the Irruptive near Milnet and more than 15 km from the Irruptive on the south (Guy-Bray *et al.*, 1966). This is in general agreement with the relationships between quartz planar features and shatter cones observed at Gosses Bluff (Cook, 1968), Charlevoix (Robertson, 1968) and Vredefort (work in progress) as illustrated diagrammatically in Figure 2. Kinking in mica is also more extensive than quartz planar features in the gneisses north and west of the Irruptive, in agreement with experience elsewhere, but in the case of mica it is more difficult to make a clear distinction between shock and tectonically induced kinking.

BRECCIAS AND IGNEOUS ROCKS

Besides the recent evidence for shock metamorphism it has long been known that all footwall rocks that predate the Irruptive are cut by an irregular network of breccias known variously as Sudbury or Levack breccias. They are somewhat more widespread (Speers, 1957) than the distribution of shock features and are characterized by extreme variability in their distribution and local geometry. Angular to subrounded blocks of the country rocks, some transported at least hundreds of metres, are held in a black, finely comminuted to locally fused matrix of the same materials. Dietz (1964) compared them to the extensively developed pseudotachylite veins and networks of the Vredefort dome (Shand, 1916; Bischoff, 1962), and

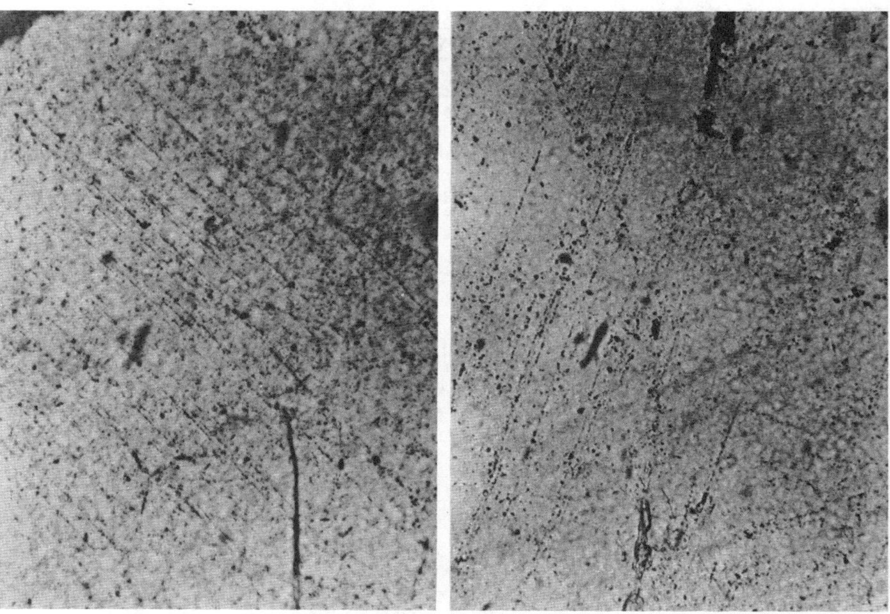

Figure 3: Two examples of decorated planar features in quartz from crystalline rocks of the North Range footwall. In both cases the features are parallel to the basal plane. Samples from west of Windy Lake (left) and south of Milnet (right).

suggested that in both cases the breccias were formed as a direct result of the action of shock waves. The similarity is indeed striking and has been strengthened by the recognition of shock effects in the clasts of both Sudbury (French, 1971) and Vredefort (Wilshire, 1971) breccias. Pseudotachylites are also known from the central regions of other large impact structures (Dence, 1965). Their location in the central uplifts of these structures makes it likely that they formed not as the direct result of shock metamorphism but from the injection of locally crushed and frictionally melted materials created by the violent movements attendant on the formation of central uplifts. If so, their abundance at Sudbury indicates that similar movements took place during the development of the Sudbury structure.

Overlying the shocked, fractured and brecciated basement rocks are the main rock units of the Sudbury Basin. These are diagrammatically compared to the rocks of a typical large impact structure in Figure 5. No one impact crater can be taken as a standard for all, so the sequence presented in the figure is a composite based mainly in the lower part on relationships observed at the moderately eroded Manicouagan and West Clearwater Lake craters and for the upper part on the less deeply eroded Lake St. Martin and Ries craters. For clarity the columns are not drawn to scale, although the relative dimensions of rock units are indicated approximately.

An overall similarity between the two columns is clearly apparent, but there are also significant differences. They have in common:

1) the general sequence of basement rocks overlain by a breccia layer, igneous rocks, additional breccias and an uppermost unit of sedimentary rocks,

2) similar deformation of the basement rocks, as discussed above,

3) relatively thin, locally discontinuous lower breccias containing clasts derived in part from the footwall and in part from deeper layers, closer to the centre of the structures,

4) substantial layers of igneous rocks overlying the basal breccias with sharp discontinuity, the bulk composition of the igneous rocks not differing greatly from that of the basement,

5) overlying breccias of two types: a lower unit comprising large, generally weakly shocked blocks of the upper levels of the country rocks and an upper unit of mixed breccias, with clasts derived mostly from deeper levels of the country rocks, many of which are moderately to strongly shocked up to and including shock melted materials,

6) gradational contacts between the mixed breccias and later sedimentary rocks, the basal units of which are derived in large part from the breccias,

7) local cross-cutting relationships between rock units, such as veins of melt cutting adjacent breccia units. However the overall conformity of the units suggests that the sequence observed from lower breccias to uppermost sedimentary rocks is in fact the order of emplacement. In impact craters the entire sequence apart from later sediments is emplaced in rapid succession so that late movements in the structure can tap unconsolidated materials forming dykes and veins of both melts and breccias. Similar conditions are inferred for the Sudbury rocks.

The more apparent differences between the two columns include:

(1) the character of the matrices of the lower breccias, which in a standard crater is clastic but at Sudbury is a combination of silicate melt (siliceous norite or quartz diorite) and sulphides (Naldrett and Kullerud, 1967; Souch *et al.*, 1969). There are no data on deformation in Sudbury sub-layer inclusions comparable to the shock metamorphism of

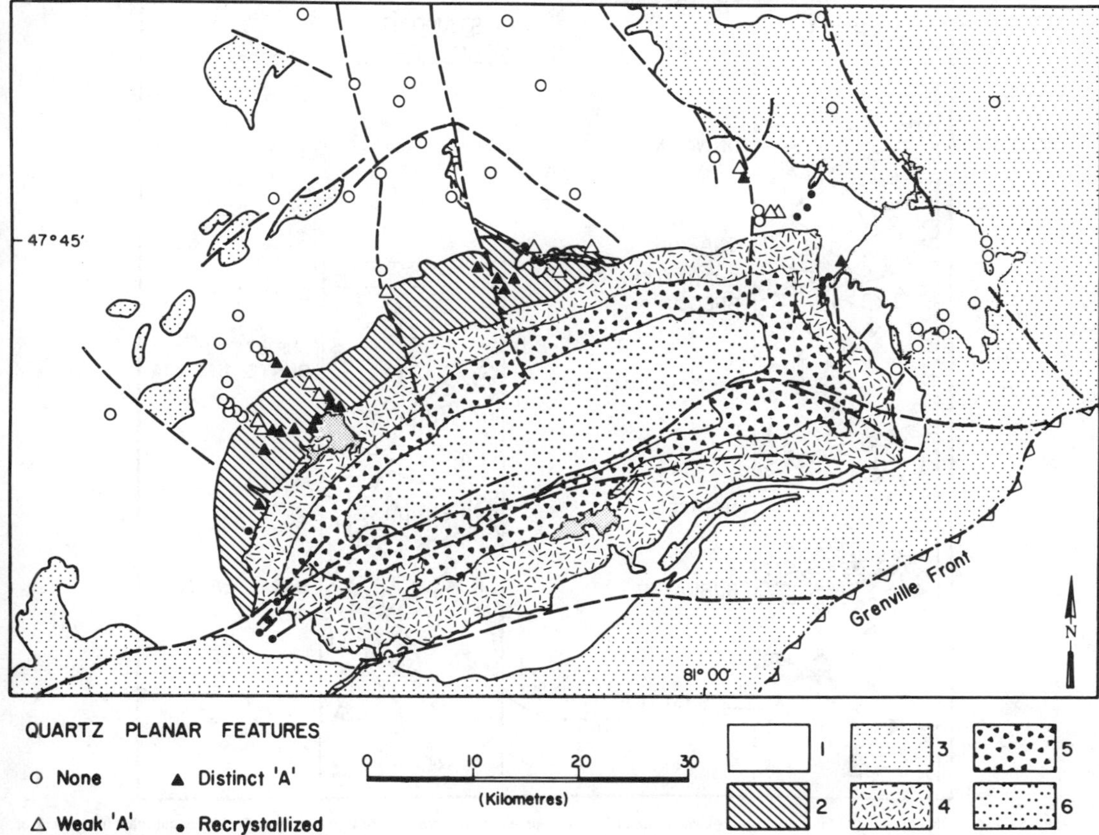

Figure 4: Distribution of planar features in quartz in footwall rocks of the Sudbury Irruptive. Footwall rocks of the South Range are all recrystallized and are not shown. Weak 'A' indicates basal planar features are present in some quartz grains, generally only at grain boundaries. Distinct 'A' indicates basal and rare omega planar features are present in most or all quartz grains, commonly throughout the grain as in Figure 3.

Rock units:

Archean and lower Proterozoic

1. Granites, granitic gneisses, greenstones
2. Levack migmatite complex

Lower Proterozoic

3. Huronian metasedimentary and basic intrusive rocks
4. Sudbury Irruptive
5. Onaping Formation
6. Whitewater sedimentary rocks

---

fragments in impact crater basal breccias, but textures resembling recrystallized maskelynite have been noted in plagioclases in some gabbroic fragments (Podolsky, personal communication),

(2) the great thickness of the igneous rocks of the Irruptive when compared with the melt rocks of standard craters. The matter of the volume of melt produced by hypervelocity impact has been discussed with regard to Sudbury by French (1971). He concluded that only the melted materials in the Onaping Formation and adjacent quartzite breccias could be attributed to impact melting. However the model he considered was of a crater 50 km in diameter, or approximately the same size as the Manicouagan crater, where there is about 400 km^3 of melt (Dence, 1971). A crater 80 to 100 km in diameter may be expected to contain 2000 km^3 of melt. Even with allowance for erosion, this volume is considerably more than that of the melt contained in the Onaping Formation. However, it is undoubtedly less than the total volume of the Irruptive which probably had a volume two to three times as great (French, 1971). It follows, as suggested by Dietz (1972), that a substantial proportion, possibly one-third to one-half of the Irruptive may be impact melt, but that a second source is required to account for the total volume,

(3) the partly clastic, partly igneous matrix of the quartzite breccias at Sudbury (Peredery, 1972) as compared with the entirely clastic matrix of the Ries Bunte breccia. This difference may be largely a matter of level of exposure, as only the peripheral parts of the Bunte breccia are exposed at the Ries. The present position of the quartzite breccias may be the

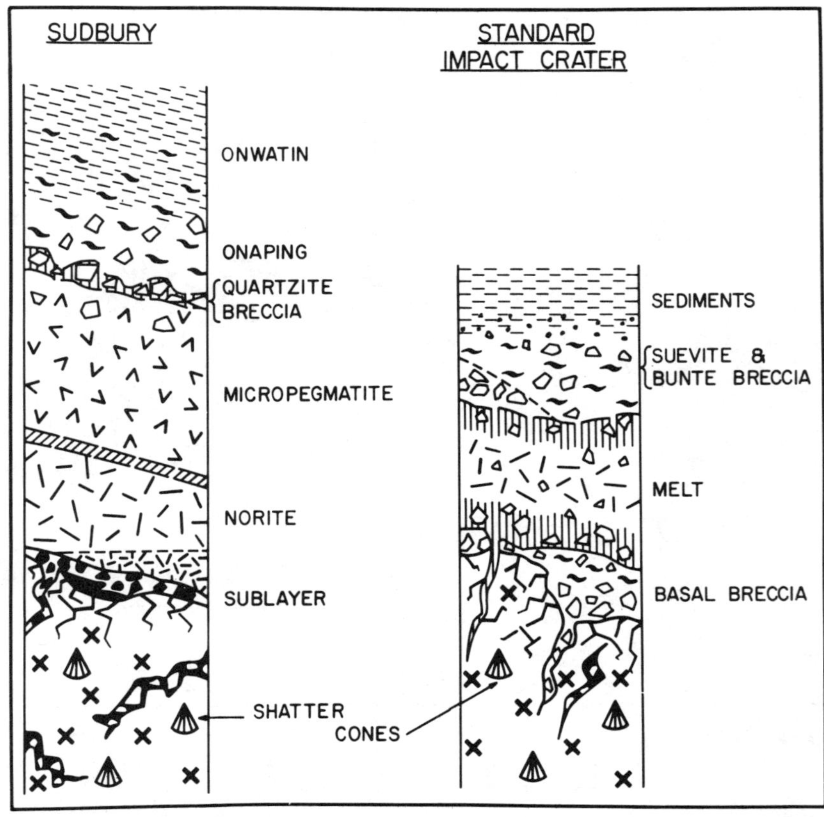

Figure 5: Main rock units of the Sudbury complex diagrammatically compared (not to scale) with units of a standard large complex impact crater. The main units are:

i) Basement crystalline rocks (crosses) mildly shocked (shatter cones), fractured, and cut by pseudotachylite breccia veins (black matrix).
ii) Lower breccia unit: Sub-layer breccia (Sudbury) of mafic inclusions with quartz diorite/norite-sulphide matrix; basal breccia (standard) of shocked basement rocks including shock melted glasses.
iii) Igneous rock unit: Sudbury Irruptive with norite at base (varying in compostion), transition zone and micropegmatite; melt rocks (standard crater) with fine-grained, inclusion-rich margins and coarser grained centre.
iv) Upper breccia unit: Quartzite breccia (Sudbury) of weakly shocked quartzite and granitic boulders in a partially molten matrix; Bunte breccia (standard crater) or equivalents of weakly shocked upper target rocks in a clastic matrix.
v) Fallout breccia unit: Onaping Formation (Sudbury) of variously shocked and melted fragments in a matrix which grades upwards into sedimentary rocks; Suevite (standard crater) breccia of variously shocked and melted fragments.
vi) Sedimentary fill unit: Onwatin slate (Sudbury); clastic or chemical sedimentary rocks (standard crater).

result of slumping from the margins of the Sudbury crater,

(4) the exceptional thickness of the Onaping Formation when compared with the Ries suevite. Again this may be in part a matter of comparing the marginal areas of the Ries with an interior section of the Sudbury structure. However, the high carbon content and crude bedding of the upper Onaping and its gradational contact with the Onwatin slate have prompted suggestions that the Sudbury area was water-covered at the time of impact. In this interpretation the Onaping consists of ejecta swept back into the crater with the turbulent flood of returning water (French, 1971).

(5) the ellipticity of the Sudbury structure. This requires distortion of original circularity by subsequent orogenic events.

In sum the similarities between the rock units of the Sudbury structure and those of typical large impact craters are those of structural position and overall texture. A common genesis is implied. The main differences, the exceptional amounts of melted rock at Sudbury, may result, as French (1971) and others suggest, from volcanic events following the impact. Alternatively, by Ockham's Razor ("multiplicity ought not to be posited without necessity"), the sufficient condition may be that of impact into a region of high heat flow, the hypothesis considered below.

HYPOTHETICAL IMPACT MODEL FOR THE SUDBURY STRUCTURE

The model presented to scale in Figure 6 is based on the requirement that at the time of impact the thermal gradient in the Sudbury area was comparable to that in present day orogenic regions, so that at a depth of 20 km the temperature was in the range of 600° to 800°C (Uyeda and Horai, 1964). A second requirement, that the area be overlain by water some hundreds of metres deep, has been discussed previously and is not shown in the diagram for reasons of scale. The existence of

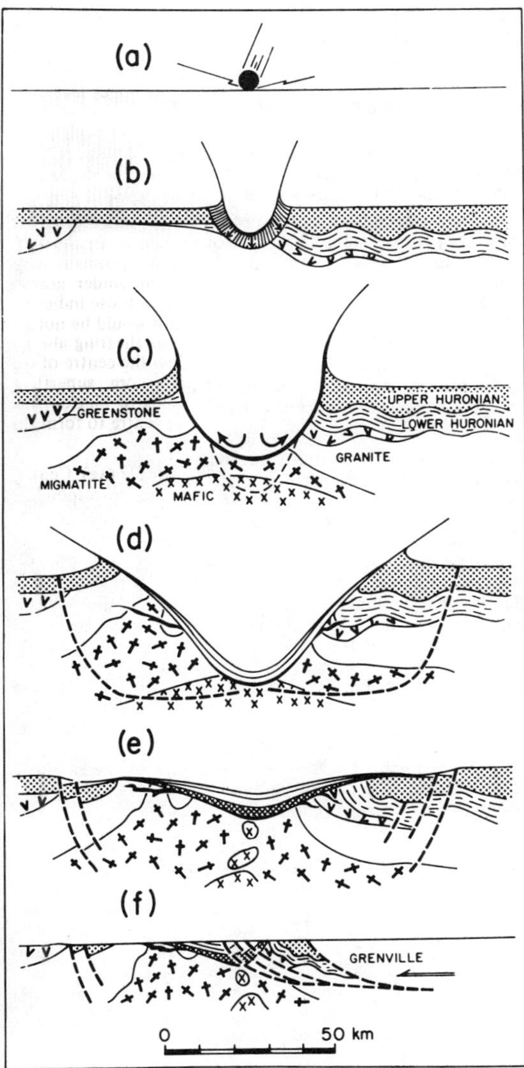

Figure 6: Diagrammatic representation presented to scale along a northwest-southeast profile of the possible sequence of events forming the Sudbury structure.

(a) Hypervelocity impact of a cosmic bolide, depicted here as a stony asteroid.

(b) Initial excavation of the crater by the high pressure shock wave and its rarefactions. At this stage the heavily shocked bolide is smeared out to line the growing cavity and the target materials behind the expanding shock front are largely molten.

(c) Continued growth of the transient cavity. The cavity lining is now a relatively thin (~1 km thick) layer of impact melt and breccia with fragments of moderately shocked country rocks. Normally the rocks underlying the cavity, having been shock metamorphosed at about 200 kb, would remain solid and there would be little further downward growth of the crater. However it is postulated that at Sudbury the initial thermal gradient was high and the rocks outlined by the dashed line were brought to the melting point, allowing excavation of an exceptionally deep cavity.

(d) Maximum dimensions of the primary crater. Addition of melt from below to initial impact melt is symbolized by a double lining of the cavity. Sub-layer breccia (black), consisting of fragments and melt derived from the deeper levels of the cavity walls, formed the lowest layer of the cavity lining and was also mobilized and forced upwards. Collapse of the uplifted rim under gravity initiated faulting along planes indicated by heavy dashed lines, with consequent modification of the crater structure by inwards and upwards movement of the cavity walls, and generation of pseudotachylites (Sudbury breccias).

(e) Form of the Sudbury crater after equilibrium was attained. The rim has collapsed to form a peripheral graben of downfaulted Huronian rocks. A subdued ring uplift inside the graben, comparable to the island ring of West Clearwater Lake (Figure 1) surrounds a central depression underlain by melt rocks which differentiated into noritic rocks (cross hatched) and micropegmatites. The entire crater is blanketed by returned ejecta (Onaping Formation), much of it redeposited in the central depression.

(f) Present form of the Sudbury structure after deposition of the Whitewater sedimentary rocks, orogenic activity involving overthrusting of the Grenville from the southeast, and general erosion.

See text for further details.

a high thermal gradient in the area at the time of impact is not unreasonable. The region was one of intermittent intrusive and orogenic activity from about 2.15 b.y. to 0.5 b.y. ago with several periods of intrusion and metamorphism between 2.15 b.y. and 1.5 b.y. in the vicinity of the Sudbury Basin and along the Grenville front (Card, 1969; Krogh, *et al.,* 1971). It is therefore likely that at least one period of folding had affected the Huronian south of Sudbury before the Sudbury event, as depicted in Figure 6, and that the thermal gradient was indeed high if the impact took place 1.72 b.y. ago, the commonly accepted age for the Sudbury Irruptive (Fairbairn *et al.,* 1968).

In Figure 6 (a) the impact of a bolide of asteroidal size is depicted. The dimensions are appropriate for a body of density about 3.5 g/cm^3 travelling at 15 km/sec. The pressure at the point of impact would be several megabars and the total energy released approximately 2 x 10^{30} ergs, as calculated by French (1971). At the initial point of contact a high velocity transient jet of fused material is ejected and widely dispersed. The main ejection process then follows as shock waves are transmitted into the ground and back into the bolide. The initial shock pressures are sufficient to vaporize the upper layers of country rocks, but with attenuation the pressure drops to levels at which melting takes place. This is the stage shown in Figure 6(b). The virtually completely molten bolide is smeared out along the lining of the growing cavity which is underlain by about 3000 km^3 of shocked and largely melted rock behind the shock front. Mainly Huronian rocks will have been melted, so the average composition of the melt would be moderately siliceous, although the proportion of extremely siliceous Upper Huronian metasediments would not be great as they would constitute much of the highly shocked initial ejecta.

Continued growth of the cavity as shown in Figure 6(c) results in melt overtaking and engulfing less strongly shocked rocks, with flow in the directions indicated by the arrows. The crater lining of mobilized materials thins to a zone about 1 to 2 km thick. At this stage the crater walls would be rocks shocked at pressures of about 200 to 300 kb. If of normal strength they would resist further downward growth of the cavity. Such, according to this interpretation, was the case at Vredefort. However the high thermal gradient postulated for Sudbury would lead to other effects. If the rocks were already at a temperature of about 700°C, a 200 degree to 300 degree C rise in temperature due to the residual heat of shock loading and decompression (Stöffler, 1971) added to the effect of the sudden unloading of more than 20 km of overburden would raise a considerable volume of rock above the melting point.

Just how much rock would melt and mix with the overlying impact melt depends on many factors, among them the composition of the country rock. Here it is taken to be mainly migmatite resembling the Levack complex (Langford, 1960) and mafic rocks such as those occurring as inclusions in the sub-layer. The regional gravity map shows that such mafic rocks may underlie much of the area southwest, east and northeast of Sudbury as well as the Sudbury Basin itself (Popelar, 1972). The suggestion that the rocks of the Levack complex originally lay at some depth is supported by the presence among them of two-pyroxene gneisses of granulite facies as well as many mafic inclusions (Card and Meyn, 1969). In this reconstruction, partial to complete melting is postulated to take place in the approximately 400 km^3 of rock outlined by the dashed line in Figure 6(c), allowing continued excavation and downward growth of the cavity as depicted in Figure 6(d).

The high particle velocities and considerable superheat still possessed by the impact melt at this point would ensure effective mixing with the lower melt, particularly in the centre of the combined molten materials. It seems possible, however, that some indication of the compositions of the two original components might be derived from the marginal facies of the Irruptive. In this regard it is of interest to compare some representative rock compositions (Table I). In column 4 of the table the composition of rocks which form much of the upper 400 m of the Irruptive is given. These rocks are considerably heavier and more basic than the main phase of the micropegmatite (column 6) or of the average of the micropegmatite (column 5). Indeed they are close to the average for the entire Irruptive at the present level of exposure (column 3). As these rocks appear to have defied the gravitational differentiation which has operated on the bulk of the Irruptive (Naldrett *et al.*, 1970), it is possible that they are a relatively unchanged portion of the original impact melt. Their mean composition and density are not unreasonable for the upper part of the crust, their density being similar to that of the lower Huronian metasediments and of the Levack complex (Popelar, 1972). From the earlier argument that the lower melt may have been derived largely from the Levack complex it might be concluded that the two melts differed little in composition, giving the mean of column 3. However, it is likely, from the gravity data, that the Irruptive has a higher proportion of norite at depth than at the surface (Popelar, 1972), and so a more basic composition for the lower melt should be considered. The silicate matrix of the offsets (column 2) is an obvious candidate.

From this assessment of the derivation of the Irruptive several corollaries follow:

(1)unlike standard basic complexes (Wyllie, 1967) there would be no large amount of ultramafic material at the base of the norite to counterbalance the granophyric upper levels, in accord with the density model of Popelar,

(2)lack of a magmatic feeder from below could explain the absence of rhythmic layering in the Irruptive, a unique feature indicating exceptionally tranquil conditions during crystallization,

(3)the sulphide ores would be derived from depth and not from the bolide, any remnants of which would be intimately mixed with the upper shock melted materials.

To return to Figure 6, the profile shown in Figure 6(d) represents the maximum dimensions of the exceptionally deep crater. From this point, collapse of the rim under gravity would initiate deep sliding along faults such as those indicated by heavy dashed lines. As the rocks at depth would be not far off the melting point the faults are shown as shearing along a zone of low strength at no great depth below the centre of the crater. Not shown on the profile are more superficial adjustments which may have included sliding of blocks from the upper margins of the crater towards the centre to form the quartzite breccias.

The final form of the crater is shown in profile 6(e). Central uplift has greatly flattened the crater with a corresponding downdrop of the margins. However, the exceptionally large depth to diameter ratio of the crater excavated results not in a central peak but in a depression, in which the various formations of the Whitewater Group were deposited. Finally, orogenic events deformed the crater, thrusting the southeast side to the northwest. Much faulting and shearing are recognized in the South Range but the North Range is commonly described as having been unaffected by these events except for a general low grade metamorphism. However, the basal breccias would have been a zone or relatively low strength and may have acted as a zone of movement during orogenesis, thus incorporating blocks of norite into the sub-layer breccias (Naldrett and Kullerud, 1967). The present day profile (Figure 6(f)) shows the results of the orogenic activity and the removal of about 5 km of rock by regional erosion to obtain the present level of exposure.

CONCLUSION

The evidence for an impact origin for the Sudbury structure is as strong as for any large, ancient, shock-metamorphosed crater. Of the three early criteria for an impact origin, the lack of circularity is explained by the effects of subsequent orogenesis, the absence of a raised rim by peripheral downdrop and erosion, the lack of a negative gravity anomaly as a result of the above average density of the regional rocks and the effects of erosion and regional metamorphism, which lead to removal or recrystallization of originally low density rocks.

The scenario sketched in this paper is but one of several that have been proposed since the original paper of Dietz (1964). Their diversity reflects the youth of the impact theory and the complexities of Sudbury geology. If Sudbury is the Gordian Knot of the Canadian Shield, the impact theory is not the Alexandrian stroke that solves all problems. It does, however, bring new facts and concepts to bear and may stimulate new attacks on this most intriguing of geological conundrums.

ACKNOWLEDGEMENTS

This manuscript has benefited from discussion with many who share the author's deep interest in Sudbury geology. They include M. J. S. Innes, J. Guy-Bray, K. D. Card, R. S. Dietz, B. M. French, J. S. Stevenson, J. Popelar, W. V. Peredery, A. J. Naldrett and others too numerous to name individually. Their encouragement in the development of the ideas presented here is greatly appreciated with the assurance that they are absolved from any responsibility for the imperfections of the proposed hypothesis.

TABLE I:

Selected Chemical Analyses of Sudbury Basin Rocks

	1.	2.	3.	4.	5.	6.	7.	8.
SiO_2	55.09	58.59	63.14	63.64	66.83	68.36	61.07	60.99
Al_2O_3	14.73	16.21	14.65	15.19	13.50	13.16	11.23	12.82
TiO_2	0.85	0.84	0.74	0.81	0.71	0.70	0.49	0.48
Fe_2O_3	3.91	2.19	1.52	1.40	1.80	1.48	1.50	0.95
FeO	5.82	5.79	5.17	5.47	4.04	4.09	7.56	7.86
CaO	7.89	5.48	4.03	3.55	2.40	2.50	4.06	1.90
MgO	4.66	4.27	2.85	2.03	1.31	1.10	4.78	4.46
Na_2O	3.04	2.88	3.20	3.64	3.29	2.97	2.71	0.75
K_2O	1.53	2.41	2.91	3.58	4.31	4.30	1.67	1.98
MnO	0.08	....	0.08	0.13	0.10	0.10	0.27	0.35
P_2O_5	0.17	0.27	0.23	0.25	0.23	0.19	0.17	0.23
H_2O	1.79	0.86	1.24	1.53	1.38	1.31	2.18	3.28
CO_2	0.24	0.03	0.16	0.04	0.06	0.06	1.20	2.32
C	.	.	.	.	.	.	0.50	1.43
S	0.20	0.24	0.08	0.06	0.04	0.04	0.98	0.49
Total	100.00	99.66	100.00	100.30	100.00	100.36	100.27	100.29
S.G.	2.82	2.86	2.77	2.73	2.69	2.68		

1. Average Norite, Levack section (Collins, 1934, p.170).
2. Average Foy offset (Collins, 1934, p.172).
3. Average Irruptive (Collins, 1934, p. 172).
4. Plagioclase-rich micropegmatite, average of samples 59W and 60W, Levack section (Collins, 1934, p.139).
5. Average micropegmatite, Levack section (Collins, 1934, p. 171, Col. I).
6. Granophyre-rich micropegmatite, Levack section, average of samples 55W, 57W, 61W, 63W (Collins, 1934, p. 139).
7. Average Lower Onaping Formation (Sadler, 1958, Table II).
8. Average Onwatin Slate (Sadler, 1958, Table XI).

REFERENCES

Beals, C. S., Ferguson, G. M. and Landau, A. (1956) – A search for analogies between lunar and terrestrial topography on photographs of the Canadian Shield; J. Roy. Astro. Soc. Can., vol. 50, pp. 203-211, 250-261.

Bischoff, A. A. (1962) – The pseudotachylite of the Vredefort Dome; Trans. Geol. Soc. S. Africa, vol. 65, pp. 207-225.

Bjork, R. L. (1961) – Analysis of the formation of Meteor Crater, Arizona, a preliminary report; J. Geophys. Res., vol. 66, pp. 3379-3387.

Borg, I. Y. and Heard, H. C. (1970) – Experimental deformation of plagioclase; *in* Paulitsch, P. (editor), Experimental and Natural Rock Deformation, Springer, New York, pp. 375-403.

Card, K. D. (1969) – Geology and geochronology of the Southern province of the Canadian shield (abs); Geol. Soc. Amer., Northeastern Section Meeting, Albany, New York, 13-15 March 1969, Program, pp. 6, 7.

Card, K. D. and Meyn, H. D. (1969) – Geology of the Leinster-Bowell Area; Ontario Dept. Mines Geol. Rept. 65.

Carter, N. L. (1971) – Static deformation of silica and silicates; J. Geophys. Res., vol. 76, pp. 5514-5540.

Chao, E. C. T. (1967a) – Impact metamorphism; *in* Abelson, P. H. (editor), Researches in Geochemistry, vol. 2, John Wiley & Sons, New York, pp. 204-233.

...... (1967b) – Shock effects in certain rock-forming minerals; Science, vol. 156, pp. 192-202.

Collins, W. H. (1934) – Life-history of the Sudbury Nickel Irruptive; Trans. Roy. Soc. Can., vol. 28, pp. 123-177.

Cook, P. J. (1968) – The Gosses Bluff cryptoexplosion structure; J. Geol., vol. 76, pp. 123-139.

Dence, M. R. (1964) – A comparative structural and

petrographic study of probable Canadian meteorite craters; Meteoritics, vol. 2, pp. 249-270.

...... (1965) – The extraterrestrial origin of Canadian craters; Ann. N. Y. Acad. Sci., vol. 123, pp. 941-969.

...... (1968) – Shock zoning at Canadian craters: petrography and structural implications; *in* French, B. M. and Short, N. M. (editors), Shock Metamorphism of Natural Materials, Mono Book Corp., Baltimore, pp. 169-184.

...... (1971) – Impact melts; J. Geophys. Res., vol. 76, pp. 5552-5565.

...... (1972) – The nature and significance of terrestrial impact structures; Proc. 24th Internat. Geol. Congress, Section 15 (in press).

Dence, M. R., Innes, M. J. S. and Beals, C. S. (1965) – On the probable meteorite origin of the Clearwater Lakes, Quebec; J. Roy. Astro. Soc. Can., vol. 59, pp. 13-22.

Dietz, R. S. (1964) – Sudbury Structure as an astrobleme; J. Geol., vol. 72, pp. 412-434.

...... (1968) – Shatter cones in cryptoexplosion structures; *in* French, B. M. and Short N. M. (editors), Shock Metamorphism of Natural Materials, Mono Book Corp., Baltimore, pp. 267-285.

...... (1972) – Sudbury astrobleme, splash-emplaced sub-layer and possible cosmogenic ores; Geol. Assoc. Can., Spec. Pap. No. 10 (this vol.).

Dietz, R. S. and Butler, L. W. (1964) – Shatter-cone orientation at Sudbury, Canada; Nature, vol. 204, pp. 280-281.

Engelhardt, W. V. and Bertsch, W. (1969) – Shock induced planar deformation structures in quartz from the Ries crater, Germany; Contr. Mineral. Petrol., vol. 20, pp. 203-234.

Fairbairn, H. W., Faure, G., Pinson, W. H. Jr., and Hurley, P. M. (1968) – Rb-Sr whole-rock age of the Sudbury lopolith and basin sediments; Can. J. Earth Sci., vol. 5, pp. 707-714.

French, B. M. (1967) – Sudbury structure, Ontario: some petrographic evidence for origin by meteorite impact; Science, vol. 156, pp. 1094-1098.

...... (1968) – Sudbury structure, Ontario: some petrographic evidence for an origin by meteorite impact; *in* French, B. M. and Short, N. M. (editors), Shock metamorphism of Natural Materials, Mono Book Corp., Baltimore, pp. 383-412.

...... (1969) – Distribution of shock-metamorphic features in the Sudbury basin, Ontario, Canada (abs); Meteoritics, vol. 4, pp. 173-174.

...... (1971) – Possible relations between meteorite impact and igneous petrogenesis, as indicated by the Sudbury structure, Ontario, Canada; Bull. Volcan., vol. 34, pp. 466-517.

French, B. M. and Short, N. M. (1968), editors – Shock Metamorphism of Natural Materials; Mono Book Corp., Baltimore, 644 pp.

Gault, D. E. and Heitowit, E. D. (1963) – The partition of energy for hypervelocity impact craters formed in rock; Proc. 6th Hypervelocity Impact Symp., vol. 2, pp. 419-456.

Gault, D. E., Quaide, W. L., and Oberbeck, V. R. (1968) – Impact cratering mechanics and structures; *in* French, B. M. and Short, N. M. (editors), Shock metamorphism of natural materials, Mono Book Corp., Baltimore, pp. 87-99.

Guy-Bray, J. and geological staff (1966) – Shatter cones at Sudbury; J. Geol., vol. 74, pp. 243-245.

Hargraves, R. B. (1970) – Paleomagnetic evidence relevant to the origin of the Vredefort Ring; J. Geol., vol. 78, pp. 253-263.

Hörz, F. (1971), editor – Meteorite impact and volcanism; J. Geophys. Res., vol. 76, pp. 5381-5798.

Innes, M. J. S. (1964) – Recent advances in meteoritic research at the Dominion Observatory, Ottawa, Canada; Meteoritics, vol. 2, pp. 219-241.

Keiffer, S. W. (1971) – Shock metamorphism of the Coconino Sandstone at Meteor Crater, Arizona; J. Geophys. Res., vol. 76, pp. 5449-5473.

Krinov, E. L. (1963) – Meteorite craters on the Earth's surface; *in* Middlehurst, B. M. and Kuiper, G. P. (editors), The Moon, Meteorites and Comets, The Solar System, vol. 4, Univ. of Chicago Press, Chicago, pp. 183-207.

Krogh, T. E., Davis, G. L. and Frarey, M. J. (1971) – Isotopic ages along the Grenville Front in the Bell Lake area, southwest of Sudbury, Ontario; Ann. Rept. Geophys. Lab. Carnegie Inst., Year Book 69, pp. 337-339.

Langford, F. F. (1960) – Geology of Levack Township and the northern part of Dowling Township; Ontario Dept. Mines Prelim. Rept. 5.

Manton, W. I. (1962) – The orientation and implication of shatter cones in the Vredefort Ring structure; Unpub. M. Sc. Thesis, Univ. Witwatersrand, S. Africa.

...... (1964) – The orientation and origin of shatter cones in the Vredefort Ring; Ann. N. Y. Acad. Sci., Vol. 123, pp. 1017-1049.

Meen, V. B. (1950) – Chubb crater, Ungava, Quebec; J. Roy. Astro. Soc. Can., vol. 44, pp. 169-180.

Millman, P. M. (1971) – The space scars of earth; Nature, vol. 232, pp. 161-164.

Murtaugh, J. C. and Currie, K. L. (1969) – Preliminary study of Manicouagan structure; Dept. Nat. Resources, Quebec, P. R. 583.

Naldrett, A. J., Bray, J. G., Gasparrini, E. L., Podolsky, T. and Rucklidge, J. C. (1970) – Cryptic variation and the petrology of the Sudbury Nickel Irruptive; Econ. Geol., vol. 65, pp. 122-155.

Naldrett, A. J. and Kullerud, G. (1967) – A study of the Strathcona mine and its bearing on the origin of the nickel-copper ores of the Sudbury district, Ontario; J. Petrol., vol. 8, pp. 453-531.

Peredery, W. V. (1972) – Chemistry of fluidal glasses and melt bodies in the Onaping Formation; Geol. Assoc. Can., Spec. Paper 10 (this vol.)

Popelar, J. (1971) – Gravity measurements in the Sudbury area; Grav. Map. Series, Earth Phys. Branch, No. 138.

...... (1972) – Gravity interpretation of the Sudbury area; Geol. Assoc. Can., Spec. Paper 10 (this vol.)

Robertson, P. B. (1968) – La Malbaie structure, Quebec – a Palaeozoic impact site; Meteoritics, vol. 4, pp. 89-112.

Robertson, P. B., Dence, M. R. and Vos, M. A. (1968) – Deformation in rock-forming minerals from Canadian craters; *in* French, B. M. and Short, N. M. (editors), Shock Metamorphism of Natural Materials, Mono Book Corp., Baltimore, pp. 433-452.

Roddy, D. J. (1968) – The Flynn creek crater, Tennessee; *in* French, B. M. and Short, N. M. (editors), Shock Metamorphism of Natural Materials, Mono Book Corp., Baltimore pp. 291-322.

Rondot, J. (1968) – Nouvel impact météoritique fossile? La structure semi-circulaire de Charlevoix; Can. J. Earth Sci., vol. 5, pp. 1305-1317.

Sadler, J. F. (1958) – A detailed study of the Onwatin Formation; Unpubl. M.Sc. thesis, Queen's Univ., Kingston, Ontario.

Shand, S. J. (1916) – The pseudotachylite of Parys (Orange Free State) and its relation to 'trap-shotten gneiss' and 'flinty crushrock'; Quart. J. Geol. Soc. London, vol. 72, pp. 198-221.

Shoemaker, E. M. (1960) – Penetration mechanics of high velocity meteorites, illustrated by Meteor Crater, Arizona; 21st Internat. Geol. Congr., Norden, Rep. Session 18, pp. 418-434.

...... (1963) – Impact mechanics at Meteor Crater, Arizona; *in* Middlehurst, B. M. and Kuiper, G. P. (editors), The Moon, Meteorites and Comets, The Solar System, vol. 4, Univ. Chicago Press, Chicago, pp. 301-336.

Short, N. M. and Bunch, T. E. (1968) – A world wide inventory of features characteristic of rocks associated with presumed meteorite impact craters; *in* French, B. M. and Short, N. M. (editors), Shock Metamorphism of Natural Materials, Mono Book Corp., Baltimore, pp. 255-266.

Souch, B. E., Podolsky, T. and geological staff, The International Nickel Company of Canada, Limited (1969) – The sulfide ores of Sudbury: their particular relationship to a distinctive inclusion-bearing facies of the Nickel Irruptive; Econ. Geol., Mono. No. 4, pp. 252-261.

Speers, E. C. (1957) – The age relation and origin of the common Sudbury breccia; J. Geol., vol. 65, pp. 497-514.

Stevenson, J. S. (1961) – Recognition of the quartzite breccia in the Whitewater series, Sudbury Basin, Ontario; Trans. Roy. Soc. Can., vol. 55 (IV), pp. 57-66.

Stöffler, D. (1971) – Progressive metamorphism and classification of shocked and brecciated crystalline rocks at impact craters; J. Geophys. Res., vol. 76, pp. 5541-5551.

Svennson, N.-B. (1971) – Probable meteorite impact crater in central Sweden; Nature Phys. Sc., vol. 229, pp. 90-92.

Thomson, J. E. (1956) – Geology of the Sudbury basin; Ontario Dept. Mines Ann. Rept. vol. 65. pt. 3. pp. 1-56.

...... (1969) – A discussion of Sudbury geology and sulphide deposits; Ontario Dept. Mines Misc. Paper 30.

Uyeda, S. and Horai, K. (1964) – Terrestrial heat flow in Japan; J. Geophys. Res., vol. 69, pp. 2121-2141.

Williams, H. (1956) – Glowing avalanche deposits of the Sudbury basin; Ontario Dept. Mines Ann. Rept. Vol. 65, pt. 3, pp. 57-89.

Wilshire, H. G. (1971) – Pseudotachylite from the Vredefort Ring, South Africa; J. Geol., vol. 79, pp. 195-206.

Wilson, H. D. B. (1956) – Structure of lopoliths; Geol. Soc. Amer. Bull., vol. 67, pp. 289-300.

Wyllie, P.J. (1967), editor – Ultramafic and Related Rocks; John Wiley & Sons, New York.

22

Reprinted from pages 19–28 of *Geol. Assoc. Canada Spec. Paper 10,* 1972

SHOCK-METAMORPHIC FEATURES IN THE SUDBURY STRUCTURE, ONTARIO: A REVIEW

Bevan M. French
Planetology Branch, NASA, Goddard Space Flight Center, Greenbelt, Maryland 20771, U. S. A.

ABSTRACT

A wide variety of shock-metamorphic features, both megascopic and microscopic, occur in the Sudbury structure and provide strong evidence for its origin by meteorite impact. The autochthonous footwall rocks exhibit megascopic shock effects (shatter cones, Sudbury breccias) around the perimeter of the basin. On the North Range, where later metamorphism has been less severe, these features are accompanied by preserved petrographic shock effects (planar features in quartz, kink bands in mica).

The Onaping Formation within the Sudbury Basin is interpreted as a fallback breccia *deposited immediately after impact in the original crater. Inclusions of basement rocks in the Onaping Formation exhibit a wider range of petrographic shock effects, from simple crushing and granulation to complete fusion. Several stages of shock metamorphism, generally reflecting increasing shock pressures, can be recognized by the presence of: (1) planar features (shock lamellae) in quartz and feldspar, (2) selective recrystallization of feldspar, (3) strong plastic deformation of feldspar with recrystallization of associated quartz, (4) local "eutectic melting" at grain boundaries, (5) complete melting of all minerals to form heterogeneous, often vesiculated, glasses, and (6) fusion, extensive flow, mixing, and accretion of glass and rock fragments to form glassy bodies analogous to the* Fladen *from the Ries Basin, Germany. The heterogeneity of the glassy fragments in the Onaping Formation and their intimate association with shocked rock fragments support the view that both types of material were produced at the same time by an impact event.*

The original Sudbury crater may have been as large as 100 km in diameter. The subsequent introduction of the Nickel Irruptive may have been accompanied by subsidence of the central part of the original crater, thereby preserving the Onaping Formation at the present deep level of erosion.

INTRODUCTION

The Sudbury structure, Ontario is an elliptical basin 60 by 27 km, elongated in an east-northeast direction. The basin is outlined, both topographically and geologically, by the outcrop belt of the Sudbury Nickel Irruptive. At the base (outer margin) of the Irruptive occur the copper-nickel sulphides that give Sudbury its great economic importance. The Sudbury structure is surrounded by older rocks. Granites and related rocks of the Superior province occur to the north and west, while Huronian and pre-Huronian metasediments and metavolcanics surround the structure on the south and east. The Grenville front passes immediately south of Sudbury and separates the older Huronian rocks from younger rocks to the south and east.

Within a few miles from the Nickel Irruptive, the older rocks have been fractured, brecciated, and locally mobilized to form a distinctive unit called the Sudbury breccia (Fairbairn and Robson, 1944; Speers, 1957). Inside the Irruptive, the Sudbury basin is underlain by rocks of the Whitewater Group (Burrows and Rickaby, 1930; Roscoe, 1969): the lowermost Onaping Formation, the Onwatin "slate," and the uppermost Chelmsford graywacke.

During the nearly ninety years since its discovery, Sudbury has been the subject of continuing controversy about the origin of the structure, the shape and emplacement of the Nickel Irruptive, and the origin of the ores (for summary and references, see Hawley, 1962). Within the last few years, interest in these problems has been renewed, partly by new field and experimental studies of the origin of the Irruptive and its ores (Naldrett and Kullerud, 1967; Stevenson and Colgrove, 1968; Naldrett, 1969; Souch *et al.*, 1969; Naldrett *et al.*, 1970), and partly by geological discoveries which support the suggestion (Dietz, 1964) that the Sudbury structure was formed by meteorite impact.

The strongest evidence for the impact origin of the Sudbury structure is the newly-recognized *shock-metamorphic features* in its rocks. In the last few years it has become generally accepted that high-velocity meteorite impacts generate intense shock waves which in turn produce unique permanent deformation features in the target rocks (for reviews, see Chao, 1967a, b, 1968; French, 1968a, 1970; French and Short, 1968). The resulting *shock metamorphism* is characterized by extremely high transient pressures (often in excess of 100 kb), extremely short reaction times, and high residual temperatures in the affected rocks.

Several of the effects of shock metamorphism are therefore sufficiently unusual to serve as unique indicators of ancient meteorite impacts, particularly: (1) the high-pressure silica polymorphs coesite and stishovite (Chao, 1967a; Stöffler, 1971a), (2) distinctive multiple planar features (shock lamellae) in quartz and feldspar (Robertson *et al.*, 1968; Engelhardt and Bertsch, 1969), (3) maskelynite, an isotropic modification of feldspar (Bunch *et al.*, 1967), (4) ultrahigh temperature decompositions such as fusion of quartz to lechatelierite or decomposition of zircon to baddeleyite (El Goresy, 1968), and (5) distinctive conical fractures (*shatter cones*) which are extensively developed in the target rocks (Dietz, 1968). The extreme conditions of shock metamorphism, combined with the short time scale of the process, produce distinctive selective mineralogical transformations (Chao, 1967a, b) and also result in the general preservation of mineralogical and chemical disequilibrium in the shocked rocks.

The conditions which produce such unique shock-metamorphic effects appear too extreme to be produced by any conventional geological process involving tectonic deformation or volcanic explosions. The exact temperature and pressure conditions indicated by any given shock effect are not precisely known, chiefly because the fabric of the target rock exerts a strong influence (Chao, 1968; Dence, 1968). However, experimental studies of rocks shocked in nuclear explosions (Short, 1968) and in laboratory experiments (Hörz, 1968; Müller and Défourneaux, 1968) indicate that the distinctive planar features (shock lamellae) develop in quartz at peak shock pressures above about 80-100 kb and occur in a pressure range of about 100-250 kb. At higher shock pressures, development of isotropic quartz and feldspar occurs in the solid state. Pressures above 400-500 kb produce extensive melting in quartzofeldspathic rocks because of the high residual temperatures (Chao, 1968; Engelhardt and Stöffler, 1968; Stöffler, 1971b). Shatter cones in granitic rocks are apparently produced by pressures below 50 kb (Roddy and Davis, 1969), but minimum and maximum pressures for their formation have not been established.

With the possible exception of shatter cones, the distinctive shock-metamorphic effects apparently require minimum shock pressures of 75-100 kb for their formation. It is generally believed that such high shock pressures cannot be produced by internal mechanisms such as volcanic explosions, for which maximum pressures of 3-5 kb have been calculated (Williams, 1954; Gorshkov, 1959; Roddy, 1968). The production of such large transient pressures by tectonic processes appears equally difficult, especially because the rapid quenching and preservation of metastable phases that is characteristic of shock metamorphism would not seem possible at great depths.

The distinctive shock-metamorphic effects at Sudbury (Dietz, 1964; Dietz and Butler, 1964; Bray *et al.*, 1966; French, 1967, 1968b, 1970) therefore provide unequivocal evidence for origin by meteorite impact. Particularly important are the petrographic effects such as shock lamellae in quartz (French, 1967), because their formation probably required peak shock pressures close to or exceeding 100 kb (Hörz, 1968). The preservation of these microscopic shock effects, despite the great age and metamorphism of the Sudbury structure, allows comparison between Sudbury and younger impact structures at which shock-metamorphic features are well preserved and evidence for a meteorite impact origin appears conclusive (see Dence, 1964, 1965, 1968; Engelhardt *et al.*, 1969).

At Sudbury, the Sudbury breccias, the shatter cones, and the petrographic shock effects (French, 1970) are regarded as primary products of the crater-forming impact. The Onaping Formation is also regarded as a result of the impact, having formed as a *fallback breccia* composed of shocked and melted material excavated by the impact and immediately redeposited in the crater (French, 1967). In this view, the glassy material in the Onaping Formation was produced by direct fusion of the target rock and is therefore *impact melt* (see Dence, 1971). W. Peredery (this volume) has indicated that larger bodies of impact melt may exist in the complex breccia zone at the base of the Onaping Formation.

Although Dietz (1964; this volume) has argued for a cosmogenic origin of the Sudbury ores, it is presently thought that both the Irruptive and the associated ores are of internal origin. A meteorite impact large enough to form the Sudbury structure would produce a significant volume of impact melt (Dence, 1971), but the volume of the Irruptive appears too large to have been produced in this way (French, 1970). Sudbury thus exhibits an apparently genetic relationship between a large meteorite impact and internal igneous activity. The existence of Sudbury as a type example for *impact-triggered volcanism* implies that other structures on the Earth and other planetary bodies may have originated in the same way.

SHOCK-METAMORPHIC FEATURES IN THE FOOTWALL ROCKS

The shock-metamorphic features at Sudbury occur in specific locations that are consistent with observed distributions of shock effects at other impact craters. Two types of occurrence may be distinguished. (1) Megascopic shock effects (shatter cones, Sudbury breccias) indicative of relatively lower shock pressures occur in an annular zone in the autochthonous older footwall rocks surrounding the basin. (2) Petrographic shock effects produced by higher shock pressures occur in inclusions of basement rock in the Onaping Formation (fall-back breccia) within the Sudbury Basin.

The Sudbury structure exposes large amounts of detectably shocked basement rocks together with a great volume of material (Onaping Formation) from the breccia lens that was deposited within the original crater. Similar associations, although not as well-developed, have been described from several other structures, especially Manicouagan, Quebec (Dence, 1964, 1965), Gosses Bluff, Australia (Milton *et al.*, 1971), and Rochechouart, France (Kraut and French, 1971). In terms of the large amount of shocked basement rock exposed, Sudbury is most comparable to Vredefort, South Africa, a structure of similar size and age which also exhibits Sudbury-type breccias ("pseudotachylite"; Shand, 1916), shatter cones (Hargraves, 1961; Manton, 1965), and petrographic shock effects in quartz (Carter, 1965).

Even before the meteorite impact origin of Sudbury was suggested, it was recognized that the extensive and unusual Sudbury breccias developed in the rocks surrounding the structure were the result of fracturing and mobilization during an intense explosive event (Fairbairn and Robson, 1944; Speers, 1957). However, the first distinctive shock-metamorphic features discovered at Sudbury were shatter cones (Dietz, 1964), that occur with Sudbury breccia in an annular zone surrounding the Sudbury basin to a distance of several miles (Bray *et al.*, 1966). Shatter cones have been observed at Nairn on Highway 17, about 13 km from the nearest norite outcrop. Sudbury breccia has been observed north of Espanola, more than 29 km from the Irruptive.

Shatter-coned inclusions have also been observed in the Onaping Formation (W. Peredery, personal communication; Dietz, this volume), strengthening the case for a simultaneous origin of the shatter-coning in the footwall rocks and the deposition of the Onaping Formation. The occurrence of shatter-coned inclusions in the fallback breccia within meteorite craters, while unusual, has also been documented at several structures, especially Sierra Madera, Texas (Howard and Offield, 1968), Rochechouart, France (Kraut and French, 1971), and the Ries Basin, Germany (W. v. Engelhardt, personal communication).

The shatter cones which occur in the footwall rocks at Sudbury, in addition to providing tangible evidence for meteorite impact, provide a tool for investigating the structural and metamorphic history of the region. According to the impact theory, the shatter cones were formed over a wide area at a single time, with the cone axes oriented in the direction from which the shock wave came (Dietz, 1968). At several shatter-coned structures, when the beds containing the shatter cones are restored to the horizontal, the cone axes point inward and upward toward a central focus, implying propagation of the shock wave from such a central point. Such relationships have been established by detailed measurements at Vredefort (Hargraves, 1961; Manton, 1965), Gosses Bluff (Milton *et al.*, 1971), and Sierra Madera (Howard and Offield, 1968).

The same measurement techniques can be profitably applied at Sudbury. Shatter cone striations were measured by the author at the excellent exposures of shatter-coned Mississagi quartzite on the south side of Kelley Lake (Figure 1) (for details of the method, see Manton, 1965). When the beds are restored to horizontal, the restored master cone axis points inward (toward the Sudbury Basin) and slightly downward, in contrast to the upward orientation observed at other impact structures. It is clearly undesirable to draw structural conclusions from one set of measurements, but if such orientations appear general, it would be possible to conclude

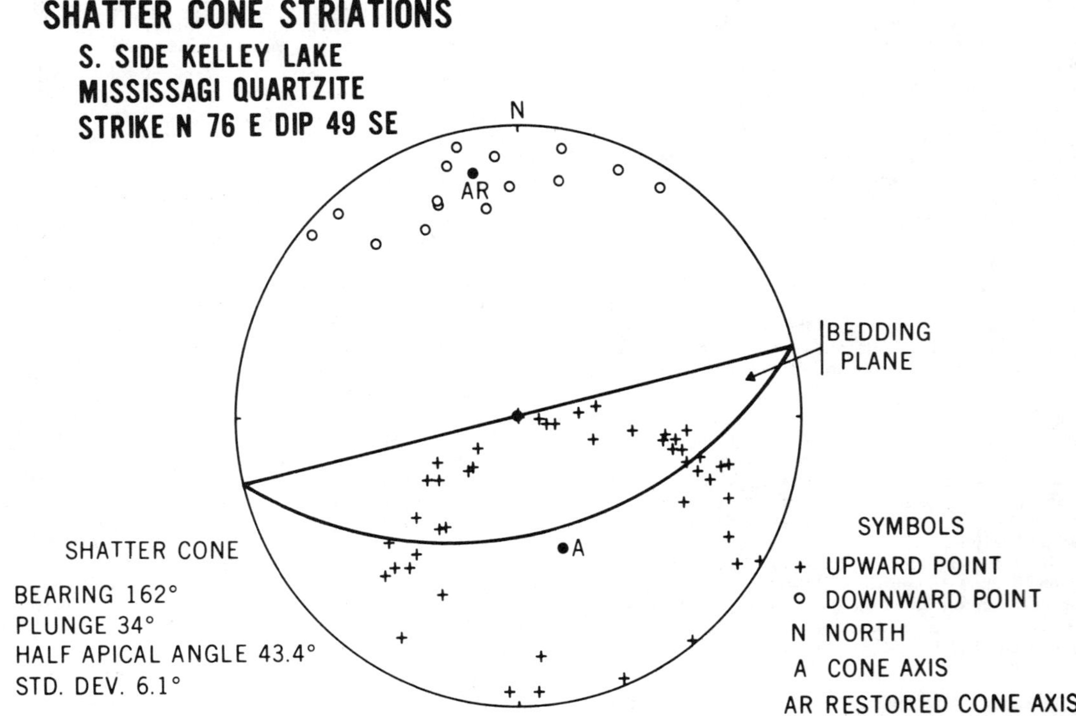

Figure 1: Stereographic projection (lower hemisphere) of orientations of shatter cone striations measured in the Mississagi quartzite on the south side of Kelley Lake, Broder Twp. The striations lie on a small circle whose centre is the projection of the "master cone" axis (Manton, 1965). When the bedding is restored to horizontal, the master cone axis points north toward the basin and slightly downward, in contrast to the upward direction generally shown by master cone axes at other structures.

that the Huronian metasediments were not horizontal at the time of impact. Further, restoring the shatter cone orientations around the entire structure to a common focus (see Milton *et al.*, 1971) might provide information on such problems as the pre-impact orientation of the Huronian section, the amount of post-impact movement, and the possible existence of post-impact differential rotation of the North and South Ranges (Sopher, 1966).

As at younger impact structures, the megascopic shock features at Sudbury are accompanied by petrographic shock effects. On the North Range, where Sudbury breccias are developed in the granitic rocks of the Superior province, weak shock lamellae in quartz and occasional kink-banding in mica are observed, in shatter cones and inclusions in the breccias. The shock lamellae in quartz usually occur as a single set with a dominantly basal orientation (Type A; Robertson *et al.*, 1968) (Figure 2A), although multiple planes (Type B; Figure 2B) are occasionally observed.

Development of kink-banding in micas, especially biotite (Figure 3), is commonly observed; the associated quartz may or may not exhibit shock lamellae. Although kink-banding can occasionally form during tectonic deformation, it is also a common shock effect (Hörz and Ahrens, 1969), and its common association with Sudbury breccias and shocked quartz on the North Range is strong evidence for its formation by shock.

On the South Range, the shocked footwall rocks have been subjected to subsequent thermal metamorphism that has caused extensive recrystallization of the breccias up to the staurolite grade (Card, 1964, 1968). No petrographic shock effects have been observed in these recrystallized rocks, although megascopic shock features such as shatter cones and Sudbury breccias are still preserved. The difference in metamorphic grade of the Sudbury breccia between the North and South Ranges is consistent with the view that the South Range rocks were more deeply buried and have been uplifted differentially by at leat 5 to 6 km to their present positions (Souch *et al.*, 1969). Further, the preservation of shatter cones and Sudbury breccia along the South Range indicates that the post-impact metamorphism has not involved intense shear deformation.

SHOCK-METAMORPHIC FEATURES IN THE ONAPING FORMATION

The numerous inclusions of recognizable basement rocks in the Onaping Formation display a wide range of distinctive shock-metamorphic effects that range from crushing and granulation to complete fusion and vesiculation (French, 1967, 1968b, 1970). The shocked rock fragments are accompanied by numerous unusual glassy bodies composed of mixtures of recrystallized flow-banded glass and rock and mineral fragments. These glassy bodies range in size from less than a millimetre to more than several metres and contain rock fragments and glass in all proportions. The occurrence of distinctively shock-metamorphosed rock fragments together with heterogeneous glassy bodies indicates that the Onaping Formation, instead of being an ash-flow tuff (Thomson, 1957;

Figure 2A: Planar features (shock lamellae) in a large quartz grain in footwall rock (North Range, Morgan Township). Crossed polarizers. The shock lamellae occur as a single set parallel to the base (0001). Note incipient development of recrystallization parallel to the lamellae. The associated feldspar (dark) has been strongly sericitized.

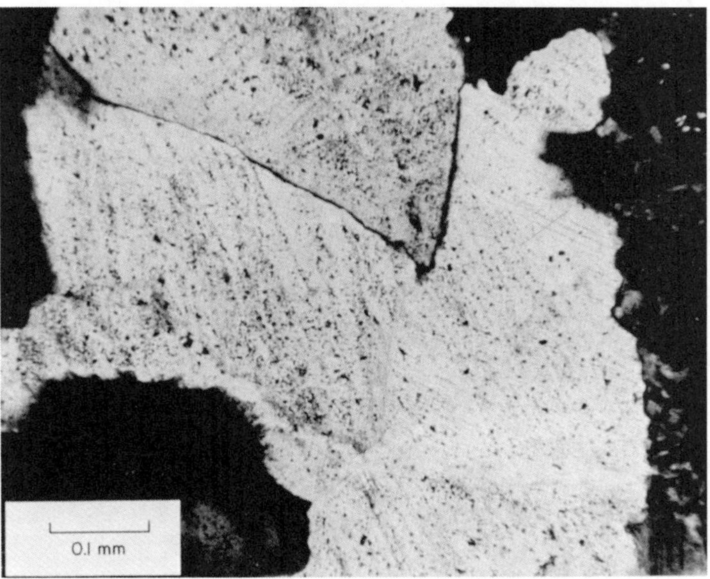

Figure2B: Multiple sets of planar features (shock lamellae) in quartz grains from a shocked inclusion in the Onaping Formation. Crossed polarizers. The lamellae are decorated with arrays of fluid inclusions, a common effect, apparently secondary, observed in older shocked rocks. At least two sets of $\omega\{10\bar{1}3\}$ lamellae are well developed in the grain at the right.

Figure3: Extensive kink-banding developed in a large biotite crystal in footwall rock (East Range, Maclennan Township). Two sets of kink bands are developed at high angles to the original cleavage (horizontal). The associated quartz (upper and lower left) does not show shock lamellae.

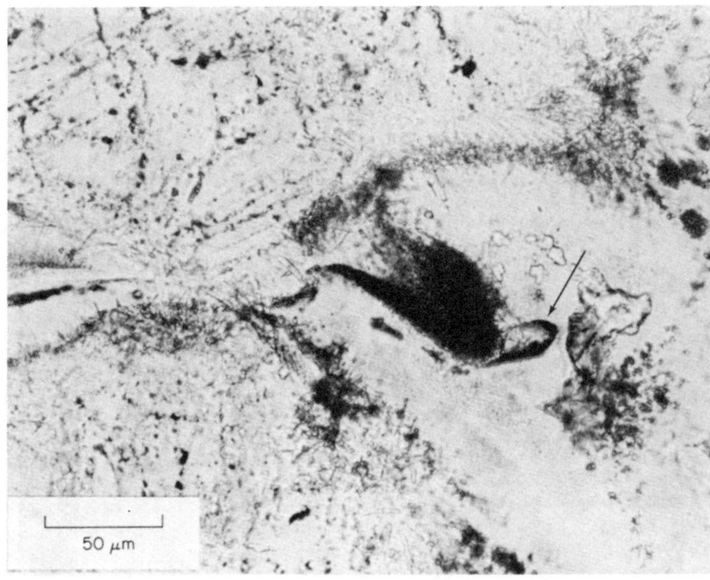

Figure 4: Incipient fusion of a grain of sphene (arrow) in a moderately shocked inclusion (Stage 2) from the Onaping Formation. Plane polarized light. The grain has partly melted to a plume of small droplets consisting either of sphene or of Ti oxides. The glassy feldspar surrounding the sphene grain (lower right) is recrystallized. Despite the local high temperatures involving the sphene (M. P. 1395°C), the associated quartz still retains visible planar features; two sets of planes can be distinguished in the grain at upper left.

Williams, 1957) is a fallback breccia similar to the suevite from the Ries Basin, Germany. The heterogeneous glassy bodies in the Onaping, in particular, are virtually identical to the fresher Fladen described from the Ries (Hörz, 1965; Engelhardt *et al.*, 1969).

Most of the inclusions in the Onaping Formation are quartzofeldspathic. They have apparently been derived both from the granites and granitic gneisses to the north and from feldspathic metaquartzites which occur as several formations in the Huronian section to the south and east. Study of more than one hundred of these inclusions has allowed recognition of *stages of shock metamorphism* based on progressive destruction of the original fabric of the rock (French, 1970). No shock grades have been established for other lithologies, chiefly greenstones and fine-grained quartzose sediments, which form a small percentage of inclusions in the Onaping Formation. These inclusions are relatively few, and classification is hampered by the fact that these lithologies are more resistant to formation of distinctive shock effects, particularly at low to moderate shock levels (Chao, 1967a; Dence, 1968).

The shock grades established for the Onaping inclusions are based on specific petrographic criteria. There are two complicating factors in the use of these criteria. First, because of the strong variability of shock-wave interactions in crystalline rocks, the specific effects may be gradational in a single specimen (Chao, 1968; Dence, 1968), and criteria for two adjacent grades may occur in the same thin section. Second, the petrographic effects observed involve both shock-produced deformation and the effects of post-shock recrystallization and metamorphism. The Onaping Formation is the oldest preserved terrestrial impact breccia known, and it has been recrystallized to a much greater degree (chlorite grade; French, 1967) than has similar material from younger structures. In particular, no isotropic material is preserved in the Onaping Formation, and it is not possible to recognize maskelynite or similar isotropic phases that might have been formed at moderate levels of shock. Despite this recrystallization, the textures observed in Onaping inclusions are comparable with classifications developed at younger impact craters (Chao, 1967a; Dence, 1968; Robertson *et al.*, 1968; Engelhardt and Stöffler, 1968; Stöffler, 1971).

The shock grades for Sudbury material (French, 1970) are based on the following petrographic criteria:

Stage 0. Crushing and granulation identical to that produced in normal metamorphism; no distinctive shock features.

Stage 1. Planar features (shock lamellae) in quartz and feldspar.

Stage 2. Planar features (shock lamellae) in quartz; recrystallization of feldspar.

Stage 3. Recrystallization of quartz; extensive plastic deformation and recrystallization of feldspar.

Stage 4. Eutectic melting at contacts of deformed quartz and feldspar grains.

Stage 5. Concurrent melting of all constituent minerals to form a physically and chemically heterogeneous glass.

Stage 6. Melting, mixing, and accretion of glass with rock and mineral fragments to form heterogeneous, inclusion-bearing, flow-banded glassy bodies.

The abundance of shocked inclusions in the Onaping Formation is remarkable. Collections of inclusions from each of several locations in relatively unmetamorphosed Onaping rocks provided specimens exhibiting the entire range of shock grades, and all stages except Stage 4 are well represented throughout the whole formation. No statistical sampling studies of shock grades have yet been made, but the distribution of ninety-one inclusions collected during 1966 and 1967 is instructive and probably typical: Stage 0, fourteen; Stage 1, twenty-two; Stage 2, eight; Stage 3, sixteen; Stage 4, one; Stage 5, ten; Stage 6, thirteen; completely recrystallized, fourteen.

The planar features (shock lamellae) in quartz are probably the most useful and widespread petrographic shock effect, and their preservation despite the age and metamorphism of the Onaping Formation is one of the most fortunate circumstances of Sudbury geology. Most of the Stage 1 and 2 inclusions exhibit two or more sets of planes, especially the distinctive ω $\{10\bar{1}3\}$ orientation (Figure 2B). At Sudbury, two general types of shock lamellae fabric can be recognized: (1) dominantly basal orientations, unaccompanied by shock lamellae in feldspar, which occur in quartz from the footwall rocks and Sudbury breccia (Figure 2A), and (2) multiple sets of shock lamellae with various orientations, associated with shock lamellae in feldspar, which are characteristic of numerous inclusions in the Onaping Formation. The multiple sets of lamellae probably reflect higher shock pressures (Hörz, 1968; Robertson *et al.*, 1968), and further subdivision of this group on the basis of number and orientation of the shock lamellae will probably be possible at Sudbury as it has been at other structures (Robertson *et al.*, 1968).

At the higher levels of shock (Stages 4 to 6), effects due to high temperature become dominant over deformational effects, because the residual temperatures produced by passage of a shock wave increase with increasing peak shock pressure. At shock pressures above about 500 kb, the resulting temperatures are high enough to produce extensive melting (Stages 5 to 6) (Chao, 1968; Engelhardt and Stöffler, 1968; Stöffler, 1971).

In a few of the Sudbury inclusions, however, high-temperature effects are anomalously associated with deformational effects suggestive of only moderate shock. One Stage 2 inclusion exhibits incipient melting and flow of an accessory sphene grain (M.P. 1395°C) (Figure 4). Furthermore the eutectic melting characteristic of Stage 4 has been observed in only one inclusion, suggesting that this texture may have resulted from an uncommon thermal effect superimposed on a moderately shocked (Stage 3) inclusion. Although textures produced by such high temperatures ($>$ 1300°C) are good evidence for meteorite impact (French, 1970), some of the high-temperature effects observed in rare Onaping inclusions may not be directly related to the shock wave itself. Post-shock thermal effects could arise from either (1) ejection of the shocked fragment through a hot gas cloud ("fireball") above the crater, or (2) burial in superheated melt or welded ejecta.

Many of the Onaping inclusions, especially the more highly shocked, are recrystallized. Because no isotropic material is observed, the term "glass" designates material that is believed to have been originally isotropic, either shock-produced isotropic mineral phases (e.g., maskelynite), or flow-banded melt material. The recrystallized feldspar (Stages 2 and 3) may have formed by annealing and devitrification of shock-produced maskelynite. The flow banded glassy material (Stage 6) now occurs as fine-grained aggregates of quartz, feldspar, chlorite, and actinolite which commonly preserve the original flow structure (Figure 5).

The degree of recrystallization of the shocked inclusions varies widely, even in specimens from the same outcrop or in the same thin section. Completely recrystallized inclusions are often associated with others that exhibit well-preserved shock features. Preserved shock effects are also observed in inclusions from the complex breccias at the base of the Onaping Formation on both the North and South Ranges (W. Peredery,

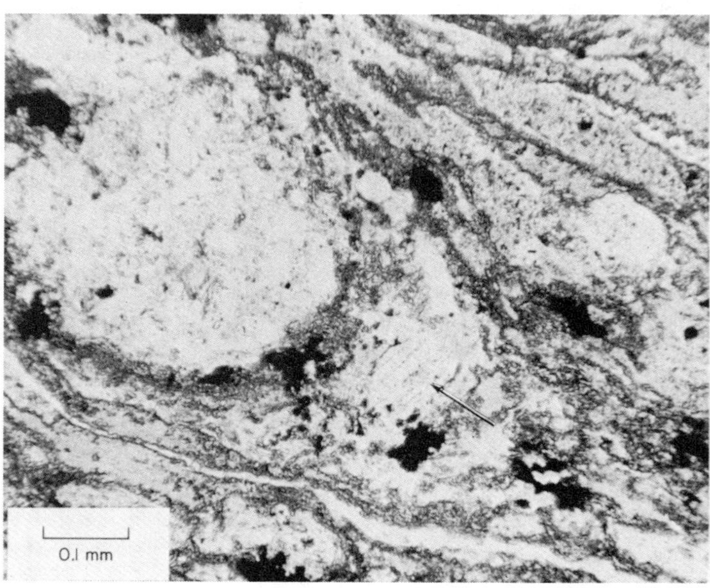

Figure 5: Photomicrograph of heterogeneous glassy inclusion from the Onaping Formation, showing preserved flow structure (outlined by differences in mineralogy) and rock inclusions. The lithic inclusion at upper left is apparently unshocked, while the small adjacent quartz grain (arrow) exhibits a set of preserved planar features.

personal communication), suggesting that the observed recrystallization is not due to contact metamorphism associated with emplacement of the Nickel Irruptive. These observations indicate that the recrystallization of the Onaping inclusions was not produced by later regional or contact metamorphism and more probably reflects conditions such as size and cooling rate of the inclusions themselves during deposition.

The heterogeneous glassy inclusions (Stage 6) are especially important for interpreting the origin of the Onaping Formation and the Sudbury structure because these bodies provided the original evidence for identification of the Onaping Formation as a tuff-breccia (Bonney, 1888; Williams, 1891) and as an ash-flow tuff (Thomson, 1957; Williams, 1957). The present interpretation is that the glassy inclusions formed in a manner identical to that of the Fladen from the Ries Basin, Germany (Hörz, 1965), i.e., by aggregation of melted target rock and rock and mineral fragments into aerodynamically shaped bodies that were redeposited in the crater as an integral part of the fallback breccia. Several lines of evidence support this view.

(1) Melting of significant amounts of target rock is an integral part of the impact process (Dence, 1968, 1971), and the volume of glassy material in the Onaping Formation compares reasonably with the amount expected from a Sudbury-size crater (French, 1970).

(2) The glassy inclusions in the Onaping Formation are virtually identical in appearance (aside from recrystallization) to the Fladen from the Ries Basin, Germany (Hörz, 1965; Engelhardt *et al.*, 1969).

(3) The inclusions in the Onaping Formation lack phenocrysts, trachytic textures, and other indications of volcanic origin that are observed in ashflow tuffs.

(4) The intimate association of glassy inclusions with shock-metamorphic effects points to a common origin for both features. One can often find, within the glassy bodies, a variety of lithic inclusions that show a range of shock effects (Figure 5). "Cored inclusions" (French, 1968b), which show a rim of heterogeneous glass surrounding a shocked rock fragment, also point to a simultaneous origin.

HISTORY OF THE SUDBURY STRUCTURE

A complete reconstruction of the history of the Sudbury structure as an impact-volcanic feature requires information about: (1) the size and shape of the original crater, (2) the shape and manner of intrusion of the Nickel Irruptive, and (3) the time interval between the impact and the emplacement of the Irruptive. Knowledge on these points is far from complete, but one reconstruction (French, 1970) proposes a three-part history (Figure 6) as follows:

(1) impact and formation of a crater about 100 km in diameter, similar to the Vredefort structure and likewise characterized by a large central uplift,

(2) instability beneath the crater, producing subsidence of the central uplift and simultaneous emplacement of the Nickel Irruptive, which intruded chiefly along the contact between the original crater floor and the fallback breccia (Onaping Formation),

(3) later metamorphism, compressive deformation and relative uplift of the South Range during the "Grenville event," and erosion to its present appearance.

This three-part history helps explain two puzzling problems in the impact origin of the Sudbury structure. The first is the clearly noncircular shape of the Irruptive outcrop belt. While some of the noncircularity is probably due to deformation during the "Grenville event," sharp bends like those in the northeast part of the basin are more apt to be original. If emplacement of the Irruptive was controlled by subsidence along fracture zones that did not exactly parallel the original

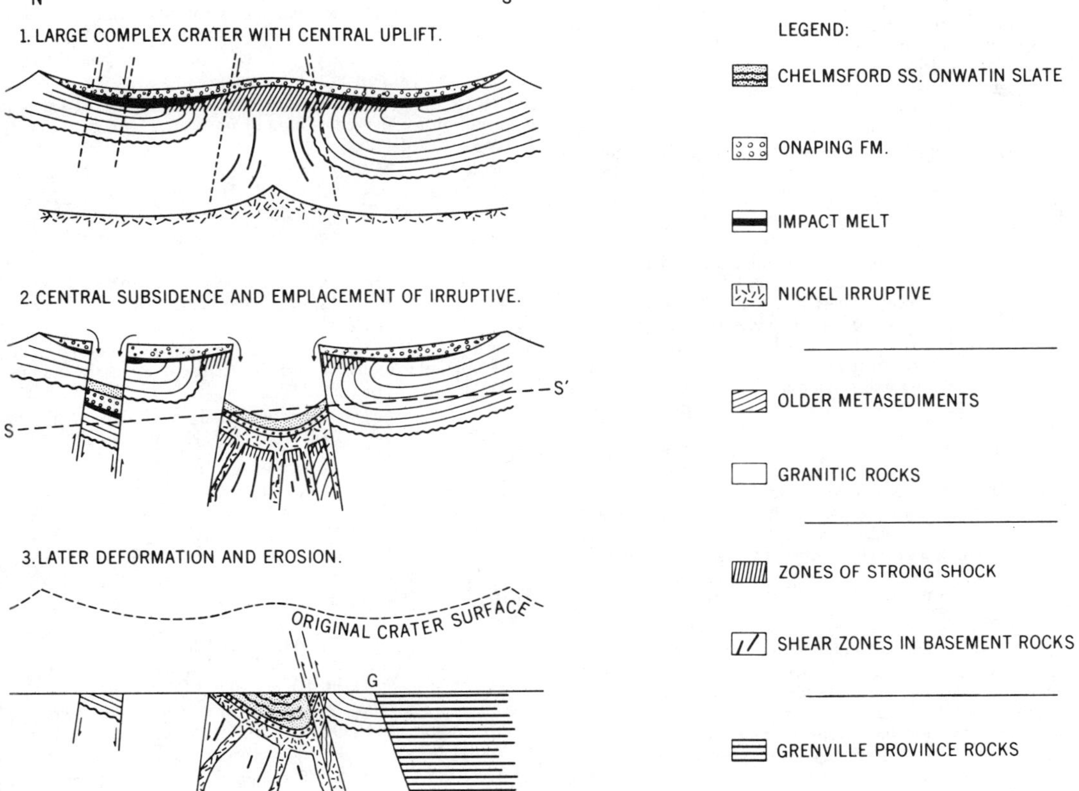

Figure 6: Schematic N-S cross section showing a suggested history of Sudbury as a composite impact-volcanic structure (from French, 1970): (1) formation of a large crater with a central uplift and a peripheral depression, (2) subsidence of the central uplift and emplacement of the Nickel Irruptive along the contact between the shocked basement rocks and the Onaping Formation (the line S-S' indicates approximately the present ground surface), and (3) deformation and erosion to the present appearance.

crater rim, a circular Irruptive outcrop is not required by the impact theory.

The second problem is the preservation of a thick section of originally near-surface crater fill (Whitewater Group) at what is now a deep level of erosion. The Onaping Formation was deposited near the surface and was probably not more than 5 km thick (Dence, 1968). A minimum estimate of the post-Irruptive erosion in the Sudbury area is 5 to 6 km (Souch *et al.*, 1968); estimates have ranged as high as 10 to 15 km. Subsidence of the central part of the Sudbury structure seems necessary to preserve the fallback breccia (Onaping Formation) at the present deep level of erosion. Subsidence of the Sudbury Basin is also suggested by the discontinuity in metamorphic grade between the Onaping Formation (chlorite grade) and the more highly metamorphosed Huronian metasediments outside the Irruptive on the South Range.

The theory suggested (Figure 6), while consistent with the present state of knowledge of Sudbury geology, is not the only possible explanation. Dietz (1964 and this volume) suggested that the original crater was bowl-shaped and about 50 km in diameter and that the Irruptive was emplaced within this bowl. Dence (personal communication) has suggested that the impact penetrated directly to the mantle, emplacing the Irruptive magma as part of the impulse that would, on a thicker crust, have formed a large central uplift similar to that at Vredefort. Until exact information can be obtained on the time interval (if any) between formation of the crater and emplacement of the Irruptive, several modifications of the impact theory will be equally satisfactory in explaining the observed features of Sudbury geology.

THE SUDBURY STRUCTURE AND IMPACT-TRIGGERED VOLCANISM

The real uniqueness and importance of the Sudbury structure is not just its identification as another large meteorite impact structure. The Sudbury structure and the Nickel Irruptive constitute the first demonstrated case of impact-triggered volcanism, a process long proposed for the origin of lunar features (for references, see French, 1970) and which may also be important in the formation of other terrestrial structures as well.

Several mechanisms have been suggested by which large meteorite impacts on planetary surfaces could promote igneous activity associated with the resulting crater: (1) direct removal of crust over zones of active magma generation, (2) partial fusion of mantle material by the pressure release produced by excavation of the crater, (3) insulation beneath the crater and development of an abnormally high near-surface thermal gradient in the crust, and (4) formation of magma channels as the result of fracturing beneath the crater. Consideration of these mechanisms (French, 1970) suggests that the production of large-scale magmatic activity within a meteorite crater requires that the impact occur in a region where there is either active magmatic activity or a high thermal gradient at depth.

If the formation of compound impact-volcanic structures requires the existence of special conditions at depth at the time of impact, then the exact time of formation of the Sudbury structure becomes particularly important. It has been suggested that the age of 1.72 b.y., determined on the Irruptive (Faure *et al.,* 1964) may be only a minimum value and that the actual age of the structure may be closer to 2 b.y. (Souch *et al.,* 1969; see also Gibbins *et al.,* this volume). This older age would be close to that of the widespread Nipissing intrusive activity dated at about 2.2 b.y. (Fairbairn *et al.,* 1969). Although the shock effects at Sudbury postdate the Nipissing-type intrusives in the area, it is possible that the impact might have occurred at the time when an anomalously high thermal gradient associated with the Nipissing volcanism was still present in the subsurface. Under these conditions, off-loading and fracturing produced by the impact may have generated an anomalously high-SiO_2 magma (in comparison with the Nipissing diabases) which formed the Nickel Irruptive.

If it can be established that the Sudbury structure and the Irruptive formed by the combination of a large meteorite impact and high-temperature subsurface conditions, it may become possible to deduce past temperature conditions in the crust by examination of meteorite impact structures formed on them. The strong contrast between Sudbury and Vredefort, two large ancient craters, only one of which shows internal magmatic effects, suggests that the different post-crater histories reflect different conditions in the crust beneath the crater. If the factors that determine the development of impact-triggered volcanism in large impact craters can be better defined, then the recognition of such structures will provide a powerful tool for investigation of the past thermal histories of the Earth, the Moon, and other planetary bodies.

ACKNOWLEDGEMENTS

Geological studies at Sudbury were greatly assisted by the cooperation and generosity of many individuals associated with The International Nickel Company of Canada, Limited, the Ontario Department of Mines, and the Federal Department of Energy, Mines and Resources. I am especially grateful to J. Guy-Bray, W. Peredery, M. R. Dence, and K. D. Card for valuable suggestions and criticisms, although the conclusions and any errors in this paper are my own. Mary-Hill French provided very helpful criticism and review of the manuscript.

REFERENCES

Bonney, T. G. (1888) – Notes on a part of the Huronian series in the neighborhood of Sudbury (Canada); Quart. J. Geol. Soc. London, vol. 44, pp. 32-45.

Bray, J. G. and Geological Staff (1966) – Shatter cones at Sudbury; J. Geol., vol. 74, pp. 243-245.

Bunch, T. E., Cohen, A. J., and Dence, M. R. (1967) – Natural terrestrial maskelynite; Am. Mineralogist, vol. 52, pp. 244-253.

Burrows, A. G., and Rickaby, H. C. (1930) – Sudbury basin area; Rept. Ontario Dept. Mines, vol. 38, pt. 2, pp. 1-55.

Card, K. D. (1964) – Metamorphism in the Agnew Lake area, Sudbury district, Canada; Bull. Geol. Soc. Amer., vol. 75, pp. 1011-1030.

...... (1968) – Geology of Denison-Waters Area; Rept. Ontario Dept. Mines, vol. 60, pp. 1-63.

Carter, N. L. (1965) – Basal quartz deformation lamellae – a criterion for the recognition of impactites; Am. J. Sci., vol. 263, pp. 786-806.

Chao, E.C.T. (1967a) – Impact metamorphism; *In* Researches in geochemistry, vol. 2, P. H. Abelson (*Editor*). pp. 204-233, John Wiley and Sons, New York.

...... (1967b) – Shock effects of certain rock-forming minerals; Science, vol. 156, pp. 192-202.

...... (1968) – Pressure and temperature histories of impact metamorphosed rocks; Neues Jahrb. Mineralogie, Abhandl., vol. 108, pp. 209-246.

Dence, M. R. (1964) – A comparative structural and petrographic study of probable Canadian meteorite craters; Meteoritics, vol. 2, pp. 249-270.

...... (1965) – The extraterrestrial origin of Canadian craters; Annals N.Y. Acad. Sci., vol. 123, pp. 941-969.

...... (1968) – Shock zoning at Canadian craters: petrography and structural implications, *In* Shock Metamorphism of Natural Materials, B. M. French and N. M. Short (Editors), pp. 169-183.

......(1971) – Impact melts; J. Geophys. Res., vol. 76, pp. 5552-5565.

Dietz, R. S. (1964) – Sudbury structure as an astrobleme; J. Geol., vol. 72, pp. 412-434.

...... (1968) – Shatter cones in cryptoexplosion structures; *In* Shock Metamorphism of Natural Materials, B. M. French and N. M. Short (Editors), pp. 267-285.

Dietz, R. S., and Butler, L. W. (1964) – Shatter-cone orientation at Sudbury, Canada; Nature, vol. 204, pp. 280-281.

El Goresy, A. (1968) – The opaque minerals in impactite glasses; *In* Shock metamorphism of natural materials, B. M. French and N. M. Short (*Editors*), pp. 531-553.

Engelhardt, W. v., and Bertsch, W. (1969) – Shock induced planar deformation structures in quartz from the Ries Crater, Germany; Contrib. Mineral. Petrol., vol. 20, pp. 203-234.

Engelhardt, W. v., Stöffler, D., and Schneider, W. (1969) – Petrologische Untersuchungen im Ries; Geologica Bavarica, vol. 61, pp. 229-295.

Engelhardt, W. v., and Stöffler, D. (1968) – Stages of shock metamorphism in crystalline rocks of the Ries basin, Germany; *In* Shock Metamorphism of Natural Materials, B. M. French and N. M. Short (Editors), pp. 159-168.

Fairbairn, H. W., Faure, G., Pinson, W. H., and Hurley, P. M. (1968) – Rb-Sr whole-rock age of the Sudbury lopolith and basin sediments; Can. J. Earth Sci., vol. 5, pp. 707-714.

Fairbairn, H. W., Hurley, P. M., Card, K. D., and Knight, C. J. (1969) – Correlation of radiometric ages of Nipissing diabase and Huronian metasediments with Proterozoic orogenic events in Ontario; Can. J. Earth Sci., vol. 6, pp. 489-497.

Fairbairn, H. W., and Robson, G. M. (1944) – Breccia at Sudbury; Rept. Ontario Dept. Mines, vol. 50, pt. 6, pp. 18-33

Faure, G., Fairbairn, H. W., Hurley, P. M., and Pinson, W. H. (1964) – Whole-rock Rb-Sr age of norite and micropegmatite at Sudbury, Ontario; J. Geol., vol. 72, pp. 848-854.

French, B. M. (1967) – Sudbury structure, Ontario: some petrographic evidence for origin by meteorite impact; Science, vol. 156, pp. 1094-1098.

...... (1968a) – Shock metamorphism as a geological process; *In* Shock Metamorphism of Natural Materials, B. M. French and N. M. Short (Editors), pp. 1-17.

...... (1968b) – Sudbury structure, Ontario: some petrographic evidence for an origin by meteorite impact; *In* Shock Metamorphism of Natural Materials. B. M. French and N. M. Short (Editors), pp. 383-412.

...... (1970) – Possible relations between meteorite impact and igneous petrogenesis, as indicated by the Sudbury structure, Ontario, Canada; Bull. Volcanologique, vol. 34, pp. 466-517.

French, B. M., and Short, N. M. (Editors) (1968) – Shock Metamorphism of Natural Materials; Mono Book Corp., Baltimore.

Gorshkov, G. S. (1959) – Gigantic eruption of the volcano Bezymianny; Bull. Volcanologique, vol. 20, pp. 77-113.

Hargraves, R. B. (1961) – Shatter cones in the rocks of the Vredefort Ring; Trans. Geol. Soc. South Africa, vol. 64, pp. 147-153.

Hawley, J. E. (1962) – The Sudbury Ores: Their Mineralogy and Origin; Univ. of Toronto Press, Toronto.

Hörz, F. (1965) – Untersuchungen an Riesgläsern; Beitr. Mineral. Petrogr., vol. 11, pp. 621-661.

...... (1968) – Statistical measurements of deformation structures and refractive indices in experimentally shock loaded quartz; *In* Shock Metamorphism of Natural Materials, B. M. French and N. M. Short (Editors), pp. 243-253.

Hörz, F., and Ahrens, T. J. (1969) – Deformation of experimentally shocked biotite; Am. J. Sci., vol. 267, pp. 1213-1229.

Howard, K. A., and Offield, T. W. (1968) – Shatter cones at Sierra Madera, Texas; Science, vol. 162, pp. 261-265.

Kraut, F., and French, B. M. (1971) – The Rochechouart meteorite impact structure, France: preliminary geological results; J. Geophys. Res., vol. 76, pp. 5407-5413.

Manton, W. I. (1965) – The orientation and origin of shatter cones in the Vredefort Ring; Annals N. Y. Acad. Sci., vol. 123, pp. 1017-1049.

Milton, D. J., Barlow, B. C., Brett, R., Brown, A. R., Glikson, A. Y., Mainwaring, E. A., Moss, F. J. Sedmik, E. C. E., Van Son, J., and Young, G. A. (1971) – The Gosses Bluff impact structure, Australia; Science, vol. 175, pp. 1199-1207.

Müller, W. F., and Défourneaux, M. (1968) – Deformationsstrukturen in Quarz als Indicator für Stosswellen: eine experimentelle Untersuchung an Quarz-Einkristallen; Zeitschr. Geophysik, vol. 34, pp. 483-504.

Naldrett, A. J. (1969) – A portion of the system Fe-S-O between 900 and 1080°C and its application to sulfide ore magmas; J. Petrology, vol 10, pp. 171-201.

Naldrett, A. J., Bray, J. G., Gasparrini, E. L., Podolsky, T., and Rucklidge, J. C. (1970) – Cryptic variation and the petrology of the Sudbury Nickel Irruptive; Econ. Geol., vol. 65, pp. 122-155.

Naldrett, A. J., and Kullerud, G. (1967) – A study of the Strathcona Mine and its bearing on the origin of the nickel-copper ores of the Sudbury District, Ontario; J. Petrology, vol. 8, pp. 453-531.

Robertson, P. B., Dence, M. R., and Vos, M. A. (1968) – Deformation in rock-forming minerals from Canadian craters; *In* Shock Metamorphism of Natural Materials. B. M. French and N. M. Short (Editors), pp. 433-452.

Roddy, D. J. (1968) – The Flynn Creek crater, Tennessee; *In* Shock metamorphism of natural materials, B. M. French and N. M. Short (Editors), pp. 291-322.

Roddy, D. J., and Davis, L. K. (1969) – Shatter cones at TNT explosion craters (abs.); Trans. Amer. Geophys. Union (Eos), vol. 50, p. 220.

Roscoe, S. M., (1969) – Huronian rocks and uraniferous conglomerates; Geol. Surv., Can., Paper 68-40.

Shand, S. J. (1916) – The pseudotachylite of Parijs (Orange Free State), and its relation to "trap-shotten gneiss" and "Flinty crush-rock"; Quart. J. Geol. Soc. London, vol. 72, pp. 198-221.

Short, N. M. (1968) – Nuclear-explosion-induced microdeformation of rocks; an aid to the recognition of meteorite impact structures; *In* Shock metamorphism of natural materials, B. M. French and N. M. Short (Editors), pp. 185-210.

Sopher, S. R. (1966) – Paleomagnetic study of the Sudbury irruptive; Geol. Surv. Canada, Bull., vol. 90.

Souch, B. E., Podolsky, T., and Geological Staff (1969) – The sulfide ores of Sudbury: their particular relationship to a distinctive inclusion-bearing facies of the Nickel Irruptive; Econ. Geol. Mono., 4, pp. 252-261.

Speers, E. C. (1957) – The age relation and origin of common Sudbury breccia; J. Geol., vol. 65, pp. 497-514.

Stevenson, J. S., and Colgrove, G. L. (1968) – The Sudbury Irruptive: some petrogenetic concepts based on recent field work; Internat. Geol. Congress, 23rd Session, Czechoslovakia, Report, sect. 4, pp. 27-35.

Stöffler, D. 1971a. Coesite and stishovite in shocked crystalline rocks; J. Geophys. Res., vol. 76, pp. 5474-5488.

...... (1971b) – Progressive metamorphism and classification of shocked and brecciated rocks at impact craters; J. Geophys. Res., vol. 76, pp. 5541-5551.

Thomson, J. E. (1957) – Geology of the Sudbury basin; Rept. Ontario Dept. Mines, vol. 65, pt. 3, pp. 1-56.

Williams, G. H. (1891) – The silicified glass-breccia of Vermillion River, Sudbury District; Bull. Geol. Soc. Amer., vol. 2, pp. 138-140.

Williams, H. (1954) – Progress and problems in volcanology; Quart. J. Geol. Soc. London, vol. 109, pp. 311-332.

...... (1957) – Glowing avalanche deposits of the Sudbury basin; Rept. Ontario Dept. Mines, vol. 65, pt. 3, pp. 57-89.

23

Reprinted from *Jour. Geology* 74:243-245 (1966)

SHATTER CONES AT SUDBURY[1]

J. GUY BRAY AND GEOLOGICAL STAFF[2]

[*Editor's Note:* Plates 1 and 2 have been omitted.]

ABSTRACT

The Sudbury structure has been interpreted by R. S. Dietz as an astrobleme. Fieldwork indicates that shatter cones are locally developed in a belt up to 11 miles wide around the outer contact of the nickel intrusive, and are confined to preintrusive rocks. Although apparently modified by folding, the cones show a strong preferred orientation toward the Sudbury Basin. Such a distribution and orientation are not incompatible with the astrobleme theory as expounded by Dietz. However, shatter-coning may not be uniquely characteristic of meteor impact, and it is hoped that further evidence will be forthcoming by which the theory may be properly evaluated.

INTRODUCTION

We are indebted to R. S. Dietz (1964; see also Dietz and Butler, 1964) for having brought to our attention the fact that shatter cones exist in the Sudbury area. Dietz and Butler regard shatter-coning as evidence of shock-induced over-pressures in rock, and therefore useful in identifying presumed astroblemes. We have conducted a survey of the more accessible parts of the area and have found that the distribution and orientation of the cones are in keeping with Dietz's astrobleme theory. Although we believe that a meteoritic impact origin for the Basin merits continuing consideration, and that it could be reconciled with the observed geological relationships without undue difficulty, we prefer to suspend judgment, having in mind that shatter cones have also been attributed to volcanic and tectonic activity, both of which are prominent in our interpretations of Sudbury geological history.

As part of the astrobleme hypothesis, Dietz also presented comprehensive speculations based on possible later consequences of meteoritic impact. We believe these speculations are incompatible with the geological facts, and they will not be dealt with in this note, which is confined to a description of the occurrence of the shatter cones.

[1] Manuscript received June 15, 1965.

[2] International Nickel Company of Canada, Ltd., Copper Cliff, Ontario.

FORM

The Sudbury shatter cones (pls. 1 and 2) range in length from under 1 inch to over 10 feet. In the bedded rocks, cone size seems to be proportional to thickness of beds. The cone surfaces are commonly shiny and micaceous, and are invariably marked by fine striations with divergent, horse-tail patterns (pls. 2, *B* and 1, *D*) but are conspicuous only in bright sunlight. Cone surfaces are generally well defined near the apex, but die out more or less irregularly toward the base, though some cones appear to be sharply terminated against joints. In most places apical angles are less than 90°.

In normal section most cones are subcircular. More elliptical or flattened cones, and particularly those having curved axes (pl. 1, *D*), may indicate imperfect development or modification by later rock deformation.

Cones typically occur in swarms (pl. 1, *A*) rather than singly. It is common to find many small parasitic cones on the flanks of a large master cone (pl. 2, *B*); cones within cones and mutually interfering cones are also seen. Thus in some places the rocks may look characteristically shattered (pl. 1, *B*) even when cone surfaces are not immediately apparent.

Comparisons with published photographs and descriptions of shatter cones from cryptoexplosion structures, etc., indicate that the Sudbury cones are excellent examples. Some of the displays in quartzite are spectacular (e.g., pl. 1, *A*).

In the Sudbury District, shatter cones are accompanied on most rock exposures by undulating striated surfaces (pl. 2, *C*). These surfaces, which are more extensive and less regular than cone surfaces, and seem to be randomly oriented, somewhat resemble slickensided fractures. However, they are marked by divergent striae similar to those on shatter cones, and in some cases appear to merge with imperfect cones.

In the field, these surfaces have proved valuable as an aid in finding shatter cones. They are more common than the cones, but are confined to the same belt around the intrusive (see below). The cones and the undulating striated surfaces seem to be related aspects of the same episode of shattering; the surfaces possibly represent incipient or attenuated coning.

DISTRIBUTION

The general distribution is shown in figure 1. To obtain reconnaissance coverage of the area, nearly all traverses were made along roads. Roads are numerous only on the south side of the Basin, but all the roads outside the intrusive intersect shatter-coned rocks, and the distribution mapped on the south appears to be representative of the whole periphery.

Shatter cones and undulating striated surfaces were found in the preintrusive footwall rocks only, in a belt up to 11 miles wide surrounding the nickel intrusive. Neither type of shattering was found on outcrops outside the belt, although traverses were made up to 60 miles away.

No shatter cones or surfaces were found in the intrusive or the overlying Whitewater rocks. In the footwall, they were seen in every main rock unit except Grenville gneiss, olivine diabase and trap dykes, and Sudbury breccia. The distribution of this shattering presents a striking parallel with the distribution of common Sudbury breccia, which also is restricted to preintrusive rocks in a belt (up to 14 miles wide) around the intrusive (see Speers, 1957).

Within the 11-mile belt, only some exposures show shatter-coning: too little rock may crop out; cone surfaces may be transected at too high an angle to be recognized; cones occur in shattered rock where weathering easily destroys them; shearing and other metamorphism obscures them north of Nairn (at High Falls) and along the Grenville Front. However, a much more important factor seems to be rock type. The best formed and most abundant cones are found in brittle, fine-to-medium-grained quartzites, for example, south of Copper Cliff (at Kelly Lake) and at Skead. The coarser granites and gneisses north of the Basin and the less brittle basic igneous rocks are generally unfavorable hosts. Nevertheless, good shatter cones have been found in all the mapped segments of the 11-mile belt.

Although the abundance of cones appears to vary irregularly on single exposures, figure 1 suggests that, on a larger scale, there are zones of maximum concentration of shattering. These may mark the areas where the relationships between the physical properties of the rock and the deforming stresses were most favorable for coning to develop.

ORIENTATION

Stereographic plotting of cone striations was not attempted: the positions of individual cone axes were estimated by eye, and their orientations were then measured with a Brunton compass, often with the aid of a pencil set parallel to the axis.

Wherever shatter cones are abundant, a

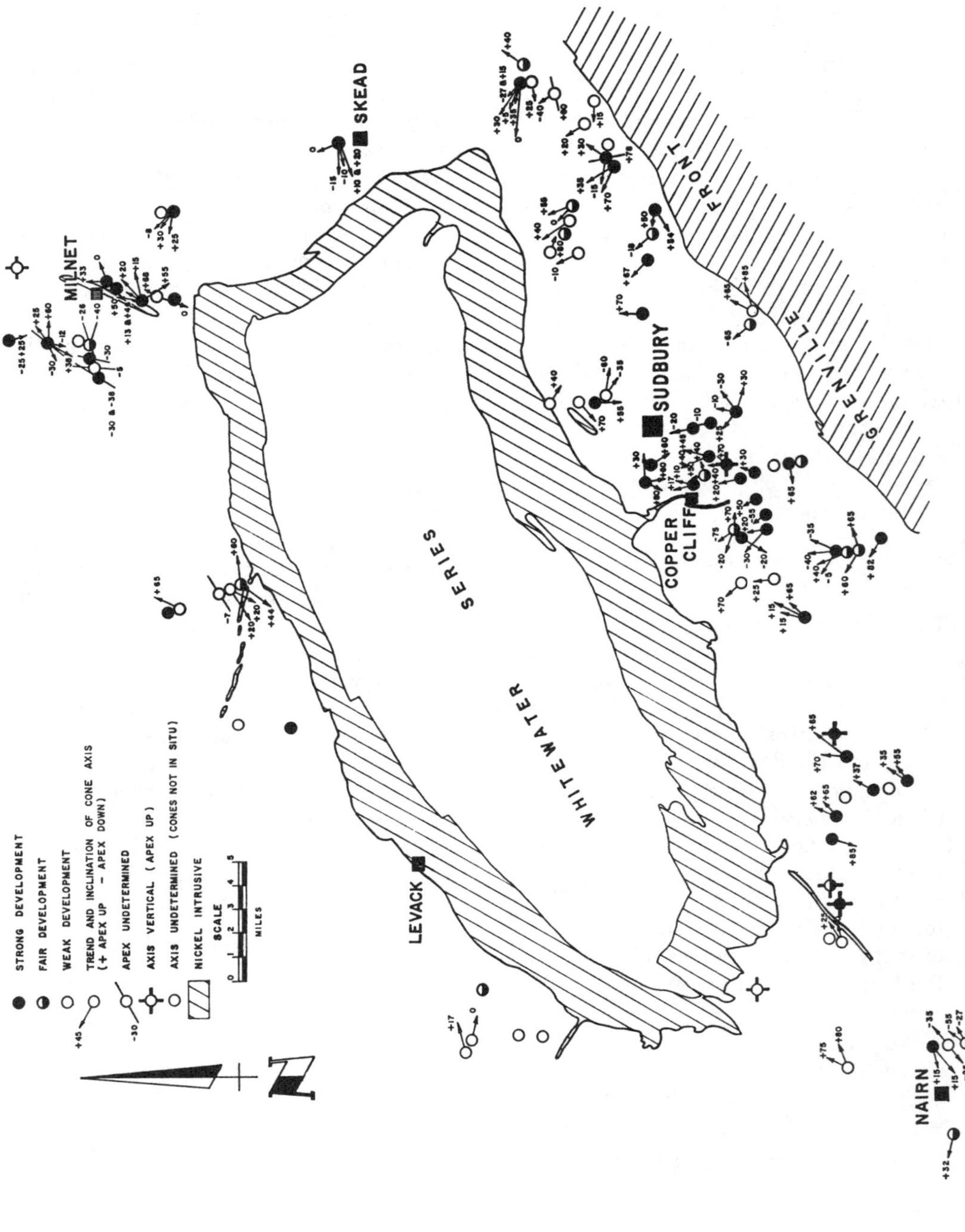

Fig. 1—Shatter cones in the Sudbury district

strong preferred orientation is observed, coinciding with the attitudes of the largest and best developed cones. Individual cones may be diversely oriented, however, and deductions from small samples can be misleading. Considering this, and the gaps in the mapping, it is remarkable that the attitudes for the whole area, plotted on figure 1, show a conspicuous general orientation of the cone apexes *toward the Sudbury Basin*, an orientation which has persisted through the folding, faulting, and metamorphism that apparently postdate the coning.

There are areas where the attitudes do not fit into this simple pattern, for example, near Nairn where there has been cross-folding and near Milnet where there may have been some overturning; additional mapping is desirable in these and other places. By contrast, in areas of lesser structural deformation where the most striking displays of shatter cones are preserved (south of Copper Cliff; north of Skead; north-northwest of Milnet), the apexes point unmistakably toward the Basin.

The attitudes of many cones appear to have been modified by folding; thus, in the steeply upturned or overturned strata close to the intrusive's southern rim, shatter cones are generally directed steeply upward. In general, the preferred orientation of cones throughout the bedded rocks is more or less parallel to the bedding; in most places these rocks now dip rather steeply, but if they are returned to the horizontal, the centripetal pattern of cone axes is strongly enhanced.

At the northeast end of the shatter cone belt, some of the cones point away from the Basin. The finding of quartzite erratics with undulating striated surfaces north of the known limit of shattering in that area may perhaps be significant in this connection. Anomalous cone orientations at the other end of the belt, near Nairn, may be related to recently reported shatter cones and Sudbury-type breccia 10 miles southwest of the belt, in Merritt Township (Dr. K. Card, Ontario Department of Mines, verbal communication).

CONCLUSION

The form, distribution, and orientation of shatter cones at Sudbury have been described above. The nature of their occurrence is not inconsistent with the predictions of Dietz's astrobleme theory. However, the precise origin and genetic significance of shatter cones in general are still open matters, and although Sudbury structure also possesses other features thought to be characteristic of cryptoexplosion structures, much additional information must be forthcoming before the theory can be satisfactorily assessed and the shatter cones used to help unravel the geological history of the Sudbury Basin. It is hoped that this note will stimulate the search for critical evidence by geologists from the Sudbury District and elsewhere.

REFERENCES CITED

DIETZ, R., 1964, Sudbury structure as an astrobleme: Jour. Geology, v. 72, p. 412–434.

——— and BUTLER, L., 1964, Shatter-cone orientation at Sudbury, Canada: Nature, v. 204, no. 4955, p. 280.

SPEERS, E., 1957, The age relation and origin of common Sudbury breccia: Jour. Geology, v. 65, p. 497–514.

24

Reprinted from pages 117–124 of *Geol. Assoc. Canada Spec. Paper 10,* 1972

EVIDENCE FOR AN IMPACT ORIGIN FOR LAKE WANAPITEI, ONTARIO

M. R. Dence and J. Popelar
*Gravity Division, Earth Physics Branch, Dept. of Energy, Mines and Resources, Ottawa

ABSTRACT

Lake Wanapitei (lat. 46° 44.7'N, long. 80° 44.6'W) is 40 km northeast of Sudbury, Ontario on the eastern margin of the Sudbury Basin. The lake has a strongly indented outline but in its north-central part contains an island-free area 8.6 km in diameter. Topographic, geophysical and petrographic data indicating this circular depression has formed by hypervelocity impact include: (i) water depths averaging 50 m and in places exceeding 100 m, (ii) a circular Bouguer gravity anomaly of −15 mgal coinciding with the circular part of the lake, (iii) absence of any significant local disturbance of the magnetic field and, (iv) evidence from glacial drift collected around the southern margin of the lake, that the lake bottom is underlain by friable red sandstone, coarse mixed breccia, fresh glass and vesicular igneous rocks. The breccias consist of fragments of Huronian quartzite and Nipissing diabase, similar to formations cropping out around all but the western margin of the lake, but showing all grades of shock metamorphism, including planar features and isotropic states in quartz and feldspar, the high pressure mineral coesite, kinking of mafic minerals, and completely melted materials which include lechatelierite. The age of the crater has yet to be determined but its state of preservation suggests formation in the late Paleozoic or Mesozoic. Certainly the crater is younger than and unrelated to other events in the Sudbury area. Its relationship to the Sudbury Basin is compared to that between the lunar craters Theophilus and Theophilus B.

INTRODUCTION

In the course of a detailed study of the gravity field in the Sudbury area of Ontario (Popelar, 1971, 1972), attention was directed to Lake Wanapitei, the large circular lake northeast of the Sudbury Basin. The general topographic similarity of the lake to others such as Lac Couture, Quebec, (Beals *et al.*, 1967) and Deep Bay, Saskatchewan, (Innes *et al.*, 1964) for which a hypervelocity impact origin has been demonstrated, prompted early speculation that Lake Wanapitei may have had a similar origin. Reservations about this proposition came from two considerations: (i) the apparent relationship between the west side of the lake and the East Range of the Sudbury Irruptive which suggested some common structural control, and (ii) the coincidence of the lake being located at the margin of the Sudbury structure, for which an impact origin had previously been advanced (Dietz, 1964; French, 1967). Nonetheless, as the investigations proceeded in 1969 and 1970, the evidence, as outlined below, swung more and more towards the thesis presented here.

The essential steps in the investigation have been the acquisition of water depth measurements, kindly provided by the Ontario Department of Lands and Forests in 1968; the publication of the aeromagnetic maps of the area (Geological Survey of Canada, 1965); the gravity observations on the ice-covered lake in the winter of 1969 by J. Popelar; and the discovery by M. R. Dence in 1970 of a rich variety of breccias in the glacial till south of the lake. Together, these data permit a reasonably complete interpretation of the structure of the lake floor and provide strong evidence for origin by hypervelocity impact. The main results of these preliminary investigations are reported here.

TOPOGRAPHY AND REGIONAL GEOLOGY

Lake Wanapitei lies 40 km northeast of the city of Sudbury and 15 km east of Capreol. In outline, the lake (Figure 1) resembles a molar tooth with an arcuate northern shore and deep embayments to the south, southeast and southwest. In the northern part of the lake there is an area 8.6 km in diameter that is devoid of islands. One major stream, the Wanapitei River, flows in from the north and discharges to the southeast, but in general there is little drainage into the lake. Indeed, the predominant pattern of the streams and lakes within 5 km of the lake is approximately concentric with the circular part of the lake, in similar style to the arcuate lakes around Deep Bay, Saskatchewan (Innes *et al.*, 1964). Other factors influencing the drainage near Lake Wanapitei are regional linear structures which follow northwest-southeast, north-south, and northeast-southwest trends, and the direction of last glaciation during which the ice sheets moved towards the south-southwest.

The lake level is 267 m (876 ft.) above sea level. In general, the shoreline is of moderate relief, rising in most directions to heights of 70 m or more above the lake. The highest point in the vicinity, a peak 130 m above the lake level, lies 2 km north of the northern shoreline. The water depth measurements (Figure 2) show a gradual increase from the northern shoreline to depths of about 100 m, with a deep channel near Massey Bay in the southern part of the lake where depths of more than 120 m have been recorded. Most of the island-free area is more than 50 m deep; this part of the lake has a mean relief of approximately 120 m.

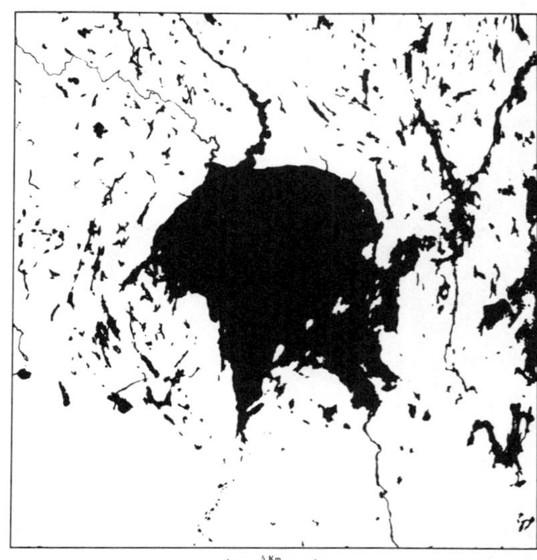

Figure 1: Outline map of the Lake Wanapitei area showing the pattern of drainage around the lake.

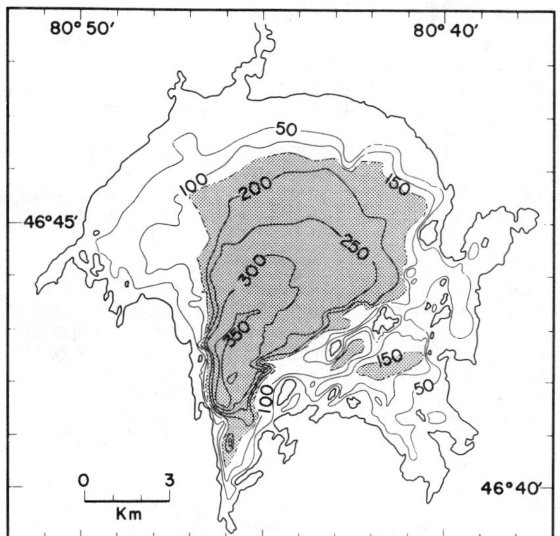

Figure 2: Contour map of water depths in Lake Wanapitei prepared from data supplied by the Ontario Department of Lands and Forests. Depths are given in feet. Water deeper than about 150 feet is considered to be underlain by crater breccias in the circular part of the lake and by tectonic breccias in the southwest arm of the lake.

The geology of the Lake Wanapitei region was first studied comprehensively by Quirke (1922), who described the Huronian rocks of the area in some detail and commented on the strongly disrupted character of the rocks exposed along the west and south shores of the lake. More recently, reports have been published by the Ontario Department of Mines covering a number of townships south, east and northwest of the lake (Thomson, 1961; Thomson and Card, 1963; Meyn, 1970, 1971) although these areas do not include much of the northern part of the lake itself. The main features of the geology of the lake are depicted on Ontario Department of Mines maps 2170 (1969) and 2188 (1971).

The rocks cropping out around the lake include Archean gneisses, Huronian sedimentary rocks and younger intrusions. On the west side of the lake are granites, gneisses and mafic rocks which are generally assigned to the Superior province. These rocks form the footwall of the East Range of the Sudbury Irruptive which on average lies 4 km from the west shoreline of the lake. The footwall rocks are strongly disturbed by brecciation of the Sudbury type and by later faulting and shearing which also cut the Irruptive. Close to the Irruptive, the footwall rocks are extensively recrystallized, but farther out some granitic rocks show weak planar features in quartz shocked by the Sudbury event (Dence, 1972).

To the south, east and north the rocks seen in outcrops nearest to the lake are mainly of the lower Huronian Mississagi Formation or upper Huronian Gowganda Formation. They also include Nipissing and younger diabase. Quartzites of the Lorrain Formation underlie the prominent ridge north of the lake but are not recorded at the lake shore. Much of the northern shore of the lake is underlain by thick sandy drift which forms plains extending 2 to 6 km to the north. On the east and south sides, the rocks exposed nearest to the center of the lake are either Nipissing diabase or Mississagi quartzite. Some of the latter contain shatter cones similar to those recorded elsewhere around the Sudbury Basin (Guy-Bray, personal communication 1970), but are otherwise no more strongly deformed than bedrock farther from the lake. There is thus little indication in the rocks *in situ* that the lake has an unusual origin.

The rocks of the area have been subjected to a complex sequence of tectonic events, including initial episodes that affected the Archean rocks and later events which followed the deposition of the Huronian sediments. For the most part, the Huronian rocks are folded about northwest-trending axes with dips flattening towards the northeast, although more complex structures are indicated south of the lake. Faults and lineations are prominent, the most conspicuous being the Wanapitei fault which is followed for a considerable distance by the upper Wanapitei River. Late Precambrian olivine diabase dikes are offset by this fault. A second important fault system of north-south trend is exposed as a broad mylonite zone at the south end of Massey Bay. The rocks of the Maclennan Mine to the south lie along the trend of this system, whereas to the north the zone strikes towards the deep water channel in the bay. This suggests that the channel is the result of deep glacial gouging along a zone of weakness. A third set of faults and lineaments trends northeast and may control, in part, the form of the bays and islands in the southern part of Lake Wanapitei. In sum, the complex pattern of lithological units and structural features in the vicinity of the Wanapitei basin comes from movements connected with the Penokean, Grenvillian and later tectonic events, all of which predate the formation of the depression now filled by the lake.

MAGNETIC DATA

Aeromagnetic maps at the standard scale of one inch to one mile were published for the Wanapitei area in 1965. The magnetic field in the vicinity of the lake is dominated by a broad, strong magnetic high of unknown origin centered about 30 km northeast of the lake. Conspicuous local magnetic anomalies are associated with iron formations (in the greenstone belt northwest of the lake) and several of the northwest-trending olivine diabase dikes, the most prominent of which passes 2 km north of the north shore of the lake. Other magnetic features are associated with the Sudbury Irruptive rocks and mafic rocks in the Archean gneisses.

The magnetic field over Lake Wanapitei, as given in the aeromagnetic maps, is relatively uniform and featureless. It shows a steady rise from south to north with an average gradient of about 80 gammas per km towards the dominant magnetic high. There is thus no sign of any significant magnetic disturbance which might be related to a geological body underlying the lake.

GRAVITY MEASUREMENTS

The gravity anomaly field of the Lake Wanapitei area was established from regional gravity observations in the Sudbury area supplemented by additional field work in 1969. During the winter of 1969, thirty-eight gravity stations were observed on the ice-covered lake surface along two parallel north-east-southwest profiles at intervals of 500 m. Another thirty-eight gravity stations were established on the islands and shore of the lake. Together with the regional stations, these measurements provide a clear picture of the gravity anomaly field of the area. The accuracy of the gravity anomaly values obtained for the detailed stations is somewhat better than that for the regional stations, for which a mean square error of ±0.61 mgal has been calculated (Popelar, 1971). All stations within or along the shore of Lake Wanapitei have been corrected for the effect of the body of water; the corrections being calculated exactly from the contoured bathymetric map. The corrections range from a few tenths of a mgal for shoreline stations to 6.9 mgal for stations above the deepest part of the lake. No terrain corrections were introduced.

The resulting Bouguer gravity anomaly map (Popelar, 1971) indicates a local, almost circular gravity depression of about 15 mgal centred over the circular part of the lake. The strong horizontal gradient of more than 4 mgal/km outward from the centre, as well as the maps of the first and second vertical derivatives of the gravity field (Popelar, 1972) all suggest a superficial mass deficiency with a regular, circular plan. There is a close resemblance to gravity anomalies associated with craters attributed to impact, notably that at Deep Bay, Saskatchewan (Innes, 1961; Innes *et al.*, 1964).

Consideration of the mean rock densities of the surrounding formations (Popelar, 1972) strengthens this comparison. The rocks in the vicinity of Lake Wanapitei are mainly upper Huronian quartzites, with a mean density of 2.63 g/cm^3, the lowest density rocks in the Sudbury area, or lower Huronian metasedimentary rocks and Archean metamorphic rocks both of which have mean densities of about 2.73 g/cm^3. The higher density gabbros and diabase (mean density 2.94 g/cm^3) form relatively minor intrusive bodies with little gravitational expression. It is clear then that the rocks underlying Lake Wanapitei must be of significantly lower density than any of the rocks of regional distribution. This is a common characteristic of rocks from impact craters elsewhere in Canada (Beals *et al.*, 1963; Dence, 1965).

DESCRIPTION OF BRECCIAS AND RELATED ROCKS

Encouraged by the analysis of the topographical and geophysical data, a search for unusual rock types was undertaken in the extensive glacial deposits on the southern, down-glaciation shore of the lake. The search was successful. The first discoveries in June 1970 by M. R. Dence were of pebbles of breccia on the beach near Wanapitei Lodge north of Skead. Subsequently, richer deposits were located by J. V. Guy-Bray in a reconnaissance survey of the shoreline of the lake (Guy-Bray, personal communication 1970). He showed that the breccia boulders were concentrated in the Massey Bay area and were increasingly rare to the north and south. This result is in agreement with the interpretation that the breccias had been scoured from the floor of the lake during the most recent period of glaciation. It was also immediately evident that most of the breccia boulders are only weakly compacted and rapidly disintegrate on exposure and handling, again indicating a local source.

The main collections taken to date have been from the east shoreline of the narrow bay north of Skead and from east of the Maclennan Mine. In these thick deposits of sand and boulders, all rock types of the area to the north are represented. The exotic breccias make up possibly ten per cent of the boulder population in the richest localities, but are generally considerably less abundant. Nonetheless, it has been possible to collect several hundred samples with relatively little effort, from which the following general description is derived.

The boulders considered to be derived from the bottom of Lake Wanapitei are almost entirely breccias of mixed lithology. They range in size from sub-rounded or well-rounded pebbles to boulders as large as 50 cm in maximum dimension (Figure 3). Some boulders, up to 20 cm across, are of a single lithology and appear to be large clasts derived from the breccias. Boulders of fine-grained to glassy igneous rocks are less abundant and small cobbles of friable, thinly bedded siltstone and sandstone are extremely rare. Most boulders are light to medium grey, with rare greenish clasts of mafic rocks, or dark grey to black masses of glass. Some, including the sedimentary rocks, are oxidized to yellow-brown or buff hues.

The breccias may be subdivided into two main groups: those containing glass, and those consisting entirely of rock and mineral fragments. The rock fragments are angular to subrounded and are mostly quartzite, arkosic quartzite, siltstone and diabase. Granitic rocks and greenstones are rare. Although many fragments are deformed and altered, making it difficult to determine their origin, it appears that most of the fragments have been derived from the lower Huronian Mississagi Formation and the Nipissing diabase which intrudes it. These are the rocks exposed along the south and east shoreline of the lake, and it may be inferred that they also underlie the central part of the lake bottom.

Evidence for shock metamorphism of many fragments in the breccias is clearly seen in thin sections, and is also manifest in hand specimen. Most conspicuous are quartzite fragments which have frothed and vesiculated so that they resemble vesicular, strongly shocked Coconino sandstone samples from Barringer Meteor Crater (Chao, 1967; Kieffer, 1971). Others, less strongly affected, have the chalky-white appearance typical of moderately shocked rocks in which quartz grains contain numerous fractures and planar features. Glass is present either as small, highly contorted fragments or in masses up to 30 cm across with a strongly layered texture due to the entrainment of numerous rock fragments. The resemblance of the glass-bearing breccias to breccias of the suevite type (Engelhardt, 1971) is strong. As at the Ries crater, the samples from Lake Wanapitei contain both fresh black glass and dull grey, partly or completely devitrified glass, as well as some fine grained, crystalline, vesicular igneous rocks in which inclusions are thoroughly recrystallized.

Densities were obtained for a representative group of the breccias and vesicular quartzite samples. The latter form a distinct group with a mean density of 1.98±0.07 g/cm^3, whereas the breccias have a mean density of 2.46±0.13 g/cm^3. The spread of measurements for the breccias results from variations in porosity and in the proportions of glass, quartzite and mafic rock fragments. Clearly these density values satisfy the material requirements for a model compatible with the gravity anomaly over the lake.

MICROSCOPIC CHARACTERISTICS

Under the microscope, features typical of shock metamorphosed rocks are apparent. Breccias found at impact craters can be grouped into two main types: breccias with weakly to moderately shocked clasts, and suevites which contain strongly shocked materials including glassy masses (Engelhardt, 1971). Both categories are readily identifiable in the Wanapitei samples, although there are few representatives of the monomict autochthonous breccias typical of the rims of craters. Most of the samples are well-mixed breccias both in regard to lithologies and degree of shock metamorphism of the clasts.

Representative rock textures (Figure 4) illustrate the banded flow texture of the large glassy masses (Figure 4 left) and the poorly sorted, well-mixed texture most commonly seen in the breccias. At low stages of shock metamorphism the original rock textures of clasts are well preserved (Figures 4 right, 5 left). At higher levels, progressive melting, flow and vesiculation of components such as micas, clay minerals and feldspars have produced increasing textural distortion and, at the highest levels, glasses containing many inclusions and schlieren of less strongly shocked phases. The strongly banded texture of the glasses is typical of that produced by impact, because, in such events, shock melted materials with high particle velocities overtake and engulf less strongly shocked rocks (Dence, 1971).

Indications of shock metamorphism are, as usual, most readily apparent in quartz, which is the most abundant mineral

Figure 3: Large boulder representative of the most common variety of mixed breccia in the glacial deposits found south of Lake Wanapitei, in this case east of the Maclennan Mine. Large clasts of white to cream quartzite showing different degrees of shock deformation are clearly visible. Smaller dark clasts are either fragments of siltstone, Nipissing diabase or irregular masses of altered glass. The matrix is clastic and composed of moderately compacted fragments of the same rock types.

in these breccias. All stages of shock metamorphism of quartz (Robertson *et al.*, 1968; Stöffler, 1971 b) are represented including simple fracturing, cleavage, planar elements of various types with characteristic orientation, diaplectic quartz glasses and lechatelierite. Planar elements (Figure 5 right) in grains at stage I of shock metamorphism in most cases are decorated with innumerable minute cavities or inclusions, giving some grains a clouded appearance. Many silica glasses are annealed in part or whole to fine-grained mosaics of secondary quartz. However, the presence of some fresh diaplectic silica glasses encouraged the hope that high pressure phases might also be found. This hope was realized with the detection of coesite (Dence *et al.*, 1972) in association with silica glass of refractive index 1.460(5) and other glasses of variable composition. According to the scheme of Stöffler (1971 a,b), this association of coesite and diaplectic glass is representative of stage III of shock metamorphism which corresponds to shock pressures of 450 to 550 kb. Other samples which consist of inhomogeneous rock glass of the type considered by Stöffler to be produced by complete melting at shock pressures exceeding 550 kb, contain pure lechatelierite in fantastically contorted forms (Figure 6). An analysis of this glass (Table I) shows that it is highly siliceous and could be formed from somewhat arkosic quartzites, which are the predominant clasts in the breccias, probably with some addition of material from highly shocked siltstones and diabases.

Shock effects in other minerals are commonly obscured by later alteration. However, examples have been recorded of

Figure 4: Photomicrograph in plane light showing the textures of two of the main rock types in the Wanapitei float: Left, the fabric of a large mass of clear, fresh glass which is colourless in thin section and black in hand specimen. A strong flow banding is evident due to the entrainment of fragments, mainly quartzite, showing all stages of shock metamorphism and disintegration. Fine schlieren in the glass matrix are commonly of lechatelierite. An analysis of the glass is given in Table I.

Right, a typical mixed breccia similar to that in Figure 3. Angular fragments of quartzite (light), siltstone (dark) and altered glass (grey, irregularly fractured) are held in a matrix of quartz and other minerals of the country rocks.

Figure 5: Photomicrograph in plane light of details of a quartzite fragment in a mixed breccia from the Wanapitei float. At medium magnification (left) the original texture of the quartzite is apparent with no sign of significant deformation. A detailed view (right) of the largest grain at left reveals abundant development of closely spaced planar features in at least two sets parallel to members of the $(10\bar{1}3)$ crystallographic directions in the grain, and more broadly spaced, irregular rhombic cleavage. The grain is heavily charged with extremely fine cavities or inclusions, as are commonly developed after slight annealing of shocked quartz.

Figure 6: Photomicrograph in plane light of a highly contorted fragment of lechatelierite in a glass matrix similar to that shown in Figure 4 (left). While the shape of this inclusion is suggestive, it can hardly be taken to prove that the Wanapitei crater formed during the age of dinosaurs!

Figure 7: Photomicrograph, left in plane light, right in crossed polarizers, of portion of a shocked fragment of Nipissing diabase in a Wanapitei breccia. The plagioclase grain in the centre of the picture has been largely converted to the isotropic form maskelynite. Some plagioclase grains also show planar features. A mosaic texture is developed in coexisting pyroxene.

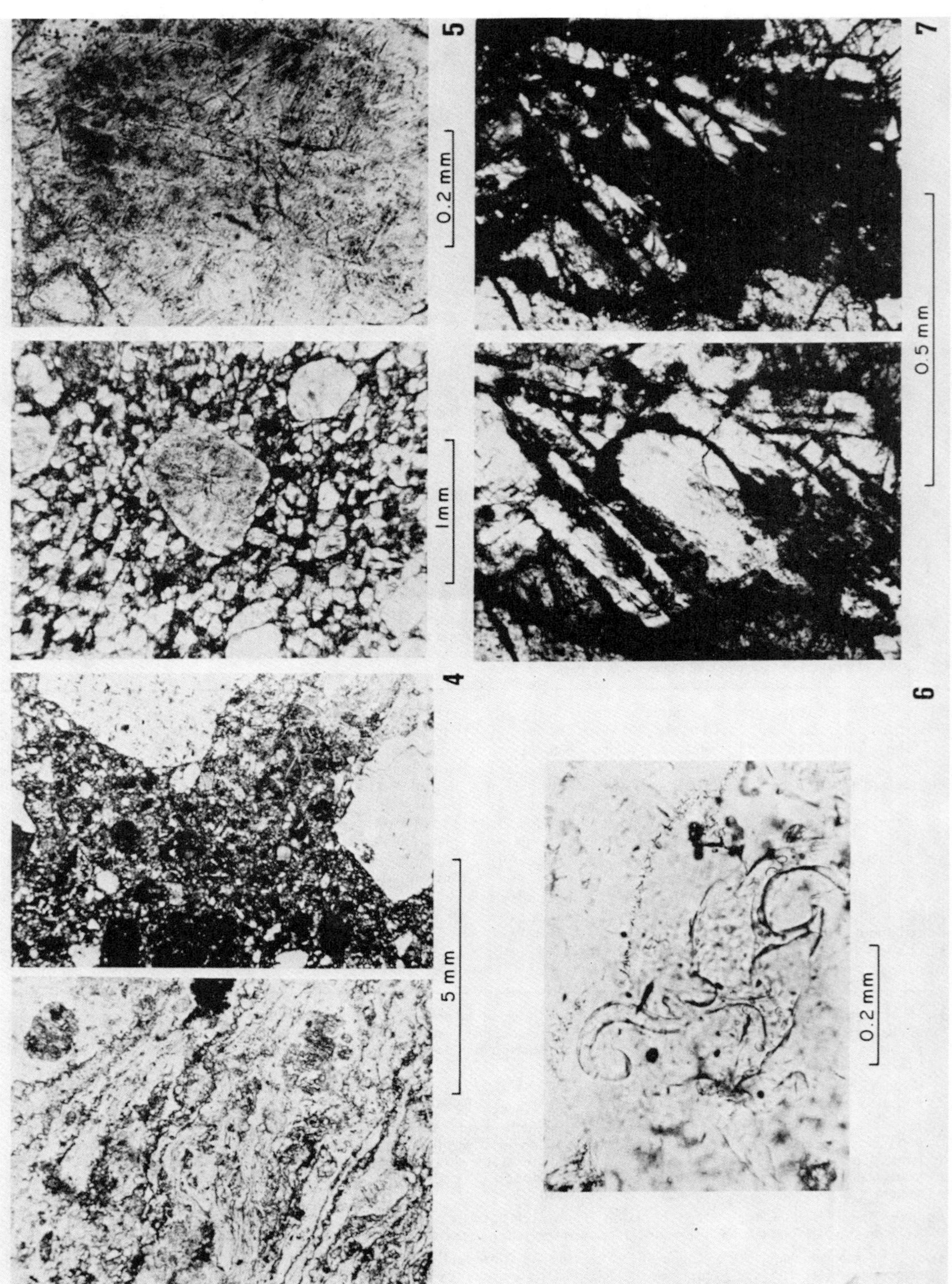
4
5
6
7
5 mm
1mm
0.2 mm
0.5 mm
0.2 mm

TABLE I

Partial major oxide analysis of matrix of glass containing lechatelierite from Lake Wanapitei, Ontario

	Wt. %	*Norm.*	
SiO_2	75.1	Quartz	42.2
Al_2O_3	9.3	Orthoclase	19.0
FeO (total Fe)	2.3	Albite	21.5
MgO	1.2	Anorthite	4.6
CaO	1.1	Hypersthene	6.7 (En_{42})
Na_2O	2.5	Diopside	0.8
K_2O	3.2		
	94.7		94.8

Refractive Index 1.5050 ± .0005

Analysis by electron microprobe methods by P. B. Robertson.

kinking in mica and hornblende, of planar features in plagioclase and potash feldspar and of the transformation of plagioclase into maskelynite (Figure 7). Other minerals derived from the country rocks such as pyroxene and epidote have mosaic textures which also are probably the result of moderate to strong shock metamorphism.

INTERPRETATION OF RESULTS

There can be little doubt that the low-density, fragile breccias in the glacial drift south of Lake Wanapitei come from the bottom of the lake. The petrographic evidence for shock metamorphism in the breccias and glasses is identical in all essential aspects with shock effects recognized at other Canadian impact craters (Robertson *et al.*, 1968) and at similar craters elsewhere (Chao, 1967; Stöffler, 1971 a). The most direct comparisons can be made with craters of approximately the same size, such as those at Deep Bay (Innes *et al.*, 1965; Dence *et al.*, 1968), Lac Couture (Beals *et al.*, 1967) and Nicholson Lake (Dence *et al.*, 1968). Of these, Lac Couture most closely resembles Lake Wanapitei, as both have a present relief of about 100 m, and in both cases glaciation has scoured shock metamorphosed breccias from the lake bottom. However, Deep Bay, by virtue of the drilling results available, and Nicholson Lake, where the floor has been exposed by erosion, provide a more complete basis for constructing a crater model for the Wanapitei structure. From these analogies the most likely model is one in which a central uplift is surrounded by a relatively thick lens of mixed breccias. The variety of breccias in the Wanapitei float is compatible with differential erosion of such a breccia lens, some varieties being similar to upper level breccias of the suevite type, others to deeper layers richer in recrystallized igneous rocks. The minor amount of apparently younger sedimentary rocks in the float suggests that such rocks have been almost completely eroded from the basin.

Using the density results, a quantitative interpretation of the circular negative gravity anomaly has been carried out. In the first stage of the interpretation, the simple formulas for the anomaly due to a vertical cylinder, as well as the method of Paul (1972) for bodies with circular symmetry, have been employed. These indicated that the best estimates lie in the range 3.5 to 5.0 km for the radius and 1.5 to 3.0 km for the depth of the structure whereas the corresponding density contrast with the country rocks varies between -0.15 and -0.30 g/cm^3. A two-layered model of a circular basin of radius 3.75 km was also devised, the upper layer having a density contrast of -0.25 g/cm^3 and the lower of -0.15 g/cm^3 (Figure 8). The model was adjusted on a 1 km by 1 km grid so that the mean square difference between the computed and measured gravity anomalies was less than 1 mgal. The results of the adjustment indicate maximum depths of about 1.6 and 3.3 km for the upper and lower layers under the centre of Lake Wanapitei. These values give the maximum possible depths of the rocks related to the crater as the density measurements indicate that some breccias are of even lower density than those assigned to the model. Considerable thicknesses (20 to 40 m) of glacial drift with density less than 2.0 g/cm^3 may also be present in the lake, and may contribute to the negative gravity anomaly.

An alternative explanation of the negative gravity anomaly at Lake Wanapitei by a stock-like intrusion is highly unlikely because it would require intrusive rock of substantially lower density than the granites of the area which are of comparable density (2.63 g/cm^3) to the Huronian quartzites. Moreover the magnetic field does not indicate any characteristic features of intrusive bodies, such as a highly variable magnetic field related to the margins of intrusions or other changes in magnetic patterns.

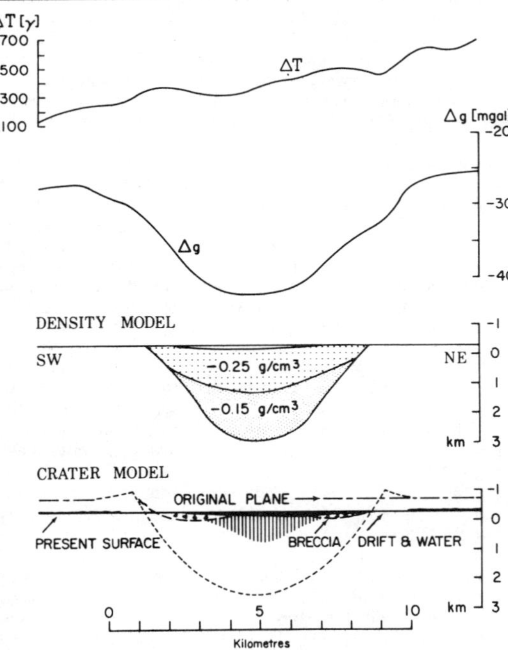

Figure 8: Northeast-southwest profile across the center of Lake Wanapitei:

ΔT - Total magnetic intensity

Δg - Bouguer gravity anomaly

Density model – The two layers of indicated density contrast with the county rocks are adjusted to fit the gravity anomaly. Depth scale given relative to mean sea level.

Crater model – The structure and degree of erosion are based on data from other Canadian craters. Estimated distribution of weakly to moderately shocked rocks in the inferred central uplift is shown by vertical lines. Dashed line, indicating limit of disturbed rocks involved in central uplift movement, corresponds approximately to the base of the lower layer in the density model.

DISCUSSION AND CONCLUSION

The gravity model agrees well with a structural interpretation based on results for the Deep Bay crater and others where a relatively modest central uplift is present. The upper layer of the model would correspond to the average mixed breccias as sampled in the drift and to moderately shocked and probably fractured and altered country rocks, as are found in the central uplift at Deep Bay (Dence *et al.*, 1968). The second layer would represent weakly shocked and fractured quartzites and gneisses, grading outwards and downwards into undeformed country rocks.

In this interpretation of the crater structure, the crater has a present diameter of about 7.5 km and is centred at latitude 46° 44.7' N, longitude 80° 44.6' W. The margin of the crater breccias corresponds approximately to the 150 foot water depth contour in the northern part of the lake (Figure 2). The gravity anomaly of -15 mgal corresponds to a mass deficiency of $(1.8 \pm 0.4) \times 10^{16}$g (cf. Innes, 1961). The comparison with the Deep Bay crater suggests that erosion has lowered the mean elevation of the countryside by some 300 m since the Wanapitei crater was formed, and that the original crater diameter was about 8.5 km.

The circumferential drainage pattern around the lake suggests that the country rocks were disturbed over an area with a diameter of about 18 km. The corresponding, more clearly defined aureole at Deep Bay is 20 km across (Innes *et al.*, 1964). From structural studies of craters involving flat-lying sedimentary rocks, e.g. the Ries crater (Dennis, 1971), it is likely that, apart from superficial breccias, rocks in the aureoles underwent only simple tilting with vertical displacements of a few hundred meters at most; horizontal displacements would be even smaller. It follows that the rocks within the Wanapitei aureole, including the Sudbury Irruptive, were probably only mildly disturbed by the Wanapitei event. The present arcuate configuration of the Irruptive near Lake Wanapitei is similar to other arcuate patterns elsewhere around the Sudbury Basin and, moreover, the Huronian rocks to the northwest and east of Lake Wanapitei show little displacement, as expected. Thus the structure of the East Range Irruptive must basically be the result of the complicated tectonic events which modified the initial Sudbury structure prior to the Wanapitei event.

The age of the Wanapitei crater has yet to be determined though it is clearly much younger than the Sudbury Basin or any of the late Precambrian tectonic or igneous events in the area. The state of preservation of the glasses and breccias is comparable to that in samples from West Clearwater Lake (Dence, 1965) and Lake Mistastin (Taylor and Dence, 1969), suggesting that Lake Wanapitei may also be of late Paleozoic or early Mesozoic age. If so it post-dates the Sudbury event by about 1.5 billion years.

Several apparently simultaneous doublets are known among terrestrial craters, notably the pair at Clearwater Lake (Dence *et al.*, 1965) and the Ries-Steinheim doublet in southern Germany (Dennis, 1971). However, the combination of a relatively small young crater superimposed on the margin of a larger, older companion has no parallel on Earth, although several comparable examples are known from the Moon. The closest comparison in terms of relative size and age among nearside lunar craters can be made with Theophilus, a 100 km crater similar in age to Copernicus (Wilhelms *et al.*, 1965), on the rim of which is Theophilus B, a younger 8.7 km crater. Other, though somewhat older examples of similar relationships are found at the lunar craters Arzachel and Posidonius. Thus, the Sudbury – Wanapitei relationship marks a rare but not unique combination of impact events recorded within the last three billion years of the Earth-Moon system.

ACKNOWLEDGEMENTS

We thank J.B. Boyd and D. Halliday for assistance with the gravity survey on the ice-covered surface of Lake Wanapitei, Michael Marchand and J.V. Guy-Bray and his colleagues for generous help with the location and collection of samples at Lake Wanapitei, and P.B. Robertson, R. Wirthlin, L.W. Warren, B. Draper and E. Gelinas for valuable time given to the analysis of samples and preparation of diagrams and photographs.

REFERENCES

Beals, C.S., Dence, M.R. and Cohen, Alvin J. (1967) – Evidence for the impact origin of Lac Couture; Publ. Dom. Obs., vol. 31, No. 10, pp. 411-426.

Beals, C.S., Innes, M.J.S. and Rottenberg, J.A. (1963) – Fossil meteorite craters; *in* Middlehurst, B.M. and Kuiper, G.P. (editors), The Solar System, vol. 4, Univ. Chicago Press, Chicago, pp. 235-284.

Chao, E.C.T. (1967) – Shock effects in certain rock-forming minerals; Science, vol. 156, pp. 192-202.

Dence, M.R. (1965) – The extraterrestrial origin of Canadian craters; Annals New York Acad. Sci., vol. 123, Art. 2, pp. 941-969.

...... (1971) – Impact melts; J. Geophys. Res., vol. 76, pp. 5552-5565.

...... (1972) – Meteorite impact craters and the structure of the Sudbury Basin; this volume.

...... Innes, M.J.S. and Beals, C.S. (1965) – On the probable meteorite origin of the Clearwater Lakes, Quebec; J. Roy. Astron. Soc. Canada, vol. 59, pp. 13-22.

...... Innes, M.J.S. and Robertson, P.B. (1968) – Recent geological and geophysical studies of Canadian craters, *in* French, B.M. and Short, N.M. (editors), Shock metamorphism of natural materials, Mono Book Corp., Baltimore, pp. 339-362.

...... Robertson, P.B. and Wirthlin, R. (1972) – Coesite from the Lake Wanapitei crater, Ontario; (in preparation).

Dennis, J.G. (1971) – Ries structure, southern Germany, a review; J. Geophys. Res., vol. 76, pp. 5394-5406.

Dietz, R.S. (1964) – Sudbury structure as an astrobleme; J. Geol., vol. 72, pp. 412-434.

Engelhardt, W. von (1971) – Detrital impact formations; J. Geophys. Res., vol. 76, pp. 5566-5574.

French, B.M. (1967) – Sudbury structure, Ontario: some petrographic evidence for origin by meteorite impact; Science, vol. 156, pp. 1094-1098.

Geological Survey of Canada (1965) – Geophysics papers 1511 and 1512, Capreol and Milnet, Ontario.

Innes, M.J.S. (1961) – The use of gravity methods to study the underground structure and impact energy of meteorite craters; J. Geophys. Res., vol. 66, pp. 2225-2239.

Innes, M.J.S., Pearson, W.J. and Geuer, J.W. (1964) – The Deep Bay crater; Publ. Dom. Obs., vol. 31, No. 2, pp. 19-52.

Kieffer, Susan W. (1971) – Shock metamorphism of the Coconino sandstone at Meteor Crater, Arizona; J. Geophys. Res., vol. 76, pp. 5449-5473.

Meyn, H.D. (1970) – Geology of Hutton and Parkin townships; Ontario Dept. Mines, Geol. Rept. 91.

Meyn, H.D. (1971) – Geology of Roberts, Creelman and Fraleck townships; Ontario Dept. Mines, Geol. Rept. 91.

Ontario Department of Mines (1969) – Sudbury Mining Area, Map 2170.

Ontario Department of Mines (1971) – Sudbury–Cobalt sheet; Geol. Compilation Series, Map 2188.

Paul, M.K. (1972) – Interpretation of the gravity anomaly

over a causative body with circular symmetry; Geophys. Prospecting (in press).

Popelar, J. (1971) – Gravity measurements in the Sudbury area; Earth Physics Br., Gravity Map Series No. 138.

...... (1972) – Gravity interpretation of the Sudbury area; this volume.

Quirke, T.T. (1922) – Wanapitei Lake map area; Geol. Surv. Canada, Summ. Rept., 1921, Part D, pp. 34-50.

Robertson, P.B., Dence, M.R. and Vos, M.A. (1968) – Deformation in rock-forming minerals from Canadian craters; *in* French, B.M. and Short, N.M. (editors), Shock metamorphism of natural materials, Mono Book Corp., Baltimore, pp. 433-452.

Stöffler, D. (1971a) – Coesite and stishovite in shocked crystalline rocks; J. Geophys. Res., vol. 76, pp. 5474-5488.

...... (1971b) – Progressive metamorphism and classification of shocked and brecciated crystalline rocks at impact craters; J. Geophys. Res., vol. 76, pp. 5541-5551.

Taylor, F.C. and Dence, M.R. (1969) – A probable meteorite origin for Mistastin Lake, Labrador; Can. J. Earth Sci., vol. 6, pp. 39-45.

Thomson, J.E. (1961) – Maclennan and Scadding townships; Ontario Dept. Mines, Geol. Rept. No. 2.

Thomson, J.E. and Card K.D. (1963) – Kelly and Davis townships; Ontario Dept. Mines, Geol. Rept. No. 15.

Wilhelms, D.E. and Trask, N.J. (1965) – Compilation of geology in the lunar equatorial belt; *in* Astrogeol. Studies Ann. Prog. Rept., July 1964 – July 1965, pt. A: U.S. Geol. Survey open-file report, pp. 29-34 and map supp.

Part IV

STRUCTURES OF CLASS IV IN THE CENTRAL UNITED STATES

Some 23 cryptoexplosion structures are known from the central plains of the United States and their margins. Not all of these structures have been attributed to impact, but 12 of them are included by Short and Bunch (1968) in a worldwide inventory of shock deformation of "unique" type, believed to be restricted to sites of impact explosion. Seventeen of the twenty-three structures have, at one time or another, been attributed to impact. Bucher (1936) suggested that some of these structures are analagous to the Rieskessel and Steinheim Basin (Papers 1 and 2), which he regarded as cryptovolcanic. Boon and Albritton (1936) suggested impact origin. Upheaval Dome, Utah, and Magnet Cove, Arkansas, were listed by Bucher, but the first is apparently a salt diapir and the second an alkaline-carbonatite ring complex. These twenty-three structures comprise four geographic groupings:

1. A linear east-west chain extending from southern Illinois, through Missouri to Kansas (Figure 2).
2. An elliptical grouping, elongated parallel to the Appalachian trend, in Ohio, Kentucky, Tennessee, and Mississippi. (See Figure 1 in Paper 26 for distribution. The structure not shown, Kilmichael Dome, is 250 miles (402 km) southwest of Wells Creek Basin.)
3. A grouping of four structures in Iowa, Indiana, and northern Illinois. (See Figure 2 and Figure 1 in Paper 26 for distribution. The structure not shown, Des Plaines, is 70 miles (113 km) due north of Kentland.)
4. Two structures in Texas.

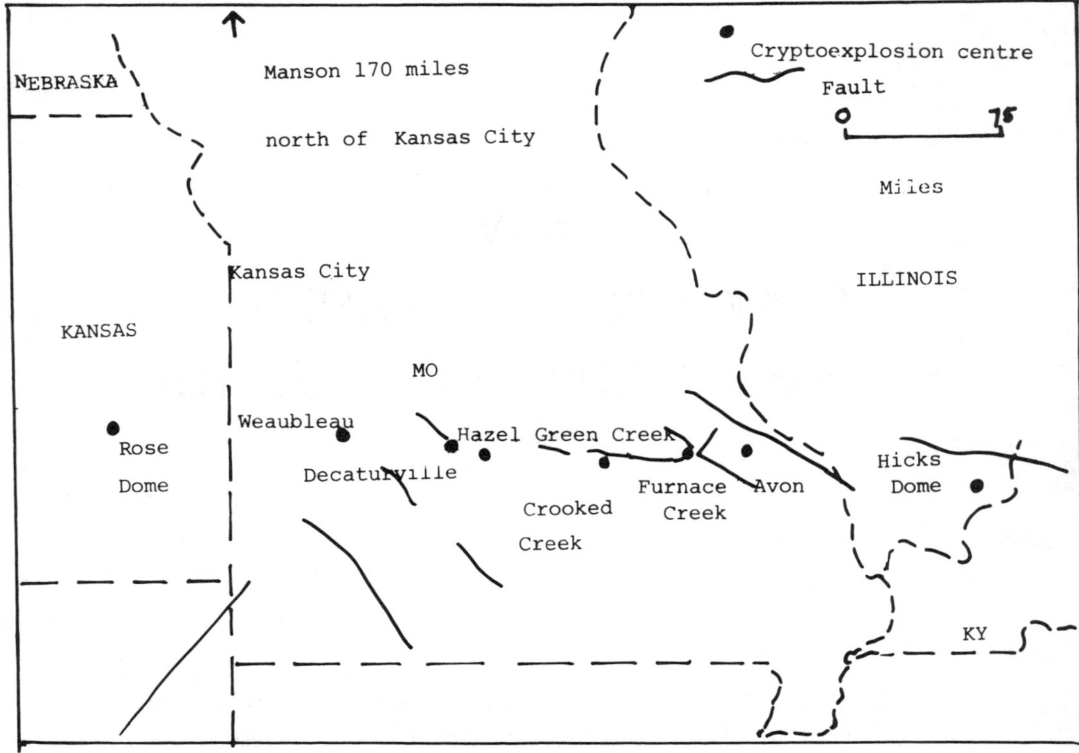

Figure 2. Distribution of the cryptoexplosion structures along the Illinois-Missouri-Kansas axis (after Snyder and Gerdemann, 1965).

All of these structures are listed in Table 9. Six structures in the second group, three structures in the third group, and one of the two structures in the fourth group are not included in the papers in this volume. These are briefly considered below.

GROUP 2 STRUCTURES

Howell

Born and Wilson (1939) described intensely deformed Black River Group and Trenton Series rocks, uplifted by 30 m above their normal level in a circular area (diameter 2.4 km) occupied by a jumble of limestone blocks embedded in a matrix of shatter breccia. Ordovician Fernvale Shale was deposited in a graded crater with moderately high dips. Mild renewal of deformation succeeded this sedimentation. Silurian and Mississippian rocks are relatively undisturbed. Breccias consist of meter-size blocks, themselves locally brecciated, together with shot-sized fragments in a matrix of powdered limestone. Stringers of injected

Table 9 Crytoexplosion structures of class IV in the Central United States

	Name	*Class*	*Diameter (km)*	*Age*
(1)	*Crooked Creek, Missouri	IVa/3	5.3	Early Ordovician to Pennsylvanian
	*Decaturville, Missouri	IVa/3	5.5	Post-Silurian
	Furnace Creek, Missouri	‡	2.4	Early Cambrian
	*Hazel Green Creek, Missouri	‡	?	Early Cambrian
	Hicks Dome, Illinois	-	16	Post-Pennsylvanian
	Avon Diatremes, Missouri	‡	17	Post-Devonian and pre-Mississippian
	Rose Dome, Kansas	-	3	Post-Pennsylvanian
	Weaubleau, Missouri	-	5	Post-Mississippian and pre-Pennsylvanian
(2)	*Dycus, Tennessee	?	(insufficient information)	
	*Flynn Creek, Tennessee	IVa/3	3	Post-Ordovician and pre-Mississippian
	*Howell, Tennessee	IVa/2?	2.4	Ordovician
	*Jeptha Knob, Kentucky	IVa/3	3.2	Early-Middle Silurian
	*Kilmichael Dome, Mississippi	IVb	8	Post-Eocene and pre-Pliocene
	*Middlesboro, Kentucky	IVa/3	6.4	Post-Pennsylvanian
	*Serpent Mound, Ohio	IVa/3	6.3	Post-early Mississipian
	*Versailles, Kentucky	IVa/3	1.6	Late Ordovician or younger
	*Wells Creek Basin, Tennessee	IVa/3	9–13	Post-Mississippian
(3)	*Des Plaines, Illinois	IVa/2?	>1.6	Post-Pennsylvanian
	*Glasford, Illinois	IVa/2	4	Late Ordovician
	*Kentland, Indiana	IVa/2 or 3	5	Post-Middle Ordovician
	*Manson, Iowa	IVa/3	32	Post-Early Cretaceous
(4)	*Sierra Madera, Texas	IVa/3	13	Post-Early Cretaceous
(5)	*Willbarger Dome, Texas	Va	3	Post-Permian

‡These appear to be simple vents.
- These are not classified.
* Structure which has been attributed to impact.

powdered breccia cut limestone blocks from five formations. Deformation was mainly post-Catheys Formation and pre-Fernvale Shale. Slight uplift preceded this latter deposition. The Fernvale Shale continuously mantled the highly deformed rocks. The structure is situated on the south flank of the westward swing of the Nashville Dome, in an area where regional dips are negligible. It occupies the crest of an anticline on a group of northwest trending flextures. The breccia matrix has flowed around the blocks and participated in dike intrusion. The authors found evidence of polyphase deformation difficult to reconcile with meteoritic origin.

Jeptha Knob

Bucher (1963) described an isolated hill rising 100 m above the surrounding country as a nearly circular disturbance 3.2 km in diameter. Eden and Trenton Series beds are uplifted to 120 m above their normal level in an angular, central area (diameter 1.6 km), the shales being wrinkled to irregular dips, not generally steep. Steep dips occur at the edge of this central area, and there is evidence of movement on normal faults in this irregular zone that borders a ring depression in which shaley limestone (Waynsville) and beds of the fossiliferous Liberty Formation are exposed, depressed by 60 m. This ring syncline passes outward into a zone of small synclines and anticlines. Undisturbed Silurian strata overlie the disturbance. This structure is also discussed by Seeger (1966).

Serpent Mound

Bucher (1963) described a nearly circular disturbance (diameter 6.4 km), in the center of which Ordovician limestone and shale has been uplifted 100 m above their normal level, in an angular uplift. Wrinkles of small amplitude have produced disordered steep dips of variable strikes in the shales. Limestone blocks of the Eden Formation have been carried upward by as much as 290 m. There is evidence of explosive deformation, and both overturning and overthrusting are evident at the edge of the central area. A string of small oval basins surrounds this area; these are broad synclines which lower the enclosed beds by up to 140 m with little accompanying faulting. The outer limit of the disturbed areas is defined by a fault or sharp flexure. There are some peripheral anticlines beyond the ring depression. The youngest rocks affected by the disturbance are Mississippian. Heyl and Brock (1962) reported zinc mineralization in shatter breccia. They note evidence of meteoritic origin—the identification of shatter cones and coesite in the Lilley Dolomite of the uplift. The zinc is epigenetic and is affected by only minor renewals of deformation, to which it is unrelated. Bucher favored cryptovolcanic origin, and Heyl and Brock did not favor impact origin because of evidence of multiphase deformation.

Middlesboro Basin

Englund and Roen (1962) described an alluviated, circular depression within rugged Appalachian plateau terrain. Fringing mountains rise to 600 m above a 6.4 km diameter expanse of low rolling hills. Intensely deformed rocks of the Pennsylvanian Lee Formation, massive quartzose sandstones and quartz pebble conglomerates, and—in the Brethitt Group—shales, siltstones, sandstones, and coal underlie the depression, the periphery of which is relatively undisturbed. A shallow syncline surrounds the basin and locally low anticlines separate it from the disturbance. In the center, dips are vertical or overturned and strikes irregular. Further out, strikes are arcuate. A core of uplifted rocks of the Lee Formation is bounded by steep normal faults. Intense shattering has produced parallel sets of fractures in quartz grains. There is no silica glass or meteoritic material. The structure was interpreted as an impact scar. Dietz (1972) recognized shatter cones at Middlesboro.

Versailles

Black (1964) described a circular structure, 1.5 km in diameter, within blue-grass country of deep soil and rare outcrop. More than 100 outcrops do, however, occur within the disturbance, the physiographic expression of which is restricted to a row of sink holes. Country rocks are Middle to Upper Ordovician carbonates and shales. There is a central uplift, and assymetrical dome of coarsely brecciated limestone, including blocks several meter across. Uplift is relative to the enclosing depression, but not the surrounding country. The annular depression is outlined by normal faults, occupied by breccias, and lowering the trough by 30 m in relation to the undisturbed peripheral rocks. Rocks within the depression and bordering it are folded. The central limestone breccia is like that at Sierra Madera (Paper 29), but it is finer and possesses a finer matrix of fragments of several rock types. The age of the structure is Late Ordovician or younger, rocks of that age being the youngest disturbed. Neither shatter cones nor coesite have been recognized. The similarity to nearby disturbances led Black to favor impact origin, though he did not reject cryptovolcanic origin.

Kilmichael Dome

Priddy (1946) described a sharp uplift in the hilly region of Mississippi, on the west limb of the Mississippi embayment, where the strata normally dip gently west. The form is that of a hat-shaped dome, with a central uplift (5–8 km diameter), raised 400 m above the normal level, a narrow graben encircling it, and a low horst beyond this graben. The rocks affected are Eocene, Wilcox Beds in the uplift and Claiborne Beds in the graben (both outcrop in the marginal horst). All are so poorly consolidated and incompetent that there is no topographic expression. Drilling revealed thickened and faulted Cretaceous and Commanchean sediments overlying slightly metamorphosed Pennsylvanian shales. The structure is mantled by Pliocene–Pleistocene gravels. The age is Eocene or younger, not younger than Pliocene. No brecciation or shock effects are described, and the structure is allocated to class IVb.

GROUP 3

Glasford

Buschbach and Ryan (1967) record the encounter, in a deep well drilled on the crest of a 4 km diameter dome, of Maquoketa Shale (Upper Ordovician), at first not disturbed, but disturbed beneath over a 500-m interval. Cambrian blocks have been uplifted as much as 330 m above their normal level. Brecciated dolomite, mylonitic sandstone, and contorted red and green shale were encountered, and dips up to vertical measured, through dips decreased with depth. A violent explosion in Cincinnatian times apparently produced this disturbance. Meteoritic impact was favored. There has, however, been a gradual and continual development of the dome since, producing thinning and arching from late Ordovician to at least Pennsylvanian, and possibly up to the present. The structure has a circular outline and central uplift. It is complicated by complex faulting and other deformation. Within the uplift there is much sheared, brecciated, and powdered rock. Mixed breccias appear identical in character to those at Kentland (Paper 28), consisting of angular dolomite fragments together with chert, sand, and shale, set in a powdery grey matrix. These mixed breccias grade into shatter breccias of a single lithology. Deformation decreases again downward. There is no igneous rock representation. The thinning of the overlying beds is attributed to the noncompactibility of the breccia core. Gradual and continued uplift may have also contributed, being itself attributed to delayed and continuing rebound after impact or deep-seated regional tectonic forces. The evidence favors the second alternative, uplift gradually waning in intensity.

Manson

Hoppin and Dryden (1958) described a raised zone of crystalline rocks, measuring 4 km^2, situated less than 30 m beneath a glacial mantle and surrounded by a depression covering 32 km^2 and filled with a thick section of strongly disturbed Lower Cretaceous sediments. Contacts of the crystalline core dip gently outward. The sediments are contorted and brecciated calc-shales, displaced downward by up to 360 m on peripheral faults. Chloritized breccia occurs in the upper part of the core, gneiss occurres at depth. Pink granite, phyllonite, syenite, and diabase are also recorded. Breccias range from those including meter-scale fragments to microbreccias. Alteration to chlorite, kaolin, epidote, calcite, pyrite, and sericite occurs in replacement zones and fracture fillings. Mylonite zones traverse the gneiss. A magnetic high is centered over the core. Hoppin and Dryden favored cryptovolcanic origin, such structures being not uncommon in flat-lying sediments or crystalline basement of cratonic regions. Dence (1972) attributed this disturbance to meteorite impact.

Des Plaines

Emrich and Bergstrom (1959) reported a 5-km^2 area of disturbance centered on the city of Des Plaines. Flat-lying to gently warped lower Paleozoic rocks enclose the disturbed area, in which beds ranging from Middle Ordovician to Pennsylvanian age underlie glacial cover. There are many fault blocks in the area of maximum disturbance, separated by high-angle faults. Mississippian and Pennsylvanian rocks are in juxtaposition with Ordovician. The younger rocks occur 80 km north of their main outcrop belt. The vertical throw of the stratigraphy amounts to 215 m. The faulting is both pre- and post-Pennsylvanian. There is a negative-gravity anomaly over the disturbance, which, together with abnormal thickening of beds, lack of overturning of blocks, and the high angle of the faulting, were not considered by Emrich and Bergstrom to be consistent with cryptovolcanic hypotheses. Dence (1972) listed this disturbance as a possible astrobleme.

GROUP 4

Willbarger Dome

Described by Hughes (1932), who related it to salt domes of the region, the Willbarger Dome is often attributed to impact (Freeberg, 1966). An uplift of 70 m of covers 5 km^2. Permian shales and limestones have been disturbed over a depth of 400 m, though outside the structure they form a northwest dipping homocline with only minor deformations. The Leuders Limestone has been raised 65 m above its normal level and deformed to dips of as much as 68°. Only the top of the structure is exposed, but core drilling has outline disturbance over an area of 3 km^2. There is no record of brecciation of unusual shock deformations.

REFERENCES

Black, D. F. B. 1964. Cryptoexplosion Structure near Versailles, Kentucky. *U.S. Geol. Survey Prof. Paper 501B,* pp. 9-12.

Boon, J. D., and C. C. Albritton, 1936. Meteorite Craters and Their Possible Relationships to Cryptovolcanic Structures. *Field and Lab.* **5**:1-9.

Born, K. E., and C. W. Wilson. 1939. The Howell Structure, Lincoln County, Tennessee. *Jour. Geology* **44**:815-835.

Bucher, W. H. 1936. Cryptovolcanic Structures in the United States. *Internat. Geol. Cong., 16th, Proc.* **2**:1055-1084.

Bucher, W. H. 1963. Cryptoexplosion Structures Caused from Within or Without the Earth (Astroblemes or Geoblemes?) *Am. J. Sci.* **261**:567-649.

Buschbach, T. C., and R. Ryan. 1967. Ordovician Explosion Structure at Glasford, Illinois. *Am. Assoc. Petroleum Geologists Bull.* **47**:2015-2022.

Dence, M. R. 1972. The Nature and Significance of Terrestrial Impact Structures. *Internat. Geol. Cong., 24th, Proc.* **15**:77-89.

Dietz, R. S. 1972. Shatter Cones (Shock Fractures) in Astroblemes. *Internat. Geol. Cong., 24th, Proc.* **15**:112-118.

Emrich, G. H., and R. E. Bergstrom. 1959. Intense Faulting at Des Plaines, Northeastern Illinois. *Geol. Soc. America Bull.* **70**:1596.

Englund, K. J., and J. B. Roen. 1962. Origin of the Middlesboro Basin, Kentucky. *U.S. Geol. Survey Prof. Paper 450E,* pp. 20-22.

Freeberg, J. H. 1966, 1969. Terrestrial Impact Structures: A Bibliography. *U.S. Geol. Survey Bull. 1220,* 91pp. and *1320,* 39pp.

Heyl, A. V., and M. R. Brock. 1962. Zince Occurrences in the Serpent Mound Structure in Southern Ohio. *U.S. Geol. Survey Prof. Paper 450D,* pp. 95-97.

Hoppin, R. A., and J. E. Dryden. 1958. An Unusual Occurrence of Precambrian Crystalline Rocks Beneath Glacial Drift near Manson, Iowa. *Jour. Geology* **66**:694-699.

Hughes, U. B. 1932. Shallow Salt Structures in Permian North Central Texas. *Am. Assoc. Petroleum Geologists Bull.* **16**:577-583.

Priddy, R. E. 1946. Kilmichael Dome, an Unusual Uplift in North Central Mississippi. *Geol. Soc. America Bull.* **57**:1273.

Seeger, C. R. 1966. Origin of the Jeptha Knob Structure. Ph.D. thesis, University of Pittsburgh.

Short, N. M., and T. E. Bunch. 1968. A Worldwide Inventory of Features Characteristic of Rocks Associated with Presumed Meteorite Impact Craters. In *Shock Metamorphism of Natural Materials,* B. M. French and N. M. Short, eds. Baltimore: Mono Book Corp., pp. 255-266.

Editor's Comments on Papers 25, 26, and 27

25 SNYDER and GERDEMANN
Abstract from and Review of *Explosive Igneous Activity Along an Illinois-Missouri-Kansas Axis*

26 RODDY
Excerpts from *The Flynn Creek Crater, Tennessee*

27 STEARNS et al.
The Wells Creek Structure, Tennessee

GROUP 1 AND GROUP 2 STRUCTURES

The first group is covered here in a single account by Snyder and Gerdemann (Paper 25); which is represented by the abstract and a review by this editor because of its length. This strongly linear group joins the more diffuse second group in Kentucky. Three of the structures in this group were included in the classic paper by Bucher (1936) and six are included by Short and Bunch (1968); Dence (1972) included the other three in a list of possible impact craters, so all in Group 2 have been attributed to impact.

Flynn Creek

The Flynn Creek structure is one of the two very important structures in the second group. It is considered here in extracts from a paper by Roddy (Paper 26). An early paper by Wilson and Born (1936) records that the Ordovician Black River and Maysville Groups, together with the Trenton Series, are involved in a disturbance of steeply dipping, intensely brecciated rocks occupying a circular area of 3 km in diameter. A central, circular uplift stands 65 m above an enclosing depression and a ring of breccia containing blocks of every Ordovician country rock formation. In the outer parts of the ring, the less disrupted blocks "join with their respective formation in situ." Dips on the flanks of the structure are steep, inward and outward. The Chattanooga strata are unusually thick, being deposited in a circular depression

100 m deep. The Fort Payne Chert forms a local overlay to them, being undeformed. Limestone breccias abound over an area of 10 km^2. Pea-sized fragments to large blocks are set in a powdered limestone matrix, which is pulverized, recrystallized, and chertified. Mixed strata include shatter breccia of limestone blocks in comminuted limestone. They are rounded due to attrition. Such breccias occur in the center of the structure, where injected powder breccia forms dikes and mush-like injections into fissures. Talus breccia shows slight rounding of blocks, suggesting talus wash. Bedded breccias are stratified parallel to the overlying Chattanooga Shale. Coarse to fine grading suggests lacustrine deposition. A lake may have existed at Flynn Creek in post-Leipers pre-Chattanooga time.

Meteorite impact, karst collapse, salt doming, expansion of hydrated anhydryte, natural gas explosion, and cryptovolcanic explosion were entertained by Wilson and Born as the cause of the Flynn Creek crater, the latter being favored because the uplift was thought to be incompatible with impact or cavern collapse. Roddy (Paper 26) reexamined the structure, recognizing shatter cones in the central peak. He interpreted breccias within the surrounding depression mainly as a breccia lens, 40 m deep, associated with some authigenic breccia. The outer ring of breccia was interpreted as highly deformed crater rim strata. Roddy recognized what he believed to be an almost entirely eroded-off ejecta blanket locally superimposed on it, of material similar to that of the supposed breccia lens. Drilling revealed that the breccia lens overlies folded and faulted limestone, displaced vertically by less than 50 m. The deformation was found to have waned almost completely at a depth of 100 m below the base of the breccia lens, the strata being flat-lying and relatively undisturbed. Roddy believed that this indicates a surface-generated energy source. He allowed only a very small amount of erosion of the original surface, in contrast to Stearns et al. (Paper 27) in the case of the nearby Wells Creek Basin.

Roddy believed that cometary impact produced the Flynn Creek structure. He suggested a 24×10^6 erg event, the energy produced in the shock wave being concentrated in the top 300 m of strata. The central peak was attributed to the special condition of cometary impact. Such features have been produced in TNT cratering experiments in Alberta. Hypervelocity impact with a penetration of 150 m by a low-density body that volatilized on impact is supposed to have propagated a cylindrically symmetrical rarefaction wave along the line of penetration, aiding rapid tensional release of highly compressed media. The form of the central peak may owe something to symmetry anomalies associated with layering, orientation of layers, and an oblique angle of impact.

The supposed configuration of the breccia lens, ejecta blanket, and deformed crater ring appears to be derived mainly by analogy with artificial cratering experiments, and the configuration of the structure and its history may yet prove to be somewhat different. There is scope for more detailed study of the two Tennessee structures, Flynn Creek and Wells Creek Basin, particularly in relation to the amount of erosion that has occurred since they were formed, the sequence of their formation, and the nature of the folding and faulting that is an integral part of them.

Wells Creek Basin

The Wells Creek Basin, described by Stearns et al. in Paper 27, is an annular, inner depressed fault ring that deforms Mississippian limestones and cherts, as well as older Paleozoic rocks (back to the Knox dolomite). Bucher (1936) suggested cryptovolcanic origin, Boon and Albritton (1937) and Dietz (1947) proposed meteoritic origin. An early paper by Wilson (1953) mentions also three small craters to the north and one inside the main structure. Of these satellite craters, Indian Mound is 30 m deep and contains a central knoll 650 m in diameter; Cave Springs Hollow is 1.6 km in diameter; and Austin is 120 m in diameter and 12 m deep. Little Elk, in the northwest quadrant of the main basin is reported to be 500 m in diameter. The main structure was described as having a diameter of 9.6 km and consisting of a central uplift that brings to the surface several underlying formations and an outer cuesta, facing inward, from which the beds dip mainly outward, though in one sector they dip steeply inward. Despite this, the outline of the cuesta is circular. The uplift stands 25 m above the basin and is composed of Knox Dolomite. Shatter cones were reported to have profuse development in the top 30 m of drill-core, becoming rare, incomplete, and poorly formed below. In the first 60 m, they are horizontal, except for a few facing obliquely upward. No azimuthal orientation was determined. The drill core commenced and terminated in Knox Dolomite, 650 m below. "The deep fingers of injection dikes and the intense, ever-changing patterns of brecciation and deformation (in the core) are awe inspiring" (Wilson, 1953). The injected breccia consists of a matrix of pulverized rock containing fragments of chert, limestone, and dolomite "up to half an inch across." It is darker than the host rocks. Calcite veins, vugs, dolomitization, silicification, pyrite, manganese dendrites, and chalcedony rings around fragmental nuclei are abundant. The injection veins, most less than 1 cm wide, are poorly developed between 1067 and 1743 ft (325–531 m), but reappear in force, continuing to the bottom. There is a progressive dip out from the uplift. Wilson (1953) believed that deformation came from above and was produced by a group of objects approaching from the south. He believed that the five structures were more or less contemporary. The structure postdates the Mississippian St. Louis Limestone. Wilson referred it to pre-Wilcox (Eocene) and post-Eutaw (Cretaceous). This data is based on the age of the filling of the small craters, not necessarily a valid argument. However, the alternative, that the craters are not contemporary with the main structure, seemed only compatible with endogenic theory, unless there was a remarkable overlap of impacts. If the Little Elk structure is a crater, it would present a major problem in terms of impact theory for it is clearly absurd to suppose that a small contemporaneous crater could be superimposed in a deeply eroded structure such as the Wells Creek Basin.

Bucher (1936) recognized two concentric depressed zones separated by a continuous ring of upfaulted blocks, a partial anticlinal ring less uplifted than the central core. He noted similarities to the Rieskessel-Steinheim-Urach group in Germany, and suggested (Bucher, 1963) that the five structures at Wells Creek line up on a north-northwesterly trend corresponding to that of the dikes at Hicks Dome (Paper 25), of similar character to the Urach bodies (Papers 1 and 2).

In Paper 27, Stearns et al. provide important new data. They recognize:

1. A central cylindrical block, uplifted by about 3000 ft (900 m) in the center, but not displaced from its normal level at its edge. Of this, a central zone 1 mi (1.6 km) in diameter is megabrecciated.
2. An inner, annular graben up to 2 mi (3.2 km) across, downthrown by 150–400 ft (46–120 m).
3. An annular horst 0.5 mi (0.8 km) across in which beds are close to the normal regional level, except for five localities near the edge where uplift has amounted to up to 400 ft (120 m).
4. An outer annular graben of the same width, downfaulted by up to 200 ft (60 m).

Stereometric plotting of shatter cones, with bedding restored to horizontal, suggests that they point up and to the center, though there is some ambiguity—they could point down and outwards. Away from the center and down at 2500 ft (762 m) in a drill core they are more incomplete and less abundant. The top of the Knox Dolomite appears to have been "ground zero" for coning, at least 2500 ft below the original land surface when the event occurred.

Breccias are of varied texture, but lack high-temperature minerals. Some show little rotation. These "crackle" breccias are similar to those of the east Tennessee zinc province and may be preexisting sedimentary breccias. Brecciated fragments occur within breccias of the Knox Dolomite, suggesting more than one episode of shock. Narrow vein and dike-like breccia bodies are common.

Stearns et al. remark that debates as to the origin of such equivocal structures often center on which side must bear the burden of proof and which criteria are taken to be unequivocal. For the endogenic protagonists, may be advanced the breccia characteristics—which leaves open the possibility of repeated activity—and the symmetrical location with respect to a gravity pattern, as well as the perfect tangent relationship to a major fault, the Carlisle Fault, part of a regional system. The location of the largest structural high in the region, perhaps a preevent feature, at the structure, could also be cited. However, Stearns et al. conclude that arguments based on regional structural symmetry do not bear critical evaluation. They believe that the shatter cone orientation yields the only independently supported data, and thus favor meteoritic origin for the Wells Creek structure. No mention is made in Paper 27 of the smaller craters reported by Wilson (1953). If these are related to the structure, it is difficult because of their smaller size, to reconcile them with a contemporaneous larger explosion 2500 ft below the existing land surface, for much smaller scale impacts such as those would have fragmentated at no significant depth and the traces of their impact would have been obliterated by erosion. It is probable that the Little Elk crater does not exist, but the others certainly do. They are either fortuitously related to the main basin, or must be explained in any hypotheses of the Wells Creek origin.

Planar deformations forming sets in quartz grains were reported from this structure by Short and Bunch (1968).

REFERENCES

Boon, J. D., and C. C. Albritton. 1937. Meteorite scars in Ancient Rocks, *Field and Lab.* **5**:53–64.

Bucher, W. H. 1936. Cryptovolcanic Structures in the United States. *Internat. Geol. Cong., 16th, Proc.* **2**:1055–1084.

Bucher, W. H. 1963. Cryptoexplosion Structures Caused from Within or Without the Earth (Astroblemes or Geoblemes?). *Am. Jour. Sci.* **261**:567–649.

Dence, M. R. 1972. The Nature and Significance of Terrestrial Impact Structures. *Internat. Geol. Cong., 24th, Proc.* **15**:77–89.

Dietz, R. S. 1947. Meteoritic Impact Suggested by Orientation of Shatter Cones at the Kentland, Indiana, Disturbance. *Science* **105**:42–43.

Short, N. M., and T. E. Bunch. 1968. A Worldwide Inventory of Features Characteristic of Rocks Associated with Presumed Meteorite Impact Craters. In *Shock Metamorphism of Natural Materials,* B. M. French and N. M. Short, eds. Baltimore, Md.: Mono Book Corp., pp. 255–266.

Wilson, C. W. 1953. Wilcox Deposits in Explosion Craters in Stewart County, Tennessee, and Their Relations to the Origin of the Wells Creek Basin Structure. *Geol. Soc. America Bull.* **64**:753–768.

Wilson, C. W., and K. E. Born. 1936. The Flynn Creek Disturbance, Jackson County, Tennessee. *Jour. Geology* **44**:815–835.

25

Reprinted from page 465 of *Am. Jour. Sci.* **263**:465–593 (1965)

EXPLOSIVE IGNEOUS ACTIVITY ALONG AN ILLINOIS-MISSOURI-KANSAS AXIS

F. G. SNYDER and PAUL E. GERDEMANN

ABSTRACT. Eight distinct geologic events, characterized by intrusive or extrusive igneous activity and/or intense localized deformation, are aligned in a 400-mile east-west trending structural zone across southern Illinois, Missouri, and eastern Kansas. The geologic features defining this zone are, from east to west: Hicks dome in Illinois; Avon diatremes, Furnace Creek volcanic crater, Crooked Creek disturbance, Hazel Green volcanics, Decaturville disturbance, and Weaubleau fault zone in Missouri; and the Rose dome area in Kansas.

The Furnace Creek structure in Washington County, Missouri, represents a period of extrusive vulcanism in Upper Cambrian time. Extensive core drilling revealed a crater approximately 7000 feet in diameter and at least 400 feet deep in Lamotte sandstone. The crater is filled with basic volcanic ejecta and fragments of basement rocks. Surrounding the crater is a thin layer of volcanic ash. The ash bed is in the lower part of the Bonneterre formation, and ash and crater are overlain by a normal Bonneterre section.

The Hazel Green volcanics in Laclede County, Missouri, represent a similar period of volcanic activity in early Lamotte time.

Considered as isolated, unrelated phenomena, the individual features along this structural axis have been interpreted as cryptovolcanic explosions, meteorite impact scars, igneous intrusives, and complex fault structures. Details of structure and lithology, subsurface information from core drilling, the remarkable alignment of the features, and the frequent association with basic igneous rocks suggest that these structures are closely related in mode of origin and that they represent intermittent deep-seated faulting and intrusion through a long period of time.

REVIEW*

Meteoritic origin has been advanced for Decaturville and Crooked Creek (Short and Bunch, 1968), but Snyder and Gerdemann find the evidence from the entire group incompatible with such an origin. The eight structures have an almost perfect linear alignment, but are of diverse age and character. Some are sites of explosive potash-alkaline igneous activity, dominated by mica-peridotite of the meimechite-kimberlite kindred (characteristic of explosion pipes). Four are situated along the Rough Creek-Shawneetown-Palmer fault, a fact strongly suggesting tectonic control: this east-west chain of disturbances appears to extend the basement fracture into Kansas, where there is no expression of it in the sedimentary cover. All the structures lie on the fault line or its inferred extension. The basement fracture is believed to have supplied the channel way for cryptovolcanic uprise. Individual sites are believed to be controlled by intersections of the master zone with transverse second-order fractures. Crooked Creek and Decaturville display shock metamorphisms of the so-called "unique" type, lamellations in quartz grains and shatter cones.

Crooked Creek: An elliptical disturbance bounded by high angle faults, this has a diameter of 5.3 km. It is a complex ring of Cambrian and Ordovician sediments with a central core. An outer ring syncline compliments the ring anticline which

*This review was prepared by the volume editor.

brings to the surface the Upper Cambrian *Bonneterre Formation.* The core exposes *Potosi Dolomite* in a collapsed zone. The structure is surrounded by normal stratigraphic sections, capped by Ordovician *Gasconade* and *Roubidoux Formations.* In the ring anticline, the *Bonneterre* is 330 m above its normal level: *Potosi* has been elevated by 160 m in the core: in contrast, rocks of the *Jefferson City Formation* are 80 m below their normal level in the syncline. The entire stratigraphic sequence of the disturbance is brought to the surface in a tight structure 1.5 km across.

Breccias are of two types, the first of broken, finely comminuted rock of a particular formation as a monomict breccia, confined between the formations above and below and composed of randomly orientated, shattered fragments: the second forming tabular bodies less than 1 m thick of mixed angular fragments: these polymictic breccias contain fragments derived from both above and below, and include intrusive breccia dikes and bedding-plane crush zones. No fragments from the basement (situated only 230 m below the *Bonneterre Formation*) are contained in the polymict breccias, which only occur toward the inside of the ring anticline.

In the north-south trending dog-leg horst at the center, *Davis Shale* is uplifted by 50 m from its normal position. It is vertically disposed and strikes parallel to the faults which bound this horst, an integral part of the disturbance. Shatter cones of diverse orientation, some doubly laminated, are abundant in the *Potosi Dolomite* at the focus. Zinc mineralization is evident in the dolomite, and lead-barite mineralization also occurs in the structure. Drilling in the ring anticline revealed intense brecciation and complex deformation. The basement appears to be depressed beneath this zone. Thickening of the sequence is evident, though stratigraphic continuity is preserved. The lack of basement fragments in the breccias suggests that the structure did not penetrate the Precambrian substratum. Its age is early Ordovician (*Jefferson City*) to Pennsylvanian. The elongation of the structure and most of the horst coincides with the trend of the 65-km long northwesterly Cuba Fault, which intersects the Palmer Fault at this point. The dog-leg horst is not a later tectonic superimposition and is difficult to reconcile with impact origin.

Decaturville: This disturbance has the form of a 5.5-km diameter ring with an 1.6-km diameter uplifted core. The outer zone of the ring consists of downfaulted younger formations. It is situated on the line of the oblique Proctor Anticline and associated faults. Granite pegmatite is exposed at the focus. The structure has been interpreted as a basement high, a granite intrusion, a cryptovolcanic structure, and as an astrobleme. Rocks of the Lower Ordovician *Gasconade* and *Roubidoux Formations* surround the disturbance, the normal section above the Precambrian basement being 450 m. Trenching has revealed a pegmatite body 16 m wide: the rock is fresh, of graphic texture and much fractured; drilling revealed sediments in contact with it, separated by a sericitic zone. This occurrence appears to be a block, torn off the basement by explosion. Similar detached blocks were revealed by drilling nearby and at another site 2.4 km away. Radiometric dating (K/Ar 1290 m yr; Rb/Sr 1445 m yr; Tilton et al., 1962) confirmed the basement origin.

Brecciation of sediments pervades the central uplift, being more intense than at Crooked Creek. There is little mixing of blocks derived from different stratigraphic units, breccia beds being contained between adjacent units of the *Davis Shale, Eminence Formation,* and *Gunter Sandstone.* Dips are steep away from the focus just outside the intensely brecciated core, in a zone possibly separated from it by a fault. Great blocks, up to 30 m long and 3 m thick, form a megabreccia: Shatter cones are abundant in bleached, dead-tan dolomite of the Cambrian *Derby-Doerun Formation* at the center. Breccia dikes occur near the pegmatite outcrop, cutting *Davis*, *Derby-Doerun*, and *Potosi*. Their contacts are irregular, fragments are centimeter-scale pieces of angular dolomite, and the matrix is black, mud-like rock-flour. These breccias closely resemble those of the second type in the Avon diatremes (see below). Iron sulphides abound in the intrusive breccia, and lead plus zinc may assay up to 3% in hand-picked specimens. Sulphides occur as individual fragments and also as coatings to fragments. The outer part of the disturbance consists of downfaulted and folded sediments (*Jefferson City* to Silurian *Bainbridge Formation*). The latter took part in the disturbance, being 160 m above its normal level.

None of the other structures in this group has been attributed to impact except Hicks Dome. Yet it is difficult to believe they are not congenetic; some are manifestly of igneous origin, and several of these show structural or deformational similarity to the two described above. In particular, individual breccia types appear identical. Like the two structures described above, the remainder are sited on intersections of master and second-order lineaments. The almost perfect linear relationship to the Palmer Fault zone and the diversity of age would seem difficult to reconcile with anything but endogenic origin.

Furnace Creek: This is a late Cambrian crater in sandstone, filled with 105 m of volcanic ejecta, which bottoms on granite. The overlying beds are not disturbed, but an unusually thick section of *Bonneterre Ash* fills a funnel which tapers downwards from a diameter of 2.5 to .65 km and originated just with the Precambrian granite basement, encountered 21 m below its normal level. Tuffisite with fragments of basement, Cambrian sandstone, and dolomite has a matrix of volcanic material, sand grains, and white carbonate. Fine ash extends in a thin layer for several kilometers from the disturbance. Alteration includes chlorite and carbonate replacement, and devetrification producing cracks and spherulites. Iron sulphides and barite occur in the tuffsite; and iron sulphides and lead occur in the Cambrian sediments close to the crater. Fracture intersection clearly controlled the site of this structure.

Hazel Green Creek: Here, drilling has revealed ash below the early Cambrian *Lamotte Sandstone*—angular and irregular quartz grains from that formation are shattered within this ash. This structure resembles the Furnace Creek disturbance, but is older. Serpentine veinlets traverse the underlying Precambrian granite, which appears to have undergone stresses of a type unusual in Missouri. No detailed study of quartz deformation appears to have been carried out. Such a study would be of critical significance.

Hicks Dome: This is a 16-km diameter circular uplift, cored by Devonian shales and ringed by Mississippian and Pennsylvanian sediments. Bucher (1963) recognized a northwest elongation of the dome, which is partly produced by ring faulting and occupies the termination of an anticline on the same trend. Drilling revealed a normal section underlain at 520 m by limestone breccia containing dolomite and shale fragments, and sand grains, but no igneous fragments. Veins of calcite, quartz, fluorite, and sulphides traverse the breccia. Fragments of *Maquoketa Shale* occur 330 m below their normal level, and fragments of the *St. Peter Formation* occur from

130 m above their normal level to the bottom of the drill hole over a 360 m interval. Pipes and dikes of breccia occur on the flanks of the dome. These intrusion breccias consist of angular to subangular fragments set in a matrix of rock-flour, and include felspar, quartz, and granite fragments carried up 2600 m from the basement. In these pipe conduits, movement has occurred both up and down.

Numerous small peridotite and lamprophyre bodies north-northwest and south-southeast of the dome follow the orientation of the dome and that of the nearby lineament. Olivine, phlogopite, augite, magnetite, apatite, perovskite, titanite, and garnet mineral assemblages indicate a potash-alkaline igneous association, which is characteristic of explosion pipes. Though these dikes contain sedimentary fragments, breccias are not recorded. A rapid two-stage intrusion process is envisaged by Snyder and Gerdemann.

The fluorite and sulphide mineralization occurred during the Cretaceous (radiometric evidence indicates a mineralization age of 90–100 m yr): however, it may well not have been congenetic or contemporary with the peridotite intrusion of deformation. The peridotite has yielded a Permian radiometric age.

Avon Diatremes: This east-west trending, oval cluster of 78 pipes, 18 km long, includes basic dikes without associated explosive effects (Bucher, 1963), breccias of country rocks devoid of igneous content, and breccias containing igneous material. These pipes are clustered at the south end of a northwest trending anticline, and are mainly situated on faults. Precambrian igneous rock fragments occur 330 m above their normal level. Devonian rocks occur 1000 m below their normal level. Olivine, augite, phlogopite, melilite, magnetite, perovskite, chromite, and apatite mineral assemblates again indicate potash-alkaline igneous rocks, like those of Hicks Dome. The mixed breccias are composed of lapilli of both igneous and country rocks set in a sand/rock flour matrix. Lead, zinc, copper, and iron sulphides occur in contact metamorphosed dolomite marginal to the dikes (Singewald and Milton, 1930), but no genetic connection is established with nearby major ore deposits. Fossils in breccias indicate a post-Devonian age for the deformation and intrusion (Tarr and Keller, 1933). There are no Mississippian fragments, despite exposure of rocks of this age nearby, and the structures are therefore probably pre-Mississippian.

Rose Dome: Three domes with an individual diameter of up to 9.4 km are clustered over a distance of 34 km along a northeast trending lineament. Granite blocks on the surface lie amid Pennsylvanian outcrop. Contact metamorphism is recorded at all three sites. This suggests intrusive origin, but a radiometric age (1220 ± 90 m yr:Rb/Sr) shows that the granite is Precambrian. It must be either erosional remnants or blocks carried up from the basement in an event associated with the basic igneous rocks exposed nearby. The apparent contact effects (bleaching, sericitization, silicification) are common in other cryptic structures in Missouri, and must be referred to other causes. The basic igneous rocks were encountered in drill holes 300 m north of the granite and southeast of the dome. They are mica peridotites, composed of olivine, augite, phlogopite, amphibole, melilite, magnetite, apatite, and perovskite (Franks, 1959, unpublished paper). Radiometric dating on micas (K/Ar) has yielded a Paleocene age (but these results are of questionable reliability). Field evidence only indicates a post-Pennsylvanian age for the intrusions.

Weaubleau: This disturbance has a diameter of 5 km. Faulting and brecciation affect the Lower Ordovician *Jefferson City* and Mississippian *Kinderhookian For-*

mations. An intensely upfaulted zone measuring 11 × 5 km is surrounded by a larger area of Mississippian breccias and conglomerates covering 40 km^2. Low-angle thrusting and minor normal faulting are integral to the disturbance. The breccias include coarse associations of both formations. Jumbled blocks several meters across recall similar coarse breccia development at Crooked Creek, Decaturville, and Avon. Conglomerates of uniform composition and grain size and white to tan color occur over a wide area. Massively bedded, they consist of rounded chert, sand, and limestone clasts, set in a dolomite or dolomitic limestone matrix. They contain Mississippian fossils. Over wide areas there are no fragments larger than 1 cm in diameter. These conglomerates appear to be water laid. Such rocks are unique in this region. Galena occurs in the structure, and barite veins cut overlying Pennsylvanian rocks. No center of explosive activity is recognized and no igneous rocks are recorded. This structure is unique and unexplained. Bucher (1963) noted a resemblance between the conglomerates and grit-sized breccias of the Rieskessel (paper 1), and compared the thrust patterns at the two localities. Weaubleau has strong affinities with Decaturville and Crooked Creek, the two supposed astroblemes in this group, yet the disturbance has marked affinities with the Avon occurrence, the most obviously endogenic of the group.

REFERENCES

Bucher, W. H. 1936. Cryptovolcanic Structures in the United States. *Internat. Geol. Cong., 16th, Proc.* **2**:1055–1084.

Bucher, W. H. 1963. Cryptoexplosion Structures Caused from Within or Without the Earth (Astroblemes or Geoblemes?). *Am. Jour. Sci.* **261**:567–649.

Short, N. M., and T. E. Bunch. 1968. A World-Wide Inventory of Features Characteristic of Rocks Associated with Presumed Meteorite Impact Craters. In *Shock Metamorphism of Natural Materials,* B. M. French and N. M. Short, eds. Baltimore, Md.: Mono Book Corp., 255–266.

Singewald, J. T., and C. Milton. 1930. An Alnöite Pipe, Its Contact Phenomena and Ore Deposition near Avon, Missouri. *Geology* **38**:54–66.

Tarr, W. A., and W. D. Keller. 1933. A Post-Devonian Igneous Intrusion in the Southeast Missouri Lead District. *Jour. Geology* **41**:805–823.

Tilton, G. R.; G. W. Wetherill; and G. L. Davis. 1962. Mineral Ages from the Wichita and Arbuckle Mountains, Oklahoma and St. Francois Mountains, Missouri. *Jour. Geophys. Research* **67**:4011–4019.

26

Reprinted from pages 291-309 and 317-322 of *Shock Metamorphism of Natural Materials,* B. M. French and N. M. Short, eds., Baltimore, Md.: Mono Book Corp., 1968

THE FLYNN CREEK CRATER, TENNESSEE[1]

DAVID J. RODDY

Branch of Astrogeologic Studies, U.S. Geological Survey, Flagstaff, Arizona 86001

The Flynn Creek structure was formed in north central Tennessee approximately 360 million years ago as a shallow crater, 3.6 km in diameter and about 150 m deep. A large central hill, highly deformed rim strata, and a breccia lens were produced during formation of the crater. Over 2 km^3 of flat-lying Middle and Upper Paleozoic limestone and dolomite were brecciated and mixed to a depth of about 200 m, and at least half of the breccia was ejected from the crater. A lens of chaotic brecciated limestone and dolomite, averaging 40 m in thickness, remained in the crater and contains fragments ranging in size from less than a meter up to megabreccia blocks 100 m in length; fragments lie in random orientation in a carbonate matrix. Drill cores indicate that the limestone and dolomite beds immediately below the base of the breccia lens are highly faulted and folded, but deformation decreases downward and the rocks are nearly flat-lying and relatively undisturbed about 100 m beneath the breccia lens.

In the middle of the crater, a sequence of steeply-dipping, folded, faulted, and brecciated Middle Ordovician limestone and dolomite of the Stones River and Knox Groups has been structurally uplifted to form a hill that rises about 100 m above the crater floor. Knox strata in the central uplift are raised about 350 m above their normal stratigraphic position and locally contain shatter cones.

In the rim surrounding the crater, flat-lying Middle and Upper Ordovician limestone and dolomite have been uplifted 10 m to 50 m, and are moderately to tightly folded into doubly plunging anticlines and synclines that have axes concentric to the crater walls. In parts of the rim, folds concentric to the crater have produced radial shortening as great as 35 percent. Normal and reverse faults with displacements of a few meters are also common in the rim strata and usually are concentric to the crater walls.

In the southeastern rim, a large thrust block forms the crater wall and has been moved out from the crater and uplifted about 50 m. The thrust block partly overrides a tilted graben that has down-dropped about 100 m and moved toward the crater. Limestone breccia overlying the graben is identical in lithology and character to the breccia in the crater and apparently is the remains of an ejecta blanket that once surrounded the crater.

In some parts of the crater, the deformed rocks in the crater walls are in sharp contact with the breccia lens; along other parts of the walls, an irregular authigenic breccia zone is present. Fracturing of the wall rocks and abundant calcite twin and microtwin lamallae are common in an irregular zone in the rim strata adjacent to the crater walls.

Post-crater erosion removed the ejecta, except for that overlying the graben, and lowered the regional ground surface less than about 30 m. A thin deposit of marine sedimentary breccia and dolomite covered the crater floor in early Late Devonian time, and the crater was then filled by the muds of the early Late Devonian Chattanooga Sea.

Structural comparisons of the Flynn Creek crater with volcanic craters and their vents indicate an absence of structural analogs. A volcanic gas origin is also inconsistent with the field evidence at Flynn Creek, which indicates strong compressional rim folding near the original ground surface; calculations suggest that high internal gas pressure in the country rocks causes fracturing rather than folding under such low confining pressure. The absence of mineralization, gravity and magnetic anomalies, and local volcanic or tectonic zones are consistent with a non-volcanic origin.

Structural comparisons with selected meteorite impact, nuclear explosive, and chemical explosive craters, however, indicate pronounced structural similarities with the types of deformation present at the Flynn Creek crater; it is concluded that the Flynn Creek crater was produced by a shock-wave mechanism. The drill core evidence for a shallow lower boundary of the breccia lens and the decrease of deformation in the rocks below the breccia lens are consistent with an origin involving surface deformation, i.e., a meteorite or cometary impact. The thin breccia lens, the absence of mineralization, volcanic, or meteoritic materials, the types of rim deformation, and the central uplift are interpreted as consistent with the impact of a cometary body.

---

[1] Publication authorized by the Director, U.S. Geological Survey.

INTRODUCTION

The Flynn Creek crater was formed in north central Tennessee sometime near the middle of the Devonian Period, about 360 million years ago. The formation of the crater required at least 10^{24} ergs to brecciate over 2 km^3 of flat-lying limestone and dolomite and to eject about 1.3 km^3 of rock. The crater, 3.6 km in diameter and initially about 150 m in depth, contains a breccia lens that averages 40 m in thickness and overlies a sequence of folded and faulted strata; deformation in the rocks below the breccia lens decreases downward and is nearly absent about 100 m below the breccia. In the middle of the crater, a sequence of originally flat-lying limestone and dolomite beds is faulted and folded and has been uplifted as much as 350 m to form a central hill that once rose about 120 m above the crater floor. In the crater rim, limestone and dolomite which were originally flat-lying are greatly fractured, folded, and faulted, and have been forced outward and uplifted as much as 50 m. Post-crater erosion lowered the regional ground surface less than about 30 m and removed all of the surrounding ejecta blanket overlying the rim, except for that preserved locally within a rim graben. In early Late Devonian time, deposition of muds in the Chattanooga Sea filled the crater and prevented further erosion. More recently, regional uplift and erosion have produced excellent exposures of this once buried crater.

The Flynn Creek crater belongs to a controversial class of structures that number at least 50 throughout the world and have been variously termed cryptovolcanic, cryptoexplosion, or meteorite impact craters. Safford (1869) first noted the unusual structural relationships in the Flynn Creek area and suggested a tectonic origin that involved folding and faulting. The possibility of a meteorite impact was considered by Wilson and Born (1936), but they rejected the idea in favor of a cryptovolcanic origin because they felt the central uplift was inconsistent with an impact process. Boon and Albritton (1937) argued instead that the central uplift at the Flynn Creek crater was the product of an elastic rebound caused during a meteorite impact. Wilson (1953) on the basis of similar types of structural deformation observed at the Wells Creek structure, Dietz (1960) on the basis of shatter cones found at Flynn Creek, and Shoemaker and Eggleton (1961) on the basis of reconnaissance studies, all interpreted the Flynn Creek crater to be the result of a meteorite impact. Bucher (1963), however, repeated his argument (Bucher, 1936) that the Flynn Creek crater was the product of a buried volcanic explosion, mainly because he felt that the central uplift could not be produced during an impact.

The possibility that these unusual structures were produced by the same general types of cratering mechanisms that have operated on other planetary surfaces has stimulated re-examination of a number of the terrestrial structures, including a detailed field and laboratory study of the Flynn Creek crater (Roddy, 1965, 1966). Field work at Flynn Creek has included detailed geologic mapping of 100 km^2 at a scale of 1 in = 500 feet (1:6000), a gravity and magnetic study of 320 km^2, and core drilling of 6 holes in the crater. Part of this work, including laboratory results, is summarized in this paper. These studies indicate that the Flynn Creek crater has the same structural features that are found in the larger meteorite impact craters and in the larger man-made craters formed by surface and near-surface nuclear and chemical explosions. The results of the field and laboratory studies and the structural comparisons indicate that the formation of the Flynn Creek crater is consistent with a cometary impact.

GEOLOGY OF THE FLYNN CREEK CRATER

Regional Setting

The Flynn Creek area is located in north central Tennessee in the extreme northeastern part of the Nashville Basin (Fig. 1). Within this part of the state, the Cumberland River and its tributaries have developed a complex dendritic drainage system of narrow, steep-sided valleys with an average relief of 150 m. Good exposures along the many valley walls and floors provide an excellent three-dimensional view of the upper part of the crater structure.

No tectonic or volcanic zones are observed in

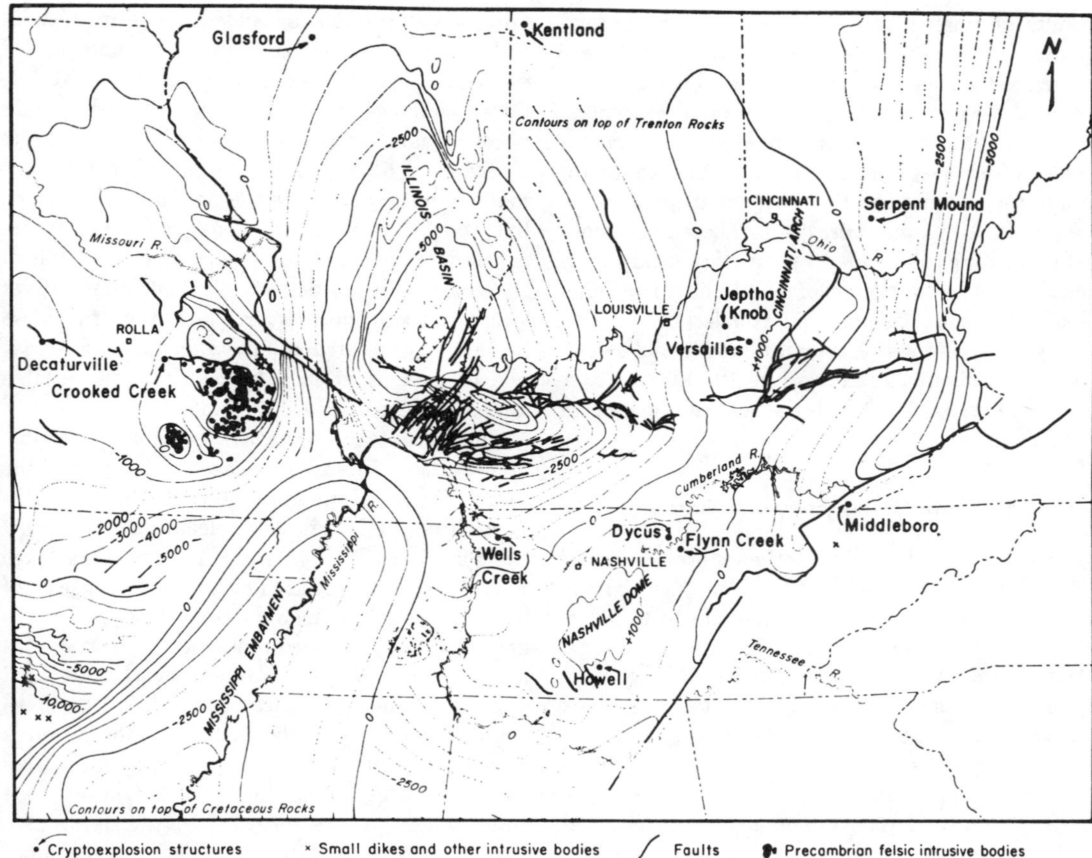

Fig. 1. Generalized tectonic map of the southern interior lowlands of the United States, showing locations of the Flynn Creek crater and other "cryptoexplosion" structures. Structure contours are drawn on the top of Trenton (Middle Ordovician) rocks, except along the lower Mississippi River, where structure contours are drawn on top of Cretaceous rocks. The structure contour interval is 500 feet. (Modified from the *Tectonic Map of the United States, 1962*).

central Tennessee (Fig. 1). The nearest exposed intrusive rock of Paleozoic(?) age is a small dike of mica peridotite in the Valley and Ridge Province in east Tennessee, at a distance of 150 km from Flynn Creek. The Nashville Dome is the dominant structural feature in central Tennessee and occurs as a broad, flat, structural arch trending northeast; the Flynn Creek area lies on the northeastern flank about 50 km east of the crest of this dome. Paleozoic limestones and dolomites dip gently away from the dome at an average slope of 2 to 4 m/km and are modified only by broad, gently plunging anticlines and synclines that cover many square kilometers. Superimposed on these larger structures are numerous smaller folds that include domes, basins, and gently-plunging anticlines and synclines. A set of vertical joints trends northeast in the Flynn Creek area and is approximately parallel to the axis of the Nashville Dome. A second, less well-developed, set of vertical joints, trends approximately northwest.

Several northwest-trending faults, up to 25 km long and with vertical displacements as large as 100 m, occur in the extreme southern part of central Tennessee, and scattered faults trending approximately east-west occur in the extreme northwest part of central Tennessee (Fig. 1). Most faulting in central Tennessee, however, appears to be limited to isolated, short, high-angle faults of small throw that trend either approximately northeast or northwest. Jewell (1947) has described a small number of fault zones in north central Tennessee which range from a few meters to about 10 m in width and trend generally northeast. These zones cannot be traced in length for more than a few tens of meters, and displacements appear to be mainly horizontal. These fault zones are usually accompanied by breccia, locally mineralized with various combinations of calcite, barite, fluorite, galena, and sphalerite. A small body of limestone breccia with fluorite and calcite is located about 25 km south of the Flynn Creek crater (Jewell, 1947) and is the nearest mineralized zone.

Pre-Crater Stratigraphy and Structure

Rocks exposed outside of the deformed area of the Flynn Creek crater range from the Cannon Limestone of Middle Ordovician age to the Fort Payne Formation of Early Mississippian age. Within the crater, however, older units are exposed; these include the upper part of the Knox Group, which is of Early Ordovician age, and the Stones River Group and the Hermitage Formation of Middle Ordovician age (Fig. 2). Only rocks of pre-early Late Devonian age are involved in the structural deformation of the crater. In the sedimentary column overlying the crystalline basement, the strata have a total thickness of about 1700 m and are nearly flat-lying, except for the deformed area associated with the crater.

The top of the Knox Group normally lies about 300 m below the present valley floors in the Flynn Creek area; however, deformed units from at least the upper 30 m of the Knox Group are exposed in the central hill within the crater. In these outcrops the upper part of the Knox consists of tan to gray, usually fine-grained, dolomite with some interbedded limestone. Large, irregular, gray to white chert masses are common in the upper thick-bedded units. The Knox Group is probably about 1000 m thick in the Flynn Creek area. The Wells Creek Dolomite of Early or Middle Ordovician(?) age lies disconformably on the Knox Group and averages about 1 to 2 m in thickness in central Tennessee. In the central hill of the crater, the Wells Creek Dolomite, at least 0.5 m thick, overlies the top of the Knox Group and consists of folded, light-greenish to yellowish-gray, microcrystalline to fine-grained dolomite. The Stones River Group of Middle Ordovician age conformably overlies the Wells Creek Dolomite and averages about 250 m in thickness. The top of the Stones River is normally about 50 m below the present valley floors, but about 170 m of the Stones River Group are exposed as deformed units in the central hill of the crater (Fig. 2, Section d). The Stones River strata exposed in this area are thin to thick-bedded and are predominately dark to light-gray, fine to medium-grained limestone, with variable amounts of interbedded dolomite.

The Hermitage Formation, Cannon Limestone, and Catheys Limestone successively overlie the Stones River Group and together form the Nashville Group of Middle Ordovician age as defined by Wilson (1949, 1962) (Fig. 2). The Hermitage Formation is about 23 m thick and consists of light-brownish to medium-gray, fine to coarse-grained, thin to medium-bedded, fossiliferous limestone and dolomitic limestone. Interbedded silty limestone and calcareous shales are dominant in some areas. The top of this formation is normally about 30 m below the present valley floors, but large thrust blocks on the southeastern crater rim expose the upper part of the Hermitage. The Cannon Limestone, which conformably overlies the Hermitage Formation, averages about 27 m in thickness and is exposed along the floors of the main valleys in the Flynn Creek area. The characteristic lithology of the Cannon strata is medium-light to dark-gray, cryptocrystalline to medium-grained, fossiliferous, thin to thick-bedded limestone. The Catheys Limestone conformably overlies the Cannon Limestone, averages about 50 m in thickness, and is exposed along the lower valley walls in the Flynn Creek area. The lithology of the lower Catheys is similar to the Cannon, except for a difference in fauna and for the occurrence of several thick-bedded, silty, ledge-forming units in the Catheys. The upper part of the Catheys has an increase in silty, dolomitic limestone and argillaceous zones.

The Leipers Limestone of Maysville (Late Ordovician) age averages about 35 m in thickness and is exposed along the upper valley walls in the Flynn Creek area. The Leipers consists of gray to brown, argillaceous, fossiliferous, fine to coarse-

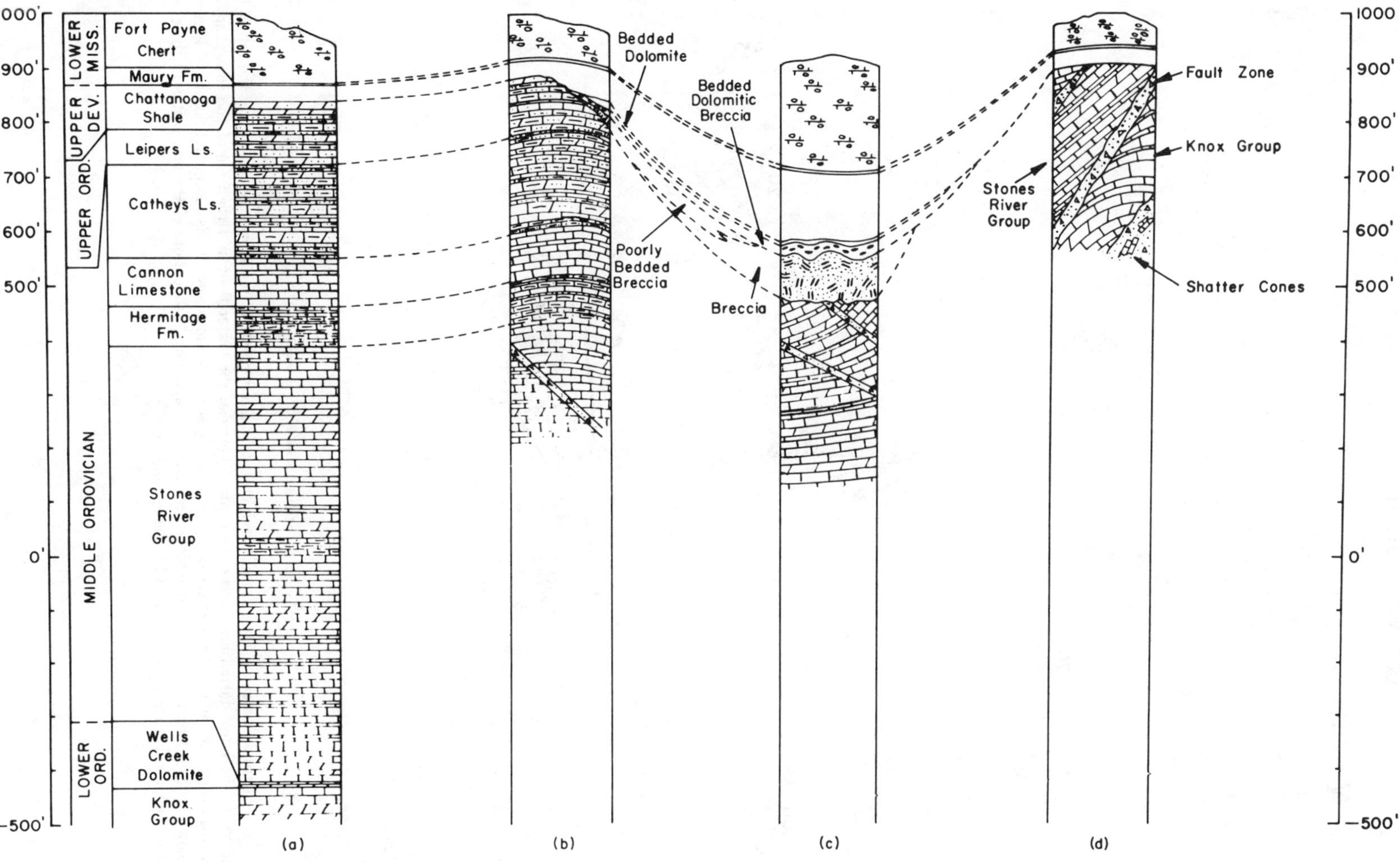

Fig. 2. Generalized stratigraphic columnar sections from the western side of the Flynn Creek crater. Column (c) is based on results of core drilling. The location of each column is shown in Figure 3.

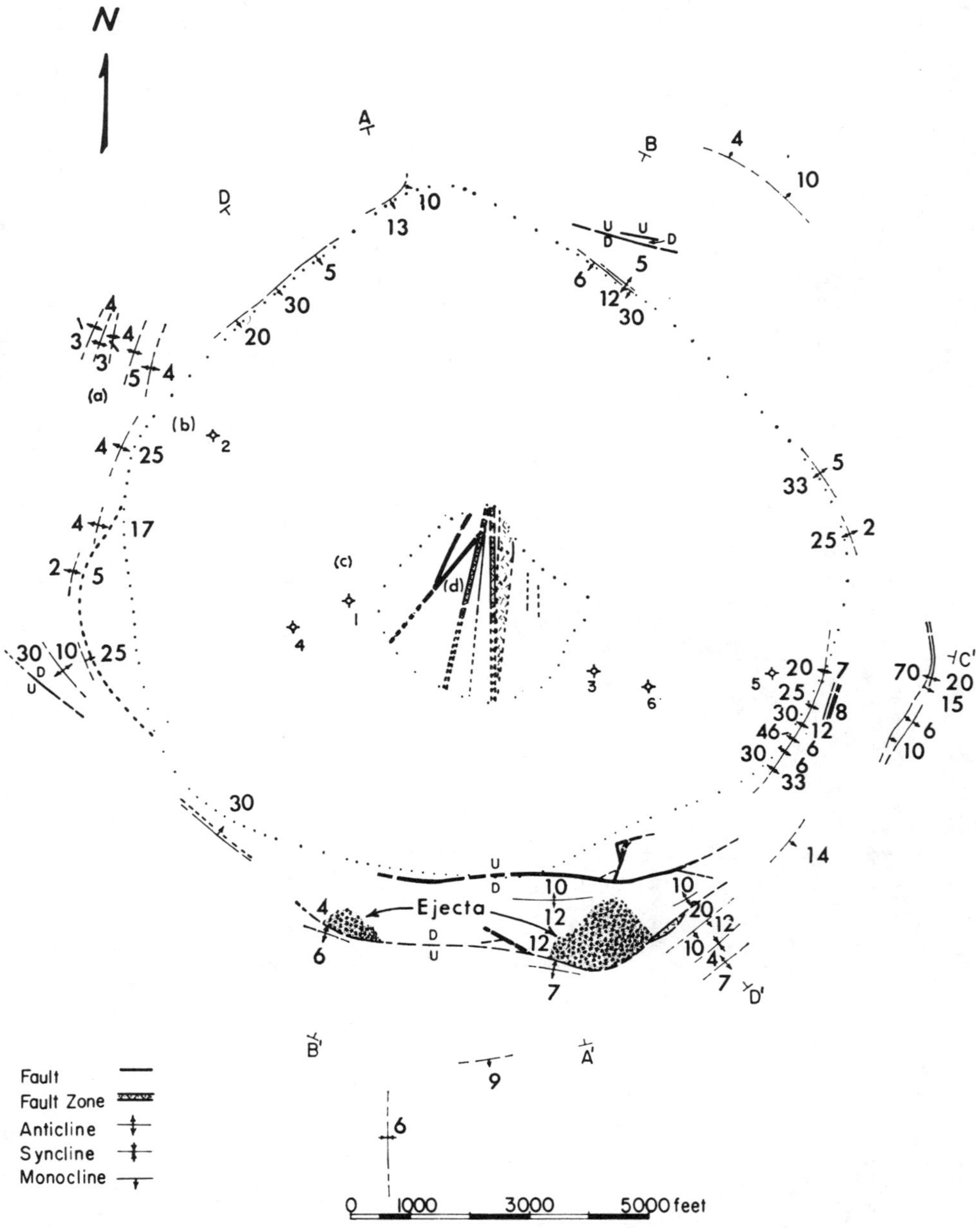

Fig. 3. Major structural elements of the Flynn Creek crater. The outer dotted line shows the location of the top of the crater walls; the inner dotted line shows the contact between the crater floor and the base of the central uplift. Exposed sections are indicated where the dots are closely-spaced. The locations of the columnar sections of Figure 2 are shown by ( ); locations of the cored wells are shown by symbol -ǫ-. Locations of the end points of the cross-sections shown in Figure 4 are indicated by the symbol ⊣.

grained, thin to thick-bedded limestone with some dolomitic beds. The upper 10 m is brown, coarse-grained, irregularly-bedded dolomite. The Leipers strata lie disconformably on the Catheys Limestone in central Tennessee (Wilson, 1962).

The Sequatchie Formation of Richmond (Late Ordovician) age normally overlies the Leipers Formation in extreme north central Tennessee and consists of a 2 m to 12 m thickness of greenish-gray to green, laminated to non-bedded, calcareous and dolomitic mudstone and "sugary" dolomitic argillaceous limestone. The nearest outcrops of these rocks lie about 23 km northwest of the crater, but breccia fragments of green dolomite which are probably Sequatchie strata have been recognized in the bedded dolomitic breccia and in the fault zone of the graben on the southern crater rim (Roddy, 1966). No Silurian or Lower or Middle Devonian strata have been recognized in the Flynn Creek area; the nearest exposures of these rocks are located about 50 km to the northwest.

Crater Stratigraphy and Structure

Immediately after its formation, the Flynn Creek crater had the form of a large, shallow, bowl-shaped depression with a central hill and apparently was surrounded by an ejecta blanket overlying the rim (Figs. 4, 19). The crater averaged 3.6 km in diameter and was less than 150 m deep, measured from the level of the pre-crater ground surface. The crater breccia lens averages about 40 m in thickness; the thickness of the folded, faulted, and brecciated strata below the base of the central hill is at least 350 m. A mass of limestone and dolomite breccia overlying a graben in the southern rim is still preserved and is the first indication from any of the cryptoexplosion structures that an ejection of crater breccia definitely occurred. The major structural elements are shown in Figures 3 and 4. In contrast to the faulted and folded rim strata, the surrounding area exhibits only a pattern of gentle folds superimposed on a shallow regional dip of 0.25° to the east. Dips on the flanks of these folds which are not associated with the crater average less than 3°.

In the crater rims and walls, beds ranging from the Cannon Limestone up to the Leipers Limestone are exposed and exhibit different types of deformation, including rim uplift, folding, faulting, fracturing, and local brecciation (Figs. 3, 4, and 5). The major part of this deformation occurs within about 500 m of the crater walls, but has an irregular distribution, and in some areas extends out to nearly 750 m. Uplifts of 10 m to 50 m occur locally in the rim strata along crests of anticlines and monoclines adjacent to the crater walls. Shortening of strata in directions radial to the crater has been as great as 35 percent in some anticlines. Most of the normal faults, thrust faults, and asymmetric anticlines, synclines, and monoclines trend approximately concentric to the crater walls. The concentric folds have crests that plunge irregularly along the traces of the folds (Fig. 5). Dips on the limbs of folds in the rim strata are commonly as high as 35°, with the steeper limbs dipping towards the crater.

Faults in the rim often grade into tight anticlinal folds or into authigenic breccias or jumbled zones. Displacements on the larger faults are as great as 100 m. Part of the southeastern rim has several large, complex, high-angle thrust blocks that have moved up and out from the crater (Fig. 4). One thrust block exposes Hermitage strata raised about 30 m along the base of a thrust plate.

A large mass of chaotic limestone breccia with a limestone matrix occurs immediately south of the thrust blocks and overlies the graben in the southeastern rim (Figs. 3 and 4). Although breccia in this mass is 50 m above the crater floor, it is identical in most lithologic respects to the breccia found in the crater, except for a crude inversion of stratigraphy, and it is interpreted as ejecta. Scattered fragments of Sequatchie dolomite occur at the base of this breccia and are present in abundance in the fault zone bounding the southern edge of the graben. A small talus deposit of dolomite from the upper Leipers overlies part of the ejected breccia. Part of the original, pre-crater ground surface is present at the base of the ejected breccia overlying the graben; below this surface are about 50 m of the

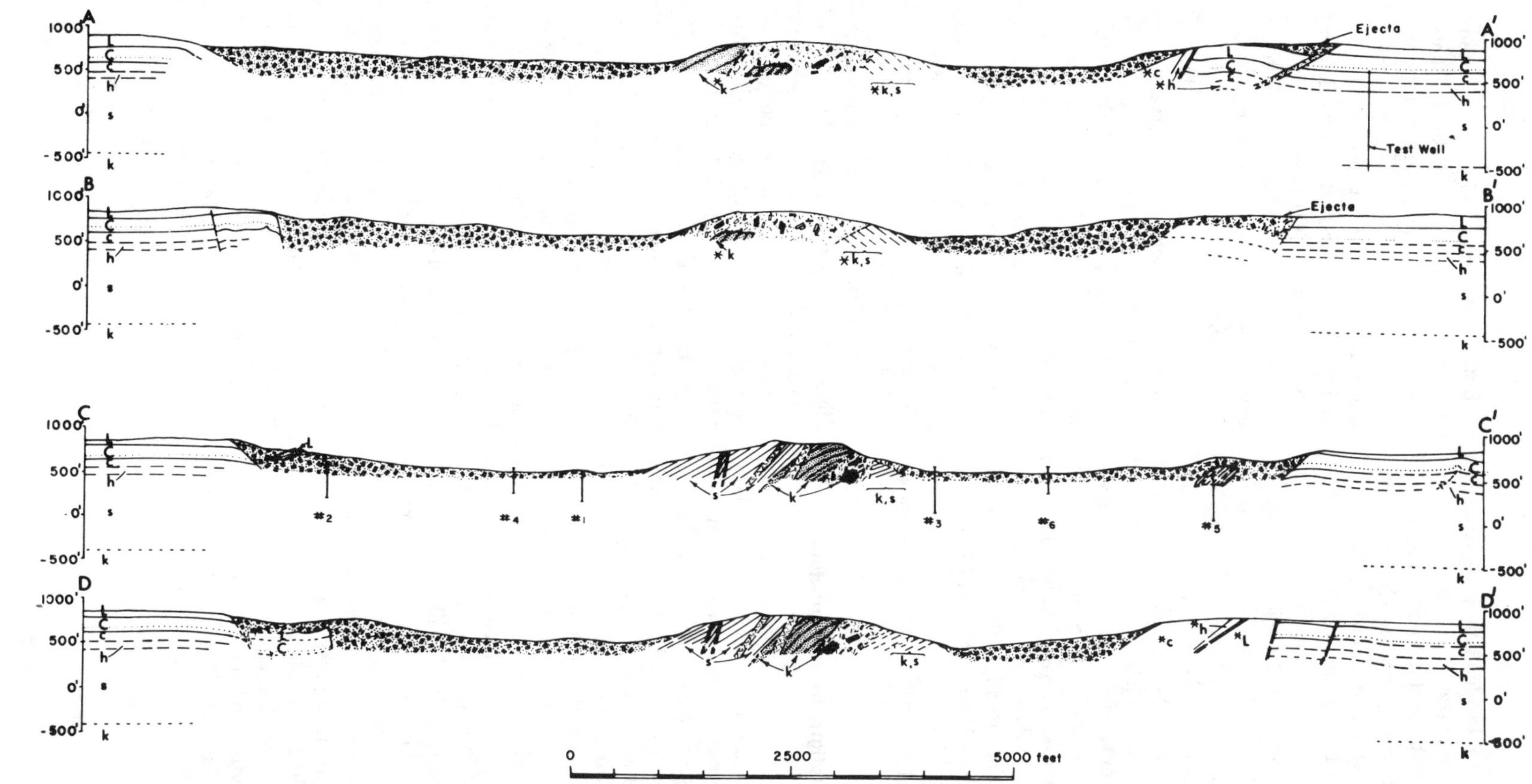

Fig. 4. Generalized geologic cross-sections of the Flynn Creek crater, showing the structure shortly before deposition of the Chattanooga Shale in early Late Devonian time. Cross section C–C′ shows positions of cored wells projected into the plane of the cross section. Locations of the cross-sections are shown in Figure 3. The symbol * indicates rocks which strike nearly parallel to the plane of the cross-section; k = Knox Group; s = Stones River Group; h = Hermitage Formation; c = Cannon Limestone; C = Catheys Limestone; L = Leipers limestone.

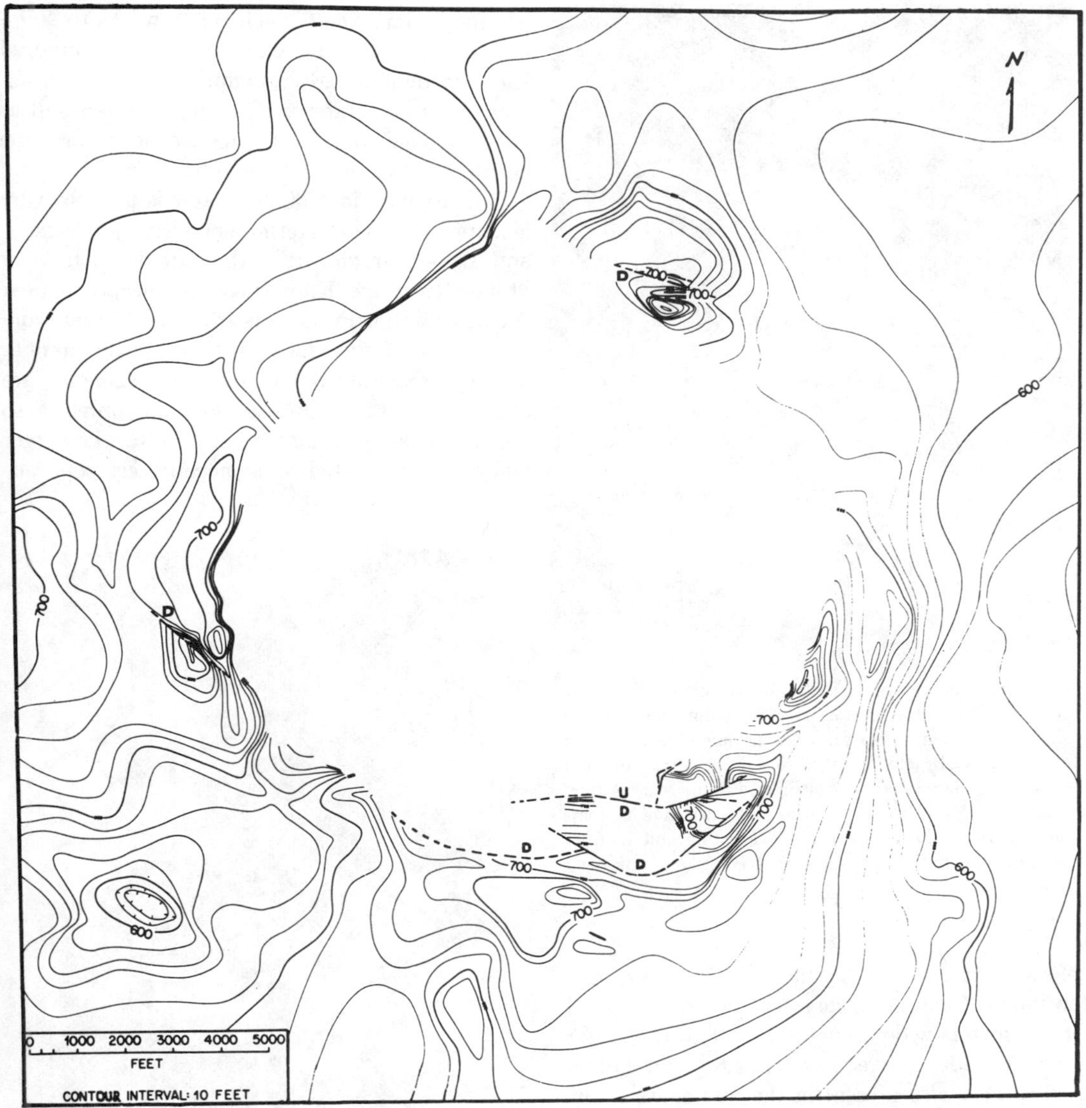

Fig. 5. Structure contour map of the Flynn Creek crater; structure contours are drawn on the top of the middle Catheys unit. Contours end at the crater walls, where rocks were removed during formation of the crater. The graben and the thrust blocks in the southern rim have not been contoured because of the lack of subsurface information. The symbols D (down) and U (up) have been placed only on the sides of faults that show a large absolute displacement.

upper part of the Leipers Limestone which, everywhere else, were eroded from this area after the crater was formed. These structural relations demonstrate that 50 m or less of upper Leipers strata were removed regionally after the crater was formed, and that all the structural deformation now exposed in the valley walls and floors was formed within less than 150 m of the precrater ground surface.

In the crater walls the deformed rock is locally in sharp contact with the crater breccia; along other parts of the crater walls an irregular authigenic breccia zone a few meters to several tens of meters wide separates the crater breccia and the

Fig. 6. Photomicrograph of a Cannon Limestone fragment collected from the crater breccia along the western side of the Flynn Creek crater. A mosaic of large calcite crystals fills a brachiopod shell (lower), which is bounded by a dark-colored matrix of argillaceous, microcrystalline limestone. Moderate to intense microtwinning is common in both the coarser calcite crystals and in fine-grained calcite in the dark matrix. Plane polarized light.

deformed rim rock. Locally, an irregular, discontinuous fractured zone a few meters to several tens of meters wide occurs adjacent to the crater walls. Abundant, irregular fractures and microfractures are also present in the fragments of the crater breccia and occur locally in rocks in the central hill in the middle of the crater. The fracture zones are especially common where a gradational, poorly brecciated, transition exists between the deformed rim strata and the crater breccia. Abundant calcite twin and microtwin lamellae are also common in the deformed limestone adjacent to the crater wall and in many fragments in the crater breccia (Figs. 6 and 7). The *e* lamellae in 100 μ to 400 μ diameter crystals increase two to four times in number as compared with calcite in the surrounding undeformed limestones. Microtwin lamellae (Conel, 1962) in 20 μ to 400 μ diameter crystals increase ten to one hundred times in number. Microtwinning becomes prominent near the crater wall in crystals as small as 20 μ, and slightly bent lamellae occur in some calcite crystals larger than 100 μ (Fig. 7).

The original floor of the crater is underlain by a chaotic breccia of Hermitage, Cannon, Catheys, and Leipers fragments with limited amounts of Stones River and Knox fragments occurring near the central hill. Fragments from the Cannon and lower part of the Catheys dominate in number in most exposures. The breccia fragments are angular and range in size from microbreccia to blocks as large as 100 m in length; the larger blocks are restricted to a megabreccia zone ad-

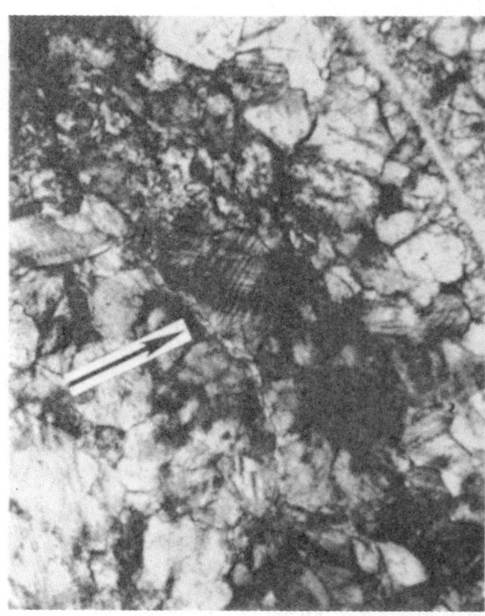

Fig. 7. Photomicrograph of a Cannon Limestone fragment collected from the crater breccia along the western side of the Flynn Creek crater. The specimen consists of a mosaic of calcite crystals in a dark, argillaceous matrix of microcrystalline limestone. The two thin, light-colored bands are chert. Moderate to intense twinning and microtwinning are common in both coarse- and fine-grained calcite. The development of large numbers of microtwinned lamellae has caused darkening in parts of some calcite crystals. The calcite crystal near the center of the photomicrograph (arrow) has a zone of bent twin lamellae. Plane polarized light.

jacent to the crater walls. Fragments exposed on the crater floor lie in all orientations in a matrix of light to medium-dark gray, locally argillaceous, dolomitic limestone.

Core drilling studies indicate that the crater contains a very shallow, bowl-shaped lens of breccia, averaging about 40 m in thickness and that it is immediately underlain by faulted and folded rocks of the upper part of the Stones River Group. Six NX (5.4 cm diameter) cores, ranging in depth from 84 m to 181 m, were drilled along an east–west line within the crater (Figs. 3 and 4). Two of the holes were core-drilled near the flanks of the central hill in areas which were originally the deepest parts of the crater floor; no lake beds were identified, and the marine bedded breccia deposits described later in this paper are underlain by chaotic crater breccia. Cores from the upper part of the breccia consist mainly of Cannon and lower Catheys fragments, set in a shaly dolomite to shaly limestone matrix. Lower in the breccia, the size of the fragments increases up to several meters in diameter, with Hermitage fragments being most abundant. The Hermitage is the lowest formation to be completely brecciated and apparently acted as an extremely incompetent, plastic unit whose limy shale beds form the matrix for most of the lower part of the crater breccia. No mineralized, volcanic, or meteoritic materials were observed in the cores.

The upper part of the Carters Limestone, which forms the highest unit of the Stones River Group, lies immediately below the breccia lens and is highly faulted and folded. The base of the crater breccia lens extends irregularly about 10 m into the upper part of the Carters, but in none of the cores was the breccia lens found below this level. The strata are repeated between faults in several of the cores, and both normal and reverse faults occur. The more intensely faulted and folded units that occur immediately below the breccia lens extend about 30 to 40 m deeper on the eastern side of the crater than on the western side. The existence of a structural high, possibly a ring anticline, is suggested about halfway between the center of the crater and the eastern and western crater walls. A zone of fine breccia composed of Stones River and Knox fragments occurs in the core immediately east of the central hill as a horizontal sheet 3 m thick and 15 m below the base of the breccia lens; this horizon was originally occupied by a one meter thick bentonite bed which was apparently forced out of the way by the injection of the breccia. Except for the region below the central uplift, the deformation decreases downward below the breccia lens, and the rocks are nearly flat-lying and relatively undisturbed below about 100 m beneath the base of the breccia lens.

In the center of the crater, Stones River, Wells Creek, and Knox strata occur as folded, faulted, and brecciated limestone and dolomite which form a hill whose top is about 120 m above the original crater floor (Figs. 3 and 4). The central hill has the general structure of a domed megabreccia block. The top of the Knox strata has been raised about 350 m, and at least 30 m of the upper Knox are exposed as folded and brecciated units. In a few blocks of the Knox breccia, groups of shatter cones occur with several different orientations. About 0.5 m of Wells Creek Dolomite strata is exposed along the folded contact at the top of the Knox strata. The overlying Stones River Group has several omissions of parts of the section caused by high and low angle faulting; bedding plane faults are also common. Most of the strata now exposed in the middle of the central hill dip between 24° and 60° to the west and northwest. Large breccia zones cut the sequence at different angles and trend generally north. The intersection of the base of the hill with bedded breccia on the crater floor measures about 900 m in diameter. The original flanks of the hill, before burial beneath the Chattanooga Shale, had an average slope of between 10° and 15° with some slopes as steep as 30°.

Post-Crater Stratigraphy and Structure

The post-crater, pre-Chattanooga surface has the same general attitude as the underlying Middle and Upper Ordovician strata except for the area that includes the crater and inner rim (Fig. 5). The local relief on the pre-Chattanooga surface surrounding the crater is about 10 m; dips of the strata average 1° to 3°. Along the crater rim many small, irregular valleys were

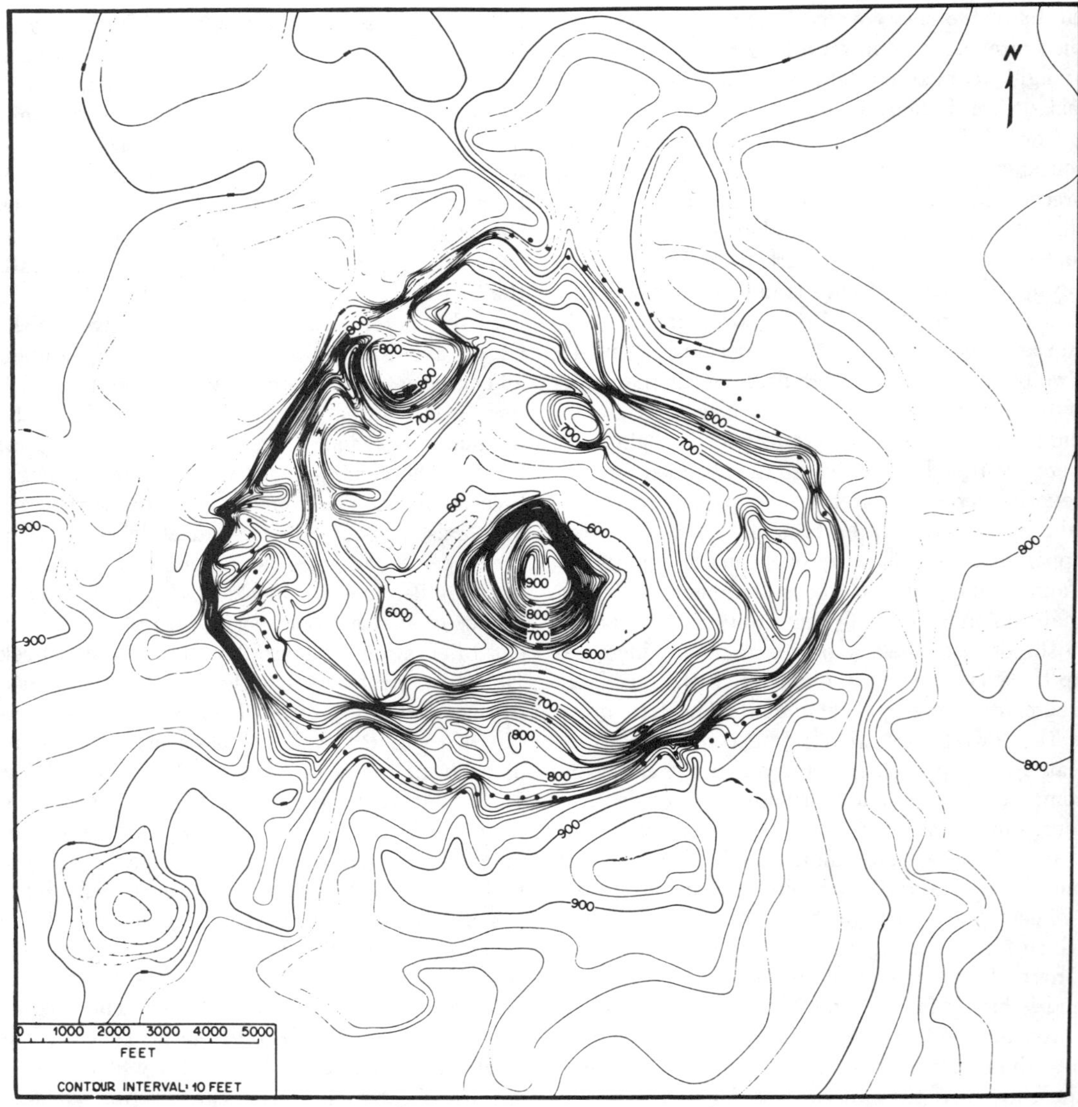

Fig. 8. Structure contour map of the Flynn Creek crater; structure contours are drawn on the base of the Chattanooga Shale. The contours represent the surface that was present immediately before deposition of the Chattanooga Shale. The tops of the crater walls are shown by a dotted line in areas where thick deposits of ejecta were washed into the crater from the rim, or where significant rim down-faulting and subsidence have occurred to modify the original crater shape.

formed before burial beneath the Chattanooga deposits, but none of the valleys was eroded deeply enough to expose the crater to external drainage. Steep cliffs up to 70 m high occurred locally along the crater walls, and large masses of ejecta that had been washed back into the depression are present immediately within the crater at the mouths of the rim valleys.

The outline of the crater is generally circular, although, locally, segments of the crater walls are relatively straight along the northeastern (N45°W) and northwestern (N45°E) rim for about 1500 m

(Fig. 8). Regional joint sets, major fault zones, and veins in the northern part of the Nashville Basin (Jewell, 1947) and local joint sets (Roddy, 1966) also have a northeastern trend of about N45°E. A minor set of joints lies in a northwest direction.

The outer part of the crater floor exhibits an irregular surface on the top of the breccia (Fig. 8). Near the northwestern and western crater walls, a continuous mass of breccia stands high within the crater and is underlain by megabreccia blocks as large as 100 m in length. These megabreccia blocks and part of the ejecta washed back into the crater locally produce a terraced effect along the crater walls (Figs. 8 and 9). Many small erosion channels and mounds of breccia are scattered throughout the inner part of the crater floor.

A nonbedded breccia cemented with a tan dolomitic matrix locally overlies much of the main breccia within the crater, but is quite irregular in distribution and thickness. This nonbedded dolomitic breccia occurs in areas where the overlying bedded breccias of early Late Devonian age are thickest, which suggests that these breccias with a dolomitic matrix are basal nonbedded parts of the reworked crater breccias and of ejecta washed back into the crater. The fragments in this breccia consist mainly of Cannon, Catheys, and Leipers strata. A typical exposure consists of yellowish to dark-gray, commonly nonbedded, unsorted, angular fragments ranging in size from microbreccia to blocks several meters long. The matrix is a yellowish-gray to yellowish-brown, argillaceous limy dolomite identical to the matrix in the overlying bedded dolomitic breccias which are of early Late Devonian age. The upper few meters of this breccia are locally very poorly bedded and gradational into the overlying bedded breccias. The lower contact of the nonbedded dolomitic breccia with the crater breccia is irregular and locally gradational over a few meters.

Within the crater, a sequence of bedded dolomitic breccia and bedded dolomite, both of early Late Devonian age (Huddle, 1963), successively overlie the crater breccias (Fig. 2). The bedded dolomitic breccia, present in most parts of the crater, is the lower of these two units and

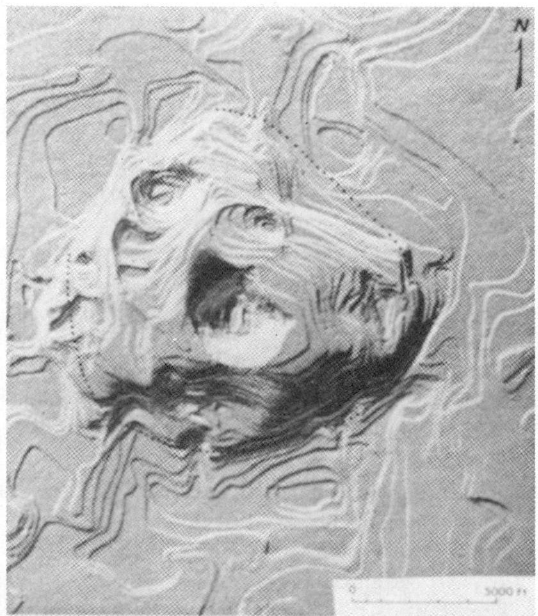

Fig. 9. Model of the Flynn Creek crater, constructed from the structure contour map, showing the crater shortly before deposition of the Chattanooga Shale in early Late Devonian time. The large hills near the outer parts of the crater are underlain by megabreccia blocks derived from the crater walls. The dotted line indicates the position of the top of the crater wall in areas where large volumes of ejecta have been washed back into the crater, modifying the original crater shape. No vertical exaggeration. Lighting is from south.

thins out high on the crater walls and on the lower flanks of the central uplift. Near the crater walls, this unit consists of predominately angular to poorly rounded fragments of dolomite from the upper part of the Leipers, enclosed in an argillaceous dolomitic matrix. The bedded dolomitic breccia is commonly underlain by a poorly-bedded dolomitic breccia that contains angular fragments derived from the Cannon, Catheys, and Leipers strata with occasional Hermitage fragments.

This bedded dolomitic breccia is usually gradational, over a few millimeters, into the overlying unit of bedded dolomite and is conformable with it. The bedded dolomite averages about one meter in thickness and usually thins out lower in elevation on the crater walls than does the underlying bedded dolomitic breccia. Both units may total up to 10 m in thickness but average

2 m. Huddle (1963) has reported conodonts of early Late Devonian age in the bedded dolomitic breccia and in the overlying bedded dolomite, indicating a marine environment of deposition.

Throughout the northeastern part of the Nashville Basin the Chattanooga Shale averages about 7.5 m in thickness, but in the Flynn Creek crater this black shale is about 55 m thick. A five-fold stratigraphic division has been applied to the Chattanooga Shale by Conant and Swanson (1961), and each unit has been recognized within the crater (Conrad *et al.*, 1957; Roddy, 1966). The lowest unit thickens from about 2 m outside the crater to 48 m inside the crater and is the only unit greatly thickened.

The Maury Formation conformably overlies the Chattanooga Shale, and Hass (1956) concluded from conodont studies that in most areas the entire Maury Formation is of Kinderhook (Early Mississippian) age.

The Maury strata are in turn overlain conformably by the Fort Payne Formation of Early Mississippian age. The Fort Payne lithology varies laterally and vertically from limestone to cherty limestone to chert and averages about 45 m in thickness in the Flynn Creek area. The upper surface is deeply weathered, and the resistant cherts act as persistent ridge formers.

Unconsolidated materials of Quaternary and possible Tertiary age are present throughout the Flynn Creek area and consist mainly of residuum and colluvium on the hillsides and alluvium along the valley floors.

STRUCTURAL COMPARISONS

Introduction

Shoemaker and Eggleton (1961) classified the Flynn Creek crater as a "buried crater with the form and structure of a meteorite crater." Bucher (1963) referred instead to this crater as a cryptoexplosion structure and suggested that the type of deformation was not consistent with an impact origin. The absence of both meteoritic and volcanic material has prevented the general acceptance of either origin. However, comparisons of structural deformation between Flynn Creek, accepted meteorite impact craters, and man-made explosion craters have shown marked structural similarities that indicate the Flynn Creek crater is an impact structure. Other lines of evidence, including petrologic, geochemical, and geophysical studies, and the absence of structural similarities between the Flynn Creek crater and volcanic explosion craters are also consistent with an impact origin.

Cryptovolcanic Hypothesis

The "cryptovolcanic" hypothesis has been most strongly advocated by Bucher (1936, 1963), although the concept was first introduced by Branco and Fraas (1905) for the Steinheim Basin in Germany. Bucher (1963, p. 642) argued that gas, under enormous pressures, derived from a rapidly crystallizing, water-rich magma would penetrate the overlying strata, "... crushing and partly pulverizing the rock in gouging out irregularly shaped pathways to the surface... explosive action may ultimately exhaust itself at depth, or end in an unsuccessful attempt to eject the rock material (a "cryptoexplosion structure") or in a successful blowout, an explosion crater." Bucher (1963) agreed that certain features of the "cryptoexplosion" structures could be accounted for by meteorite impact, but stated that impact craters should be:

1. Randomly distributed, e.g., not demonstrably related to structures of purely terrestrial origin;
2. Independent of magmatic activity in the region;
3. Structurally of a nature that is comprehensible in terms of instantaneous impact and its immediate consequences.

The Flynn Creek crater appears to fit ideally the criteria for a "random distribution"; it occurs as an isolated structure in otherwise flat-lying, undisturbed sedimentary strata (Fig. 1). The possible existence of a shallow, buried, igneous intrusive body was examined by a geophysical study (Biehler, 1964), but no gravity anomalies on the level of one milligal are associated with the crater (Fig. 10), and no large magnetic anomalies are present (Fig. 11). On the basis of regional geologic studies, no known tectonic zones or areas of volcanic activity appear to be related in time or space to the Flynn Creek crater.

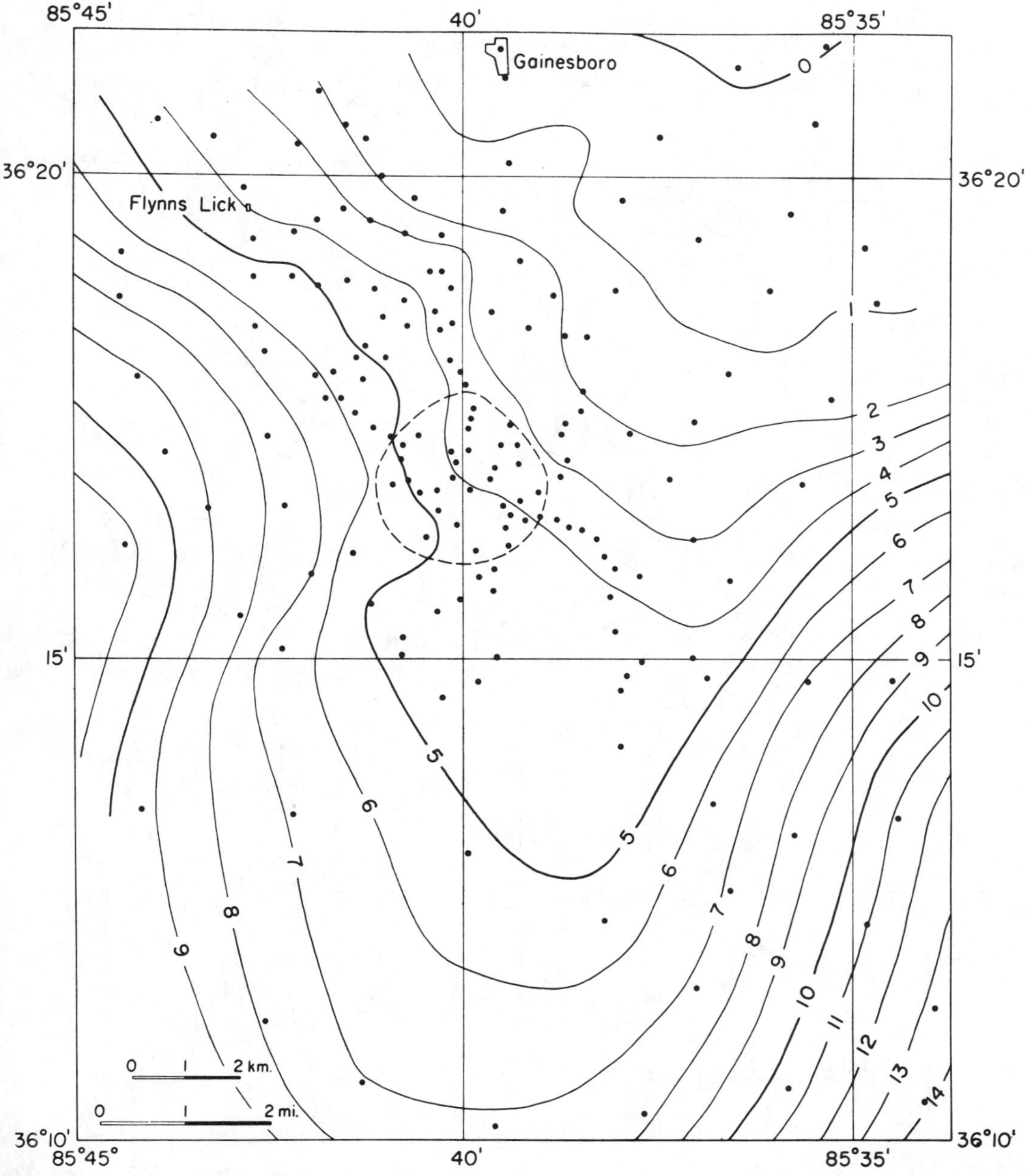

Fig. 10. Complete Bouguer anomaly map of the Flynn Creek area with a contour interval of 1 milligal. The locations of the gravity stations are indicated by dots. The location of the Flynn Creek crater is shown by a dashed circle.

Comparison of the structural deformation at the Flynn Creek crater with volcanic structures of similar size, such as maars and diatremes, that were produced by violent gas eruptions in sedimentary strata, has shown that very few structural similarities exist. The maar and diatreme fields used in these comparisons include: Eastern Fife in Scotland (Geikie, 1902), Swabian Alb in Germany (Cloos, 1941), Eifel District in Germany (Hopmann *et al.*, 1957), Hopi Buttes in Arizona

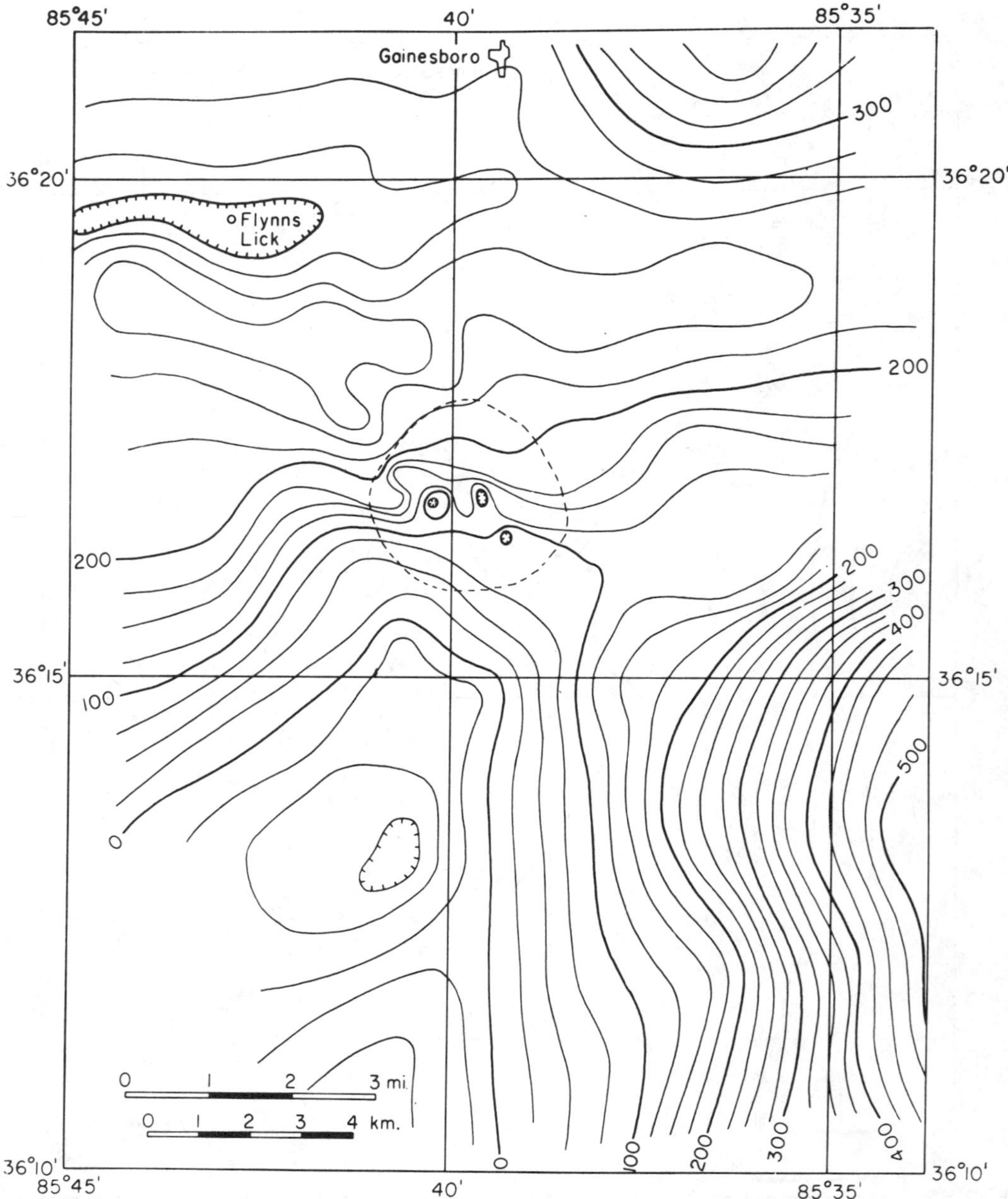

Fig. 11. Total intensity magnetic anomaly map of the Flynn Creek area with a contour interval of 25 gammas. The magnetic stations have the same locations as the gravity stations shown in Figure 10. The location of the Flynn Creek crater is shown by a dashed circle.

(Williams, 1936; Shoemaker, oral communication, 1966), Pinacate Craters in Sonora, Mexico (Jahns, 1959), the Kimberlite District in South Africa, the Riñinahue volcanic field in southern Chile (Müller and Veyl, 1957) and the Ubehebe volcanic field in northern Death Valley, California (Roddy, 1967). Marked similarities in the structural deformation of their respective surrounding strata exist in each of these volcanic features, and it is necessary to summarize only the more important comparisons with the Flynn Creek crater.

Comparisons of the shape, size, and location indicate marked differences between these volcanic structures and the Flynn Creek crater. Maars commonly occur in groups; the individuals range in diameter from a few tens of meters to nearly 4 km, and they are usually circular to elliptical, bowl-shaped craters as deep as 250 m. Diatremes, the vents or pipes of maars, are commonly circular to elliptical in plan, and may extend in depth to many kilometers. Neither maars nor diatremes usually exhibit a well-defined polygonal outline such as is locally observed along the crater walls at Flynn Creek. The Flynn Creek crater occurs as an isolated structure, unlike maar and diatreme fields which are usually associated with other volcanic and tectonic features.

In general, maars and their vents appear to form mainly by local brecciation of a column of strata with very little, if any, lateral compression of their walls. By contrast, the rim of the Flynn Creek structure is faulted, uplifted, and generally tilted away from the crater, with many concentric, doubly-plunging, asymmetric anticlines and synclines. The width of the deformed rim zone at Flynn Creek averages about 500 m. The walls of maars and diatremes commonly have sharp contacts with the surrounding country rock, and exhibit little or no folding in the rim strata. Where structural deformation is present in the rims of maars and diatremes, it is commonly restricted to a narrow band a few meters or less in width. Some diatremes have a monoclinal fold with a steep inward dip adjacent to the vent wall, but the zone of folded strata is rarely wider than 5 m and averages one meter. Increasing the size of the maar or diatreme does not appear to proportionally increase the width of the very narrow, inward-dipping monocline or of the fracture zone. A narrow band of fractured rock with occasional brecciation is also present in some of the vent walls. The same zone which is fractured is commonly indurated to some extent, and mineralization may be present. Slumping within the craters and concentric normal faulting in the rims are present both in maars and in the Flynn Creek structure. Large-displacement thrust faulting has not been reported in maar rims, but is present in the Flynn Creek structure.

The Flynn Creek crater has a large central hill or central uplift of older strata, with some units containing shatter cones, raised nearly 300 m. Some maars have central or non-central small volcanic cones in the crater, but large blocks forming a central uplift and the presence of shatter cones have not been reported from maars or diatremes.

The breccias in the crater of the Flynn Creek structure were formed from the same lithologic units as are exposed in the surrounding country rock. Deformation in the individual breccia blocks is often severe and includes fractures, folds, and intense twinning in the calcite. The agglomerates in maars and diatremes, on the other hand, particularly those in structures the size of the Flynn Creek crater, consist of a variable assemblage of volcanic tuff and breccia derived from both local and deeper horizons. Maars commonly have an ash and agglomerate ejecta blanket overlying the rim. The Flynn Creek structure had an ejecta blanket consisting only of sedimentary rock and displaying a crude inversion of the stratigraphy, part of which is now preserved overlying a rim graben.

No volcanic material has been observed at the Flynn Creek crater, and mineralization related to hydrothermal or volcanic processes has not been recognized in the Flynn Creek area. Examination of drill cores, and trace element studies of the rocks, including breccia fragments and breccia matrix in the crater and deformed rim strata, have shown no anomalous mineralogical or chemical variations (Roddy, 1966).

From the preceding outline, it is clear that few structural similarities exist between the Flynn Creek structure and the structures of maars and diatremes, but it is recognized that this lack of comparison does not uniquely rule out some other type of volcanic gas origin whose mechanism of eruption is as yet unexplained. Indeed, Bucher (1936) considered maar craters as end products of the volcanic gas "explosion" process. Wilson and Born (1936) also described the structure at Flynn Creek as the product of a "relatively rapid, deep seated volcanic explosion accompanied by a gas explosion near the surface."

Apparently Bucher (1936, 1963) considered this a possibility when he said that "the explosion, shallow and strong enough, blows out a shallow, more or less circular explosion basin filled with a jumble of disordered blocks and surrounded by a zone of materials blown or pushed out from it."

In the earlier papers of Branco and Fraas (1905), Bucher (1936), Wilson and Born (1936), and others, the physical concept of a "buried volcanic explosion" and how it could occur was not made clear. Bucher (1963) restated the process in terms of very high gas pressures derived from rapidly crystallizing, water-rich magmas, and equated "fluidization" and comminution of the rock to an "explosion." More recently, Snyder and Gerdemann (1965) have also discussed high volcanic gas pressures, accompanied by fracturing and structural disturbance, as the process which produces "cryptoexplosion" structures. They suggest that, when the sedimentary cover is thin, as at the Crooked Creek structure in Missouri, gas pressures are able to dome the rocks until an "explosion" occurs and the strata collapse as a breccia into the crater. It would seem very unlikely, however, that this process could explain the types of rim folding at the Flynn Creek crater, in which rim strata have been moved out from the crater with radial shortening locally as great as 35 percent; horizontal compression of considerable magnitude appears to have operated on much of the rim strata around the crater.

This aspect, rim compressional folding, can be examined briefly by considering the minimum mechanical energy necessary to form the crater, and by relating this energy to gas pressures that might accompany the process. The gas pressures and their relationship to the folding in the rim strata can then be discussed qualitatively. A minimum volume of about 2×10^{15} cm^3 of brecciated rock can be calculated for the crater, breccia lens, and central uplift region. The minimum energy necessary to fracture this volume of rock is estimated to be about 10^{23} ergs, based on reported comminution studies (Bergstrom, 1963). A minimum energy of 10^{23} ergs is also calculated for the work done in raising the central uplift and the large thrust blocks in the rim, and for ejecting the volume of rock originally present in the crater. If the energy expended in heating and friction are neglected, the *minimum* total energy required to form the Flynn Creek structure is between 10^{23} and 10^{24} ergs. It should be emphasized that this energy was concentrated in a column of rock which was about 200 m deep and 3.6 km in diameter.

The *minimum* volume of gas (volcanic gas and/or steam from ground water) that would provide energies on the order of 10^{24} ergs can be estimated if an adiabatic expansion and ideal gas behavior are assumed. A maximum temperature of 250°C is also assumed for the calculation, although no thermal metamorphic effects have been noted in the Flynn Creek structure, either in the field or in petrographic studies, and temperatures were probably below 200°C. Using these assumptions, the *minimum* volume of gas that will provide 10^{24} ergs of work in an adiabatic expansion at 250°C is about 5×10^{14} cm^3. This volume is equivalent to a 10 percent porosity in the rock column before brecciation. Very similar volumes are calculated using Goguel's (1951) value of 2.5×10^9 erg/cm^3 for an adiabatic expansion at shallow depths. The pressure computed for gas (steam) under these *P-V-T* conditions is between 1000 and 2000 bars, which is at least 50 to 100 times the pressure necessary to produce tensile fracturing in limestone under the same conditions.

The relationship that these gas pressures have to the folding in the rim strata of the Flynn Creek crater can only be discussed qualitatively. The level of the pre-crater ground surface can be determined from the structural relations in the rim graben; this information allows calculation of the lithostatic pressures that existed at various depths. When the crater was formed, these pressures were about 50 bars or less at the level of the deformed strata in the crater rim and about 150 bars at the lowest level of known brecciation. Under these low confining pressures at temperatures less than 200°C, laboratory studies (Robertson, 1955; Handin and Hager, 1957, 1958; Handin *et al.*, 1963) have shown that brittle fracture of limestone and dolomite will occur in the region of a few thousand bars compressional differential pressure. Ductile behavior or shortening is commonly less than a few percent. In systems under tensile stress, failure commonly occurs at differential pressures 50 to 100 times lower than for compressional stress. Consequently, if a gas is introduced under high pressure from depth, failure of rocks near the surface by brittle fracture is to be expected. Such a process is capable of explaining the formation of the breccia, but it does not explain the large-scale rim folding, during which the major stress component was in a horizontal direction. It seems unlikely that gas pressures could build up to more than a few hundred bars near the ground surface before fracturing the rocks and releasing the pressure; little or no folding would be expected under these conditions.

Impact Hypothesis

INTRODUCTION

An impact origin for the Flynn Creek crater was first discussed by Wilson and Born (1936), but the idea was rejected because of the raised strata in the central uplift. Boon and Albritton (1937) suggested that the argument of Wilson and Born overlooked the point "... that elasticity of rocks would cause a strong rebound following intense compression produced by impact and explosion." During a study of the deformation at the Wells Creek structure, Tennessee, Wilson (1953) revised his opinion on the origin of the Flynn Creek crater and referred to it as a possible impact crater. Dietz (1960) reported shatter cones from the central uplift at Flynn Creek and also argued for an impact origin.

Shoemaker (1960, 1963) presented the first careful structural comparisons between Meteor Crater, Arizona and nuclear craters, and showed that shock mechanisms play a dominant role in the formation of impact craters. No attempt was made in the earlier work to compare the crater at Flynn Creek with either impact or volcanic craters because structural studies had not been made. It is now possible to show that close structural similarities exist between the Flynn Creek crater and several of the different types of "shock-wave-produced" craters, including meteorite impact craters and nuclear and chemical explosion craters.

[*Editor's Note:* Material has been omitted at this point.]

DISCUSSION

A number of marked structural similarities exist between the Flynn Creek crater and the chemical explosion craters. Crater breccia and ejecta were formed at each of the craters. The Flynn Creek and 500-ton TNT craters are shallow, flat-floored depressions formed in flat-lying strata; both of these craters, including the 100-ton TNT crater, have well-defined central uplifts with similar internal structures. The deepest horizons raised in the central uplifts of the Flynn Creek and the 500-ton TNT craters have a similar ratio of scaled vertical distances of structural uplift compared with the ratio of crater diameters; identical values exist between the ratios of scaled shear strengths and the ratio of crater diameters.

Low-angle thrust zones and high-angle faults and folds that are concentric to the crater walls are common in both the rims and walls of the 100-ton and 500-ton TNT craters and the Flynn Creek crater. Local concentric anticlinal folding, immediately adjacent to the crater walls of the 100-ton and 500-ton TNT craters, is equally common to the Flynn Creek, Odessa, and Jangle U craters. At least one thrust fault locally offsets the top of a ring anticline in the crater wall at the 500-ton TNT crater; identical structural deformation is exposed at the larger Odessa crater. The upper plates of thrusts at the 500-ton TNT, Odessa, and Flynn Creek crater have moved out from the craters. Several concentric, discontinuous fracture zones around the 500-ton TNT crater occur as concentric faults or folds at equivalent scaled distances at the Flynn Creek crater. The southeastern rim graben at the Flynn Creek crater is equivalent, in scaled distance, to one of the inner, irregular fracture zones at the 500-ton TNT crater; large subsidence may have occurred with time at the Flynn Creek crater as it did at the 500-ton TNT crater.

CONCLUSIONS

Three meteorite impact craters, three nuclear explosion craters, and two chemical explosion craters exhibit pronounced structural similarities to the Flynn Creek crater, although they were formed in different types of rock materials and vary in size from $\frac{1}{3}$ to $\frac{1}{35}$ the diameter of the Flynn Creek crater. In particular, the Jangle U and the 100-ton and 500-ton TNT craters have structural deformation remarkably similar to that observed at the Flynn Creek crater, especially in the types of rim folding. A central uplift and concentric fractures occur at both the 100-ton and 500-ton TNT craters, and are present at the Flynn Creek crater. The three ratios of diameters vs. shear strengths, diameters vs. distances to concentric fracture zones, and diameters vs. depths to deepest horizons exposed in the central uplifts, are nearly identical for both the Flynn Creek crater and the 500-ton TNT crater.

By contrast, structural comparisons of the Flynn Creek crater with volcanic explosion craters and their vents have demonstrated a notable lack of similarity. The volcanic gas origin appears equally inconsistent with the field evidence at Flynn Creek, which indicates that the rim deformation, including compressional folding, occurred within 150 m of the pre-crater ground surface; calculations indicate that the introduction of high internal gas pressure into the country rock should produce mainly fracturing rather than folding in very shallow buried strata under low confining pressures. The drill core evidence of a shallow, lower boundary of the crater breccia and the decrease in deformation in the rocks below this breccia lens strongly indicate an origin involving surface deformation.

On the basis of the structural analogues found in meteorite impact craters and in nuclear and chemical explosion craters, it seems reasonable to conclude that the crater at Flynn Creek was also produced by a shock-wave mechanism, in this case by the hypervelocity impact of an extraterrestrial body. The conclusion is strengthened by the absence of regional tectonic and volcanic zones or lineaments. The absence of both lake deposits and an identifiable fallback zone, and the occurrence of bedded dolomitic breccia with marine conodonts suggest that the first shallow waters of the Chattanooga Sea were present very near the time of crater formation; an early Late Devonian age is thus implied.

The possibility exists that the crater at Flynn Creek was produced by an impacting body other than an iron or stony meteorite. The shallow depth of the crater, the pronounced anticlinal rim folding, and the central uplift, indicate that a large amount of deformational energy was concentrated within 200 m of the surface. Shoemaker (1963) has pointed out that the energy of an impacting body is distributed along the line of penetration, and he has calculated that a large iron meteorite, impacting under hypervelocity conditions, can penetrate four to six times its diameter. The resulting deformation is similar to that produced by the explosion of a buried nuclear device, such as at the Teapot Ess crater, which has structural deformation nearly identical to that at Meteor Crater (Shoemaker, *this vol.*, p. 627).

The "center of gravity" of energy at the Flynn Creek crater was relatively shallow, considering the types of deformation and their similarity to the Jangle U and to the 100-ton and 500-ton TNT craters which were formed by near-surface and surface explosions respectively. Calculations, modified from Shoemaker's (1963) study, indicate that a cometary body would penetrate less than one quarter of the distance that an iron meteorite would penetrate, at impact velocities on the order of 15 km/sec., and it is suggested that shock conditions similar to a very shallow or surface explosion could be generated by a comet impact. Shallow penetration and large energies also appear necessary to form and preserve a central uplift; a large iron meteorite with an impact velocity of between 10 and 30 km/sec. apparently penetrates deeply enough to continue cratering well within the region producing the central uplift. A cometary origin, however, explains the absence of chemical, mineral, and magnetic anomalies, since cometary bodies are composed primarily of frozen volatiles. A very high velocity meteorite or stony meteorite, as suggested by Dr. G. H. S. Jones (oral communication, 1967), is a possible alternative to a comet impact, but the absence of meteorite materials in the crater breccias and ejecta remains unexplained.

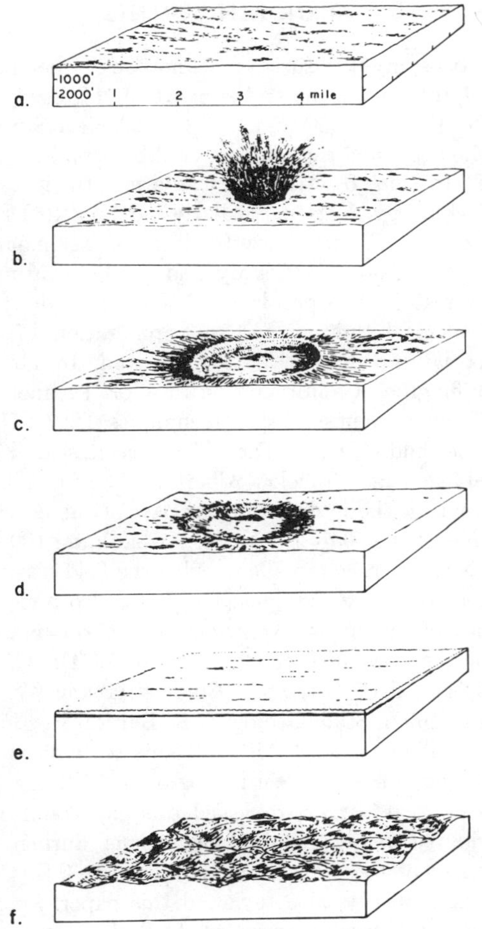

Figure legend in next column

Fig. 19. Diagrammatic sequence of events during formation and subsequent history of the Flynn Creek crater. (a) The Flynn Creek area, shortly before formation of the crater, about 360 million years ago, was apparently a rolling lowland or coastal plain exposing rocks of Leipers (Upper Ordovician) age. (b) Formation of the Flynn Creek crater by a comet impact about 360 million years ago, as interpreted by the writer. The area may have been inundated by the first shallow waters of the Chattanooga Sea. (c) The final crater, immediately after formation, was 3.6 km in diameter and about 150 m in depth; it contained a central hill that stood about 120 m above the crater floor. Rocks in the central hill were uplifted as much as 350 m from their original position; brecciated Knox dolomite contains shatter cones. An ejecta blanket surrounded the crater and overlay the highly deformed strata in the crater rim. The large rim thrust blocks and rim faulting and folding occurred when the crater was formed. (d) A period of erosion followed formation of the crater; many small valleys were formed on the crater rim. The ejecta blanket was removed, much of the ejecta being washed back into the crater. Subaqueous erosion and marine deposition of bedded breccias and bedded dolomite occurred in the crater during early Late Devonian time, in the first shallow waters of the Chattanooga Sea. (e) The Chattanooga Sea filled the crater with black muds in early Late Devonian time; the Flynn Creek area remained under water through at least Early Mississippian time, when the Fort Payne sediments were deposited. (f) Uplift and erosion during Quaternary (and probably Tertiary) time has produced a highly dissected region of steep-sided hills and valleys with an average relief of 150 m. The structure of the Flynn Creek crater is now well-exposed along the valley floors and walls.

Scaling laws from nuclear and chemical explosion craters indicate that a minimum of between 10^{24} and 10^{25} ergs were necessary to form the Flynn Creek crater; comminution and mass movement calculations also give 10^{24} ergs for the minimum energy needed to form the crater. For illustrative purposes, if a cometary body with $\rho = 1.0$ gm/cc and an impact velocity of 15 km/sec. produced the crater at Flynn Creek, the body would have had a diameter of about 250 m. The interpretation of the sequence of events that occurred in the Flynn Creek area is shown diagrammatically in Figure 19.

The genetic significance of the intense twinning in calcite and of the shatter cones has not yet been established, although both features apparently require a high pressure history (Roddy, 1966; Dietz, 1963a, 1963b). Shatter cones have yet to be found at proven meteorite impact craters (craters with associated meteoritic iron); however, Shoemaker, *et al.* (1963) have produced by hypervelocity impact small, poorly developed, striated surfaces very similar to shatter cones. Johnson and Talbot (1964) have presented a theoretical treatment indicating that a high stress environment is necessary for shatter cone formation. As far as the author is aware, neither shatter cones nor the particular type of intense twinning of calcite found at Flynn Creek have been reported from any type of volcanic explosion structure.

High pressure phases, such as coesite and stishovite, have not been identified at the Flynn Creek crater; however the maximum amount of detrital quartz in the carbonate rocks is less than one percent. Detrital quartz grains may not be greatly affected by a shock wave in a matrix which deforms as easily as calcite.

Preliminary results from core drilling have indicated that no anomalous minerals are present in the breccias yet examined, and trace element studies of the surface rocks have detected no volcanic or meteoritic materials in the Flynn Creek area. It seems likely that a crater as large as the Flynn Creek structure, with its extensive brecciation, would contain some trace of volcanic material or mineralization if it were related to a volcanic event, particularly since fault breccias within 20 km of the Flynn Creek area exhibit mineralization apparently related to hydrothermal solutions. If volcanic materials had been initially present in the breccias, even in small amounts, it would be difficult to explain their present absence by weathering processes, since such materials have remained in similar environments for equal lengths of time. The same argument can be made for the silicate phases of a stony meteorite. Fragmental material from an iron meteorite, however, most probably would not survive the weathering processes that have operated since middle Paleozoic time. A cometary body, on the other hand, presumably would leave no mineralogical or chemical evidence of impact and is considered, at present, as the most likely type of impacting body.

ACKNOWLEDGEMENTS

It is my pleasure to acknowledge the many helpful discussions and support of this study by Dr. E. M. Shoemaker, U. S. Geological Survey, who suggested the problem, and by Drs. C. Allen, H. Brown, B. Murray, California Institute of Technology. I am particularly indebted to Dr. S. Biehler, Massachusetts Institute of Technology, for the geophysical study and for his continued interest in the problem; to J. Alderman, U.S. Geological Survey, for the preparation of the special topographic base map; and, to Dr. F. Buffington, California Institute of Technology, for guidance in solid state techniques. Dr. G. H. S. Jones and C. H. H. Diehl, Defence Research Establishment, Suffield, Alberta, Canada, kindly provided the information for part of the descriptions of the 500-ton TNT (Snowball) and 100-ton TNT craters and made possible the field observations of the excavations; Dr. Jones also reviewed part of this paper. Appreciation is also expressed for the assistance and cooperation of Dr. C. W. Wilson, Jr., Vanderbilt University, and W. D. Hardeman, State Geologist, R. Barnes, R. Floyd, H. Milhous, and R. Miller, Tennessee Division of Geology. Drs. H. Wilshire and D. Milton, U.S. Geological Survey, reviewed this paper and contributed many helpful suggestions during the course of the work; Dr. K. Howard, U.S. Geological Survey, also reviewed this paper. Special appreciation is extended to A. Roddy for assistance in preparation of this paper and for part of the geophysical data reduction. This work was supported by NASA Traineeships and NASA Grant NsG 56-60 at the California Institute of Technology and the Branch of Astrogeology, U.S. Geological Survey.

REFERENCES

Bergstrom, B. H., Energy and size distribution aspects of single crushing, *Rock mechanics*, ed. by C. Fairhurst, p. 152–172, New York, Macmillan Co., 1963.

Biehler, S., Recent investigations of the Flynn Creek structure, by D. J. Roddy, with a section on Geophysical Studies, by Shawn Biehler and D. J. Roddy, *U.S. Geol. Survey, Astrogeologic Studies Ann. Prog. Rept., July 1, 1964–July 1, 1965*, Part B, pp. 163–180, 1964.

Boon, J. D., and C. C. Albritton, Jr., Meteorite scars in ancient rocks, *Field and Lab.*, *5*, *(2)*, 53–64, 1937.

Branco, W., and E. Fraas, Das kryptovulkanische Becken von Steinheim, *Akad. Wiss. Berlin, Phys.-math. Kl., Abh. 1*, 64 p., 1905.

Bucher, W. H., Cryptovolcanic structures in the United States, *Internat. Geol. Cong., 16th, Washington, D. C., 1933, Rept., 2*, 1055–1084, 1936.

Bucher, W. H., Cryptoexplosion structures caused from without or from within the earth? ("astroblemes" or "geoblemes"?), *Am. J. Sci.*, *261*, 597–649, 1963.

Bunch, T. E., and A. J. Cohen, Shock deformation of quartz from two meteorite craters, *Bull. Geol. Soc. Am.*, *75*, 1263–1266, 1964.

Carter, N. L., Basal quartz deformation lamellae—a criterion for the recognition of impactites, *Am. J. Sci.*, *263*, 786–806, 1965.

Cloos, H., Bau und Tätigkeit von Tuffschleten: Untersuchungen an dem schwabischen Vulkan, *Geol. Rundschau*, *32*, 709–800, 1941.

Conant, L. C., and V. E. Swanson, Chattanooga shale and related rocks of central Tennessee and nearby areas, *U.S. Geol. Survey Prof. Paper 357*, 91 p., 1961.

Conel, J. E., Studies of the development of fabrics in some naturally deformed limestones, unpub. Ph.D. thesis, Calif. Inst. of Tech., 257 p., 1962.

Conrad, S. G., R. T. Elmore, Jr., and S. W. Maher, Stratigraphy of the Chattanooga black shales in the Flynn Creek structure, Jackson County, Tennessee, *J. Tenn. Acad. Sci.*, *32*, 9–18, 1957.

Diehl, C. H. H., and G. H. S. Jones, A tracer technique for cratering studies, *J. Geophys. Res.*, *70*, 305–309, 1965.

Diehl, C. H. H., and G. H. S. Jones, The Snowball Crater, general background information, Defence Research Establishment, Suffield, Canada, *Suffield Tech. Note No. 187*, unclassified, 23 p., 1967a.

Diehl, C. H. H., and G. H. S. Jones, The Snowball Crater, profile and ejecta pattern, Defence Research Establishment, Suffield, Canada, *Suffield Tech. Note No. 188*, unclassified, 30 p., 1967b.

Dietz, R. S., Meteorite impact suggested by shattercones in rocks, *Science*, *131*, 1781–1784, 1960.

Dietz, R. S., Cryptoexplosion structures: a discussion, *Am. J. Sci.*, *261*, 650–664, 1963a.

Dietz, R. S., Astroblemes: ancient meteorite-impact structures on the earth, in *The Solar System, v. 4, The moon, meteorites, and comets*, B. M. Middlehurst and G. P. Kuiper, eds., p. 285–300, Chicago, Univ. of Chicago Press, 1963b.

Evans, G., Investigation at the Odessa, Texas meteor craters, in *Proc. Geophys. Lab./Lawrence Radiation Lab. Cratering Symp.*, Lawrence Radiation Lab. Rept. *UCRL-6438*, pt. 1., paper D, 11 p., 1961.

Geikie, A., The geology of eastern Fife, *Geol. Survey Scotland, Mem.*, 421 p., 1902.

Goguel, J., Le régime thermique de l'eau souterraine, *Annales des Mines, Ministere de l'Industrie es de l'Energie, v. 10*, 3–32, 1951.

Handin, J. W., and R. V. Hager, Experimental deformation of sedimentary rocks under confining pressure: tests at room temperature on dry samples, *Am. Assoc. Petroleum Geol. Bull.*, *41*, 1–50, 1957.

Handin, J. W., and R. V. Hager, Jr., Experimental deformation of sedimentary rocks under confining pressure: tests at high temperature, *Am. Assoc. Petroleum Geol. Bull.*, *42*, 2892–2934, 1958.

Handin, J. W., R. V. Hager, Jr., M. Friedman, and J. N. Feather, Experimental deformation of sedimentary rocks under confining pressure: pore pressure tests, *Am. Assoc. Petroleum Geol. Bull.*, *47*, 717–755, 1963.

Hass, W. H., Age and correlation of the Chattanooga shale and the Maury formation, *U.S. Geol. Survey Prof. Paper 286*, 47 p., 1956.

Hopmann, M., J. Frechen, and G. Knetsch, *Die vulkanische Eifel*, 2nd ed., Bonn Germany, Wilhelm Stolfuss Verlag, 143 p., 1957.

Huddle, J. W., Conodonts from the Flynn Creek cryptoexplosion structure, Tennessee, *U.S. Geol. Survey Prof. Paper 475 C*, C55–C57, 1963.

Jahns, R. H., Collapse depressions of the Pinacate volcanic field, Sonora, Mexico, Ariz. Geol. Soc., *Southern Arizona Guidebook II*, 165–184, 1959.

Jewell, W. B., Barite, fluorite, galena, sphalerite veins of Middle Tennessee, *Tenn. Div. Geol. Bull. 51*, 114 p., 1947.

Johnson, G. P., and R. J. Talbot, *A theoretical study of the shock wave origin of shatter cones*, M.S. thesis, Air Force Inst. of Technol., Wright-Patterson AFB, Ohio, 92 p., 1964.

Jones, G. H. S., and C. H. H. Diehl, Bosumtwi: an African meteorite crater? *Sky and Telescope*, *30*, No. 1, 15, July, 1965.

Milton, D. J., Structural geology of the larger Henbury Craters, *U.S. Geol. Survey Astrogeologic Studies Ann. Prog. Rept., July 1, 1964–July 1, 1965, Part B*, 26–49, 1965.

Müller, G., and G. Veyl, The birth of Nilahue, a new maar type volcano at Riñinahue, Chile, *Int. Geol. Congr.*, 20th, *Mexico City, Rept.*, 1, 375–396, 1957.

Robertson, E. C., Experimental study of the strength of rocks, *Bull. Geol. Soc. Am.*, *66*, 1275–1314, 1955.

Roddy, D. J., Recent geologic and laboratory investigations of the Flynn Creek structure, Tennessee, *U.S. Geol. Survey Astrogeologic Studies Ann. Prog. Rept., July 1, 1964–July 1, 1965, Part B*, 50–59, 1965.

Roddy, D. J., *The Paleozoic crater at Flynn Creek, Tennessee*, unpub. Ph.D. thesis, Calif. Inst. of Tech., 232 p., 1966.

Roddy, D. J., Minimum energy of formation of Ubehebe Crater, Death Valley, California (abs.), Geol. Soc. Am. Meeting, New Orleans, La., Nov. 20–22, 1967, program, pp. 187–188, 1967.

Rooke, A. D., and T. D. Chew, Operation Snow Ball Project 3.1, crater measurements and earth media determinations, interim rept., *Miscellaneous Paper No. 1-764*, U.S. Army Eng. Waterways Exp. Sta., Vicksburg, unclassified, 61 p., 1965.

Safford, J. M., *Geology of Tennessee*, S. C. Mercer, Nashville, 124 p., 1869.

Shoemaker, E. M., Penetration mechanics of high velocity meteorites illustrated by Meteor Crater, Arizona, *Internat. Geol. Cong., 21st, Copenhagen, 1960, Rept., 18*, 418–434, 1960.

Shoemaker, E. M., Interpretation of lunar craters, Chap. 8 of *Physics and astronomy of the moon*, Z. Kopal, ed., pp. 283–380, New York, Academic Press, 1962.

Shoemaker, E. M., Impact mechanics at Meteor Crater, Arizona, in *The solar system, v. 4, the moon, meteorites, and comets*, B. M. Middlehurst and G. P. Kuiper, eds., pp. 301–336, Chicago, Univ. of Chicago Press, 1963.

Shoemaker, E. M., Structure of the Jangle U and Teapot Ess nuclear explosion craters (abs.), *this vol.*, p. 627.

Shoemaker, E. M., and R. E. Eggleton, Terrestrial features of impact origin, in *Proc. Geophys. Lab./Lawrence Radiation Lab. Cratering Symp.*, Lawrence Radiation Lab. Rept. UCRL-6438, pt. 1, paper A, 27 p., 1961.

Shoemaker, E. M., D. E. Gault, H. J. Moore, and R. V. Lugn, Hypervelocity impact of steel in Coconino sandstone, *Am. J. Sci.*, *261*, 668–682, 1963.

Short, N. M., A comparison of features characteristic of nuclear explosion craters and astroblemes, *Ann. N.Y. Acad. Sci.*, *123*, 573–616, 1965.

Snyder, F. G., and P. E. Gerdemann, Explosive igneous activity along an Illinois-Missouri-Kansas axis, *Am. J. Sci.*, *263*, 465–493, 1965.

Williams, H., Pliocene volcanoes of the Navajo-Hopi country, *Bull. Geol. Soc. Am.*, *47*, 111–172, 1936.

Wilson, C. W., Jr., Pre-Chattanooga stratigraphy in Central Tennessee, *Tenn. Div. Geol. Bull. 56*, 407 p., 1949.

Wilson, C. W., Jr., Wilcox deposits in explosion craters, Stewart County, Tennessee, and their relation to origin and age of Wells Creek Basin structure, *Bull. Geol. Soc. Am.*, *64*, 753–768, 1953.

Wilson, C. W., Jr., Stratigraphy and geologic history of Middle Ordovician rocks of central Tennessee, *Bull. Geol. Soc. Am.*, *73*, 481–504, 1962.

Wilson, C. W., Jr., and K. E. Born, The Flynn Creek disturbance, Jackson County, Tennessee, *J. Geol.*, *44*, 815–835, 1936.

27

Reprinted from pages 323–337 of *Shock Metamorphism of Natural Materials*, B. M. French and N. M. Short, eds., Baltimore, Md.: Mono Book Corp., 1968

THE WELLS CREEK STRUCTURE, TENNESSEE

RICHARD G. STEARNS, CHARLES W. WILSON, JR., HERBERT A. TIEDEMANN[1], JOHN T. WILCOX, AND PHYLLIS S. MARSH[2]

Vanderbilt University, Nashville, Tennessee

The Wells Creek structure is an irregularly concentric series of annular fault blocks with throws of 50 to 600 feet, superimposed on a structural "background" of north-trending 50- to 150-foot highs and lows. Its north edge is at the apparent south limit of the Western Kentucky fault system, represented here by an east-northeast-trending normal fault downthrown about 200 feet on the north side. The entire structure is about 9 miles in diameter.

From the center outward, the structure consists of: (1) a central cylindrical block uplifted more than 2500 feet in its center, but not displaced from the normal regional level at its periphery (at the center of this block is a circular zone of megabreccia a mile in diameter); (2) an inner annular graben downthrown 150 to 400 feet; (3) an annular "horst" ring in which beds are generally at about normal regional position, except for five places near its edge where local uplift exceeds 400 feet; and (4) an outer annular graben downfaulted 50 to 200 feet.

The outermost fault is nearly vertical. The next annular fault is steep or dips outward, in places as little as 30°. The inner two annular faults bounding the inner graben are steep normal faults. Radial faults occur in all parts of the structure, but none was mapped as cutting more than two blocks. The most (17) radial faults are mapped in the central cylinder, where 12 tangential faults also are mapped. Some of these tangential faults, perhaps all 12, are outward-dipping thrusts.

Breccias occur near all the annular faults. Breccias of Mississippian rocks of the outer graben appear to have been formed while the rocks were soft. This conclusion implies a Late Mississippian or Early Pennsylvanian age for the structure, whose previously determined age was known only to be from Late Mississippian to pre-Middle Cretaceous. Extensive breccia in the Ordovician Knox Dolomite may be an Early Ordovician karst deposit, but it may be also of local tectonic origin, which leaves open the possibility of earlier deformational movements in this structure during the Early Ordovician.

The structure is symmetrically located with respect to gravity patterns. Regional gravity lows trend just outside the east and west periphery. Regional gravity highs bracket the periphery to the north and south. A small gravity high occupies the center of the structure, expressing uplift of the relatively dense Knox Dolomite. The exterior highs have magnetic anomalies, but the central high does not. The presence of a dense or magnetic intrusion at depth is not suggested by the geophysical data.

The most likely origin of the Wells Creek structure is believed to be meteorite impact, because of the shatter cone orientation and distribution. When orientation of cones and bedding are plotted on stereonets, and bedding is restored to a horizontal position, the shatter cones point upward and toward the center. Cones from deep parts of a 2500-foot core hole and from the surface farthest from the center are incomplete and less abundant. The original "ground zero" relative to shatter cones appears to have been at the top of the Knox Dolomite, which was buried at least 2500 feet below the original land surface when the event occurred.

INTRODUCTION

The Wells Creek structure is named from Wells Creek in Houston and Stewart Counties, Tennessee, in the northwestern part of the state (Fig. 1). The Wells Creek Basin is a small circular valley through which Wells Creek flows northward to the Cumberland River. The structure is about 70 miles northwest of Nashville and 30 miles southwest of Clarksville, and may be reached via State Highways 13 and 149 from Clarksville, or via State Highway 149 from Erin.

[1] Now with NASA, Manned Spacecraft Center, Houston, Texas.

[2] Now a graduate student at the University of Montana, Missoula, Montana.

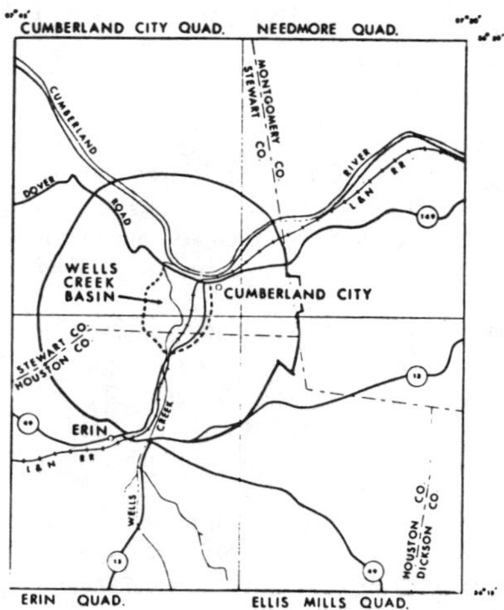

Fig. 1. Maps showing the location of the Wells Creek structure, Tennessee.

This intensely deformed circular structure is striking, because it is surrounded by flat-lying strata which characterize this interior plateau region. The structure occupies parts of the Cumberland City, Needmore, Erin, and Ellis Mills $7\frac{1}{2}$-minute quadrangles which have been mapped in detail and are currently being jointly published by the Tennessee Division of Geology and the Tennessee Valley Authority (see Stearns *et al.*, 1968a, 1968b; Wilson *et al.*, 1968; and Tiedemann *et al.*, 1968).

Several theories have been advanced as to the origin of this unusual geologic feature, which has been studied by many geologists over a period of at least 75 years. During the present investigation, much information has been accumulated, and this paper mentions only a few of the observations that may lead to the ultimate answer. The theories of origin suggested here are based mostly on structural interpretations, gravity and magnetic data, breccia characteristics, and shatter-cone orientation.

PHYSIOGRAPHY AND GENERAL GEOLOGY

The Wells Creek Basin, a small circular valley which averages about $2\frac{1}{2}$ miles in diameter, has been eroded into the northwestern Highland Rim, the general upland plateau of central Tennessee. The Highland Rim level has an elevation of about 600 to 800 feet above sea level in this area. The elevation of the floor of the Wells Creek Basin averages 360 to 380 feet, and Central Hill in the middle of the basin averages 450 feet in elevation.

The intensely deformed megabreccia core of the Wells Creek Basin, underlying Central Hill, is composed of early Ordovician Knox Dolomite, which represents the oldest rocks in the basin (Fig. 2). The rocks underlying the lower slopes of the basin, and the floor of the basin surrounding Central Hill include formations of Middle Ordovician to early Silurian age, and consist of alternating shales and limestones. Silurian and Devonian limestones and shales underlie the walls of the basin, and siliceous Mississippian limestone (the Fort Payne Formation) caps the rim. A total thickness of about 2800 feet of strata are exposed. On the Highland Rim and in the outer parts of the structure all the rocks are of Mississippian age (the Fort Payne, Warsaw, and St. Louis limestones).

PREVIOUS INVESTIGATIONS

The first geologic map of the Wells Creek Basin was made by James M. Safford (1869, insert on State Geologic Map). The first detailed geologic map (Fig. 3) of the entire structure was made between 1889 and 1893 by James M. Safford, first professor of geology at Vanderbilt University, and his student assistant, W. T. Lander (Wilson and Stearns, 1966). This is perhaps the earliest detailed geologic map made of a cryptoexplosion structure in the United States. The map was lost for many years, and was discovered about 1947 by the late Josiah Bridge of the U.S. Geological Survey, who presented it to Vanderbilt University.

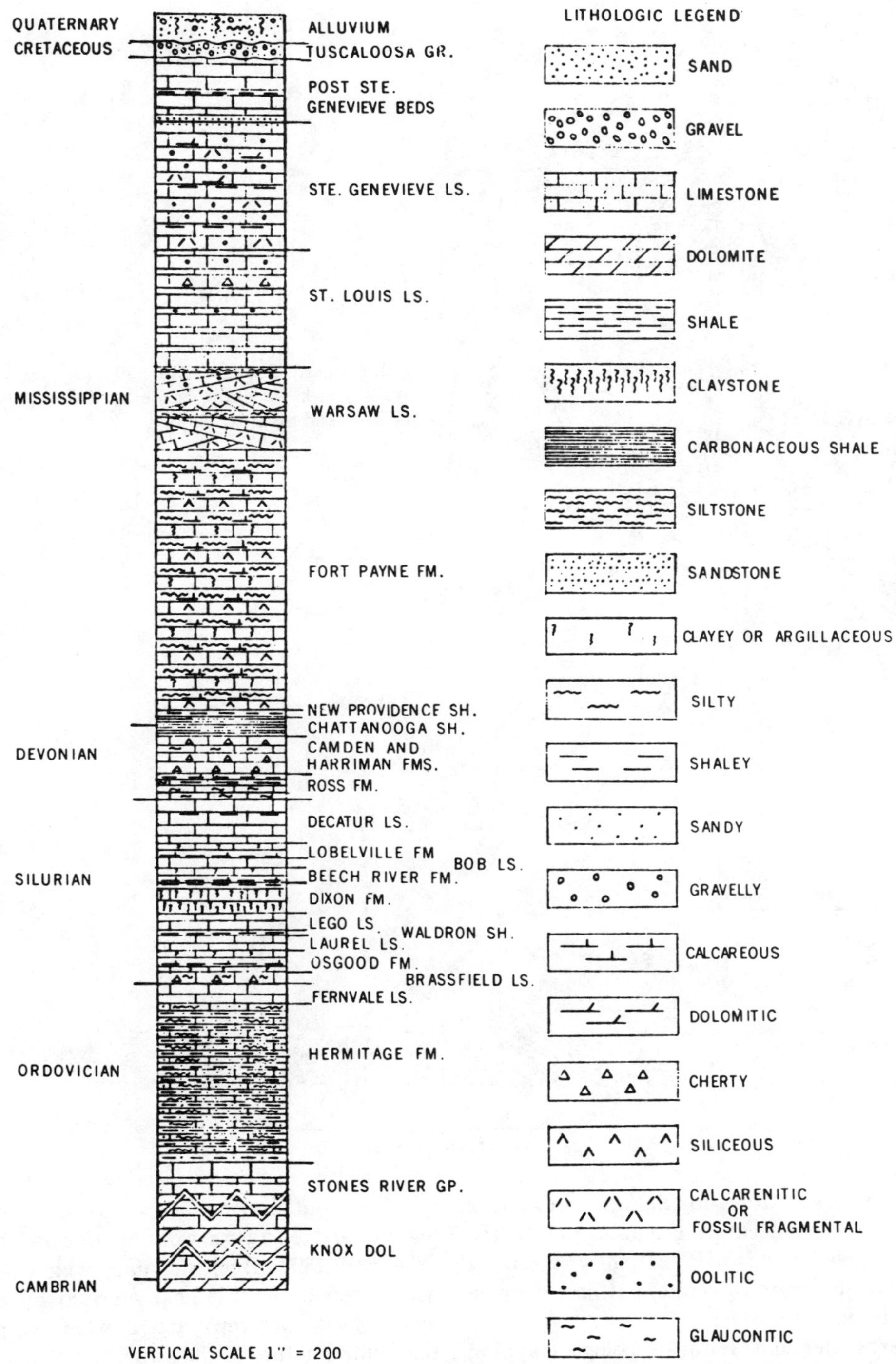

Fig. 2. Stratigraphic column of formations at the Wells Creek structure, Tennessee.

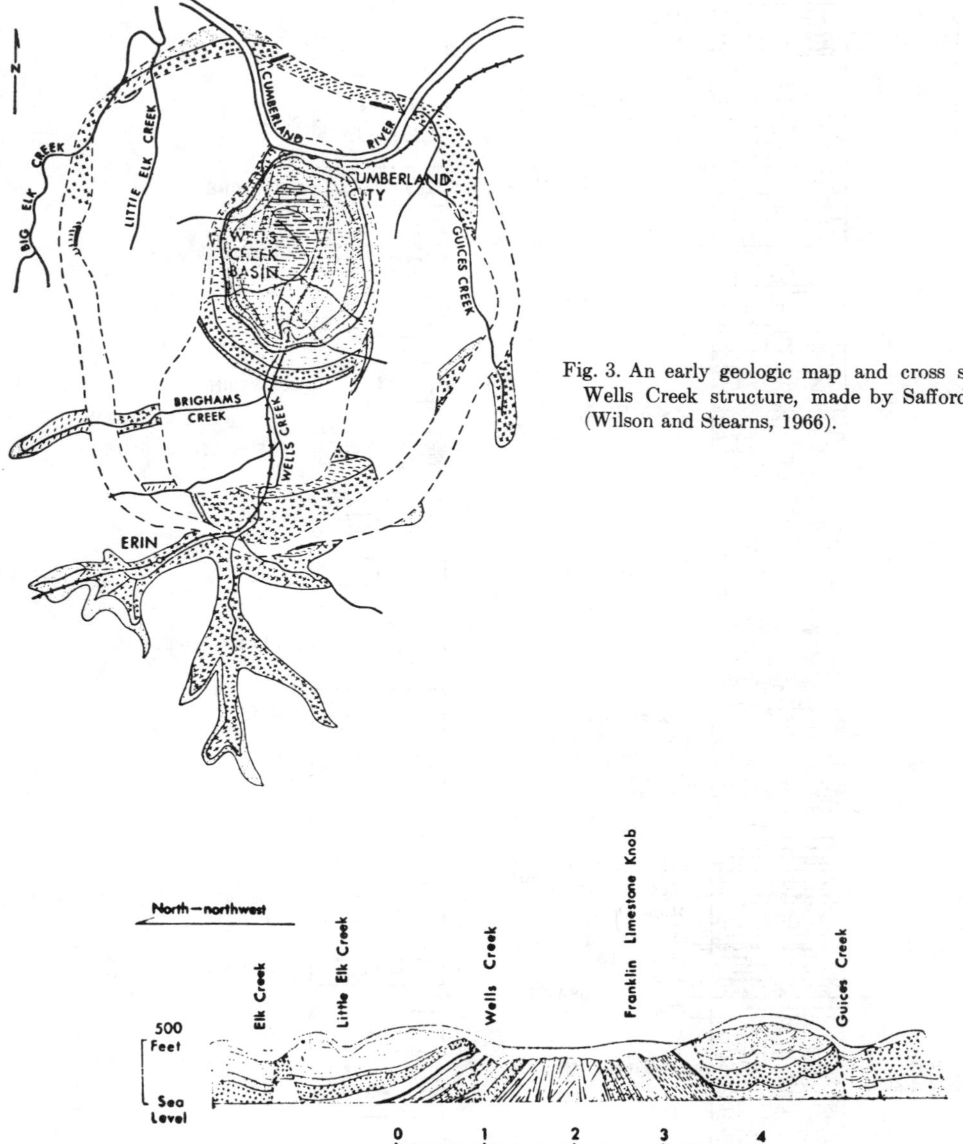

Fig. 3. An early geologic map and cross section of the Wells Creek structure, made by Safford and Lander (Wilson and Stearns, 1966).

The cross section drawn by Safford and Lander (Fig. 3) is a fairly accurate generalization of the structure as we know it today. It shows a central uplifted area of Knox Dolomite, but only a single outer fault ring or graben.

The next major contributor to geologic mapping of the Wells Creek area was Bucher (1932; 1936, pp. 1066–1070, 1074). His work was apparently done without having access to the results of Safford and Lander's work, performed between 1889 and 1893. Bucher's map, which has been widely republished, is characterized by a prominent blocky outcrop pattern which emphasizes the faulted nature of the structure. Bucher recognized and mapped the radial faults more clearly than the ring faults.

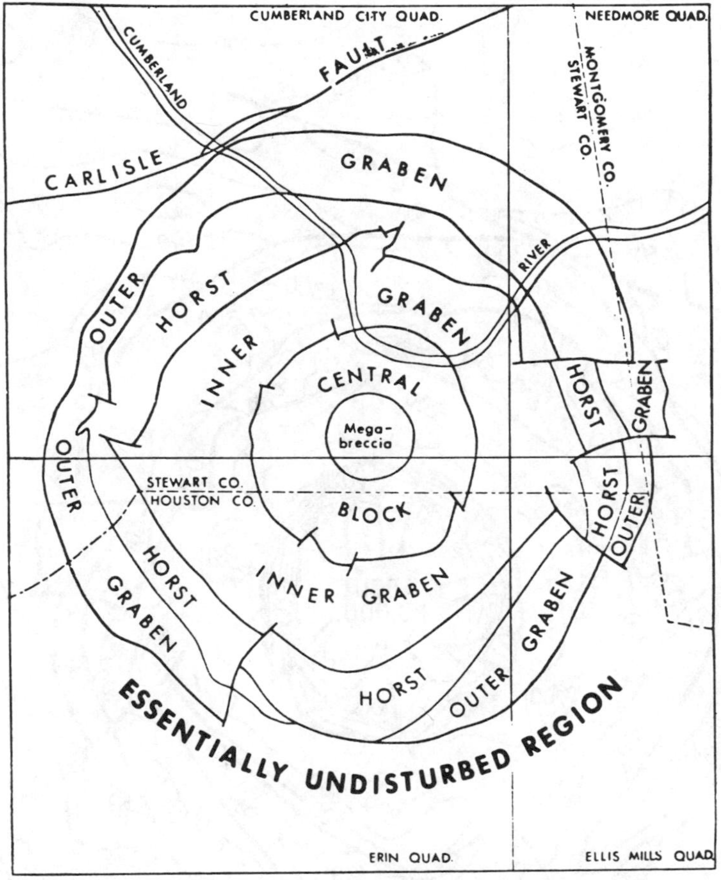

Fig. 4. Generalized geologic map of the Wells Creek structure.

Considering the relatively short times that both Safford and Bucher spent in studying this structure, it is remarkable that each of these men had such an accurate understanding of the geology and of the structural problems.

Both Safford and Bucher believed that this structure had been produced by unspecified internal forces. Safford thought that it was caused by an uplift from "subterranean forces," and Bucher called it a "cryptovolcanic" structure of underground origin.

The first suggestion that the Wells Creek Basin was a "cryptoexplosive" feature, possibly caused by meteoritic impact, was advanced by Boon and Albritton (1936). This idea has also been favored by some later investigators, including Wilson (1953), Dietz (1959, 1960), and Beals *et al.*, (1963). Kellberg (1959), however, assigns to the structure a tectonic origin.

RECENT INVESTIGATIONS

Structure

Field mapping and geophysical investigations during the current project have considerably refined the structural geology over that mapped by earlier workers. Figure 4 is a generalized geologic map that illustrates the major structural features, as derived from the present investigations. In general, it may be seen that the core of intensely deformed Knox Dolomite is surrounded by four roughly circular faults that are, in turn, offset by radial faults. The radial faults are mainly parallel to joint sets recognizable outside the structure.

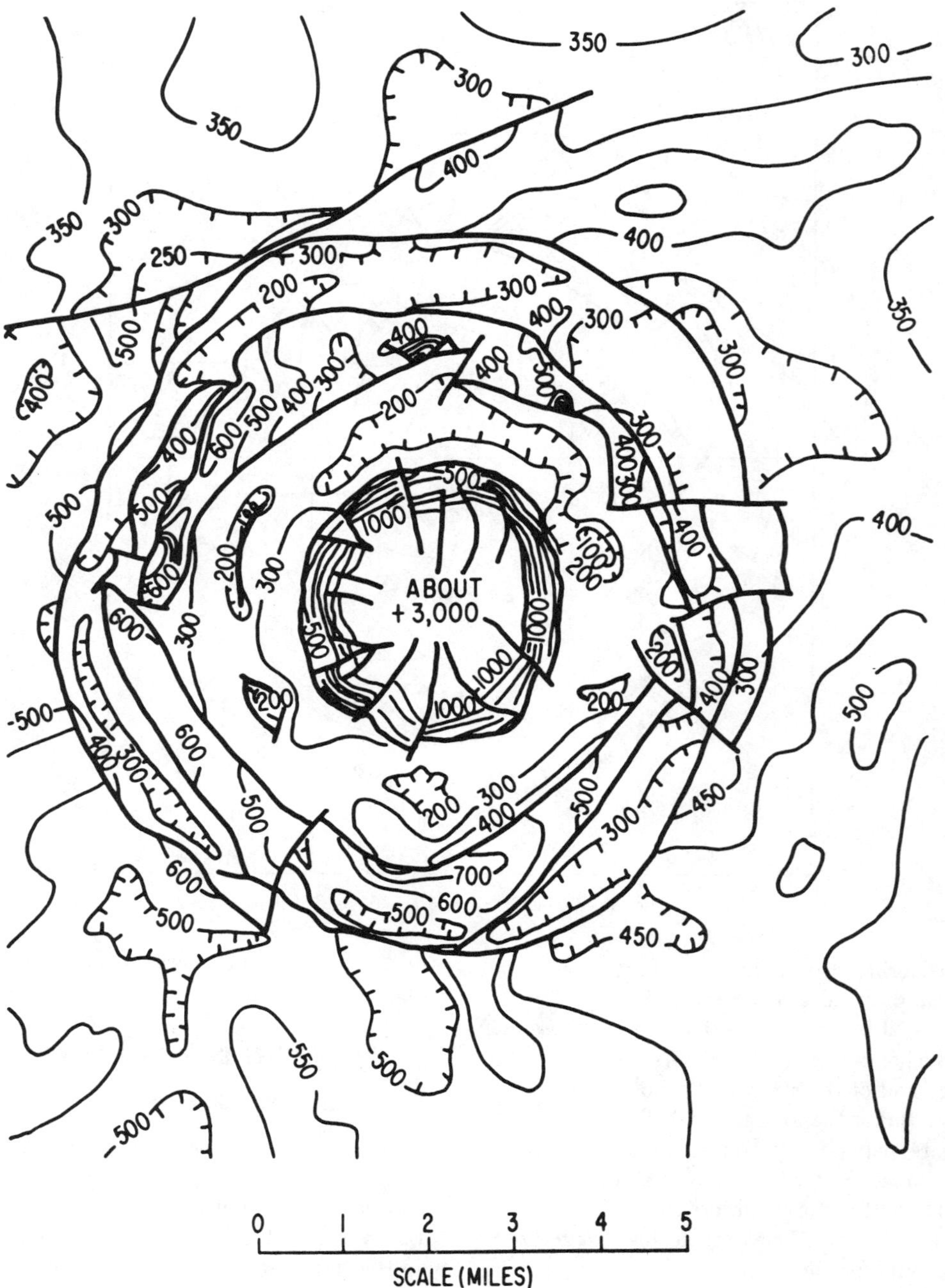

Fig. 5. Structure contour map drawn on the top of the Mississippian Fort Payne Formation, showing the detailed structure of the Wells Creek basin. Scale is in miles.

Fig. 6. Structure of the Wells Creek area as it would have been prior to formation of the Wells Creek basin. This map is constructed by projecting structure contours on the Fort Payne Formation through the structure, using areas believed not to have been moved vertically by the deformation. Scale is in miles.

A 100-foot structure contour map on top of the Fort Payne Formation (Fig. 5) can conveniently be used to demonstrate the relationships of the various fault blocks. Proceeding from the outside inward, one passes abruptly from undeformed rocks into brecciated, downfaulted, and twisted St. Louis rocks of the *outer graben*, which are downfaulted an average of about 150 feet. This outer graben completely encircles the structure, although it is interrupted by radial faults. The graben is variable in width but averages about $\frac{1}{2}$ mile wide. The next inward feature is a *horst* composed of Warsaw and Fort Payne formations which are inferred to be in virtually their normal regional position; this horst is about $\frac{1}{2}$ mile wide. Inward from the horst is the *inner graben*, composed of St. Louis Limestone which has been downthrown an average of about 250 feet. The inner graben averages about $1\frac{1}{2}$ to 2 miles wide.

Inside the inner graben is the *central block* in which a core of Knox Dolomite has been uplifted more than 3000 feet above its normal regional position. The strongly deformed rocks in the central block exhibit both shatter cones and a great variety of breccias. Relatively undisturbed units ranging from Stones River upward to Fort Payne (and locally to Warsaw) formations surround the central core of Knox Dolomite and extend outward to the inner graben. On the outer periphery of the central block, these rocks are believed to be approximately in their normal regional position.

Radial faults, generally parallel to joint sets outside the structure, offset the beds in the central block and also offset the annular inner graben, the horst, and the outer graben.

The outermost fault is nearly vertical. The next annular fault is steep or dips outward, in places as little as 30°. The inner two annular faults bounding the inner graben are steep normal faults. Radial faults occur in all parts of the structure, but none was mapped as cutting more than two blocks. The most (17) radial faults are mapped in the central cylinder, where 12 tangential faults also are mapped. Some tangential faults, perhaps all 12, are outward-dipping thrusts.

Breccias occur near all the annular faults. Breccias of Mississippian rocks of the outer graben look as if they were stirred up while they were soft. This implies a Late Mississippian or Early Pennsylvanian age for the structure, otherwise known only to be from Late Mississippian to pre-Middle Cretaceous.

The structural "grain" of the region surrounding the Wells Creek Basin consists generally of north–south-trending highs and lows which have an average amplitude of about 100 feet, except near the Carlisle fault where the trend is almost east–west. Owing to the 50-foot contour interval chosen, these regional structures are somewhat obscure in Figure 5.

By considering the horst ring and the perimeter of the central block to be essentially in normal regional position, we can construct a contour map showing the structure prior to formation of the Wells Creek basin by projecting structure contours on the Fort Payne Formation from outside the structure across the grabens to connect with similar contours on the horst and on the periphery of the central block. Such a map (Fig. 6) shows that the regional north–south highs and lows continue into the structure, and that the most pronounced high in the region is inside the structure near its south edge.

The structure of the faulted rings themselves can best be illustrated by subtracting the regional structure map (Fig. 6) from the present contour map of the entire structure (Fig. 5). This produces a map (Fig. 7) which shows only the deformation due to the event that produced the Wells Creek structure. The outer graben can be seen to have been downthrown an average of about 150 feet and locally as much as 300 feet, while the inner graben has been downthrown an average of about 250 feet and locally as much as 500 feet. Neither the horst ring nor the perimeter of the central block is believed to have been shifted significantly from normal regional position, but local highs, uplifted from 300 to 400 feet above their normal position, are present in the horst ring. In these local highs the Chattanooga Shale (along with a few older units) crops out, whereas normally it would be about 500 feet below the surface (see Fig. 2). At present, we do not have a satisfactory explanation for these unusual local uplifts.[3]

Geophysics

The current project included a gravity study of the area. The resulting gravity map (Fig. 8), prepared from about 800 station points, with the Wells Creek structure as its center, covers an area about 18 by 28 miles. The fault pattern superimposed on the gravity map shows some interesting relationships. Regional gravity lows trend down both the east and west sides of the structure, and regional gravity highs bound the structure to the north and south. The gravity high in the center is small and is believed to reflect the uplift of the relatively denser Knox Dolomite. No dense intrusion at depth is suggested by the gravity data. The dashed line along the northwest side of the structure is the Carlisle fault, the southernmost member of the West Kentucky fault system, which is tangent to the Wells Creek structure. The Carlisle fault has a throw of about a hundred feet down on the north side.

North–south gravity and magnetic profiles through the center of the structure (Fig. 9) are similar, except that the small positive gravity anomaly in the center of the structure is not duplicated on the magnetic profile. This lack of correspondence substantiates the belief that neither an igneous intrusion nor a buried meteorite is present. Of course, an intrusion that is both light and non-magnetic could be present without being detected with the present techniques.

Breccias

The Wells Creek structure is noted for a variety of breccias. The most extensive brecciation occurs

[3] The local uplifts, and many other structural features, are consistent with the interpretation that the Wells Creek structure consists of a series of collapsed roughly circular rings, which display movement toward the center of the structure such that each block has moved more than the next exterior block. A full discussion of the detailed structural evidence cannot be given here, but the interpretation is developed in detail in Wilson, C. W., Jr., and R. G. Stearns, *Geology of the Wells Creek Structure, Tennessee*, which is soon to be published as a Bulletin of the Tennessee Division of Geology.

Fig. 7. Structure of the Wells Creek area due to the cryptoexplosive event itself. This map exhibits the residual structure when Figure 5 (present structure) is subtracted from Figure 6 (regional structure). Scale is in miles.

Fig. 8. Bouguer gravity and structure map of the Wells Creek area. Note the south-trending gravity "lows" east and west of the circular fault limits, and the gravity "highs" north and south of the structure.

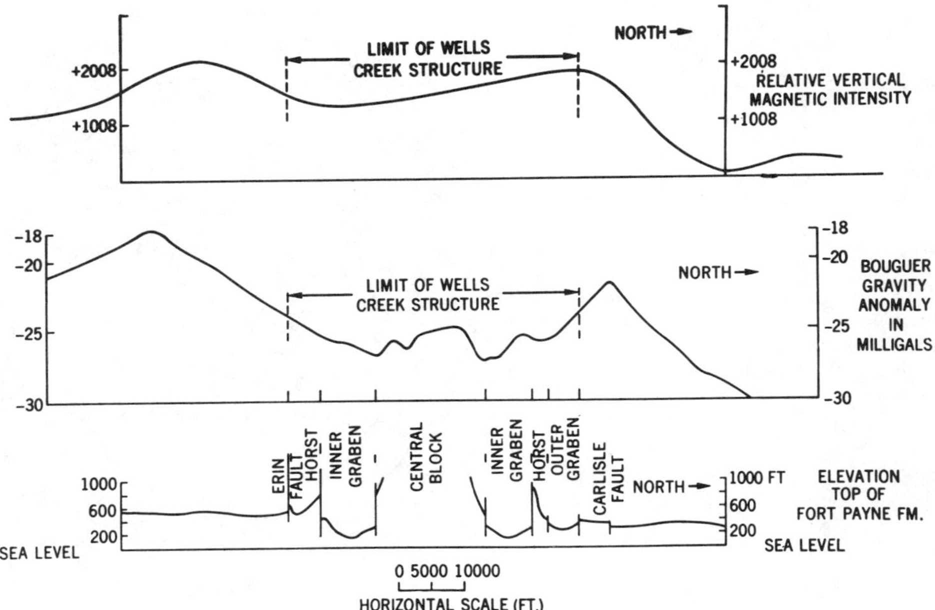

Fig. 9. Gravity and vertical component magnetic profiles across the Wells Creek structure. Note that the small positive gravity anomaly at the middle of the structure, which is probably caused by uplift of denser dolomite at the surface, is not present on the magnetic profile.

in Knox Dolomite in the central block; however, breccias are also found in the outer parts of the structure in younger stratigraphic units, particularly adjacent to the ring faults where resistant brecciated cherts facilitate mapping. In the Central Hill area, large zig-zag structures show that most of the Knox outcrop is actually a megabreccia with blocks up to 100 yards or more in length.

Although the breccia textures are extremely varied, mineral composition is not. All the breccias examined in thin section appear to consist of varying amounts of calcite, dolomite, quartz (including chert), clay minerals, and occasional pyrite. Neither of the dense silica minerals, coesite or stishovite, was found in the samples studied, and other minerals generally considered indicative of elevated temperatures (e.g., calc-silicates such as wollastonite or diopside) are also apparently absent.

In some of the breccias, the rock has been extensively fragmented, but the fragments themselves have undergone little rotation. This so-called "crackle breccia" is similar to much material in the East Tennessee zinc district. In other breccias, fragments of carbonate, quartz, and chert have been thoroughly mixed to yield a heterogeneous or polymict breccia. In this type of breccia, the lithic fragments are of rock types which are also seen in apparently unbrecciated form, either in outcrop or in core samples. The two types of breccia are commonly seen in contact (Fig. 10).

Heterogeneous breccias are often found as vein- or dike-like bodies commonly from a few inches to a few millimeters across (Fig. 11). These bodies often show pronounced lineation parallel to the walls, which might be attributed to injection under explosive force, but which could also have resulted from some less violent crevice-filling process.

A number of geologists have called attention to similarities in appearance between various breccias from the Wells Creek basin and those at the top of the Knox Dolomite in East Tennessee. The implication of this comparison is that some,

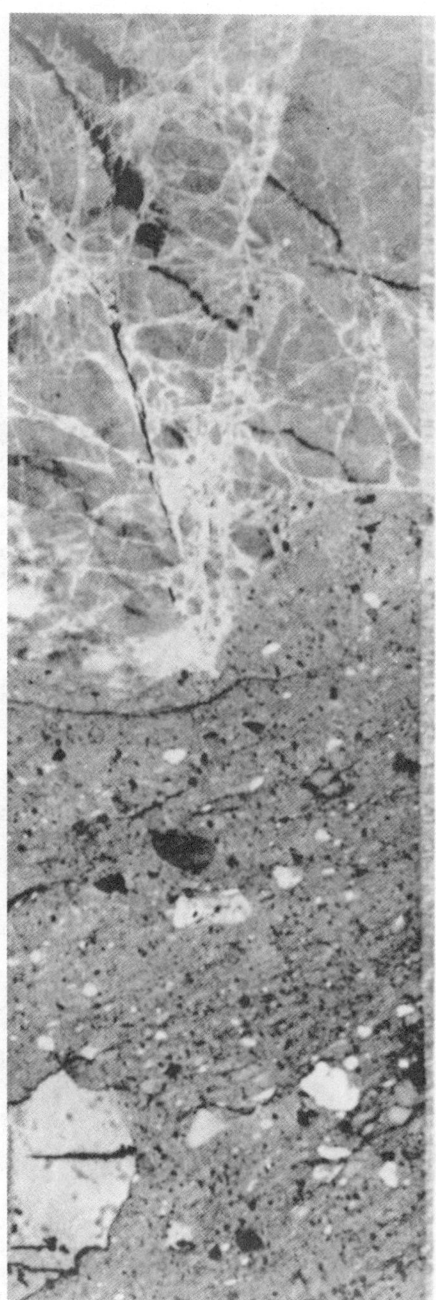

Fig. 10. Split and polished core sample from Knox Dolomite in the central uplift, showing both "crackle" type breccia (above) and heterogeneous breccia (below). Note the lineation in the heterogeneous breccia. The specimen is 1.6 inches wide.

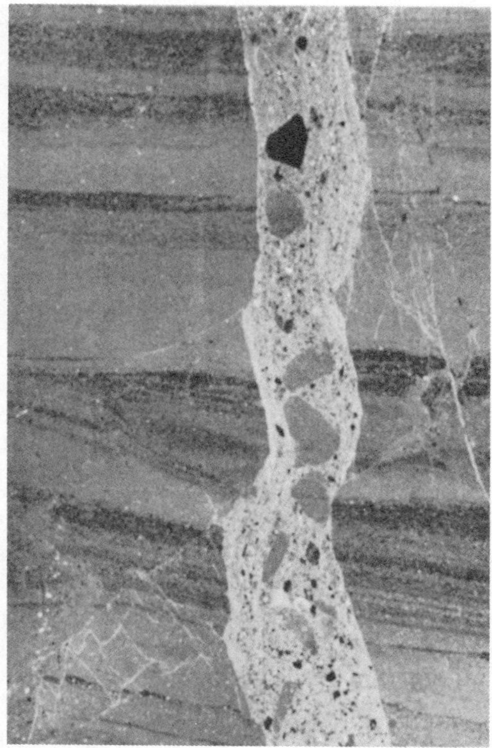

Fig. 11. Vein of heterogeneous breccia, 3 to 5 millimeters across, cutting thin bedded sandy Knox Dolomite in a core sample from a deep drill hole in the central uplift. The breccia consists of dolomite and chert fragments and quartz grains in a dolomite matrix.

at least, of the Wells Creek breccias in the Knox Dolomite must have existed long before the structure was formed. This possibility of several discrete and separated stages of brecciation is supported by the occurrence of breccias in which some of the fragments themselves are breccias (Fig. 12), and also by veinlets of heterogeneous breccia cut by carbonate veins which are in turn involved in "crackle type" brecciation (Fig. 13). In other specimens, relatively unbroken stringers of shale may be seen to be bounded by breccia on either side (Fig. 14). Assuming that a violent origin for this breccia would have caused greater disruption of the intervening shale, a sedimentary origin appears more likely.

Small-scale breccias are abundant in the Knox Dolomite, present, although rare, in upper

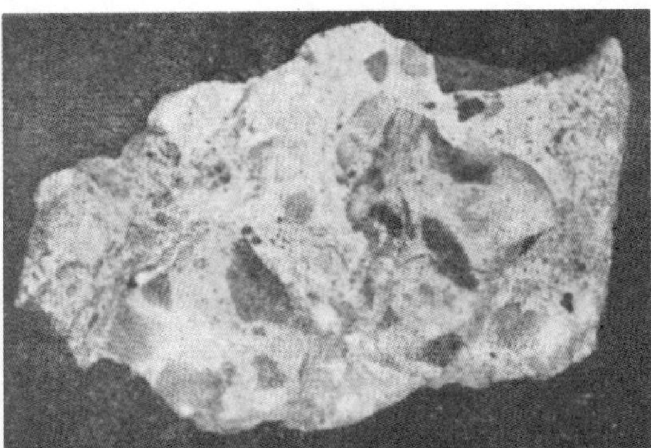

Fig. 12. Knox Dolomite breccia from an operating quarry in the central uplift. Some of the larger fragments consist of preexisting breccia, indicating either that two or more periods of movement occurred during development of the structure, or else that some (perhaps most) of the breccia is unrelated to the major structural event. The specimen is 3.5 inches long.

Ordovician and Silurian units, and common in the St. Louis Limestone adjacent to ring faults. Breccias of Mississippian rocks in the grabens appear to have been formed when the rock was soft. This conclusion implies a late Mississippian or early Pennsylvanian event for the structure, whose age has previously been known to lie between Late Mississippian and Middle Cretaceous. If this discontinuous stratigraphic distribution of breccias means that the structure has undergone

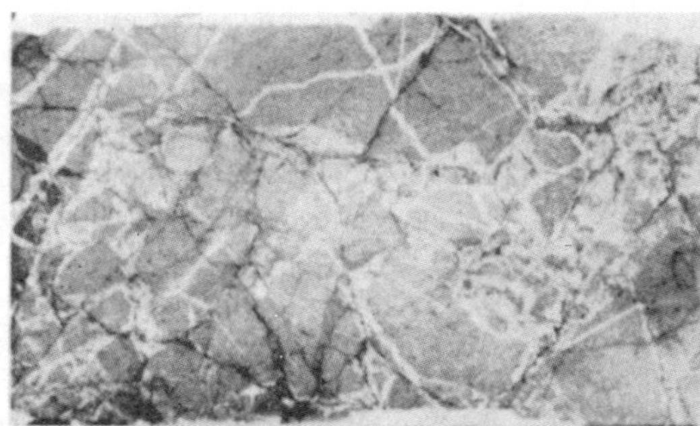

Fig. 13. Thin section of brecciated Knox Dolomite from a deep drill hole in the central uplift. A heterogeneous breccia at left (dark gray) is cut by carbonate veins (light gray to white). The carbonate veins themselves are offset by an intricate fracture system which has imparted a "crackle" breccia appearance to the rock. Section is 1.5 inches by 0.8 inches.

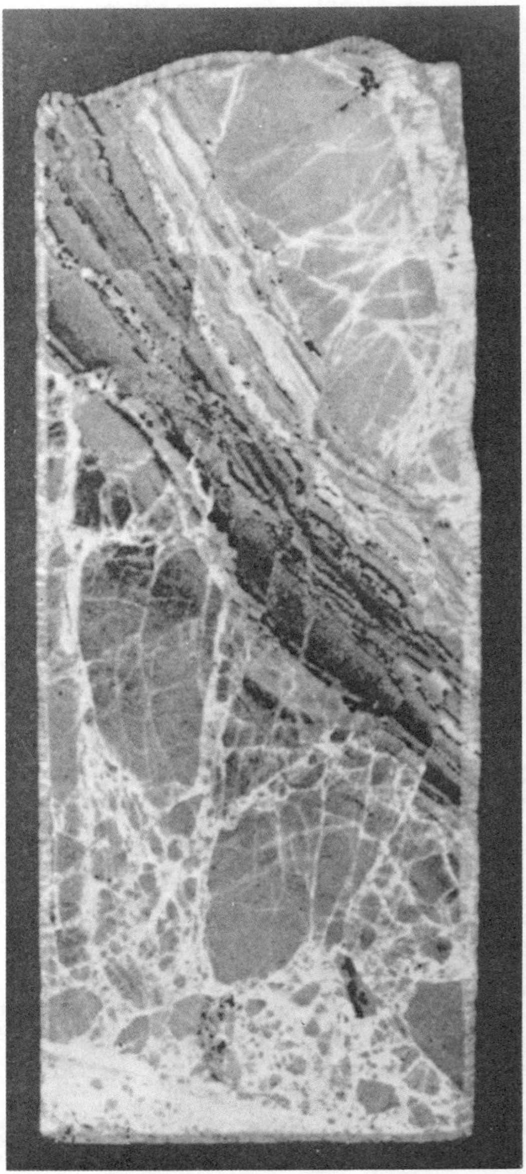

Fig. 14. Brecciated Knox Dolomite from a deep core drill hole in the central uplift. Shale stringer (dark diagonal band) is relatively unbroken, in contrast to brecciated dolomite on either side. A violent origin for the breccia adjacent to the shale would probably have caused more complete disruption of the shale also. A sedimentary origin therefore appears more likely. Surface is 1.5 inches by 4 inches.

several major episodes of deformation, an origin by meteorite impact would appear less likely.

Bedding plane attitudes in the Road Materials Co. quarry in the central uplift suggest that the quarry is located entirely within one, or perhaps within a few, discrete megabreccia blocks. Inasmuch as the largest blocks in "ordinary" breccias seen in the quarry are on the order of a few feet across, there appears to be a large discontinuity in fragment size between megabreccia and the material referred to simply as breccia. It seems likely, therefore, that these are of two different origins; the megabreccia is more likely related to the origin of the entire structure, while a large proportion of the smaller-size breccia represents an earlier stratigraphic breccia. This interpretation of the origin of the small-scale Knox breccia at the center makes the lesser abundance of small-scale breccia away from the center consistent with a single-event origin such as meteorite impact.

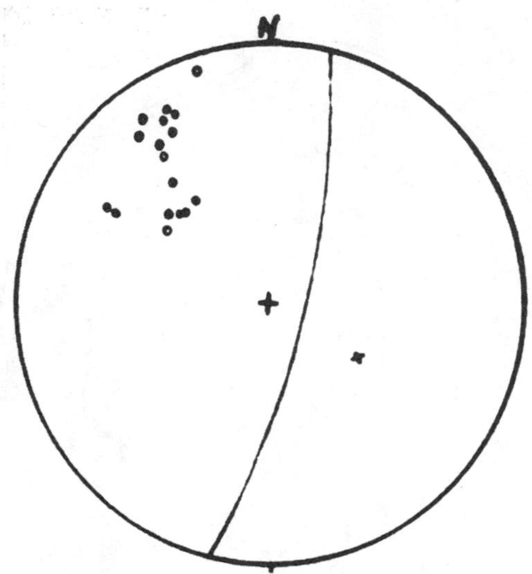

Fig. 16. Stereogram of an outcrop at the Old Quarry on Central Hill. Bedding, which dips steeply westward, is drawn as a great circle on the upper hemisphere. Shatter cones, which point steeply upward toward the northwest, are drawn as small circles on the upper hemisphere. One opposite-pointing cone is plotted as a small "x" on the lower hemisphere. Original orientation of shatter cones can be estimated by rolling the plot to the east-southeast until bedding is horizontal. The shatter cones will then point upward toward the northeast, which is toward the center of the structure.

Shatter Cones

The Wells Creek structure has, at its center, a remarkable development of shatter cones (Fig. 15), particularly in and near the abandoned quarry on Central Hill where an abundance of shatter cones occur with diverse breccias.

Fig. 15. A typical group of shatter cones, developed in Knox Dolomite, from Central Hill in the Wells Creek basin.

If we believe the hypothesis advanced by Dietz (1959, 1960), that the shatter cones were formed by shock waves fractions of a second before the beds were deformed, and that they formed pointing toward the direction from which the shock waves came, the direction from which the shock waves came can be estimated. This is done by making stereo plots of bedding with shatter cones. By rotating the beds back to horizontal, the restored cone apex directions can be plotted in a fashion similar to the restoration of shatter cones around the Vredefort Ring by Manton (1965).

During this investigation, the orientations of more than 1,000 shatter cones were measured at several exposures of shatter-coned Knox Dolomite at different localities in the central uplift. The resulting data (an example is shown in Fig. 16) indicate that the apices of the restored cones generally point upward and toward the center of

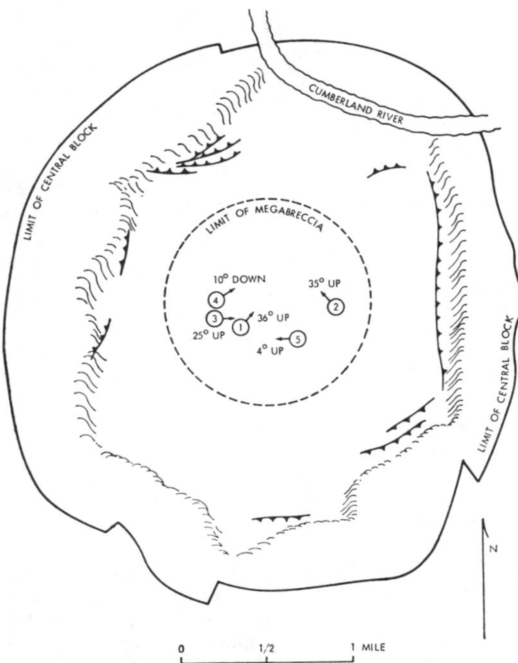

Fig. 17. Map showing shatter cone localities and median orientations of the cones relative to bedding planes which have been restored to the horizontal. The data for the shatter cones from the outcrop at Locality No. 1 are shown in Figure 16.

the structure (Fig. 17), to a position almost at the top of the Stones River Group near the top of the Ordovician section, whose normal position is about 1500 feet underground in this vicinity. The assumed "ground zero" for the shatter cones was thus at least 1500 feet underground (and somewhat above the present cone locations) when the cones were formed. How much deeper underground the inferred explosion occurred depends on the amount of regional erosion that has occurred since formation of the Wells Creek structure. Evidence from the studies of shatter cone orientations thus leads to the conclusion that either (1) a meteorite penetrated in such a manner that shock pressures were generated from at least 1500 feet underground, or (2) that an internal explosion occurred at that depth or deeper.

Distribution and changing character of cones also supports these conclusions. Cones from the deepest portion of a 2500-foot core hole and from surface exposures farthest from the center are incomplete and less abundant.

In spite of considerable effort to determine the attitude of the beds which contain the shatter cones, we must admit that we are not sure whether or not they are overturned. We have assumed that the beds studied were not overturned, but if they are, the shatter cones must have pointed downward and outward. The cones could not have pointed downward and inward—only upward and inward (which we assume to be the case) or downward and outward (if all the beds examined were overturned).

CONCLUSIONS

A main purpose of this project was to try to find data or relationships which would bear on the problem of the origin of the Wells Creek structure; the following summary is directed to "volcanists" and "impacters" alike. Unfortunately, debates as to the origin of such equivocal structures often center on which side must bear the burden of proof, and on what criteria are unequivocal.

For those who favor an internal origin, such as a volcanic explosion, we can cite some fairly convincing evidence. Breccia characteristics leave open the possibility of recurrent activity. The Wells Creek structure is symmetrically located with respect to the gravity pattern; it is bracketed to the north and south by highs and to the east and west by lows. The structure is also precisely tangent to a major fault (the Carlisle fault) of a regional system, the West Kentucky fault system. Furthermore, one can restore the structure by projecting contours from outside to connect with those inside the structure; this restoration reveals that the largest structural high in the region (perhaps a pre-event feature) is located within the structure itself.

It is, however, also true that structural, geophysical, and lithologic data can be satisfactorily rationalized by those in favor of a meteoritic impact origin. For example, the Knox breccias are at the center of the structure and St. Louis breccias are adjacent to ring faults. The various regional structural symmetry arguments generally do not bear critical examination; the projected high, although within the structure, is not located near the central "ground zero" area.

At present, the only independently supported data are the shatter cone orientation results. Therefore, if we consider that shatter cones are characteristic features developed by shock waves of an intensity impossible to generate by terrestrial forces, we must lean heavily toward an origin by meteoritic impact.

ACKNOWLEDGMENTS

This project, which has been active for 2½ years, has been conducted by the Geology Department of Vanderbilt University, under the sponsorship of the National Aeronautics and Space Administration (Grant NsG 465). Many organizations and individuals deserve thanks for contributing to the effort. The Tennessee Division of Geology and the U.S. Geological Survey provided base maps and geologic data. The Tennessee Valley Authority made available cores obtained from drilling near the center of the structure for a steam plant foundation. They also prepared a special large-scale base map of the complex central area for our project. The U.S. Bureau of Mines furnished a gravity meter. Numerous geologists gave advice and assistance, and to each the writers are deeply grateful.

REFERENCES

Beals, C. S., M. J. S. Innes, and J. A. Rottenberg, Fossil meteorite craters, in *The solar system, v. 4, The moon, meteorites, and comets*, B. M. Middlehurst and G. P. Kuiper, eds., pp. 235–284, Chicago, Univ. of Chicago Press, 1963.

Boon, J. D., and C. C. Albritton, Jr., Meteorite craters and their possible relationship to "cryptovolcanic structures," *Field and Lab.*, *5*, 7–9, 1936.

Bucher, W. H., Wells Creek Basin, Tennessee, a typical cryptovolcanic structure (abs.), *Bull. Geol. Soc. Am.*, *43*, 147–148, 1932.

Bucher, W. H., Cryptovolcanic structures in the United States, *Internat. Geol. Cong., 16th, Washington, D. C., 1933, Rpt.*, *2*, 1055–1084, 1936.

Dietz, R. S., Shatter cones in cryptoexplosion structures (meteorite impact?), *J. Geol.*, *67*, 496–505, 1959.

Dietz, R. S., Meteorite impact suggested by shatter cones in rocks, *Science*, *131*, 1781–1784, 1960.

Kellberg, J. M., Recent subsurface investigations in the Wells Creek structure, Stewart County, Tennessee (abs.), *Bull. Geol. Soc. Am.*, *70*, 1765, 1959.

Manton, W. I., The orientation and origin of shatter cones in the Vredefort Ring, *Ann. N.Y. Acad. Sci.*, *123*, 1017–1048, 1965.

Safford, J. M., *Geology of Tennessee*, Nashville, Tennessee, 550 p., 1869.

Stearns, R. G., H. A. Tiedemann, and C. W. Wilson, Jr., Geologic map of the Needmore Quadrangle, Tennessee, *Tenn. Div. Geol. Geol. Map G. M. 38 NE*, 1968a.

Stearns, R. G., H. A. Tiedemann, and C. W. Wilson, Jr., Geologic map of the Erin Quadrangle, Tennessee, *Tenn. Div. Geol. Geol. Map G. M. 38 SW*, 1968b.

Tiedemann, H. A., C. W. Wilson, Jr., and Richard G. Stearns, Geologic map of the Cumberland City Quadrangle, Tennessee, *Tenn. Div. Geol. Geol. Map G. M. 38 NW*, 1968.

Wilson, C. W., Jr., Wilcox deposits in explosion craters, Stewart County, Tennessee, and their relations to origin and age of Wells Creek Basin structure, *Bull, Geol. Soc. Am.*, *64*, 753–768, 1953.

Wilson, C. W., Jr., and R. G. Stearns, Circumferential faulting around Wells Creek Basin, Houston and Stewart Counties, Tennessee—a manuscript by J. M. Safford and W. T. Lander, circa 1895, *Tenn. Acad. Sci.*, *41*, 37–48, 1966.

Wilson, C. W., Jr., R. G. Stearns, and H. A. Tiedemann, Geologic Map of the Ellis Mills Quadrangle, Tennessee, *Tenn. Div. Geol. Geol. Map G. M. 38 SE*, 1968.

Editor's Comments on Papers 28 and 29

28 **DIETZ**
Excerpt from *Shatter Cones in Cryptoexplosion Structures (Meteorite Impact?)*

29 **WILSHIRE et al.**
Excerpts from *Geology of the Sierra Madera Cryptoexplosion Structure, Pecos County, Texas*

GROUP 3

The third grouping of cryptoexplosion structures in the central United States is diffuse and poorly defined but seems to extend southward both the tectonic lineament that controls Lake Michigan (Des Plaines, Kentland) and the lineament that controls Lake Winnipeg (and is occupied, fortuitously or otherwise, by several cryptoexplosion structures in Canada). The Manson and Glasford structures appear to be situated on an extension of this lineament. All the structures in the third group have been attributed to impact (Short and Bunch, 1968; Dence, 1972).

Kentland

The Kentland structure is discussed by Dietz in Paper 28. It is of obscure overall form, even though it provided, in quarries, some of the best material used by Dietz in his pioneer studies of shatter cones. Bucher (1936) and Shrock (1937) regarded the structure as cryptovolcanic. Cohen et al. (1961) recorded coesite in the Middle Ordovician St. Peter's Sandstone of the central uplift, a network of megashatter cones up to 50 m long. Such giant cones are composed of small parasitic cones. Cohen et al. believe that the summit may have once had a small crater formed by a gaseous jet produced at maximum penetration of the meteorite, and they consider the structure to have possible connection with lunar rays. Cohen (1962) suggested that central uplifts are produced by metallic meteorites, a conclusion in striking conflict with Roddy's (Paper 26) invocation of low-density, cometary impacts and also in conflict with the delayed isostatic-response hypothesis. The other three structures in the group are briefly summarized here.

GROUP 4

Sierra Madera

The fourth group includes only two structures, but the Sierra Madera (Paper 29) is among the most complex and best documented of cryptoexplosion structures. It is situated at the tail end of the immense internal plains of the United States, separating the Appalachians and Cordillieras. All the north American structures are, in fact, situated within this tract, for the Canadian shield extends it northward where the Paleozoic cover, here thin, has been removed by erosion. This tract contains the majority of cryptoexplosion structures known throughout the world.

The complex dome of the Sierra Madera, with its overturned folds on the flanks, was mapped by King (1930), who noted that it is broken by radial faults and partly enclosed by smaller, concentric folds. Meteoritic origin was suggested as early as 1937 by Boon and Albritton, and shatter cones were recognized by Dietz (1960). Eggleton and Shoemaker (1961; Shoemaker and Eggleton, 1964) attributed this structure to impact. The important paper on the Sierra Madera by Wilshire et al. (Paper 29) has been slightly abridged for inclusion here. The somewhat full stratigraphic descriptions have been omitted, together with a number of line diagrams and photographic illustrations not suitable for reproduction in the format of this book. Omissions also include a table listing all the quartz lamellation planes recorded and a section suggesting analogy with lunar structures. Megabreccia outcrops in the core, being composed of blocks of carbonates several hundred meters in maximum dimension, set in a matrix of angular fragments and grading down to microbreccia. Both monomict and polymict breccia are represented, the latter grading into monomict breccia, which in turn grades into shattered in situ country rock. Drill cuttings from the upper 1000 m of the Elsinore No. 1 well revealed microbreccia with a light-colored matrix and with a dark matrix, also unfractured dolomite and limestone. In the dark matrix, there may be some carbonaceous and pyritic material. Shatter cones of basal diameter greater than or equal to 20 cm are present in this core material. The breccia is supposed by Eggleton and Shoemaker to form a great lens, 2.4 km across and 900 m thick, nested in a collar of steeply upturned and overturned beds, a crater 3 km in diameter being supposedly formed by impact. Wilshire et al. describe a much larger bowl or funnel

13 km in diameter and up to 2700 m deep. Shoemaker and Eggleton are considered to have studied only the central part of this much larger structure. A shock wave of hemispherical form, of decreasing intensity outward and downward from the focus, is considered to have produced the structure. Wilshire et al. believe that only a surface impact of an extraterrestrial object could produce this strain pattern, and they do not believe that it could have been produced by a deep-seated, subsurface explosive geologic event.

Sierra Madera seems to be unique in an otherwise normal geologic environment and to bear no relationship to any other nearby geologic structure. The mechanics of the uplift remain unclear, but it resembles productions in TNT detonation experiments. It is supposed to have protruded through a 13 km diameter crater, the rocks comprising it coming up from beneath the crater floor. The structure is of post-early Cretaceous age, that of the youngest rocks deformed. The uplift is of the same order of magnitude as many lunar crater uplifts. Lowman (1965) noted a magnetic high under the southeast sector of the uplift, attributing it to a hidden syenite intrusion. Van Lopik and Geyer (1963) attributed it to a meteorite mass. The deformation dies out at between 2000 and 2750 m according to oil drilling records and the breccias only extend to 1000 m. The conglomerates and sandstones of the uplift are more mobile under deformation than dolomites, suffering crushing, contortion, and faulting, with pods moving out of their normal position. In contrast, dolomite accomodated by convolution. Thus, folding is disharmonic and there is crushing and breaking of folds. The base of the Cretaceous sandstone is highly mobile and tends to be injected, the injections themselves being later folded and faulted. Cross-bedding in sandstone has been reconstituted due to shock deformation of quartz grains. In general, though the movement inward and outward is comparable with that in salt domes, the medium involved is more rigid than salt and the deformation is likely to have been more rapid. The deformation in the uplift has produced many quaquaversal domes and basins, with steep circumferential and radial axial plunges. The authors consider two possibilities: that the ring faulting and down-dropping of the depression caused the uplift, or that the uplift caused the depression by removing rock material in depth. This structure was formed during mild deformation in the Glass Mountains, and both Zimmerman and Amstutz (1965) and Lowman (1965) believed that the structure forms part of the Tertiary Trans-Pecos igneous province, and is of endogenic origin. They believed that it is aligned with the Solitario Dome and breccia pipes of the Terlingua District. No obvious connection exists between the structure and these manifestations of igneous activity.

Wilshire et al. favored impact because of lack of integration in the regional geology, the shape of the deformation, the geometry of the uplift, and the intesity of the shock deformation. They mention the possibility that the impacting mass was of low density, of cometary character as suggested by Roddy, 1968a,b). They also mention the alternative possibility that the structure was produced by delayed collapse inward on inward-dipping shear surfaces, in the manner proposed by Dence (1968).

REFERENCES

Boon, J. D., and C. C. Albritton. 1937. Meteorite Scars in Ancient Rocks. *Field and Lab.* **5**:53–64.

Bucher, W. H. 1936. Cryptovolcanic Structures in the United States. *Internat. Geol. Cong., 16th, Proc.* **2**:1055–1084.

Cohen, A. J. 1962. Central Uplifts and Lunar Craters: II. Megashatter Cone Mechanism for Ray Formation. *Jour. Geophys. Research* **67**:1632.

Cohen, A. J.: T. E. Bunch; and A. M. Reid. 1961. Coesite Discoveries Establish Cryptovolcanics as Fossil Meteorite Craters. *Science* **134**:1624–1625.

Dence, M. R. 1968. Shock Zoning at Canadian Craters: Petrography and Structural Implications. In French and Short, 1968, pp. 169–184.

Dence, M. R. 1972. The Nature and Significance of Terrestrial Impact Structures. *Internat. Geol. Cong., 24th, Proc.* **15**:77–89.

Dietz, R. S. 1960. Meteorite Impact Suggested by Shatter Cones in Rock. *Science* **131**:1781–1784.

Eggleton, R. E., and E. M. Shoemaker. 1961. Breccia at Sierra Madera. *U.S. Geol. Survey Prof. Paper 424D,* pp. 151–153.

French, B. M., and N. M. Short, eds. 1968. *Shock Metamorphism of Natural Materials.* Baltimore, Md.: Mono Book Corp.

King, P. B. 1930. Geology of the Glass Mountains, Texas, Part I: Descriptive Geology. *Texas Univ. Bull. 3038,* 167pp.

Lowman, P. D. 1965. Magnetic Reconnaissance of Sierra Madera, Texas, and Nearby Igneous Intrustions. *New York Acad. Sci. Annals* **123**(article 2):1182–1197.

Roddy, D. J. 1968a. Comet Impact and Formation of Flynn Creek and Other Craters with Central Peaks. *Am. Geophys. Union Trans.* **49**:272 (abstr.)

Roddy, D. J. 1968b. The Flynn Creek Crater, Tennessee. In French and Short, 1968, pp. 291–322.

Shoemaker, E. M., and R. E. Eggleton, 1964. Re-examination of the Stratigraphy and Structure of Sierra Madera, Texas. In *Astrogeologic Studies Annual Progress Report, Aug. 25, 1962–July 1, 1963,* U. S. Geol. Survey Open Field Report, Part B, pp. 98–106.

Shrock, R. R. 1937. Stratigraphy and Structure of an Area of Disturbed Ordovician Rocks near Kentland, Indiana. *Am. Midland Naturalist* **18**:471–531.

Van Lopik, J. R., and R. A. Geyer. 1963. Gravity and Magnetic Anomalies of the Sierra Madera, Texas, Dome. *Science* **142**:45–47.

Zimmerman, R. A., and G. C. Amstutz. 1965. The Polygonal Ring Structure at Decaturville, Missouri: New Tectonic Observations. *Neues Jahrb. Mineralogie Monatsh.* **9**:288–307.

28

Reprinted from pages 502–503 of *Jour. Geology* **67**:496–505 (1959)

SHATTER CONES IN CRYPTOEXPLOSION STRUCTURES (METEORITE IMPACT?)

Robert S. Dietz

[*Editor's Note:* In the original, material precedes and follows this excerpt.]

KENTLAND DISTURBANCE, INDIANA

A large quarry near Kentland, Indiana, exposes the center of a well-known localized area of intensely deranged Paleozoic beds in a region of flat-lying strata. Most writers agree (e.g., Bucher, 1933; Shrock, 1937) that the structure was formed by a violent natural explosion which they presume to be cryptovolcanic. Violent shock is indicated by the jumbling of the strata, the shattering of the limestones, and the pulverization of the St. Peter Sandstone into a rock flour resembling the comminuted Coconino sandstone at Meteor Crater. The Kentland structure is mainly exposed only in the large McCray quarry, so that the over-all form of the structure is unknown. However, Pullen's (1953) radio-wave transmission investigation suggests that it has a roughly circular shape and is about 3 miles across, with the McCray quarry in the middle.

Shatter cones are prominently developed in the excellently exposed limestones (pl. 5). Along the north face of the quarry, the ver-

PLATE 5

Shatter cones developed in the limestone quarry which exposes the central portion of the Kentland cryptoexplosion structure.

A, Two large shatter cones, both oriented normal to the vertically dipping Ordovician strata. The apexes point toward the tops of the beds.

B, A close-up showing the surface details of a large shatter cone about 6 feet in length. This is an exceptionally large example of such a cone and, to the writer's knowledge, is the largest ever photographed. Note that the surface is covered with small striated and imbricated parasitic half-cones which point in the direction of the primary cone.

C, A group of about five small shatter cones pointing in the same direction on a loose limestone block.

Shatter cones

tically dipping limestone strata appear to be only moderately fractured, but, when tapped with a hammer, they tend to break into fragments of shatter cones. Although usually only 10–15 cm. long, some of the cones attain lengths of 2 meters (pl. 5, *B*). If well developed, they consist of a large primary cone with a grooved surface and with small secondary, overlapped, and imbricated half-cones with their apexes pointing toward the tip of the primary cone.

On a visit to the quarry in 1945 the writer noted a preferred orientation—the apexes were pointed normal to the bedding and toward the top of the beds. It was therefore suggested (Dietz, 1946) that the deforming stress must have come from above, as in a meteorite impact, rather than from below, as would be expected from a cryptovolcanic explosion. (It would, of course, be possible to consider that the steam explosion occurred above the shattered rocks observed, but this would not account for their uplifted position.) However, such preferred orientation has subsequently been questioned by some of the writer's colleagues (e.g., R. Gutschick, personal communication) and is thus open to some doubt. It should be pointed out, however, that the determination of the orientation requires more than a cursory examination, since there are many pitfalls. Upon the examination of a loose block of limestone, the slickensided surfaces appear to run in all directions. These surfaces are due to complex overlapped and interlaced large cones, so that little can be definitely established from a hand specimen. Many of the cones observed on a hand specimen may be secondary ones, i.e., parasitic on the surface of a large cone; their orientation changes with respect to their position on the primary cone (pl. 5, *B*). Further, shatter cones are a cup-and-cone feature, consisting of a mold and a cast. The mold or cup surfaces, which require careful scrutiny for certain identification, can easily be interpreted as pointing in a direction opposite from the true orientation of the cone. Such a false impression of cone orientation can be seen on plate 3, *C*.

[*Editor's Note:* Plate 3 is not reproduced here.]

29

Reprinted from pages H1-H4, H6, H9-H19, H21-H23, H25-H26, H29-H30, H32-H33, H36-H39, H40, H41-H42 of *U.S. Geol. Survey Prof. Paper 599H,* 1972, 42pp.

GEOLOGY OF THE SIERRA MADERA CRYPTOEXPLOSION STRUCTURE, PECOS COUNTY, TEXAS

By H. G. Wilshire, T. W. Offield, Keith A. Howard, and David Cummings

ABSTRACT

The Sierra Madera structure is a bowl- or funnel-shaped body, about 8 miles in diameter and 6,000 to 8,000 feet deep, of intensely deformed sedimentary rocks at the southern edge of the Val Verde basin in west Texas. The deformed rocks are Permian and Lower Cretaceous shelf-facies carbonate strata. The structure is composed of three main parts: a central uplift, about 5 miles across, in which the oldest rocks have been raised 4,000 feet above their normal position; a surrounding structural depression, about ½ to 1 mile wide, that is floored mainly by Lower Cretaceous strata; and a concentric structurally high rim, about half a mile wide, in which Lower Cretaceous rocks are locally folded and cut by concentric normal faults downthrown toward the center.

The intensity of folding and faulting increases inward from the flanks of the central uplift toward a central zone, about a mile in diameter, where dips and fold plunges are near vertical or overturned. Individual beds are repeated by folding and faulting, so that their total strike length is greater than the length of the perimeter on which they lie, indicating both inward and upward movement of the strata forming the uplift. This movement pattern is substantiated by a thickened section in the center where only 1,200 feet of the oldest strata with generally steep dips fill an area at least a mile wide. Drill data show that this thickening is not the result of domical arching of the oldest beds but is instead, caused by repetition of beds resulting from centripetal movement.

The structural depression probably resulted from tectonic thinning accompanying the inward movement of rocks to form the central uplift. Folding and uplift in the outer rim may have been produced by outward-directed compression early in the deformational event.

Monolithologic and mixed breccias, shatter cones, and certain types of internal structures of minerals at Sierra Madera are ascribed to shock deformation. Monolithologic breccias, composed of shattered but unmixed rocks, are abundant on the central uplift. Generally, they show few signs of shock deformation, but there are indications that some shatter cones formed concurrently. When beds containing shatter cones are restored to horizontal, the cones point inward and upward above the center of the structure. Mixed breccias form intrusive bodies in which fragments from beds as much as 1,700 feet apart occur together. Quartz grains in the mixed breccias have multiple planar elements and cleavages, dominantly parallel to $\{0001\}$, $\{10\bar{1}3\}$, $\{10\bar{1}2\}$, $\{10\bar{1}1\}$, and $\{10\bar{1}0\}$, and abnormally low refractive indices that indicate peak pressures above 200 kb (kilobars). Planar elements in quartz from rocks still in place near the center of the structure record pressures above 100 kb, and those in quartz from rocks on the flanks of the uplift record pressures near 50 kb.

The symmetry of restored shatter-cone orientations suggests that the cones were formed by a shock wave of generally hemispherical form. As the shock wave traveled downward and outward from the central focus, shock intensity progressively decreased. Only the surface impact of an extraterrestrial body could have produced such a shock wave. This interpretation is supported by the uniqueness of Sierra Madera in an otherwise normal geologic environment, its lack of relation to any regional or local structure, and the geometry of the central uplift, which could not have been produced by a deep subsurface explosive event.

Although the mechanics of formation of the central uplift are still not completely understood, its structure is similar to uplifts in craters formed by surface detonation of TNT (trinitrotoluene) in Canadian cratering experiments. Analogy with experimental craters and with other cryptoexplosion structures indicates that the uplift at Sierra Madera protruded into a crater, since destroyed by erosion, that was about 8 miles in diameter. Inasmuch as the rocks forming such uplifts are derived from below the crater floors, analogous lunar craters potentially offer samples of lunar crust uplifted from distances below the crater floor on the order of one-tenth the crater diameter.

INTRODUCTION

Sierra Madera (wooded hills) is a small group of hills about 3 miles in diameter that rise 600 feet above the plains between the Glass Mountains and Fort Stockton in west Texas (fig. 1). The hills are encircled at a distance of about 2 miles by a ring of hills that rises only slightly above the general level of the plains. Between these two features is an annular depression largely covered by alluvium. Six Shooter Draw, an intermittent stream, breaches the outer ring of hills on the southwest and northeast sides and follows the annular depression along a crescent-shaped path around the west side of Sierra Madera. The Sierra Madera structure, about 8 miles in diameter, includes the central hills from which it takes its name, the annular depression, and the outlying ring of low hills.

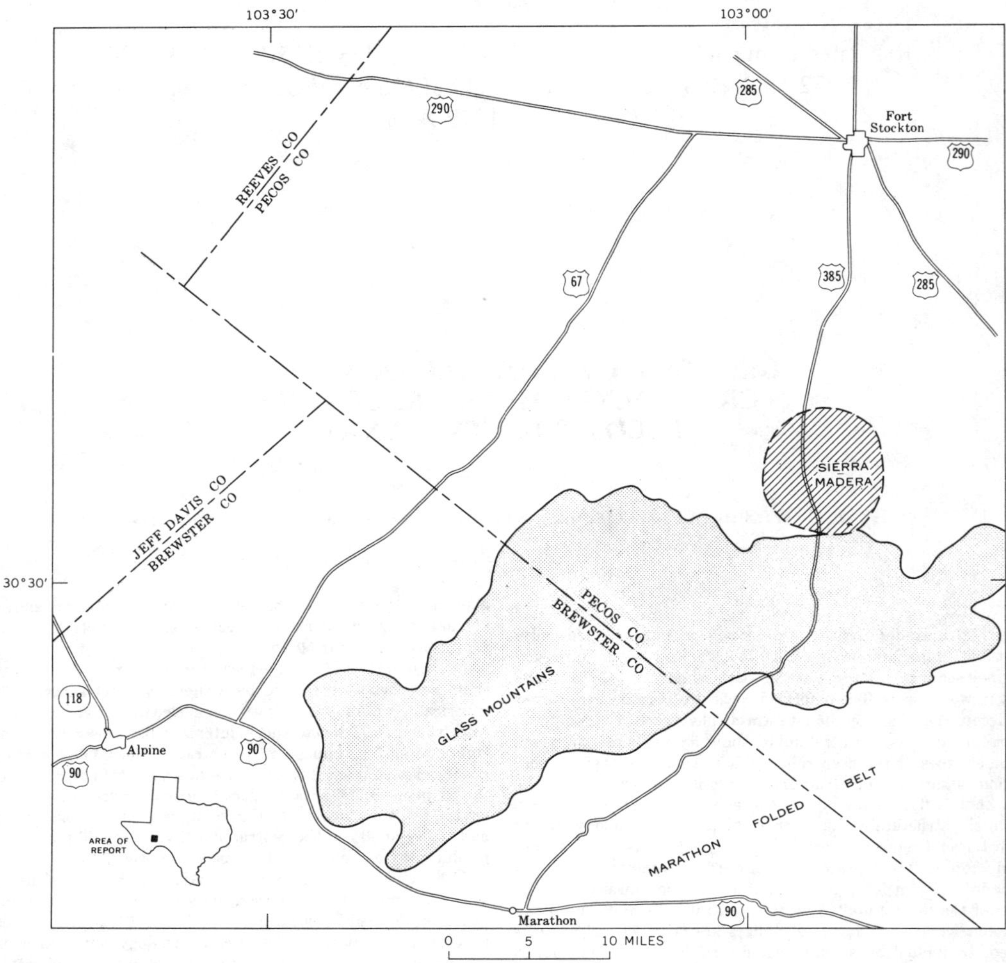

FIGURE 1.—Index map of Pecos and adjacent counties, west Texas, showing location of the Sierra Madera structure, the Glass Mountains, and part of the Marathon folded belt.

The Sierra Madera structure is one of a class characterized by roughly circular areas enclosing a central zone of uplifted and intensely deformed rocks surrounded by a ring of subsided rocks with or without an outlying zone of concentric folds. Bucher (1936) denoted these by the name "cryptovolcanic," originally used for the Steinheim structure (Branco and Fraas, 1905). Subsequent work has shown that such structures are also characterized by the presence of shatter cones (Dietz, 1968) and unusual types of mineral deformation (Short, 1966a), which support the idea, dating back to the work of Branco and Fraas (1905), that these structures formed by explosive release of energy. Whether the source of the energy was terrestrial or extraterrestrial has been the subject of lively controversy for more than three decades (Bucher, 1936; Boon and Albritton, 1936). To describe these structures without bias toward terrestrial or extraterrestrial origin, Dietz (1946) introduced the term "cryptoexplosion structure," to which Bucher (1963) later subscribed.

The current investigation of Sierra Madera was begun in 1960 by E. M. Shoemaker to provide detailed information on the geology of this well-exposed structure. The main impetus of the study was to obtain information on possible terrestrial analogs of lunar structures.

PREVIOUS WORK

The first geologic map of Sierra Madera was made by King (1930) as part of a regional study of the Glass Mountains. He recognized the unusual character of the structure (p. 123), and his comparison (p. 125) of Sierra Madera with the Vredefort ring of South Africa led Boon and Albritton (1937) to their challenging suggestion that these two structures as well as others like them are of meteorite impact origin. King (p. 124) further demonstrated the domical uplift of the center and noted evidence for severe radial compression of the uplifted rocks, both of which are confirmed by our work. His suggestion (p. 125) that igneous intrusion may have played a role in formation of the structure was subsequently disproved by holes drilled for oil and gas.

Shatter cones were discovered at Sierra Madera in 1959 (Eggleton and Shoemaker, 1961, p. 151), and were later studied in detail by Howard and Offield (1968). Shoemaker and Eggleton (1964) found, contrary to King's interpretation, that Lower Cretaceous rocks at Sierra Madera are as severely deformed as adjacent Permian rocks and concluded that the structure was formed by a single episode of deformation some time after deposition of Lower Cretaceous rocks. Eggleton and Shoemaker (1961) described the breccias in the central uplift of Sierra Madera as a cup-shaped mass resembling the breccia zones under Meteor Crater (Arizona) and the nuclear crater Teapot Ess (Nevada Test Site).

Lowman (1965) made a reconnaissance magnetic survey that revealed a magnetic high under the southeast part of the central uplift and tentatively concluded that Sierra Madera was formed by a syenite intrusion. Van Lopik and Geyer (1963) suggested that the aforementioned anomaly may be caused by meteoritic material, but the data are not yet adequate to distinguish among the possible causes of the anomaly.

In most of the foregoing studies, the structure was thought to be confined to the topographic feature of Sierra Madera, 3 miles in diameter, but it is now known to include two other structural elements—a concentric structural depression and a peripheral ring of folds and faults.

ACKNOWLEDGMENTS

This work was done on behalf of the National Aeronautics and Space Administration under contract R–66. We are indebted to R. P. Bryson and V. M. Wilmarth (NASA) for their interest and encouragement. P. B. King, D. J. Milton, D. J. Roddy, O. B. James, and E. C. T. Chao, U.S. Geological Survey, critically read the manuscript and considerably improved this final version by their suggestions. Discussion of the problems at various stages of the project with D. J. Milton, U.S. Geological Survey, was of great benefit to us. We are especially indebted to Addison Young, J. L. Williams, and C. J. Perusek, Phillips Petroleum Co.; G. L. Evans, Louisiana Land and Exploration Co.; V. O. Cook and Y. B. Newsom, Gulf Oil Corp.; and J. K. Somerville, Southern Natural Gas Co.; for their interest and cooperation. The cooperation of E. M. Shoemaker, California Institute of Technology, and R. E. Eggleton and D. P. Elston, U.S. Geological Survey, is gratefully acknowledged. Access to Sierra Madera and assistance in many other ways were graciously rendered by Mr. Peller Matthews, Elsinore Ranch.

STRATIGRAPHY

Sierra Madera is underlain by a sequence of Cambrian to Lower Cretaceous sedimentary rocks about 18,000 feet thick (fig. 2). They presumably overlie Precambrian crystalline rocks, as granite with an age of 900 million years was penetrated by wells about 23 miles east of Sierra Madera (Young, 1960, p. 89), and silicic plutonic rocks were penetrated on the Fort Stockton high to the north (Flawn, 1956). The pre-Permian Paleozoic strata, mainly carbonate and fine-grained clastic rocks, were deformed in Late Pennsylvanian and very early Permian time during the formation of the Val Verde trough. Sierra Madera is situated on the southern edge of the trough (fig. 3; Young, 1960; Oriel and others, 1967). The 7,000 to 12,000 feet of clastic sedimentary rocks of the Lower Permian Wolfcamp Series (fig. 2) near Sierra Madera were deposited in the Val Verde trough, which, in Wolfcamp time, was continuous with the Delaware basin (fig. 3; Vertrees and others, 1959). Later in Wolfcamp time, the Sierra Madera site became part of a shelf on which carbonate rocks formed. The Hess Formation (fig. 2) is transitional between basin and shelf environments and is composed of interbedded clastic and carbonate rocks. Thin, shallow-water, clastic sedimentary rocks of Triassic age unconformably overlie Permian strata near Sierra Madera but probably do not occur at Sierra Madera. Permian rocks at Sierra Madera are unconformably overlain by a conformable sequence of Lower Cretaceous rocks—the Trinity, Fredericksburg, and Washita Groups—deposited during an extensive transgression of the sea from the Gulf region (Oriel and others, 1967). Cretaceous rocks overlap progressively older Permian rocks southward in the Glass Mountains, a fact indicating pre-Cretaceous tilting of the Glass Mountains Permian sequence (King, 1930, p. 127). Lower Cretaceous and older rocks are partly covered by thin unconsolidated Quaternary deposits.

Rocks exposed at Sierra Madera range in age from Early Permian to Early Cretaceous (Comanche Series).

SYSTEM	GROUP	FORMATION		REGIONAL THICKNESS (FEET)	REGIONAL LITHOLOGY	SIERRA MADERA THICKNESS (FEET)	SIERRA MADERA LITHOLOGY
CRETACEOUS	Lower part of Washita	Georgetown Limestone		200-740		30-475	
		Duck Creek Limestone					
	Fredericks-burg	Kiamichi Formation				130-250	
		Edwards Limestone					
	Trinity	Basal Cretaceous sandstone		50-360		50-100+	
TRIAS-SIC		Bissett Conglomerate		0-500		0?	
PERMIAN		Tessey Limestone		400-1460		0-400	
		Gilliam Limestone		590-980		700-980	
		Word Formation	Vidrio Limestone Member	280-1070		200-750	
			Sandstone member			0-35	
			Limestone member			0-250	
		Cathedral Mountain Formation		0-1800		0-80	
		Hess Formation (King, 1930)		1870-2850		1200-1400	
		Wolfcamp Formation (King, 1930)		6995-12,220			
PENNSYL-VANIAN	Strawn			0-850			
MISSIS-SIPPIAN		Barnett Shale equivalent		70-450			
		Novaculite of Kinderhook(?) age		40-180			
DEVO-NIAN		Woodford Shale equivalent		70-255			
		Chert of Devonian(?) age		30-185			
SILU-RIAN		Fusselman Dolomite		15-190			
ORDO-VICIAN		Montoya Dolomite		160-590			
	Simpson			2130-2480			
	Ellenburger			1420?			
CAM-BRIAN				100?			
PRE-CAMBRIAN							

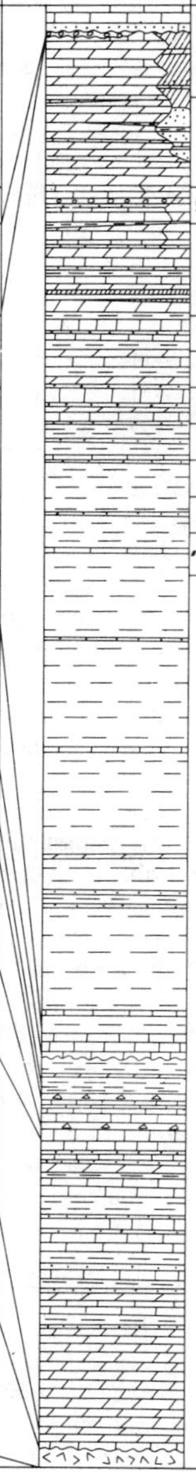

FIGURE 2.—Generalized stratigraphic column of Precambrian to Lower Cretaceous rocks at Sierra Madera (left column) and detailed stratigraphic column of rocks exposed at Sierra Madera (right column).

[*Editor's Note:* Material has been omitted at this point.]

TABLE 2.—*Petrography and paleontology of rocks exposed at Sierra Madera*

[Units arranged from youngest to oldest. Limestones classified according to Folk (1959, 1962), sandstones according to Williams, Turner, and Gilbert (1954). Fusulinids identified by C. A. Ross, Western Washington State College]

Georgetown, Duck Creek, Kiamichi, and Edwards Formations.	Biomicrite, pelmicrite, and microsparite most abundant; local fine to very coarse, sandy calcirudite. Fossils common, not identified.
Basal Cretaceous sandstone	Fine- to coarse-grained quartz arenite with finely to coarsely crystalline calcite cement, locally with microcrystalline quartz cement; quartz has bimodal size distribution with large subrounded to well-rounded grains and small angular to subangular grains; zoned and weak to moderate undulose extinction fairly common; most grains have fluid inclusions, many in planar concentrations; poorly developed cleavages uncommon; a few grains have one to three sets of thin planar discontinuities. Minor detrital constituents include micrite, chert, tourmaline, rutile, and monazite.
Tessey Limestone	Finely to very finely crystalline dolomite and pellet dolomite; variable quantities of sparry calcite, minor detrital quartz and muscovite. Some interbedded pelmicrite. Dolomites commonly brecciated with clasts in aphanocrystalline groundmass. Fossils fragmental and rare, not identified.
Gilliam Limestone	Finely to very finely crystalline dolomite and pellet dolomite, and fine to very fine grained quartz arenite with very finely to finely crystalline carbonate cement. Sandstones moderately to well sorted, quartz angular to subangular; planar concentrations of fluid inclusions common in quartz; undulose extinction uncommon; cleavage and planar discontinuities rare. Common accessories in sandstones are chert, muscovite, microcline, plagioclase, tourmaline, rutile, zircon, and opaque minerals. Fusulinids and gastropods rare, poorly preserved, not identified.
Word Formation	Vidrio Limestone Member; finely to medium-crystalline dolomite, local dolomitized microsparite, biomicrudite, pelmicrite(?). *Parafusulina sellardsi.* Sandstone Member; finely- to coarse-grained quartz arenite with finely to coarsely crystalline carbonate cement, and sandy finely to medium-crystalline dolomite. Quartz moderately sorted, tendency to bimodal size distribution, subangular to subrounded; undulose extinction uncommon; planar concentrations of fluid inclusions common; cleavage and one to three sets of planar discontinuities uncommon. Minor detrital constituents include chert, biomicrite, micrite, and opaque minerals. Limestone Member; dolomitized biomicrudite, biomicrosparite, microsparite, biosparrudite. *Parafusulina deliciasensis, P. wildei, P. sullivanensis.*
Cathedral Mountain Formation	Conglomerate and fine to very coarse grained quartz arenite with finely to coarsely crystalline carbonate cement. Conglomerate pebbles mainly chert and quartzite. Quartz moderately to poorly sorted, subangular to subrounded; undulose extinction uncommon; planar concentrations of fluid inclusions moderately common; cleavage poorly developed, moderately common; planar structures uncommon but form one to three sets where present, except in quartzite grains where one to five sets of such structures are common.
Hess Formation (King, 1930)	Variously dolomitized microsparite, biomicrorudite, biosparrudite, and finely to coarsely crystalline dolomite. *Schubertella melonica, Parafusulina durhami, P. spissisepta, P. brooksensis, P. splendens, Schwagerina hessensis, P. vidriensis.*

STRUCTURE

The Sierra Madera structure consists of (1) a central uplift (fig. 4) about 5 miles in diameter in which the oldest Permian rocks are raised about 4,000 feet above their normal position, (2) a surrounding alluviated structural depression about ½ to 1 mile wide that is floored mainly by Lower Cretaceous rocks, and (3) an outer, structurally high rim about half a mile wide of faulted and folded Upper Permian and Lower Cretaceous strata (pl. 2). Each of these zones is prominently expressed in the topography except on the outer rim at the south and southeast sides. Outside the structure, regional joints are unrelated to the circular form (fig. 5); some joints (northwest and north-northwest strikes) cut mixed breccia within the structure and so postdate it. High-altitude photographs show a nearly complete circular form that is apparently, along the south and southeast, marked by concentric drainage. However, no deformation was found along the concentric drainage channels, so that they are apparently controlled by joints that have no fault offsets. Hence, the limits of the structure are set by the mapped limits of deformation, although future mapping with finer subdivision of Lower Cretaceous strata may show that mild deformation extends beyond the presently recognized boundaries.

FIGURE 4.—View west-northwest from the outer rim of Sierra Madera showing the central uplift across the alluviated structural depression in the middle ground. Photograph, courtesy of Jeanette L. Gillette, San Angelo Standard-Times.

Deep holes drilled for oil and gas (pl. 1) show that the intense deformation observed at the surface dies out rapidly at depth and probably disappears in Upper Wolfcamp strata between 6,000 and 8,000 feet below the surface (fig. 6).

Rocks of the central uplift, especially the Permian rocks, but also locally the Lower Cretaceous strata, are pervasively brecciated. The most common type of breccia (fig. 18) consists of thoroughly shattered rock in which fragments are generally so little displaced that thin beds maintain their continuity. (See fig. 17.) A second type of breccia, occurring in generally crosscutting bodies, contains a mixture of fragments of different lithologies. Examination of cuttings from the Phillips Elsinore 1 well drilled on the center of the structure (pls. 2, 3; Eggleton and Shoemaker, 1961) suggests that both types of breccia become less common downward and are absent about 3,000 feet below the surface.

The gross structure, then, is represented by a bowl-shaped body approximately 8 miles in diameter and 6,000 to 8,000 feet deep, the central part of which has a high structural relief near the present ground surface, with intensely shattered beds that are tightly folded and cut by numerous faults. The intensity of deformation and brecciation decreases outward and downward from this area.

RIM ZONE

The outer rim of the structure consists of faulted and locally folded Upper Permian and Lower Cretaceous strata. There appears to be little or no structural displacement of Cretaceous rocks immediately outside the bounding ring faults at the southwest and south edges of the structure, but Cretaceous strata in the same position relative to circumferential faults along the north and east sides are as much as 100 feet, and perhaps more, above their normal regional level. Both the faults and the fold axes are generally concentric about the central uplift (pl. 2). Folds are best developed from the northeast part of the rim around the east to the southeast corner. Fold axes have a generally low plunge, and limbs generally dip less than 45°. Dips are as steep as 90° in places but do not persist more than a few tens

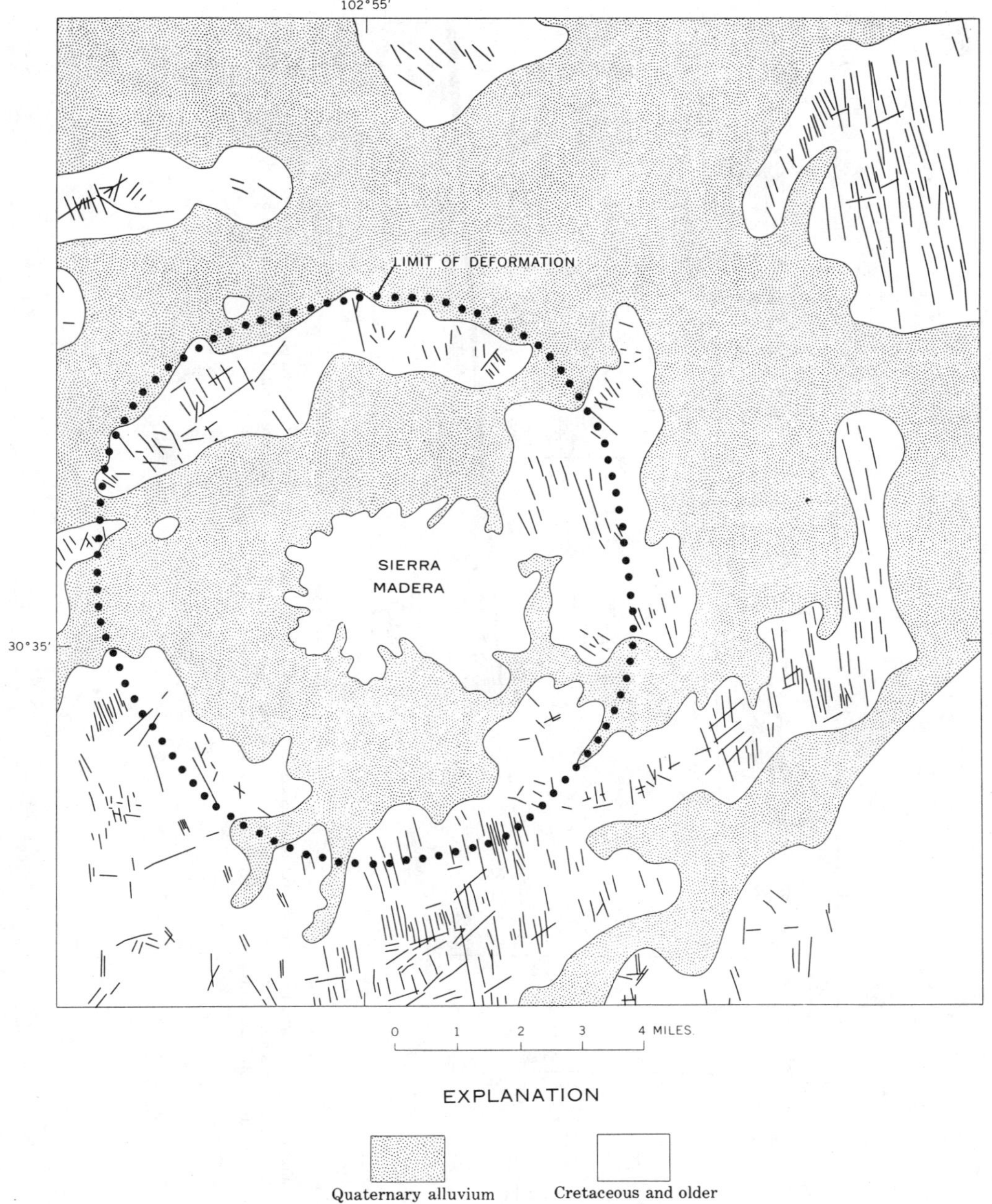

FIGURE 5.—Generalized lineaments in the vicinity of Sierra Madera. Drawn from aerial photographs.

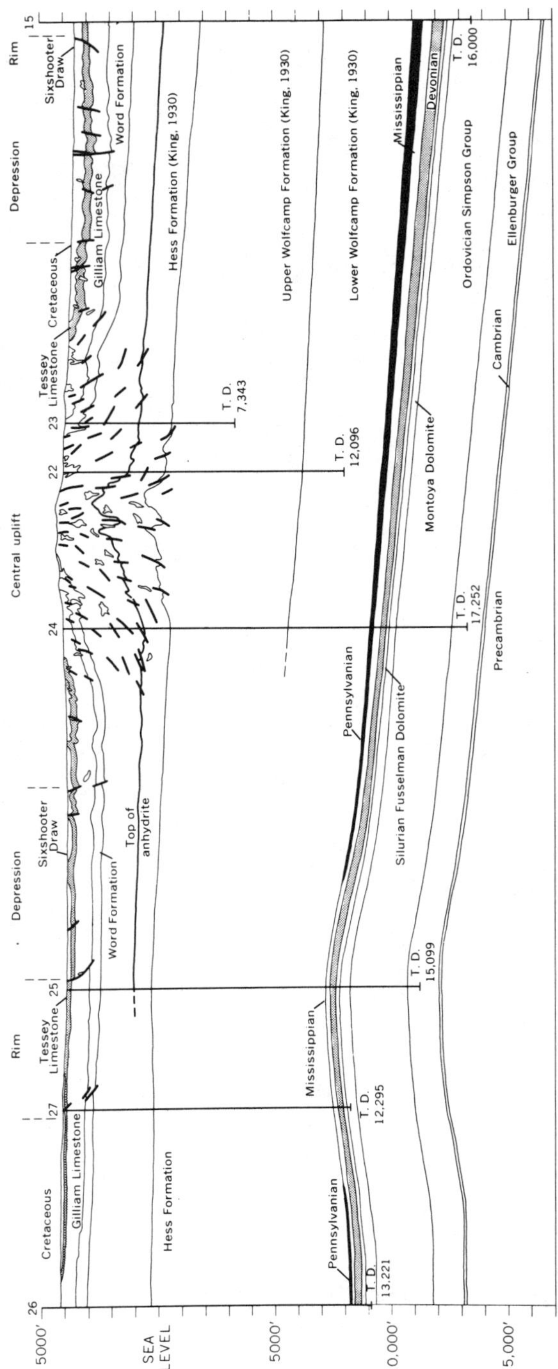

FIGURE 6.—Generalized southwest-northeast structure section across Sierra Madera. Section line connects drill holes shown on plate 2. Wells are: 15, Hunt Oil Elsinore 51; 22, Phillips Petroleum Elsinore 1; 23, Thompson (full name not known) Elsinore 1; 24, Hunt Oil Elsinore 48; 25, Hunt Oil Elsinore 52; 26, Hunt Oil, Fulk 1; and 27, Hunt Oil Elsinore 53. T.D. indicates total depth of well in feet.

of feet. Drag along faults also produced local steep dips. All the main concentric faults in the rim zone are normal faults downthrown toward the center of the structure. Displacements are generally on the order of 50 to 100 feet, but drill data (pl. 2, Hunt Elsinore 19) and exposures suggest structural relief of as much as 700 feet in some places.

The Tessey Limestone and the basal Cretaceous sandstone at their normal stratigraphic positions are exposed in the rim along its southwestern segment (pl. 2; the largest outcrop of the sandstone in the northern part of the rim may also be at its normal position), but both also form small, irregular masses along faults and other concentric fractures within Edwards Limestone. At these places blocks of Tessey dolomite (most of which contain shatter cones) and pods of overlying sandstone occur above their normal levels on both sides of the faults.

RING DEPRESSION

The ring depression separating the outer rim and central uplift is largely buried under alluvium, but drill holes (pl. 1) and exposures on the southeast side (pl. 2) show that it is floored mainly by Lower Cretaceous rocks. The largest exposures as well as the drill data reveal much disturbance, with dips up to vertical and some faulting. The depression is separated from the outer rim by concentric normal faults but is not separated from the central uplift by any significant change in style of deformation.

CENTRAL UPLIFT

STRUCTURAL FRAMEWORK

Drill-hole data and exposures indicate that strata were lifted above their normal regional levels in the central uplift across a width of 5 miles (pls. 1, 2). Inward from the ring depression the rocks are increasingly uplifted until, in the complexly folded and faulted core of the structure, the Hess Formation lies 4,000 feet above its normal level. Folds and faults are more abundant toward the center, and dips and fold plunges steepen, so that they are near vertical to overturned in a central zone about a mile wide. Despite the extreme complexity of faulting, the structure is coherent enough that the term "megabreccia" (Eggleton and Shoemaker, 1961) is inappropriate. Steeply plunging radial folds, in places broken or crushed by faults, dominate the interior. Farther out are concentric as well as radial folds; these are also broken by numerous faults.

The uplift has approximately radial symmetry. Perhaps because of the slight regional northeast dip, the Gilliam-Tessey contact outlines an oval, slightly elongate east-west, and farther in, the top of the Hess outlines an oval elongate to the northwest (pl. 3). Exposures of Hess varicolored beds in the core of the structure indicate that the greatest uplift is slightly northwest of the center of the concentric outcrop patterns. The most extreme variations in deformation around the uplift are in the complex faulting and thickening of the Gilliam on the east side (pl. 3). However, even these variations do not greatly mar the gross radial symmetry, so that the folding and faulting are classifiable into concentric zones which coincide with outcrop belts of the formations. Nevertheless, there are significant relationships between lithology and deformational style, as described below.

STYLE OF DEFORMATION

Response of the various rock units to the deformation varied according to their massiveness and competence. Massive dolomites of the Vidrio Limestone Member, Tessey Limestone, and much of the Hess Formation are slightly less convoluted than adjacent better bedded strata. The Gilliam-Tessey contact, which is fairly straight for considerable strike distances, shows only gentle undulations, and folds outlined by the Vidrio are generally larger and more open than those in the overlying Gilliam (pl. 3). Within the Tessey Limestone, structure is difficult to unravel because much of the formation is a reef breccia with high initial dips or is otherwise poorly bedded. Most observable folds in the Tessey are small and open. This is illustrated on plate 3 by the contrast of open folds in the Tessey and tight folds in the Gilliam in the same area, as at coordinates 350.2, 180.3 and 354.0, 194.2. In this report, all coordinates locating specific features refer to plate 3.

In the Gilliam, prominent bedding and numerous sandstone beds permit delineation of many small folds. Folds range from shallow basinlike synclines (348.3, 183.6; 345.5, 193.5), through somewhat rounded but chevronlike folds (349.7, 183.2), to nearly isoclinal folds (343.3, 183.5). The thickness of sandstone and dolomite beds is not changed significantly by folding, and all the mapped folds appear to have formed by flexural slip. Small-scale folds and parallel boudinage occur in dolomite in a few places. Because faults break many of the folds, one limb (346.7, 184.7; 346.1, 194.0) or the hinge itself (348.2, 196.0) may commonly be replaced. Faulting in the Gilliam is most intense on the east and west sides of the uplift.

Conglomerate of the Cathedral Mountain Formation and the sandstone member of the Word are distinctive key beds for tracing the structure around the subvertical core of the uplift. Their behavior may have been slightly more mobile than that of adjacent carbonate rocks. Both key beds are severely contorted, crushed, and faulted, as plate 3 shows, and in some places small pods occur considerably out of their normal position, as if they had been dragged or even injected. The Cathedral Mountain conglomerate forms boudins several feet across at one place (345.8, 187.7) and locally has small-scale internal folds.

Considerable disharmony in the folding is illustrated by the anticline (fig. 7) at the top of the Vidrio Limestone Member of the Word Formation (352.3, 186.3). This anticline is box-shaped in outline and becomes faulted and broader inward at the base of the Word. Another anticline (348.1, 192.5) is rounded (where not faulted) at the level of the Word sandstone but is angular and faulted below in the Cathedral Mountain and upper Hess. Most folds are crushed (347.1, 185.9) or broken, and some are transitional to faults (346.5, 191.6; 344.8, 189.3) or to mixed breccia along a fault (345.3, 191.0). Fragmentation of the Cathedral Mountain conglomerate and Word sandstone on the west limb of a syncline (344.8, 189.3) shows that some folds were sheared apart by crushing and thinning rather than by clean faulting. Many faults in the carbonate rocks of the uplift are clean breaks with neither gouge nor breccia.

The very complicated map pattern (pl. 3) indicates that the basal Cretaceous sandstone was highly mobile

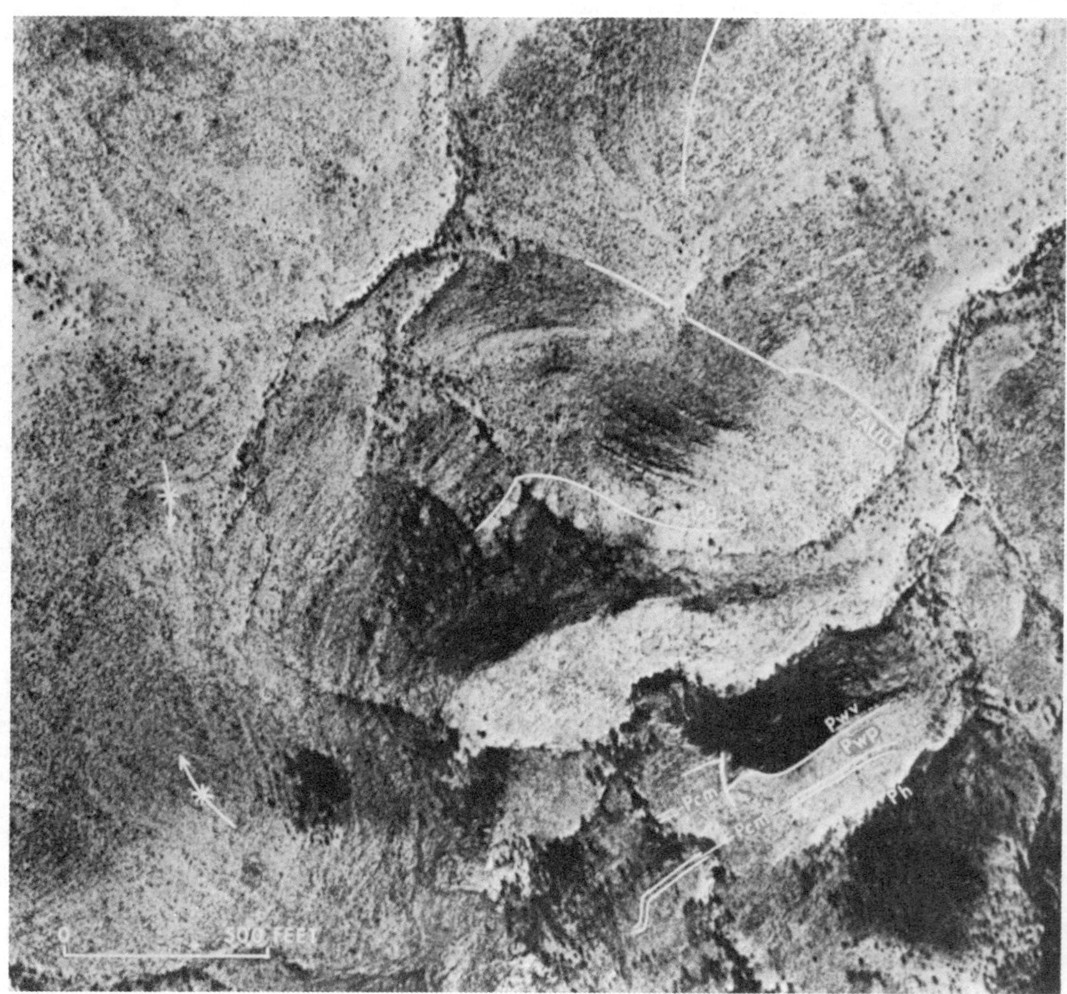

FIGURE 7.—Vertical aerial photograph (pl. 3, coordinates: 352.3, 186.3) of radially plunging anticline in the Vidrio Limestone Member of the Word Formation (Pwv) and Gilliam Formation (Pg). Traces of beds (left and top) in thin-bedded Gilliam outline other folds. Ph, Hess Formation; Pcm, Cathedral Mountain Formation; and Pwl, limestone member of Word Formation.

during the deformation. The sandstone is extensively injected (fig. 8), mainly into overlying Cretaceous limestone but also into underlying dolomite of the Tessey Limestone. It forms pods and irregular dikes from a few sand grains thick to tens of feet across, at considerable distances from its normal stratigraphic position. The sandstone map pattern, compared with those patterns of Tessey Limestone and Edwards Limestone, thus shows only the gross folds. Some injections were themselves folded and perhaps faulted after injection. Original crossbedding (fig. 9) is replaced in the intrusions by intricate flow banding (fig. 10), and isolated pods have lineated or contorted boundaries (fig. 11). Commonly the injections are tightly cemented by silica and are harder than calcite-cemented sandstone from the less-disturbed exposures. The sandstone

FIGURE 8.—Thin dike of basal Cretaceous sandstone (below and approximately parallel to hammer handle) injected into Edwards Limestone.

FIGURE 9.—Undisturbed crossbedding in basal Cretaceous sandstone. Such bedding was entirely reconstituted during injections of sand.

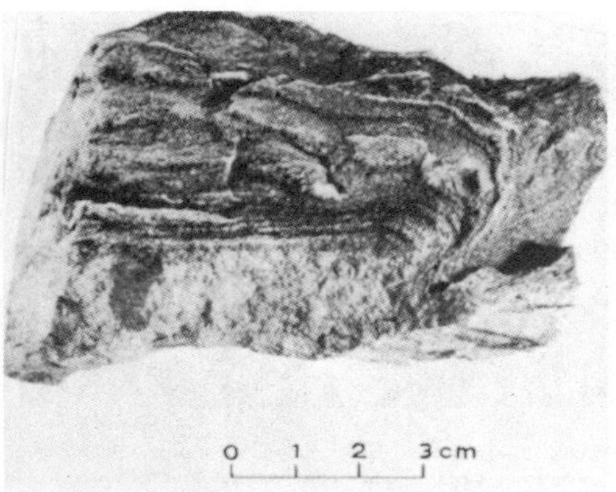

FIGURE 10.—Hand sample of basal Cretaceous sandstone from an injected mass, showing contorted flow banding.

FIGURE 11.—Irregular dike injection of basal Cretaceous sandstone showing flow foliation parallel to dike wall (top of photograph).

is missing from its normal stratigraphic position in places and is several times its normal thickness in others. Its mobility suggests that it was largely unconsolidated when the Sierra Madera structure formed.

Where the basal Cretaceous sandstone is entirely removed from its normal stratigraphic position, typical Edwards Limestone is separated from the underlying Tessey by a few inches to several feet of sandy limestone breccia. In the limestone clasts the laminae are severely contorted (fig. 12) as though the clasts were unconsolidated when incorporated in the breccia. In such places, as well as in other structurally complex areas of Tessey Limestone, basal Cretaceous sandstone, and Edwards Limestone, limestone dikes one-fourth of an inch to 2 feet wide and over 50 feet long occur, one of

FIGURE 12.—Limestone clast in breccia separating Tessey Limestone from Edwards Limestone. Laminae internally contorted in clast.

which contains shock-deformed quartz grains. The dikes and internally contorted limestone clasts in the breccia suggest that Edwards limestones were not fully consolidated at the time of deformation. This condition, combined with the distance of Cretaceous rocks from the center of the structure, may explain the absence of shatter cones in otherwise favorable parts of the Edwards Limestone.

FOLDING

Folds in Tessey Limestone and Cretaceous rocks at the margin of the uplift are mostly a few hundred feet across. Many are quaquaversal domes and basins, so that outcrop belts form concentric patterns a few hundred feet across (pl. 3). Some of the fold plunges are gentle. Others are very steep or even inverted (339.0, 196.0; 343.8, 199.7), but even so, the structural relief is not great, as only about 600 feet of Tessey and Cretaceous strata is exposed across an outcrop width of about 4,000 feet. Circumferential folds predominate in the northeast, and radial folds, in the east and southeast. However, axial trends are not uniform, and axial surfaces are mostly curved. Some folds branch (340.5, 194.9; 342.0, 193.7). The curved axial surfaces, branching folds, and domes and basins may result from interference of radial and circumferential folding, but there is no consistent sequence. Some asymmetric and rare overturned circumferential folds in the Tessey step up toward the core of the structure.

In the Gilliam Limestone most folds are either roughly circumferential or radial to the center of the uplift. Several folds have markedly curved axial traces (353.5, 186.7; 342.3, 187.1), and in several areas folds have different axial trends (351.7, 185.0; 350.0, 183.3). Most axial surfaces dip steeply, and a few folds of circumferential trend are overturned either outward or inward. The dips of axial surfaces of some adjacent synclines and anticlines differ by 20°. If the folds formed before major tilting, their axial surfaces dipped outward. Where asymmetric, the circumferential folds in Gilliam beds step up toward the uplift, suggesting relative inward movement of upper beds over lower. Uplift and tilt were almost certainly under way when the folds were being formed, but many of the folds must have acquired their shape and symmetry at an early phase before they were much tilted.

In the Cathedral Mountain and Word Formations, most folds are radial, though there are a few tight circumferential folds (345.5, 190.8). The inward transition from mainly circumferential folds in the Gilliam to radial folds is well shown on plate 3. Radial folds in the Word Formation plunge steeply, so that the map pattern approximates the down-plunge profile. Contrary to what might be expected from the northeastward regional dip, folds on the south side of the structure generally plunge more steeply than those on the north; some plunge at angles greater than vertical (345.0, 188.2; 346.9, 191.3). Broad anticlines on the south side are separated by extremely tight, faulted synclines where Cathedral Mountain conglomerate was squeezed deep toward the center of the uplift. Such folding indicates that these strata were crowded together more than rocks farther out on the flanks. A few steeply plunging, near-circumferential folds (345.3, 191.0; 347.5, 191.7; 347.1, 185.9) may reflect such crowding of folds that were originally more radial.

Structural details within the exposed Hess Formation are poorly known because of the scarcity of marker beds and the rapid facies changes. Bedding attitudes and form lines indicate considerable disharmony and complexity. Intricate repetition is such that steeply dipping beds representing only 1,200 feet of the Hess Formation fill an area at least a mile wide. Varicolored beds near the base of the exposed sequence form convolute outcrops or discontinuous fault slivers in which the direction that the beds face is mostly uncertain. The concentric outcrop patterns apparently express steep, faulted, spinelike basins and domes (as at 349.0, 187.8; 349.6, 187.8). A complexly faulted and brecciated exposure of Cathedral Mountain surrounded with steep contacts by the Hess Formation may represent one such basin, west of the center of the structure (349.5, 186.5). The spinelike folds in the core of the central uplift are perhaps analogous to those in the core of the Grand Saline salt dome that were described by Muehlberger (1959).

FAULTING

In upper Permian and Cretaceous rocks on the outer flank of the uplift faults are few in number and generally of small displacement. The intensity of faulting increases inward dramatically. It is complex in the Gilliam Limestone (pl. 3). In the core of the uplift the discontinuity of varicolored beds of the Hess Formation demonstrates relatively great displacements along steep faults, but nowhere great enough to expose the anhydrite beds that occur a few hundred feet below the varicolored beds in the undisturbed stratigraphic section. Throughout the structure many of the faults have no systematic pattern but appear to have formed irregularly as late adjustments to folding.

Most faults dip steeply. Curved fault traces in Word and older rocks mostly indicate not topography but curvature which was either original or caused by later folding. Low-angle faults occur in only a few places.

Along subradial faults, true directions of slip are generally unknown, but the displacement would be least if the slip were mainly perpendicular to bedding. Thus, strike slip is thought to characterize most of the subradial faults that cut steep beds in the uplift (with some exceptions, at 346.3, 191.2; 344.8, 190.7). The most prominent example is a fault in the northeast that offsets the Word Formation over a thousand feet right-laterally. Imbrication of Word and Cathedral Mountain Formations across subradial faults is thus ascribed, like the steeply plunging folds, to crowding caused by centripetal inward movement. If formed before much tilting, some subradial faults may initially have had mainly dip-slip displacements, now rotated into a strike-slip geometry. Faults on the uplift flank that bound folds of Vidrio protruding into Gilliam (346.8, 194.8; 346.1, 194.0) are best explained by such dip-slip displacement.

Circumferential faults are best seen in the Gilliam outcrop belt; most have dip-slip displacement. However, one with strike-slip displacement is suspected where the axial trace of a large syncline is offset (346.7, 185.6). The circumferential faults generally dip more steeply than bedding; thus, if they have been tilted, they originally dipped outward. Along most of the circumferential faults, strata are cut out rather than duplicated, and the center of the uplift is displaced upward relative to the flanks. This relationship is opposite to the sense implied by asymmetric folds. At one locality such a fault cuts out concentric isoclinal folds (345.4, 190.8) and therefore is later. Thus, repetition by folding due to inward movement of upper beds over lower appears to have been followed by upward movement of the uplift core pulling away from the flanks by faulting.

Very few thrust faults are proved, contrary to an earlier conclusion based on mapping of a small area (Shoemaker and Eggleton, 1964). One flat-lying fault displaces the Word Formation over the Hess Formation. Other thrusts displace older rocks outward over younger ones on the north (fig. 13) and east (349.0, 196.8; 347.9, 194.2) sides of the uplift. The northern thrust (fig. 13), which displaces Vidrio on overturned Gilliam beds, is bounded on the west side by a strike-slip fault that curves around behind to become a major normal fault bounding the thrust block as if it were a large outward-directed slump. The thrusts suggest that the Vidrio rose until outward movement became easier than vertical movement. The situation may be analogous to the formation of a mushroom cap when a diapiric mass completes its penetrative rise.

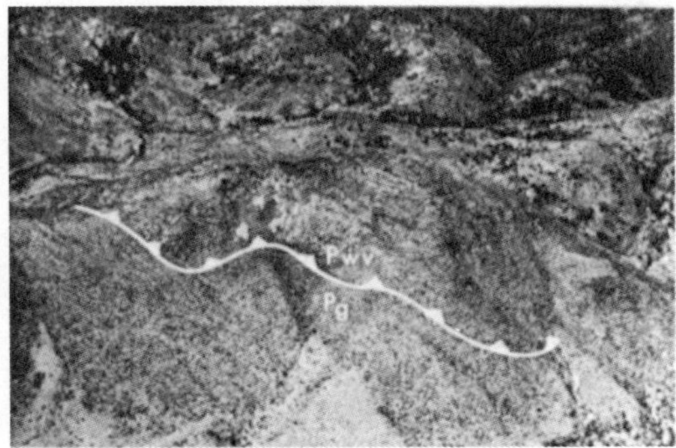

FIGURE 13.—Oblique aerial photograph showing part of low-angle thrust with the Vidrio Limestone Member of the Word Formation (Pwv) structurally overlying the Gilliam Limestone (Pg) (pl. 3, coordinates 351.4, 187.3).

SUMMARY OF CENTRAL UPLIFT

The geometry of the central uplift suggests that it formed by inward as well as upward movement of strata (Wilshire and Howard, 1968). Individual beds, exemplified by the Cathedral Mountain Formation, are duplicated by faulting and folding so that the total strike length of the beds is greater than the length of the perimeter on which they lie; the perimeter at the stratigraphic level of the Cathedral Mountain was evidently shortened about 25 percent by inward movement during doming. Moreover, the section in the center is thickened by repetition so that near-vertical beds representing only 1,200 feet of Hess strata fill an area with a minimum width of a mile, and the Phillips drill hole near the center of the structure (pl. 2) shows that the thickening persists below the surface and is thus more than superficial. The rocks are not sufficiently bulked by brecciation to otherwise account for the volume of uplifted rocks, so the fact that the structure narrows and dies out at depth requires that the uplift volume be accounted for by inward movement of the strata involved.

Faulting and radial folding are believed to be a consequence of this inward and upward motion. Just as with a napkin being pulled up through a ring or with salt flowing into a diapir (Balk, 1949), crowding together of strata by this centripetal movement produces radial folds that plunge down the flanks (fig. 14), and produces subradial faults. The intense faulting, as compared with internal structures of salt domes (Balk, 1949), surely results in part from the greater rigidity of the rocks at Sierra Madera but also suggests that deformation was rapid.

The circumferential folds and complex branching, crossing, and domical folds that dominate the flanks of the uplift suggest irregular adjustments to compression and inward movement of upper beds over lower, as if centripetal motion of the uppermost beds had dragged lower strata inward and upward. Circumferential faults then formed that have an opposite sense of movement, indicating that the center of the uplift was pulling upward away from the flanks. Spinelike faulted folds may have formed in the uplift core at this stage. The few outward-directed thrusts may have formed late in the sequence by gravitational spreading of the rising dome.

The ring depression formed in response to tectonic thinning accompanying the inward movement of rocks to form the central uplift. This feature resembles the rim synclines which surround many salt domes (Nettleton, 1955; Parker and McDowell, 1955). Thus, in several respects, Sierra Madera resembles diapiric structures, but, for reasons discussed later, diapirism is not a tenable explanation for the structure. The inward motion that formed the uplift is also recorded by the normal faults that drop rocks down from the rim into the ring depression. It is unknown, however, whether downdropping along these faults caused the inward motion and the uplift or whether the uplift caused the rim faults by withdrawal of material from the ring depression.

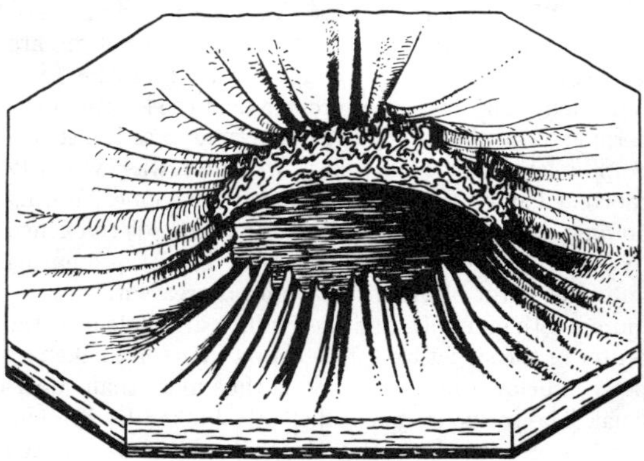

FIGURE 14.—Diagram (from Balk, 1949) illustrating radial folds on top of a salt bed formed by centripetal inward movement into a salt dome; pattern is comparable to folding in the central uplift of Sierra Madera. The cut along the hypothetical vertical cylinder shows inferred internal structure of the salt.

SHOCK DEFORMATION

Brecciation, shatter cones, and certain types of internal structures of minerals at Sierra Madera are ascribed to shock deformation. There are two types of deformational breccia: one composed of fragments of only one lithology, the other of fragments of several lithologies. Shatter cones, a very common mode of rock failure in cryptoexplosion structures, are well developed in Permian rocks at Sierra Madera. Within the area in which breccias and shatter cones occur, quartz grains commonly have cleavages and other planar elements characteristically formed by shock. Calcite and dolomite are also deformed but were not examined closely.

MONOLITHOLOGIC BRECCIAS

The more common of the two types of deformational breccia is composed of only a single lithology. The distribution of well-developed monolithologic breccias is shown in figure 20, but less obviously brecciated rock is considerably more widespread on the central uplift than illustrated. Lower Cretaceous rocks are only rarely brecciated (fig. 15), consistent with the suggestion made before that they were not fully consolidated at the time of deformation.

Except for original sedimentary breccia in the Tessey Limestone (well displayed in the Glass Mountains), brecciation at Sierra Madera suggests intense shattering and is unique to the region. In places, the deformational brecciation is clearly superposed on Tessey sedimentary brecciation.

Incipient brecciation in Permian rocks at Sierra Madera is marked by widely spaced, thin (generally

FIGURE 15.—Monolithologic breccia of basal Cretaceous sandstone.

less than 0.5 millimeter), irregular veinlets of mylonite. An electron micrograph indicates that the mylonite is composed of very tightly packed, angular grains of carbonate a few microns long. In a more advanced stage of brecciation, mylonite veinlets, less than 0.5 mm thick, form an intricate intersecting network that separates unrotated clasts as much as 2.5 cm long (fig. 16). Clasts more than a few millimeters long are internally shattered, and the cracks filled with mylonite. The mylonite in such breccias does not weather out differentially from the clasts, and the rocks generally are not recognizable as breccias in the field. The most advanced stage of brecciation is represented by rocks in which the mylonite weathers out differentially, so that the clasts are conspicuous in outcrop (fig. 17).

FIGURE 16.—Photomicrograph of intermediate stage of monolithologic brecciation showing unrotated clasts in network of mylonite veinlets.

FIGURE 17.—Monolithologic breccia of dolomite from the Gilliam Limestone. These thoroughly shattered beds grade laterally (outside the photograph) to unbrecciated dolomite with no change in form and smoothness of bedding-plane partings.

TABLE 3.—*Partial chemical analyses of clasts and matrices of monolithologic breccias*

[Analyst: J. A. Thomas. Ca and Mg by colorimetric titration, Fe by atomic absorption. Results in percent except for molal ratio]

Formation or Member	Ca	Mg	Molal ratio	Total carbonate	Fe_2O_3
Tessey:					
Clast	21.7	12.7	1.04	98.4	0.195
	22.5	12.4	1.10	99.3	.330
Matrix	21.6	12.2	1.07	96.3	.197
	22.7	11.8	1.17	97.7	.445
Gilliam:					
Clast	21.6	12.9	1.02	98.8	.130
	20.7	12.3	1.02	94.5	.175
	22.4	12.2	1.11	98.3	.225
Matrix	22.1	12.8	1.05	99.7	.182
	21.8	12.7	1.04	98.6	.210
	21.8	11.8	1.12	95.4	.225
Vidrio:					
Clast	21.8	12.2	1.08	96.8	.100
Matrix	22.7	12.2	1.13	99.1	.100

Staining of cut slabs and partial chemical analyses (table 3) of mylonite and clasts indicate very small compositional differences. In well-bedded rocks such as the dolomites in the Gilliam Limestone, pervasive brecciation had no effect on the spacing and smoothness of bedding-plane partings (fig. 17). Clast sizes range from a few centimeters down to grain size of the original rock (fig. 18), and the size distribution appears to vary randomly, both vertically and horizontally in individual beds. As in the intermediate stage of brecciation, clasts larger than a few millimeters are typically shattered (fig. 20) and the cracks are filled with mylonite. Mylonite between clasts makes up as much as 30 to 40 percent of such breccias and is commonly laminated

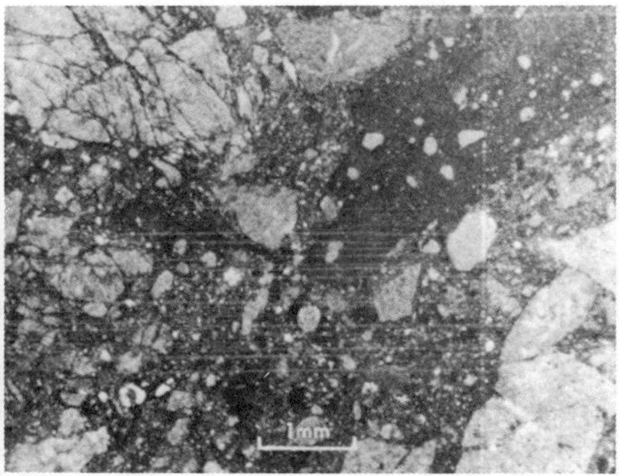

FIGURE 18.—Photomicrograph of monolithologic breccia of dolomite from the Tessey Limestone showing extreme clast-size variation, extensive internal shattering of clast (top left), and thin finger (diagonal from top right) of darker, finer grained mylonite with internal flow foliation.

parallel or subparallel to clast boundaries; small elongate rock fragments are also oriented parallel to larger clast boundaries (figs. 18, 19). Since the mylonite veins and clasts are variously oriented in a particular rock, such laminations (caused by grain-size variations) and fragment orientations must have been produced by flow of the mylonite. Preservation of original bedding-plane partings, however, indicates that only small movements are required for the size sorting and grain orientation.

Tessey sedimentary breccias are characterized by a homogeneous matrix in which detrital quartz is concentrated and by little change in grain size between clasts and matrix, so that clast boundaries are vague. The matrix generally has a moderate proportion of exceedingly fine-grained carbonate(?) along with more abundant carbonate of nearly the same grain size as in the clasts. In contrast, the deformational breccias are characteristically flow-banded and vary greatly in grain size; clasts are sharply bounded and internally shattered, and detrital quartz is uniformly distributed in clasts and matrix. These features suggest that the deformational breccias formed by shattering and cataclastic flow. Of the several types of breccias in carbonate rocks described by Blount and Moore (1969), only the fault breccias bear much resemblance to those at Sierra Madera. Most monolithologic breccias at Sierra Madera, however, are clearly unrelated to faults (fig. 20).

Monolithologic breccias (fig. 20) occur generally within the zone of shatter cones. (See fig. 30.) Generally these breccias contain only trace amounts of quartz, which is undeformed; carbonate minerals are deformed in some monolithologic breccias but not in others. The physical characteristics of the monolithologic breccias suggest tensile shattering, perhaps caused by the rarefaction phase of a shock wave.

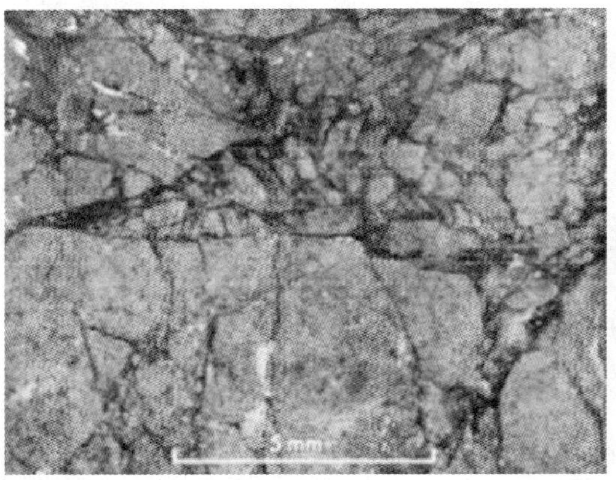

FIGURE 19.—Photomicrograph of Vidrio monolithologic breccia showing internal shattering of clasts.

MIXED BRECCIAS

The second type of deformational breccia at Sierra Madera is composed of clasts of two or more lithologies, either from the same or different formations. Clasts of monolithologic breccias also are present. Mixed breccias occur in all formations on the central uplift, but they form considerably larger masses and are more abundant in the older rocks (fig. 20). These breccias form steeply dipping tabular sheets, 1 to 150 feet thick and 100 to 1,700 feet long, and irregular masses as large as 1,100 feet across. Most of the mixed breccias appear to crosscut the country rock at steep angles, but gradations to monolithologic breccia are locally present. Some tabular sheets of breccia were clearly emplaced along faults, whereas others were emplaced along bedding planes that now dip steeply. In one place (353.1, 189.5), the top of a thin steeply dipping sheet of breccia connects with a large irregular breccia mass; the latter appears to be approximately concordant with local bedding in the Gilliam, as though it were a sill with a dike feeder. Other irregular breccia masses, however, clearly crosscut the country rock and show small-scale intrusive relations to it (fig. 21). At one locality the country rock at its contact with mixed breccia is slickensided, and striae plunge nearly down the 75° dip of the contact.

Steeply dipping flow foliation is locally conspicuous, especially in tabular sheets of breccia but also in some irregular masses. The flow foliation is expressed by orientation of elongate and slabby rock fragments and by size sorting of clasts to produce a repetitive layering parallel to the walls of the breccia masses. Locally, there is a crude size sorting of clasts, such that generally smaller ones are concentrated near the breccia margins, and one thin breccia sheet has a prominent central concentration of large clasts (fig. 22). These relationships indicate emplacement of the breccia masses as dense suspensions of clasts and finely comminuted rock debris, probably in CO_2-rich water or water vapor. The tendency for axial migration of solid particles under possibly comparable conditions of flow is well known (Simkin, 1967), and Wilshire (1961, 1969) has suggested that repetitive size-sorted layers may form by shear flow in dense suspensions of particles in fluid. In one place, very fine-grained sandstone veinlets crosscut marginally flow-foliated breccia. The veinlets may represent late injection of softened and disaggregated inclusions into fractures in the breccia. (Similar phenomena in diatremes were described by Wilshire, 1961.)

The mixed breccia is hard, dense rock, difficult to break with a hammer. The cement of the clasts is

[*Editor's Note:* Figure 20 has been omitted.]

FIGURE 21.—Small intrusion of mixed breccia (light, bottom of photograph) into steeply dipping sandstone unit of Gilliam Limestone. Two thin dark bands in sandstone are truncated by apophysis of mixed breccia. Dolomite in the Gilliam at and above hammerhead.

FIGURE 22.—Sample from thin dike of mixed breccia showing central concentration of large clasts.

typically a rust-colored, exceedingly fine-grained carbonate in which individual grains cannot be clearly resolved with the microscope. The cement may represent slightly recrystallized rock powder rather than a later chemical deposit.

Wherever mixed breccias occur, the dominant clasts were derived from the immediately adjacent formation. Many of them, however, also contain clasts derived from other formations (fig. 20). Because of the general similarity of lithologies, only fragments of such formations as the Cathedral Mountain, the sandstone member of the Word, and the Cretaceous basal sandstone could be consistently identified as foreign. Foreign fragments range from common to rare; in one example, the breccia consists almost entirely of broken red dolomite from the Gilliam derived from immediately adjacent beds with a few scattered quartz and chert pebbles from the Cathedral Mountain Formation, separated about 1,000 feet stratigraphically from this part of the Gilliam. Figure 20 shows that fragments derived from formations other than the one in which the breccia occurs have moved both upward and downward stratigraphically for stratigraphic distances of as much as 1,700 feet. The direction and true distance of movement of such fragments, however, are not known.

A common clast in mixed breccias in the Gilliam and older rocks is white silty claystone. The clay, which is optically nearly isotropic, was identified as kaolinite by X-ray. Some such clasts are deformed (fig. 23). No rocks of this lithology are exposed at Sierra Madera, and their source is unknown.

Another common constituent of mixed breccias is unbroken, commonly fossiliferous, chert nodules. That these are products of replacement of limestone clasts after brecciation is clear from the gradation between chert with small, irregular limestone cores to very thin chert rinds on limestone clasts. Associated with these chert nodules in a few mixed breccias are octahedral crystals as much as 1½ inches across, some of which show penetration twins. These crystals were probably originally fluorite but are now entirely replaced by very fine grained quartz and carbonate. Silification of limestone clasts and fluorite as well as cementation of the breccias by fine-grained carbonate indicates that the mixed breccias were considerably modified after they were emplaced. Alteration was probably accomplished by ground water or by fluids in which the comminuted rock was carried.

FIGURE 23.—Photomicrograph of deformed foliated claystone clast in mixed breccia.

Another type of chert clast, found only in the mixed breccia near the center of the structure at the Phillips well site (pls. 2, 3), is of special interest because of its well developed flow structure (fig. 24). Seventeen such specimens were collected from the breccia, some of which have poorly preserved fusulinids that may have been derived from the Word Formation (R. C. Douglass, written commun., 1969). If derived from the Word, such fragments are below their normal position in breccia composed mainly of the adjacent Hess strata. The clasts are composed almost entirely of fine-grained quartz, but voids conspicuous in some specimens are locally filled with calcite and rare fluorite. In hand specimen, the fusulinids (up to 1 inch long) range from undeformed to almost unrecognizable forms smeared out in a foliation; probably at least part of the foliation is a result of thoroughly smeared fusulinids. The foliation is generally curved, and linear streaking is also conspicuous, so that specimens resemble stretched taffy or volcanic bombs. Thin sections (fig. 25) show clearcut evidence of fluidal flow, despite complete recrystallization to fine-grained quartz mosaics. The larger quartz grains typically have a pronounced zoned or radial extinction that suggests recrystallization of spherulites. In contrast to other quartzose clasts in the mixed breccia, the quartz and calcite of these flow-foliated clasts contain no shock-formed planar elements. Taken together, these features suggest that such clasts may have been melted and subsequently recrystallized. However, these characteristics could also result from silicification or recrystallization of an unmelted mylonite, and somewhat similar textures were produced experimentally without melting (Carter and others, 1964).

Mixed breccias occur in all formations on the central uplift (fig. 20); thus their distribution roughly coincides with that of monolithologic breccias and shock-

FIGURE 24.—Strongly foliated, melted(?), and recrystallized chert clast from mixed breccia.

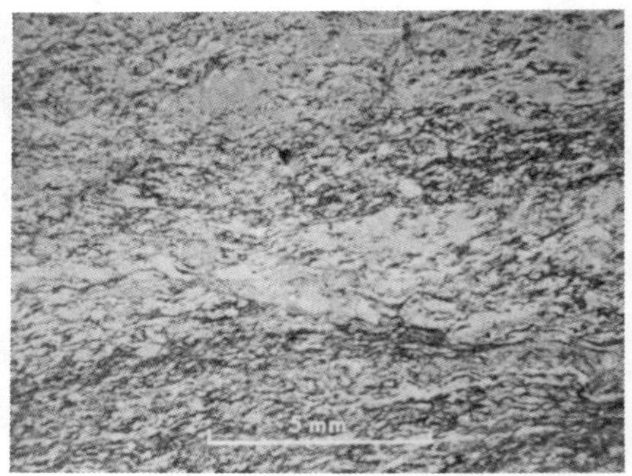

FIGURE 25.—Photomicrograph of foliated chert clast showing fluidal flow features. Light areas are more coarsely recrystallized quartz.

deformed quartz. (See fig. 31.) Unlike the monolithologic breccias, the mixed breccias contain quartz that has been subjected to intense shock deformation. The occurrence of some mixed breccia bodies along faults and the common occurrence of clasts of monolithologic breccia and broken shatter cones in these breccias suggest that much of the mixed breccia was emplaced during folding and uplift. Because of folding and faulting, foreign fragments in mixed breccias may have moved obliquely for shorter distances than the stratigraphic separation of the source bed of a foreign fragment and the formation in which the host breccia is now found. The occurrence together of foreign fragments derived from stratigraphically both above and below the formation in which the breccia is found suggests that probably very transient dilational or shear fractures were propagated over stratigraphic distances of as much as 1,700 feet. Fragments injected or spalled into such fractures may have been mixed by compressive pump action of the fracture walls if the radial inflow of material forming the central uplift were nonuniform.

SHATTER CONES

Shatter cones at Sierra Madera, as elsewhere, are conical fracture surfaces with characteristic striae that fan outward from the apex in "horsetail" fashion (fig. 26; Howard and Offield, 1968). The striae are sharp grooves between intervening rounded and broader ridges; they are straightest, most regular, and best developed in fine-grained rocks; in coarser rocks they are coarser and less detailed. Away from the apex, the surface of a cone is built up by parasitic cone segments successively overlapping like shingles. Whole cones and clusters of whole cones having a common axial orientation are found in places (fig. 26*A*), but segments or

FIGURE 26.—*A*, Whole shatter cones, 2 to 10 millimeters high, in marly dolomite of Gilliam Limestone; length of specimen 11 centimeters. *B*, Variously oriented shatter-cone segments in marly dolomite of Gilliam Limestone. Surfaces in foreground and background are parallel and have similarly oriented cone segments. Note shinglelike overlap of parasitic cone segments. Specimen 6.5 by 5.5 cm.

partial cones (fig. 26*B*), which may intersect at high angles, are more common. Cone segments may have any orientation relative to bedding; in several places, particularly in aphanitic dolomite, they are preferentially developed on joint and bedding surfaces, so that in outcrop they may be developed almost exclusively on two or three sets of such subplanar surfaces. Some whole cones are asymmetric, having one side longer than the other.

Shatter cones are found in all the exposed Permian formations. Figure 27 shows their outer limit of abundance on the central uplift. The distribution of cones is very irregular, partly because lithology is an important factor in their formation; fine-grained and brittle rocks appear to have been most favorable. Thus cones are most common in aphanitic and dense marly dolomite; are present in crystalline dolomite, siltstone, fine-grained sandstone, and calcareous chert; are least common in limestone; and do not occur in coarse-grained sandstone. They may, however, be abundant in a particular rock type at one place and not be developed at all in virtually the same lithology at another place.

Whole cones range in height from less than 2 mm to 12 cm and the sides are 2 mm to 17 cm long, but incomplete cone segments are typically larger, as much as 45 cm long. Even in a single bed with 3-cm whole cones, cone segments reach 10 cm in length. For any given rock type, cones are more abundant toward the center of the structure but show no change in range of size. Along the outer limit of abundant shatter cones (fig. 27), the cone segments typically are 5 cm long in aphanitic dolomite, up to 12 cm long in marly dolomite, and up to 20 cm long in sandy siltstone; in like rocks near the center they are comparable in size.

The rocks in which shatter cones are most abundant also have generally the smallest cones. This lithologic control of size of shatter cones is illustrated in the Gilliam Limestone where sandy layers with 20-cm cone segments and finely crystalline dolomite beds with 5-cm cone segments are in direct contact. Another example is an outcrop of Word Formation, a few meters across, where finely crystalline dolomite contains a few cone segments 9 cm long, marly dolomite layers contain abundant well-developed cone segments 5 cm long, but crystalline dolomite, limestone, and medium-grained sandstone contain no cones. Aphanitic marly dolomite of the Gilliam contains cones as small as 2 mm; the longest cone segments are in sandy dolomite of the Word Formation (up to 35 cm) and limestone of the Hess Formation (up to 45 cm).

Shatter cones are not present in Cretaceous rocks at Sierra Madera, even where these rocks are in contact with marly dolomite of Tessey Limestone containing abundant cones. The absence of cones in aphanitic to medium-grained Cretaceous limestone may be due to lack of consolidation, combined with distance from the center of the structure.

Most shatter cones are within 1 to 2 miles of the center of the structure, as figure 30 shows. However, cones also occur in the uppermost dolomite of the Tessey near the east and south margins of the map (pl. 2). At the most distal locality, the rock was driven upward as a small slice into Cretaceous strata along a fault on

[*Editor's Note:* Figure 27 has been omitted.]

the outer perimeter of the structure, about 4 miles southeast of its center.

Cones also occur at depth. Cuttings from a well drilled within about 0.3 miles of the center of the structure reveal probable cone segments to a depth of 1 mile—4,000 feet stratigraphically below the top of the Permian. Possible cone segments are found from lower cuttings throughout the rest of the hole to a depth of 12,000 feet.

Axes of whole cones in clusters have a nearly common orientation in any outcrop (fig. 26). In exposures showing only segments of cones (fig. 26*B*), measurement of many striae shows that the segments are incomplete parts of cones having a common orientation and that they define a common cone axis, even though the segments themselves are at large angles to one another.

Each striation measured at an outcrop was plotted on a lower hemisphere stereographic projection with the direction of point indicated. Most plots resulted in a circular pattern corresponding to a conical arrangement of striae. Orientations and apical angles of the cones were determined by fitting small circles to these circular patterns. Figure 28*A* shows a typical plot of striae from cone segments; for comparison, figure 28*B* shows striae measured on a single whole cone from a different locality.

On average, 26 striae per locality were measured except for three localities in which whole cones were measured directly; it was found that cone geometry could be reasonably estimated from as few as eight striae.

Cone segments in a single structural block define an average cone orientation and an apical angle at most places at Sierra Madera. However, a few measured striae in one locality fall far off the circular plot, and data from three localities were discarded altogether, because striae orientations were scattered and no unidirectional cone axis was obvious. The discrepant orientations of cone segments may be attributable to effects of inhomogeneities in the rock.

Cones in a single structural block generally point in the same direction, as well as having the same axial orientation. In a few localities, however, some segments (as many as 31 percent) point in a direction opposite to the dominant direction of the cluster.

Apical angles range from 75° to 108° and average 88.5°; thus they are significantly smaller than those reported from the Vredefort ring of South Africa, where angles range from 90° to 122° (Manton, 1965). Figure 29 suggests that, as at Vredefort, each stratigraphic unit has a characteristic range of apical angles and that apical angles may be influenced by rock type.

Rocks containing shatter cones have been faulted and folded, and no pattern is obvious when the average

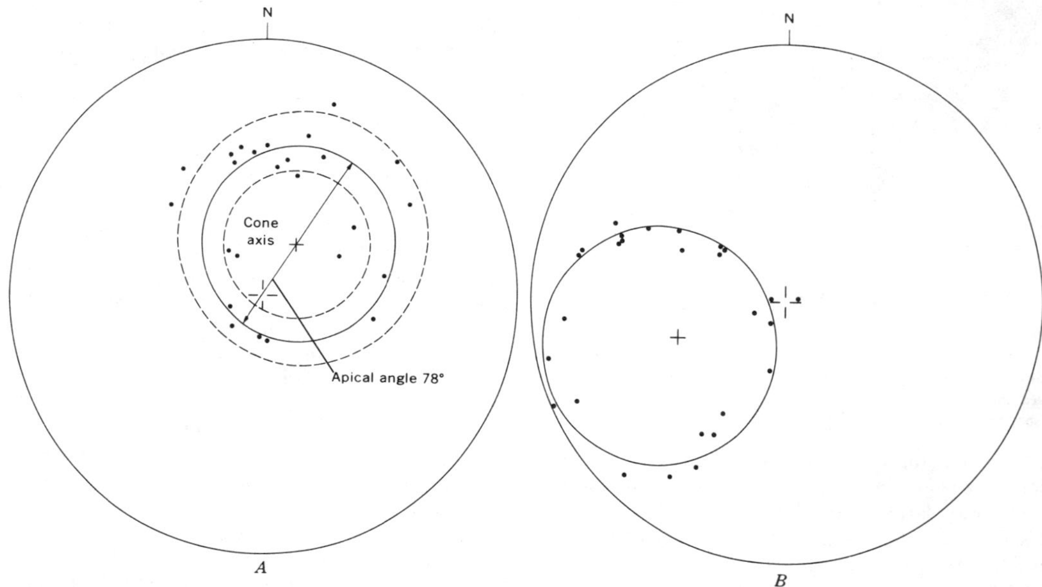

FIGURE 28.—*A*, Striae from shatter-cone segments from one locality, plotted on lower hemisphere stereographic projection (all indicate upward-pointing cones); dashed circles are drawn 10° from small solid circle selected to represent cone. *B*, Striae measured on a single upward-pointing whole cone.

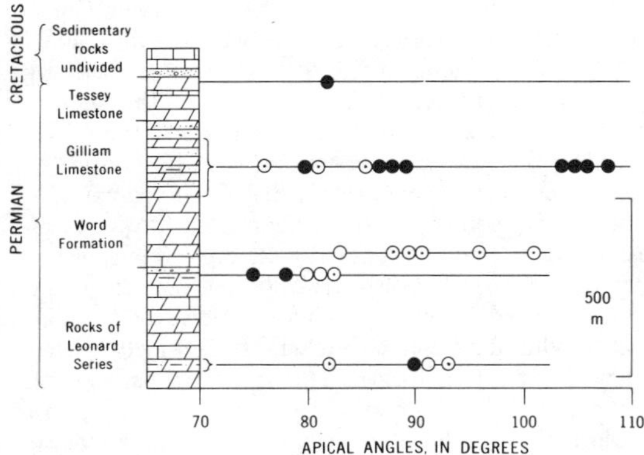

FIGURE 29.—Apical angles of shatter cones measured at various stratigraphic levels. Each measured apical angle is represented by a symbol showing rock type: dotted circles, siltstone or sandstone; solid circles, aphanitic or marly dolomite; open circles, crystalline dolomite. Rock-column symbols: rectangular brick, limestone; slanted brick, dolomite; circles, conglomerate; dots, sandy or sandstone: dashes, shaly or marly.

orientation of cones from each measured locality is plotted on a map (fig. 27). Evidence from the three northwesternmost localities, where shatter-cone orientations were measured on both limbs and the trough of a syncline, indicates that the cones formed before folding: although oriented differently at all three places, if the syncline is unrolled, the cones approach a common orientation. Likewise, some broken cones found in fault breccia show that the cones predate faulting.

The original cone orientations can be estimated but cannot be determined exactly. The paths through which they were deformed were undoubtedly complex; the initial attitudes of bedding also are unknown. Before the cryptoexplosion structure formed, Cretaceous strata lay nearly flat, but the underlying Permian rocks of the area had a small northward regional dip and gentle undulations, with dips locally as high as 20°.

Nevertheless, if the cones formed before the central uplift, their initial orientations can be approximated by assuming that the Permian beds were horizontal and that there were simple rotations. The facing direction of beds is known certainly for all but the three most central localities. Bedding orientations were rotated to horizontal about their strike lines by computer.

The resultant pattern (fig. 30) clearly shows cones pointing inward and upward, as at other cryptoexplosion structures where shatter-cone orientations have been well studied. Considering the simplifying assumptions used in restoring the beds to their initial positions, the in-and-up pattern is surprisingly good. One exception, the easternmost locality, lies in a chaotic sedimentary reef breccia where initial dips were so high that the rotation assumption is not valid. A single location of outward point, in the southwest, is representative of the minority of cones that developed along the same axes but with points opposite those of the prevailing pattern. Apart from the easternmost locality, 13 percent of all striae measured indicate cones pointing away from the center; 87 percent indicate cones pointing toward the center.

Cone-axis plunges are steepest in the center and decrease outward. A crude central focus of intersecting cone axes was found by computer after restoring the strata to a horizontal position. This approximate focus (standard deviation of 2,000 feet) occurs at the top of the Permian section directly over the center of the uplift (fig. 30).

These results generally confirm those of Manton (1965), Bray and others (1966), Stearns, Wilson, Tiedmann, Wilcox, and Marsh (1968), and Milton (1969), all of whom have shown that shatter cones in other structures point inward and upward if the beds containing the cones are restored to the attitudes they had before the structures formed.

The formation of shatter cones by shock has been demonstrated by hypervelocity impact experiments (Shoemaker and others, 1963) and at craters formed by nuclear and chemical explosions (Bunch and Quaide, 1968; Roddy, 1969a). Roddy studied craters formed by explosion of TNT spheres near ground surface. The apices of shatter cones at these craters point toward the energy source, as do those formed by the aforementioned impact experiments. The symmetry of restored shatter-cone orientations at Sierra Madera suggests that the cones were formed by a shock wave of generally hemispherical form that originated above the present ground surface and at the center of the structure.

SHOCK-DEFORMED QUARTZ

Quartz grains in many deformed rocks from Sierra Madera display multiple sets of planar structures characteristic of naturally and artificially shocked quartz (fig. 31). Carter (1968) distinguished four types of planar structures in quartz—cleavage, faults, planar features (a group of structures as yet poorly understood), and deformation lamellae. Of these, we distinguished with certainty only cleavage (fig. 32). The remaining planar structures (fig. 33) in quartz from Sierra Madera are hereafter referred to collectively as "planar elements."

Multiple sets of planar elements in individual grains of quartz were first discovered at Sierra Madera by N. M. Short (written commun., 1966). We have exam-

[*Editor's Note:* Figures 30 and 31 have been omitted.]

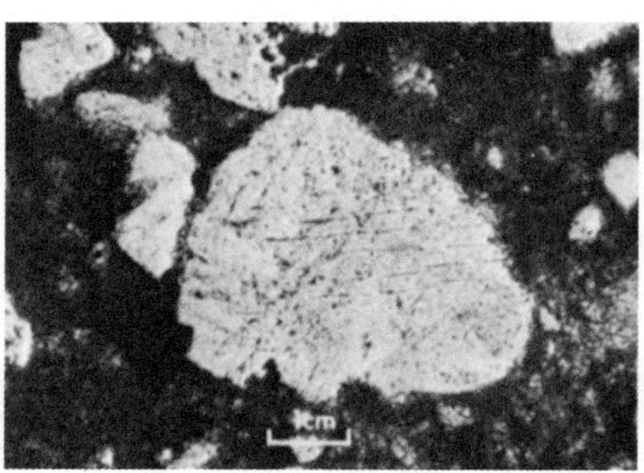

FIGURE 32.—Photomicrograph of well-developed basal cleavage (wide-spaced lines nearly east-west across grain) in quartz. Other planar elements present but not well developed. Sample is from a small pod of basal Cretaceous sandstone injected into Edwards Limestone.

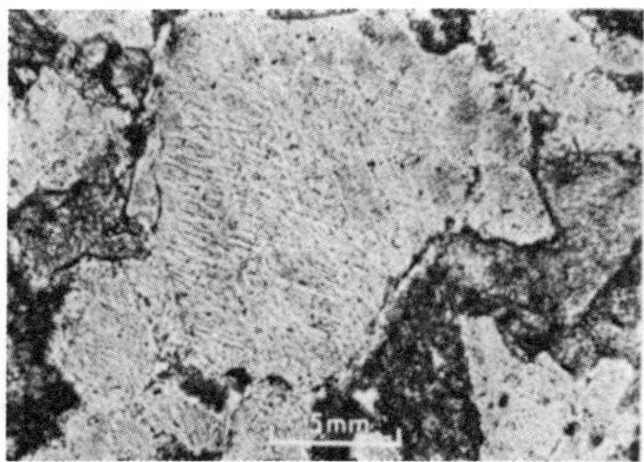

FIGURE 33.—Photomicrograph of multiple sets of planar elements in quartz. Sample is a clast of Cathedral Mountain Formation in mixed breccia.

ined several hundred thin sections of rocks from Sierra Madera and from the same formations in the Glass Mountains to determine the distribution of this type of deformation. About 1,400 sets of planar elements and cleavages were measured with a universal stage microscope in 165 quartz grains from 23 localities at Sierra Madera (fig. 31). Asterism and refractive indices of selected quartz grains were also studied. Although calcite and dolomite are also severely deformed in many rocks from Sierra Madera, they were not studied in detail.

Some types of internal structures in quartz, briefly described in table 2, were found in rocks from the Glass Mountains and presumably formed before the Sierra Madera structure. Planar elements in quartz of undeformed sandstones and conglomerates are rare, but as many as five sets in a single grain were observed. Commonly the best developed set consists of subparallel, slightly wavy discontinuities that are very difficult to bring to a recognizably vertical position on the universal stage. Other planar elements in quartz of the undeformed rocks consist of concentrations of cavities(?) that are not visibly connected. Still other planar elements, and possibly the most abundant, in quartz of unshocked rocks are not optically distinguishable, either in bright-field or phase-contrast illumination, from the shock-produced planar elements. However, unlike shock-produced elements, they tend to be widely spaced and vary greatly in length. The angular separation of such planes from the *c* axis is random, in contrast to concentrations of basal pinacoidal and rhombohedral planes among the shock-produced planar elements. Another feature, common in unshocked quartzite pebbles but not observed in densely packed quartz of shocked rocks, is planar elements that cross grain boundaries. Such planar structures are apparently irrational. The technique used by Hörz (1968, p. 249) for indexing planar elements was used and consists of matching, by rotation about the *c* axis, the largest number of poles of planar elements with a plot of poles of rational crystallographic planes in quartz. This technique gives no guarantee that an indexed plane is in fact rational; but where several planes of a particular zone can be indexed, the indication of rational crystallographic direction is strong. Using this procedure for planar structures that cross grain boundaries, it was found that in one example a planar element could be indexed as a rational crystallographic direction in one grain but not in the adjacent grain; in a second example, the planar element could not be indexed in either grain, and its angular separation from the *c* axis was quite different in the two grains.

In rocks containing the most strongly deformed quartz at Sierra Madera, virtually all quartz grains typically show multiple sets of planar elements, in marked contrast with the rarity of deformed grains in the corresponding unshocked rocks. However, gradations exist at Sierra Madera between rocks that display extensive shock damage and rocks that do not differ from their counterparts in the Glass Mountains. A qualitative estimate of the degree of shock deformation, based on flat-stage microscopic examination of the frequency of deformed quartz grains within single thin sections, is shown in figure 31. A significant feature of the distribution of shock-deformed quartz is that quartz in mixed breccias consistently shows extensive shock damage, whereas quartz in rocks still in place generally shows less shock damage and the occurrence is sporadic. Cleavages, locally the most prominent structure in quartz, are widely spaced (about 10μ to 50μ), broad, apparently open cracks that are commonly decorated by cavities or inclusions. Although they extend to edges of grains, they do not offset grain boundaries. The most prominent set of cleavages is generally parallel to the basal pinacoid, but rhombohedral and prismatic cleavages are moderately developed (table 4). A few grains in which cleavage is subordinate to other planar elements have both basal cleavage and basal planar elements.

Examination of the planar elements with a flat stage and both bright-field and phase-contrast illumination did not allow distinction of "planar features" and "deformation lamellae" (Carter, 1968) because they are so commonly decorated by inclusions and cavities. Among the less decorated planar elements, some that are oriented at a high angle to the plane of the thin section show the asymmetric light and dark edges mentioned by Carter (1968, p. 467) as characteristic of deformation lamellae. The planar elements consist of thin, closely spaced (about 1μ to 5μ) discontinuities that are equally visible in plane- and cross-polarized light. They do not offset or cross grain boundaries. Typically there are at least five planes in a single set; such sets may occur in a single, small zone in a quartz grain, in several small zones or may be equally distributed across the entire grain. Decoration by minute cavities and inclusions is very common, but also common are such irregularities as decoration of only some planes in a single set, decoration of all planes in only one set and none in another set in the same grain, and lateral changes in degree of decoration along a single plane.

The orientation of planar elements with respect to the *c* axis of the host grain was measured in 165 grains from 23 localities at Sierra Madera (fig. 31). After measurement of each grain and appropriate rotations, an attempt was made to index the planes by a best fit to a plot of poles of rational crystallographic directions in quartz (fig. 34); by accepting measured poles that plot within 5° of the poles to rational planes, 73 percent of the planar elements were indexed, and, because several members of specific zones are consistently present, most of the indexed planes are believed to be rational. The frequency of occurrence of planar elements and cleavages is shown in figure 35, but it must be emphasized that these results are biased: grains in which the *c* axis is inclined at a low angle to the plane of the thin section were specifically selected for measurement so that basal planar elements would not be missed. However, because of the blind area of the universal stage, this meant that certain other orientations, such as some rhombohedra and prisms, could not be seen. On the other hand, when two or more members of a zone were plotted by best fit, other members of the same zone were then specifically sought; such planar elements, missed in the initial measurements because of high tilts or because their traces were nearly parallel to another set, were found in a number of cases. Apparently irrational planes were not similarly sought. Although no basal planar elements were missed, this orientation is underweighted on the histogram because for each grain there is only one such orientation, whereas the other common rational directions occur in zones of six members.

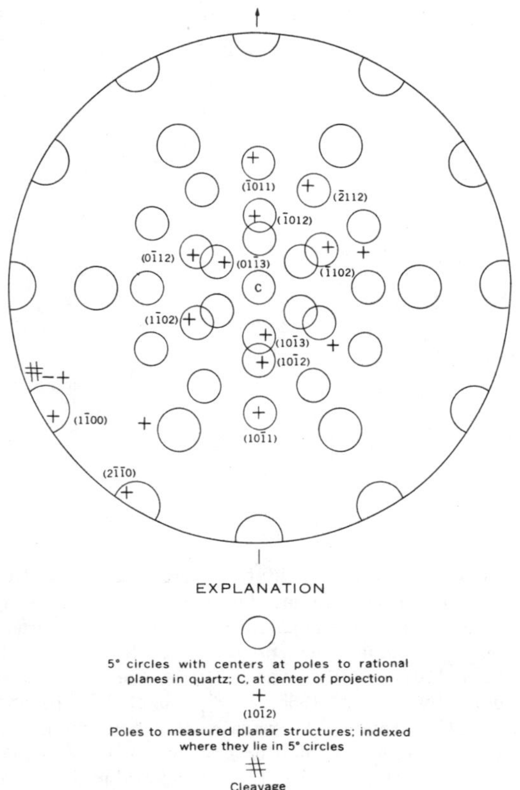

FIGURE 34.—Rotated stereogram showing best fit of measured poles (+) to sets of planar elements with poles (at center of 5° circles) to rational planes in quartz from mixed breccia.

Table 4 shows, in order of decreasing relative abundance, the most prominent rational planar elements measured in grains from the 23 localities. The number of basal orientations measured was multiplied by six to determine the position of {0001} in the table, but the number of other crystallographic orientations was not modified to account for inaccessibility or absence of some members of the zone in each grain. Accordingly,

[*Editor's Note:* Figure 35 and Table 4 have been omitted.]

table 4 gives only a qualitative picture of the relative abundance of the different planar elements. Despite the statistical flaws, it is apparent that the dominant orientations of planar elements are parallel to $\{0001\}$, $\{10\bar{1}3\}$, $\{10\bar{1}1\}$, $\{10\bar{1}2\}$, and $\{10\bar{1}0\}$ at the 23 localities. Studies by Carter (1965, 1968), Chao (1967b), and Short (1966a) have demonstrated that both the multiplicity [1] and the common orientations of these planar elements are grossly unlike those developed in tectonites at slow strain rates, and that they are the same as those formed by shock, both naturally (Bunch and Cohen, 1964; Engelhart and Stöffler, 1965; Chao, 1967a, b; and other workers) and experimentally (Short, 1965, 1966a, b, 1968a, b; Hörz, 1968; Müller and Défourneaux, 1968; and Bunch and others, 1968).

In experiments performed by Hörz (1968), single quartz crystals shock loaded at pressures of about 50 kb (kilobars) and higher developed cleavages oriented like those at Sierra Madera. Hörz' experiments showed that basal and prismatic cleavages form when the impact direction is normal to $\{10\bar{1}1\}$ and that unit rhombohedral cleavages form when impact is normal to $\{0001\}$ and $\{10\bar{1}0\}$. Planar elements form in single quartz crystals only at pressures of over 100 kb (Hörz, 1968; Müller and Défourneaux, 1968), but the orientations of the planes are independent of the impact direction. Müller and Défourneaux (1968) reported formation of planar features parallel to $\{10\bar{1}3\}$ at shock pressures above 105 kb; Hörz (1968) reported their formation at pressures above 119 kb. Planar features parallel to $\{10\bar{1}2\}$ first formed between 160 and 200 kb in Hörz' experiments and above 200 kb in experiments by Müller and Défourneaux. Above 250 kb, planar features parallel to $\{10\bar{1}2\}$ are about as common as those parallel to $\{10\bar{1}3\}$ (Hörz, 1968).

The data from measured samples and from flat-stage examination indicate that cleavages are best developed in quartz from rocks that are far from the center of Sierra Madera or from less strongly deformed rocks closer to the center. Of the measured planar elements, $\{10\bar{1}3\}$ is dominant over $\{10\bar{1}2\}$ in all samples of Cretaceous rocks, whether in place or in mixed breccia, and in the majority of Gilliam fragments in mixed breccia; $\{10\bar{1}2\}$ is as well or better developed than $\{10\bar{1}3\}$ in samples of Word and older rocks from mixed breccia. These relations indicate that, in general, rocks closer to the center of the structure record the highest pressures and those incorporated in mixed breccias were subjected to higher pressures than equivalent rocks that remained in place. Experimental data referred to above suggest that rocks incorporated in mixed breccias were subjected to pressures in excess of about 200 kb near the center of the structure, rocks still in place near the center were locally subjected to pressures above about 100 kb, and those around the edge of the central uplift were subjected to pressures of about 50 kb and higher. Although these pressures are believed to be of the right order of magnitude and are compatible with those suggested by Robertson and others (1968) for the same types of deformation, experimental data from single quartz crystals probably cannot be extrapolated directly to sandstones. Kieffer's (1969) theoretical treatment of shock-wave propagation through granular quartz (Rinehart, 1968; Chao, 1968) suggests that a considerable range of peak pressures could be recorded in different but nearby grains of a sandstone because reverberation of shock and rarefaction waves created by multiple grain impacts results in local energy concentrations not produced in similar solid material. Accordingly, an accurate appraisal of peak pressures reached at different points on Sierra Madera would require a considerably more sophisticated study of the shock deformation.

In addition to the planar elements in quartz at Sierra Madera, permanent crystal damage is indicated by pronounced asterism (see Dachille and others, 1968) and abnormal refractive indices of deformed grains. Asterism—spreading of spot reflections from single grains X-rayed during rotation in a Debye-Scherrer camera—is apparent in photographs of shocked quartz and calcite shown in figure 36. Photographs of unshocked calcite and quartz (Nos. 2 and 4) are shown for comparison. A quartz grain from the basal Cretaceous sandstone in the Glass Mountains, chosen to show

[1] The maximum number of planar element sets measured in a single quartz grain from Sierra Madera is 22, of which 14 were matched to rational directions by best fit.

that slight asterism may also be displayed by unshocked minerals. The undulose extinction of this grain indicates that asterism was produced by strain. In Sierra Madera samples, asterism ranges from slight in weakly deformed grains to patterns approaching those expected from powder mounts. Different degrees of asterism for different grains from the same sample are characteristic. The degree of asterism illustrated for quartz samples as a whole is roughly correlative with the departure of the refractive index n_O from its normal value of 1.544; thus, samples 192 and 143, which show the most marked asterism, have the lowest value of n_O. Samples 186, 159, and 112 show less abnormal refractive indices and less asterism; they also show a greater range of refractive index and asterism. The range of refractive index ($n_O=1.538$–1.544) and asterism also correlates with the degree of shock damage inferred from the orientations and relative abundances of planar elements. Thus, samples 192 and 143 are from mixed breccias in which planar elements indicate high shock pressures, sample 112 is from the Cathedral Mountain Formation in place in which moderate shock pressures are indicated, and samples 186 and 159 are from slightly shocked Cretaceous sandstone in mixed breccias on the flanks of the central uplift. Reduced refractive index and extreme asterism shown by the severely deformed quartz suggest pressures above about 200 kb. (See Hörz, 1968; Chao, 1968.)

Coesite and stishovite reflections were looked for in all the X-ray photographs, but none were found. A few other highly shocked samples were searched, also unsuccessfully, for these high-pressure silica polymorphs, but special techniques for concentration were not employed.

SUMMARY OF SHOCK DEFORMATION

Four main products of shock deformation are found at Sierra Madera: monolithologic breccias, mixed breccias, shatter cones, and deformed quartz and carbonate minerals. The general absence of shock-deformed quartz in monolithologic breccias and scattered occurrences of deformed calcite and dolomite in these breccias and in shatter-coned rocks suggest that monolithologic breccias and shatter cones formed at comparatively low pressures. All rocks emplaced as mixed breccias in Gilliam and older strata contain quartz that has been severely deformed at peak pressures of over 200 kb, whereas rocks still in place contain quartz that was subjected to pressures from about 50 kb, at the flanks of the uplift, to more than 100 kb, nearer the center of the structure. Considerably higher pressures may have accompanied shock melting of certain clasts in a centrally located mixed breccia.

The shock predated the central uplift, as shown by the rotations of shatter cones. Ideally the shock wave might be considered to have propagated as an expanding hemisphere from the approximate focus indicated by shatter cones. Shock intensity should have decreased with increasing radius of the hemisphere. After uplift, the presently exposed Hess Formation should all have been about equidistant initially from the shock origin; but outward the exposed Gilliam Limestone should have been half again as far from the shock origin, and the nearest exposed Cretaceous strata, more than twice as far as the lowest exposed Hess beds. In a crude way this is consistent with the gross distribution of shock deformation features.

ORIGIN

The keys to the origin of Sierra Madera are: (1) lack of relationship to local and regional structure, (2) the shape of the deformed zone, in which intensity of deformation diminishes laterally and downward from the center, (3) geometry of the central uplift, which closely restricts the choices for its mode of formation, and (4) shock deformation features that indicate peak pressures in excess of 200 kb and a shock source located above the center of the structure. Although certain of these features are compatible with a terrestrial origin, taken together they can be viewed only as evidence of deformation resulting from a surface impact event.

REGIONAL AND LOCAL SETTING

The Sierra Madera structure is unique in its form, style, and severity of deformation in a broad region of otherwise mildly deformed rocks of Permian and Early Cretaceous age. The structure formed some time after the Lower Cretaceous strata were deposited but probably before they were completely consolidated, thus, in Late Cretaceous or early Tertiary time. During that time the region, including the Delaware and Val Verde basins, was apparently undergoing only very mild deformation. Major late Cenozoic block faults occur in some parts of the Delaware basin (King, 1948) but only in areas remote from Sierra Madera. Although Sierra Madera is unique in this region, its general setting—a moderate thickness of shelf-facies sedimentary rocks overlying a thick section of basin-facies rocks—is common around the Permian basins (fig. 3; Oriel and others, 1967).

Sierra Madera is near the northeast end of the Glass Mountains, a gently north-dipping homocline of Permian and Lower Cretaceous strata. The main structural features of the Glass Mountains—northwest tilting and minor folding into low domes and north-trending anticlines—formed before deposition of the Lower Cretaceous rocks (King, 1930, p. 127). Lower Creta-

[*Editor's Note:* Figure 36 has been omitted.]

ceous strata were deposited on the beveled surface of Permian rocks, and the strata were later (probably near the close of the Pliocene) retilted and broken by a system of northwest-trending normal faults (King, 1930, p. 127), few of which intersect the Sierra Madera structure. Hence, the very mild deformations of the Glass Mountains appear to have bracketed the formation of Sierra Madera. A fault, possibly Tertiary in age, has about 400 feet of displacement in Lower Cretaceous rocks (pl. 1, near Hunt Elsinore 31) north of Sierra Madera but does not intersect the structure. Sierra Madera lies above the north flank of an east-west trending subsurface anticline. (See fig. 6, Hunt Elsinore 52 and 53.) The fold must have formed before Permian time because Pennsylvanian strata were eroded from its top and rocks of Permian age and younger are not affected by it.

Zimmermann and Amstutz (1965, p. 304) and Lowman (1965) have suggested, respectively, that Sierra Madera is part of a regional lineament and that it is a member of a regional association of features. Zimmermann and Amstutz implied that Sierra Madera is alined with the Solitario dome and breccia pipes in the Terlingua district, and Lowman considered it to be part of the Trans-Pecos Tertiary igneous province. However, Sierra Madera is not connected with these features by any structure expressed in rocks exposed at the surface or known to exist in the subsurface, and moreover, it is very different from all these other features.

Because Sierra Madera is unique in an otherwise common regional setting and because it is not connected either spatially or in time with local structures, we conclude that the cause of deformation is not genetically related to regional or local features.

ZONE OF DEFORMATION

Surface mapping and drill data indicate that the intensity of folding and faulting dies out rapidly from the center of Sierra Madera, both laterally and downward. Hence, the gross shape of the zone of deformation is that of a cup or funnel about 8 miles in diameter and 6,000 to 8,000 feet deep, with the most intensely deformed rocks in the upper central part. The intense shattering of rocks seen at the surface similarly disappears downward and was not observed in cuttings from the Phillips well at depths greater than about 2,800 feet (Eggleton and Shoemaker, 1961). The shape of the zone of deformation as well as the distribution of the more severely disrupted rocks accordingly indicates that the deforming forces originated high in the affected section of rocks and near the center of the structure.

GEOMETRY OF THE CENTRAL UPLIFT

Formation of the central uplift involved both radial inflow and uplift of strata. This structural pattern leads to the conclusion (Wilshire and Howard, 1968) that the uplift could not have formed by a doming process involving (1) emplacement of material, such as magma (King, 1930; Lowman, 1965), (2) diapiric intrusions of salt or shale (Lowman, 1965), or (3) solid blocks of subjacent strata or basement rock (Amstutz, 1960), beneath the uplifted rocks. All such mechanisms would require distension of the section rather than the observed compression. Igneous intrusions, saline diapirs, and reactivated basement structures are also disproved by drill data. Centripetal movement of strata in the uplift furthermore rules out an explosion at depth (Milton and Brett, 1968), perhaps the most prominent hypothesis for the terrestrial origin of cryptoexplosion structures (Bucher, 1963; Snyder and Gerdemann, 1965).

Structures of terrestrial origin that do resemble the central uplift of Sierra Madera are diapirs, and a diapiric origin for cryptoexplosion structures has been proposed by Amstutz (1964) and Zimmermann and Amstutz (1965). The form of the uplift at Sierra Madera resembles that within diapiric structures such as salt domes, so that any explanation by diapirism must seek the motivating cause within the section of rocks now exposed. The only appropriate density contrast for initiating buoyant diapirism is between the dominantly shaly lower Hess and older Permian strata and the overlying, dominantly carbonate rocks of upper Hess and younger rocks. An origin of the uplift by emplacement of a shale diapir yet unexposed runs into the same geometrical problem as formation by other intrusions, namely, the addition of new material, which would cause outward rather than inward movement. Despite the resemblance of the uplift and ring depression to salt domes, notable differences at Sierra Madera include the structurally high rim containing upward-driven slices of older rocks and the various signs of shock deformation, not known in any bonafide diapiric structure. Hence, an origin by diapirism is untenable.

Several hypotheses have been advanced for the origin of central uplifts in impact craters, such as reflection of shock waves from deep discontinuities (Dence and others, 1965), isostatic adjustment (Dence and others, 1965), and elastic rebound (Rooke and Chew, 1965). None of these, however, account for the radial inflow of rock responsible for the structural geometry of the uplifts; other strong objections have also been lodged against these mechanisms (Beals, 1965; Dence, 1968).

Uplifts with a geometry apparently similar to that of Sierra Madera have been produced in high-explosive cratering experiments [2] performed in Canada (Diehl

[2] The analogies in cratering mechanics of impacts and explosives were discussed by Shoemaker (1960). An excellent recent paper by Moore (1969) points out the similarities in subcrater deformation between craters produced by missile impact and by explosives.

and Jones, 1967; Roddy, 1968b) where the charge was detonated at ground surface rather than buried beneath it. Roddy (1968a, b; 1969b) likened this surface detonation of explosives to the low penetration capacity of an impacting comet, as opposed to an iron or stony meteorite, and suggested that craters with uplifts were produced by the impact of comets. Experimental impact of low-density projectiles with metals and water (for example, Partridge and Van Fleet, 1959; Engel, 1962; Worthington, 1963; Harlow and Shannon, 1967b) have also produced central uplifts, but the mechanics of uplift have not yet been fully explained. Dence (1968) suggested that collapse of crater walls along inward-dipping shear surfaces that extend below the crater could produce the observed geometry of the uplifts. We have no information that allows a choice between these proposals. In any case, the occurrence of uplifts in impact craters is apparently related to the scale of the event—most known and suspected terrestrial impact structures more than 1½–2 miles wide contain central uplifts and smaller ones do not. On the moon, where gravity is one-sixth that on earth, craters with central peaks are generally more than 12 miles wide.

SHOCK DEFORMATION

Two types of deformation common at Sierra Madera—shatter cones and, especially, internal structures in quartz—appear to have well-substantiated shock-wave origins. The symmetry of shatter cone orientations indicates that shock waves were generated at a central focus in the structure and at a level high in the section of deformed rocks; the shock wave passed downward and outward from this focus at a time when the strata were nearly flat lying. Local peak pressures reached during passage of the shock wave were in excess of 200 kb in rocks incorporated in mixed breccia masses. Peak pressures recorded in quartz of rocks still in place are crudely zoned (fig. 31), such that older rocks near the center of the structure were subjected to pressures in excess of about 100 kb and those on the periphery of the uplift were subjected to pressures on the order of 50 kb. This crude zoning is comparable to what might be expected from deformation, uplift, and erosion of the generalized subcrater shock zones described by Dence (1968, p. 179).

CONCLUSIONS ON ORIGIN

Of the four significant features of Sierra Madera that bear on its origin, two—the shape of the deformed zone and shock deformation—clearly indicate an origin by surface impact. One of the remaining two—lack of relationship to local structure—is consistent with an impact hypothesis. The fourth feature—geometry of the central uplift—was discussed mainly in terms of its inconsistency with other proposed mechanisms of uplift in cryptoexplosion structures. Although the mechanics of uplift have not yet been fully explained, the existence of such uplifts in craters of probable impact origin has been established (Roddy, 1968a; Engelhart and others, 1967).

DEPTH OF EROSION

Bucher (1936) recognized the structural characteristics of cryptoexplosion structures. Other features, such as shatter cones and shock deformation of minerals, have also been shown to be characteristic of such structures. (See Short and Bunch, 1968). Because these structures formed at different times, ranging from Precambrian for the Vredefort ring of South Africa to Miocene for the Steinheim Basin of Germany, and because they formed under different conditions, present exposures show a wide range of erosional level. Nevertheless, their common characteristics tie them together and allow the more deeply eroded structures, such as Sierra Madera, to be reconstructed by comparison with those that are less eroded. The structures at Flynn Creek, Tenn. (Roddy, 1968a), and Steinheim Basin (Branco and Fraas, 1905; Engelhart and others, 1967) represent the highest levels of exposure in well-studied cryptoexplosion structures. Both structures have a small central uplift surrounded and partly buried by a lens of breccia that occupies a shallow crater; the crater rim at Flynn Creek has well-developed concentric folds, and a small lens of breccia ejected from the crater is preserved in a graben (Roddy, 1968a). A somewhat lower erosional level is found at Gosses Bluff, Australia (Milton and Brett, 1968), where some highly shocked and melted fallback breccia is preserved, but the original crater walls have been destroyed by erosion. Sierra Madera represents a still lower erosional level from which all vestiges of a crater-form depression, fallback, and ejecta have been stripped. We infer from the other cryptoexplosion structures that the outer rim of Sierra Madera represents folded strata beneath the original crater rim, that, before erosion, the central uplift protruded into the crater, and that the crater was comparatively shallow. Experimental craters in which analogous folds formed beneath the crater walls include the nuclear explosion crater Jangle U (Shoemaker, 1968), the chemical explosion craters Snowball and Prairie Flat (Roddy, 1968a), and certain impact craters in sand (Gault and others, 1968, fig. 18). Scaling to the approximate dimensions of the Flynn Creek crater and various experimental craters suggests that about 2,000 feet of cover may have been present over Lower Cretaceous strata and have now been removed by erosion (fig. 37).

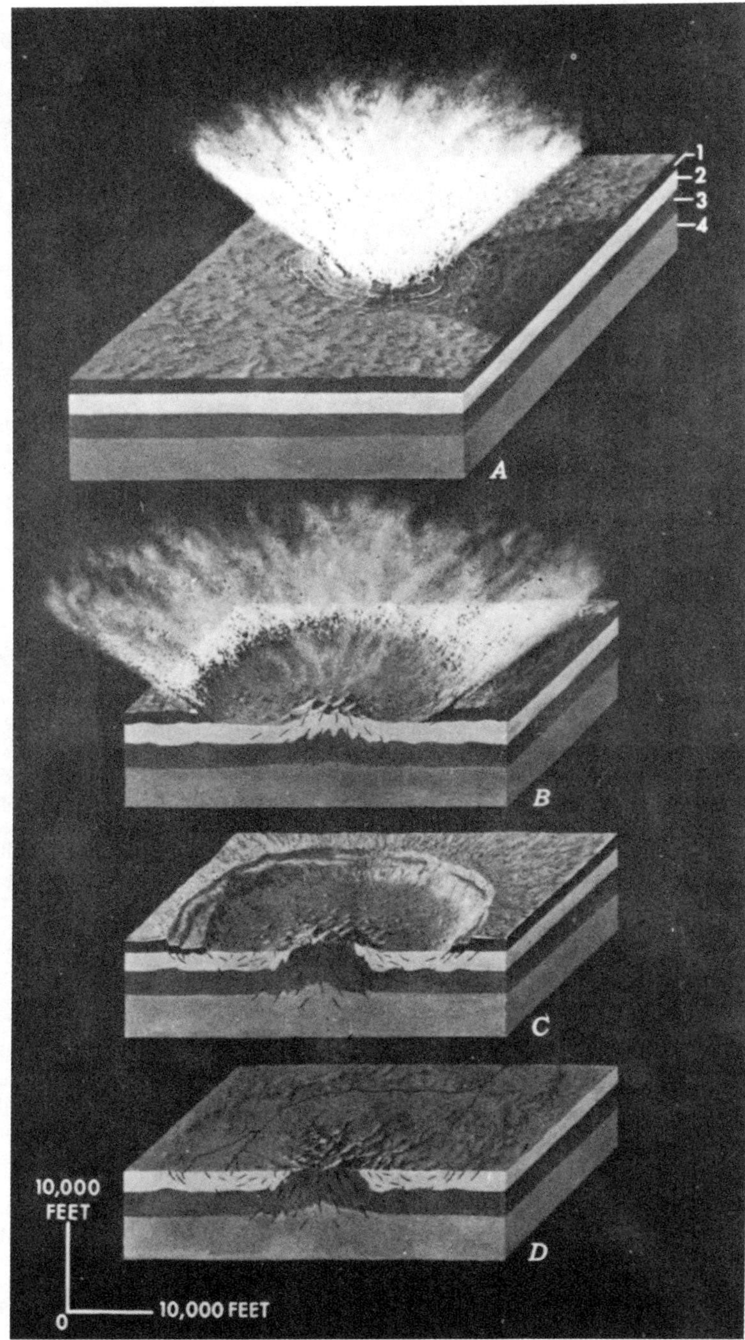

FIGURE 37

Alternatively, the Lower Cretaceous strata may have been under water.

HYPOTHETICAL RECONSTRUCTION OF THE EVENT

Figure 37 illustrates our concept of how Sierra Madera formed. Immediately after impact (fig. 37*A*), the crater grew by low-angle jetting (Gault and others, 1968), while a shock wave passed outward and downward. Figure 37*B* shows the central uplift forming during excavation of the crater, in accordance with the Canadian high-explosive cratering experiments and with the sequence inferred at Gosses Bluff (Milton and Brett, 1968). Outward thrusting may have dragged or injected slices of dolomite from the Tessey up around the periphery; the slices were then left isolated by subsequent normal faulting related to crater wall collapse. Withdrawal of material moving into the uplift may have caused collapse of the crater walls, or inward movement of collapsing wall blocks may have been responsible for the uplift. Either way, the adjustments may have continued past the excavation stage. Injection of sandstone and of highly shocked, shattered rock in mixed breccias probably began early in the deformation. If analogous to sand flows around the Snowball TNT crater (Diehl and Jones, 1967), some of the fluidized injections may have continued to flow for days. The freshly formed crater is shown complete in figure 37*C*. Figure 37*D* represents the present eroded form of the structure.

[*Editor's Note:* Material has been omitted at this point.]

REFERENCES CITED

Adkins, W. S., 1927, The geology and mineral resources of the Fort Stockton quadrangle: Texas Univ. Bull. 2738, 166 p.

Amstutz, G. C., 1960, Polygonal and ring tectonic patterns in the Precambrian and Paleozoic of Missouri, U.S.A.: Eclogae Geol. Helvetiae, v. 52, no. 2, p. 904–913.

——— 1964, Impact, cryptoexplosion, or diapiric movements?: Kansas Acad. Sci. Trans., v. 67, no. 2, p. 343–356.

Balk, Robert, 1949, Structure of Grand Saline salt dome, Van Zandt County, Texas: Am. Assoc. Petrol. Geol. Bull., v. 33, no. 11, p. 1791–1829.

Beals, C. S., 1965, The identification of ancient craters, *in* Geological problems in lunar research: New York Acad. Sci. Annals, v. 123, art. 2, p. 904–914.

Blount, D. N., and Moore, C. H., Jr., 1969, Depositional and non-depositional carbonate breccias, Chiantla quadrangle, Guatemala: Geol. Soc. America Bull., v. 80, no. 3, p. 429–442.

Boon, J. D., and Albritton, C. C., Jr., 1936, Meteorite craters and their possible relationship to "cryptovolcanic structures": Field and Lab., v. 5, no. 1, p. 1–9.

——— 1937, Meteorite scars in ancient rocks: Field and Lab., v. 5, no. 2, p. 53–64.

Branco, W., and Fraas, E., 1905, Das kryptovulkanische Becken von Steinheim: Akad. Wiss. Berlin, Phys.-Math. K1, Abh., v. 1, p. 1–64.

Bray, J. G., and others, 1966, Shatter cones at Sudbury: Jour. Geol., v. 74, no. 2, p. 243–245.

Bucher, W. H., 1936, Cryptovolcanic structures in the United States: Internat. Geol. Cong., 16th, Washington, D.C., Rept., pt. 2, p. 1055–1084.

——— 1963, Cryptoexplosion structures caused from without or from within the Earth? ("astroblemes" or "geoblemes"?): Am. Jour. Sci., v. 261, no. 7, p. 597–649.

Bunch, T. E., and Cohen, A. J., 1964, Shock deformation of quartz from two meteorite craters: Geol. Soc. America Bull., v. 75, p. 1263–1266.

Bunch, T. E., Cohen, A. J., and Dence, M. R., 1968, Shock-induced structural disorder in plagioclase and quartz, *in* French, B. M., and Short, N. M., eds., Shock metamorphism of natural materials: Baltimore, Md., Mono Book Corp., p. 509–518.

Bunch, T. E., and Quaide, W. L., 1968, Shatter cones in the Danny Boy nuclear crater, *in* French, B. M., and Short, N. M., eds. Shock metamorphism of natural materials: Baltimore, Md., Mono Book Corp., p. 285.

Carter, N. L., 1965, Basal quartz deformation lamellae—A criterion for recognition of impactites: Am. Jour. Sci., v. 263, no. 11, p. 786–806.

Carter, N. L., 1968, Dynamic deformation of quartz, *in* French, B. M., and Short, N. M., eds., Shock metamorphism of natural materials: Baltimore, Md., Mono Book Corp., p. 453–474.

Carter, N. L., Christie, J. M., and Griggs, D. T., 1964, Experimental deformation and recrystallization of quartz: Jour. Geology, v. 72, no. 6, p. 687–733.

Chao, E. C. T., 1967a, Impact metamorphism, *in* Abelson, P. H., ed., Researches in geochemistry, v. 2: New York, John Wiley and Sons, p. 204–233.

——— 1967b, Shock effects in certain rock-forming minerals: Science, v. 156, no. 3772, p. 192–202.

——— 1968, Pressure and temperature histories of impact metamorphosed rocks—based on petrographic observations, *in* French, B. M., and Short, N. M., eds., Shock metamorphism of natural materials: Baltimore, Md., Mono Book Corp., p. 135–158.

Cooper, G. A., and Grant, R. E., 1964, New Permian stratigraphic units in Glass Mountains, west Texas: Am. Assoc. Petroleum Geologists Bull., v. 48, no. 9, p. 1581–1588.

——— 1966, Permian rock units in the Glass Mountains, west Texas: U.S. Geol. Survey Bull., 1244–E, p. E1–E9.

Dachille, Frank, Gigl, Paul, and Simons, P. Y., 1968, Experimental and analytical studies of crystalline damage useful for the recognition of impact structures, *in* French, B. M., and Short, N. M., eds., Shock metamorphism of natural materials: Baltimore, Md., Mono Book Corp., p. 555–569.

Dence, M. R., 1968, Shock zoning at Canadian craters: petrography and structural implications, *in* French, B. M., and Short, N. M., eds., Shock metamorphism of natural materials: Baltimore, Md., Mono Book Corp., p. 169–184.

Dence, M. R., Innes, M. J. S., and Beals, C. S., 1965, On the probable meteorite origin of the Clearwater Lakes, Quebec: Royal Astron. Soc. Canada Jour., v. 59, no. 1, p. 13–22.

Dence, M. R., Innes, M. J. S., and Robertson, P. B., 1968, Recent geological and geophysical studies of Canadian craters, *in* French, B. M., and Short, N. M., eds., Shock metamorphism of natural materials: Baltimore, Md., Mono Book Corp., p. 339–362.

Diehl, C. H. H., and Jones, G. H. S., 1967, The Snowball crater, general background information: Defence Research Establishment (Canada), Suffield Expt. Sta., Tech. Note 187, 23 p.

Dietz, R. S., 1946, Geological structures possibly related to lunar craters: Pop. Astronomy, v. 54, no. 9, p. 465–467.

——— 1960, Meteorite impact suggested by shatter cones in rock: Science, v. 131, no. 3416, p. 1781–1784.

——— 1968, Shatter cones in cryptoexplosion structures, *in* French, B. M., and Short, N. M., eds., Shock metamorphism of natural materials: Baltimore, Md., Mono Book Corp., p. 267–284.

Eggleton, R. E., and Shoemaker, E. M., 1961, Breccia at Sierra Madera, *in* Short papers in the geologic and hydrologic sciences: U.S. Geol. Survey Prof. Paper 424–D, p. D151–D153.

Engel, O. G., 1962, Collisions of liquid drops with liquids, Part II—Crater depth in fluid impacts: U.S. Air Force Systems Command Tech. Rept. WADD–TR–60–475, 52 p.

Engelhart, W. C., Bertsch, W., Stöffler, D., Groschopf, P. and Reiff, W., 1967, Anzeichen für den meteoritschen Ursprung des Beckens von Steinheim: Naturwiss., v. 54, no. 8, p. 198–199.

Engelhart, W. v., and Stöffler, D., 1965, Spaltflächen im Quarz als Anzeichen für Einschläge grosser Meteoriten: Naturwiss., v. 52, no. 17, p. 489–490.

Flawn, P. T., 1956, Basement rocks of Texas and southeast New Mexico: Texas Univ. Pub. 5605, 261 p.

Folk, R. L., 1959, Practical petrographic classification of limestones: Am. Assoc. Petroleum Geologists Bull., v. 43, pt. 1, p. 1–38.

——— 1962, Spectral subdivision of limestone types, *in* Classification of carbonate rocks, a symposium: Am. Assoc. Petroleum Geologists Mem. 1, p. 62–84.

Gault, D. E., Quaide, W. L., and Oberbeck, V. R., 1968, Impact cratering mechanics and structures, *in* French, B. M., and Short, N. M., eds., Shock metamorphism of natural materials: Baltimore, Md., Mono Book Corp., p. 87–99.

Groves, James, 1925, Fossil charophyte fruits from Texas: Am. Jour. Sci., 5th ser., v. 10, no. 1, p. 12–14.

Harlow, F. H., and Shannon, J. P., 1967a, The splash of a liquid drop: Jour. Appl. Physics, v. 38, no. 10, p. 3855–3866.

——— 1967b, Distortion of a splashing liquid drop: Science, v. 157, no. 3788, p. 547–550.

Hörz, Friedrich, 1968, Statistical measurements of deformation structures and refractive indices in experimentally shock loaded quartz, *in* French, B. M., and Short, N. M., eds., Shock metamorphism of natural materials: Baltimore, Md., Mono Book Corp., p. 243–353.

Howard, K. A. and Offield, T. W., 1968, Shatter cones at Sierra Madera, Texas: Science, v. 162, no. 3850, p. 261–265.

Kieffer, S. W., 1969, Shock wave propagation through granular quartz [abs.]: Am. Geophys. Union Trans., v. 50, no. 4, p. 221.

King, P. B., 1930, The geology of the Glass Mountains, Texas, Part 1, Descriptive geology: Texas Univ. Bull. 3038, 167 p.

——— 1942, Permian of west Texas and southeastern New Mexico, Part 2 *of* DeFord, R. K., and Lloyd, E. R., eds., West Texas-New Mexico Symposium: Am. Assoc. Petroleum Geologists Bull., v. 26, no. 4, p. 535–763.

——— 1948, Geology of the southern Guadalupe Mountains, Texas: U.S. Geol. Survey Prof. Paper 215, 183 p.

Lowman, P. D., Jr., 1965, Magnetic reconnaissance of Sierra Madera, Texas, and nearby igneous intrusions, *in* Geological problems in lunar research: New York Acad. Sci. Annals, v. 123, art. 2, p. 1182–1197.

Manton, W. I., 1965, The orientation and origin of shatter cones in the Vredefort ring, *in* Geological problems in lunar research: New York Acad. Sci. Annals, v. 123, art. 2, p. 1017–1049.

Milton, D. J., 1969, Gosses Bluff astrobleme, Australia—shatter cones [abs.]: Am. Geophys. Union Trans., v. 50, no. 4, p. 220.

Milton, D. J., and Brett, Robin, 1968, Gosses Bluff astrobleme, Australia—the central uplift [abs.]: Geol. Soc. America, 64th Ann. Mtg., Tucson, Ariz., Program, p. 82.

Moore, H. J., 1969, Subsurface deformation resulting from missile impact, U.S. Geol. Survey Prof. Paper 650–B, p. B107–B112.

Muehlberger, W. R., 1959, Internal structure of the Grand Saline salt dome, Van Zandt County, Texas: Texas Univ., Bur. Econ. Geology Rept., no. 38, 18 p.

Müller, W. F., and Défourneaux, M., 1968, Deformationsstrukturen in Quarz als Indikator für Stosswellen: Eine experimentelle Untersuchung an Quarz-Einkristallen: Zeitschr. Geophysik, v. 34, no. 5, p. 483–504.

National Research Council's Committee on Stratigraphy, Permian Subcommittee, 1960, Correlation of the Permian formations of North America: Geol. Soc. America Bull., v. 71, no. 12, pt. 1, p. 1763–1805.

Nettleton, L. L., 1955, History of concepts of Gulf Coast salt-dome formation: Am. Assoc. Petrol. Geol. Bull., v. 39, no. 12, p. 2373–2383.

Oriel, S. S., Myers, D. A., and Crosby, E. J., 1967, West Texas Permian basin region, Chap. C, *in* McKee, E. D., Oriel, S. S., and others, Paleotectonic investigations of the Permian System in the United States: U.S. Geol. Survey Prof. Paper 515, p. C19–C60.

Parker, T. J., and McDowell, A. N., 1955, Model studies of salt-dome tectonics: Am. Assoc. Petrol. Geol. Bull., v. 39, no. 12, p. 2384–2470.

Partridge, W. S., and Van Fleet, H. B., 1959, Similarities between lunar and high-velocity-impact craters: Astrophys. Jour., v. 128, p. 416–419.

Rinehart, J. S., 1968, Intense destructive stresses resulting from stress wave interactions, *in* French, B. M., and Short, N. M., eds., Shock metamorphism of natural materials: Baltimore, Md., Mono Book Corp., p. 31–42.

Robertson, P. B., Dence, M. R., and Vos, M. A., 1968, Deformation in rock-forming minerals from Canadian craters, *in* French, B. M., and Short, N. M., eds., Shock metamorphism of natural materials: Baltimore, Md., Mono Book Corp., p. 433–452.

Roddy, D. J., 1968a, The Flynn Creek crater, Tennessee, *in* French, B. M., and Short, N. M., eds., Shock metamorphism of natural materials: Baltimore, Md., Mono Book Corp., p. 291–322.

——— 1968b, Comet impact and formation of Flynn Creek and other craters with central peaks [abs.]: Am. Geophys. Union Trans., v. 49, no. 1, p. 272.

——— 1969a, Shatter cones at TNT explosion craters [abs.]: Am. Geophys. Union Trans., v. 50, no. 4, p. 220.

——— 1969b, Similarities of 100 and 500 ton TNT explosion craters and proposed comet impact craters [abs.]: Am. Geophys. Union Trans., v. 50, no. 4, p. 220.

Rooke, A. D., Jr., and Chew, T. D., 1965, Operation Snowball, Project 3.1—Crater measurements and earth media determinations, interim report: U.S. Army Eng. Waterways Expt. Sta. Misc. Paper I-764, 61 p.

Shoemaker, E. M., 1960, Penetration mechanics of high velocity meteorites, illustrated by Meteor Crater, Arizona: Internat. Geol. Cong., 21st, Copenhagen 1960, Rept., pt. 18, p. 418–434.

——— 1962, Interpretation of lunar craters, *in* Kopal, Zdeněk, ed., Physics and astronomy of the Moon: New York, Academic Press, p. 283–380.

——— 1963, Impact mechanics at Meteor Crater, Arizona, *in* Middlehurst, B. M., and Kuiper, G. P., eds., The moon, meteorites, and comets, v. 4, *of* The solar system: Chicago, Ill., Univ. Chicago Press, p. 301–336.

——— 1968, Structure of the Jangle U and Teapot ESS nuclear explosion craters [abs.], *in* French, B. M., and Short, N. M., eds., Shock metamorphism of natural materials: Baltimore, Md., Mono Book Corp., p. 627–628.

Shoemaker, E. M., and Eggleton, R. E., 1964, Re-examination of the stratigraphy and structure of Sierra Madera, Texas, *in* Astrogeologic Studies Ann. Prog. Rept., Aug. 25, 1962–July 1, 1963: U.S. Geol. Survey open-file rept., pt. B, p. 98–106.

Shoemaker, E. M., Gault, D. E., Moore, H. J., and Lugn, R. V., 1963, Hypervelocity impact of steel into Coconino sandstone: Am. Jour. Sci., v. 261, no. 7, p. 668–682.

Short, N. M., 1965, A comparison of features characteristic of nuclear explosion craters and astroblems, *in* Geological problems in lunar research: New York Acad. Sci. Annals, v. 123, art. 2, p. 573–616.

——— 1966a, Shock processes in geology: Jour. Geol. Education, v. 14, no. 4, p. 149–166.

——— 1966b, Effects of shock pressures from a nuclear explosion on mechanical and optical properties of granodiorite: Jour. Geophys. Research, v. 71, no. 4, p. 1195–1215.

——— 1968a, Experimental microdeformation of rock materials by shock pressures from laboratory-scale impacts and explosions, *in* French, B. M., and Short, N. M., eds., Shock metamorphism of natural materials: Baltimore, Md., Mono Book Corp., p. 219–242.

——— 1968b, Nuclear - explosion - induced microdeformation of rocks: an aid to the recognition of meteorite impact structures, *in* French, B. M., and Short, N. M., eds., Shock metamorphism of natural materials: Baltimore, Md., Mono Book Corp., p. 185–210.

Short, N. M., and Bunch, T. E., 1968, A worldwide inventory of features of rocks associated with presumed meteorite impact craters, *in* French, B. M., and Short, N. M., eds., Shock metamorphism of natural materials: Baltimore, Md., Mono Book Corp., p. 255–266.

Simkin, Thomas, 1967, Flow differentiation in the picritic sills of North Skye, *in* Wyllie, P. J., ed., Ultramafic and related rocks: New York, John Wiley & Sons, p. 64–68.

Snyder, F. G., and Gerdemann, P. E., 1965, Explosive igneous activity along an Illinois-Missouri-Kansas axis: Am. Jour. Sci., v. 263, no. 6, p. 465–493.

Stearns, R. G., Wilson, C. W., Jr., Tiedmann, H. A., Wilcox, J. T., and Marsh, P. S., 1968, The Wells Creek structure, Tennessee, *in* French, B. M., and Short, N. M., eds., Shock metamorphism of natural materials: Baltimore, Md., Mono Book Corp., p. 323–337.

U.S. Geological Survey, 1967, Gosses Bluff structure, *in* Astrogeologic Studies Ann. Prog. Rept., July 1, 1966–Oct. 1, 1967: U.S. Geol. Survey open-file rept., p. 31–35.

Van Lopik, J. R., and Geyer, R. A., 1963, Gravity and magnetic anomalies of the Sierra Madera, Texas "dome": Science, v. 142, no. 3588, p. 45–47.

Vertrees, C. D., Atchison, C. H., and Evans, G. L., 1959, Paleozoic geology of the Delaware and Val Verde basins, *in* Geology of the Val Verde basin and field trip guidebook: West Texas Geol. Soc., p. 64–73.

Williams, Howel, Turner, F. J., and Gilbert, C. M., 1954, Petrography—an introduction to the study of rocks in thin section: San Francisco, Calif., W. H. Freeman and Co., 406 p. (repr. 1958).

Wilshire, H. G., 1961, Layered diatremes near Sydney, New South Wales: Jour. Geology, v. 69, no. 4, p. 473–484.

——— 1969, Mineral layering in the Twin Lakes granodiorite, Colorado, *in* Larsen, E. H., ed., Petrological and structural studies: Geol. Soc. America Mem. 115, Poldervaart Volume, p. 235–261.

Wilshire, H. G., and Howard, K. A., 1968, Structural pattern in central uplifts of cryptoexplosion structures as typified by Sierra Madera: Science, v. 162, no. 3850, p. 258–261.

Wilson, C. W., Jr., and Stearns, R. G., 1969, Lateral and vertical movements of rock in the Wells Creek structure, Tennessee [abs.]: Geol. Soc. America Southeastern Sec., 18th Ann. Mtg., p. 87.

Worthington, A. M., 1963, A study of splashes: New York, Macmillan Co., 169 p.

Young, Addison, 1960, Paleozoic history of the Fort Stockton-Del Rio region, west Texas, *in* Aspects of the geology of Texas—A symposium: Texas Univ. Pub. 6017, p. 87–109.

Zimmermann, R. A., and Amstutz, G. C., 1965, The polygonal ring structure at Decaturville, Missouri: new tectonic observations: Neues Jahrb. Mineralogie Monatsh., v. 9, no. 11, p. 288–307.

Part V
STRUCTURES OF CLASS IV IN AFRICA

The largest and most important structure in Africa is the Vredefort Dome situated in South Africa (Table 10), to the south of Johannesburg. Two other structures in South Africa have been described in a previous volume—the Class II Pretoria Salt Pan (Milton and Naeser, 1971—in McCall, 1977) and Roter Kamm (Dietz, 1965—in McCall,

Table 10 Cryptoexplosion structures in Africa discussed in this part and part VIII

Name	*Class*	*Diameter (km.)*	*Age*
Vredefort Dome, South Africa	IVa/3	160	Early Proterozoic
Kalkkop, South Africa	V	.6	Recent
Nebiewale, Ashanti	VI	2.4	?
Richat, Mauretania	IVa?	50	Post-Ordovician
Temimichat, Mauretania	V	.5	Post-Ordovician
Semsiyat, Mauretania	V	5	Post-Ordovician
Tenoumer, Mauretania	IVa or II	1.8	End-Tertiary or Quaternary
B.P., Libya	IVa/3	11.5	Post-Nubian (Cret.)
Oasis, Libya	IVa/4	2.8	Post-Nubian (Cret.)
East, Libya	V	3.5	Post-Nubian (Cret.)
*			

*Eleven structures of class V from Algeria, Mali, Mauretania, Morocco and Libya are listed in Table 2: little is known about them, and they are therefore omitted from this table.

1977). The Kalkkop structure in Cape Province is briefly described in Part VIII herein. Structures in western Africa include the Class II Bosumtwi Crater (Maclaren, 1931; Rohleder, 1936; and Littler et al., 1961–in McCall, 1977) and the Nebiewale structure in Ashanti, which, though obscure in origin, is almost certainly endogenic (Rohleder, 1937). The Richat, Tenoumer, Temimichat, and Semsiyat structures in Mauretania form a group with the previously described El Aouelloul crater (Monod, 1952–in McCall, 1977). There is a further group of three cryptoexplosion structures in Libya, the B.P., Oasis, and East structures.

MAURETANIAN STRUCTURES

Monod (1952) described these structures together with El Aouelloul (see Figs. 3 and 4). They are:

Temimichat—crater; 500 m diameter; in granite, with no lava flow association—probably an explosion crater, but origin unknown.

Tenoumer—crater; 1800 m diameter; in Precambrian rocks, but

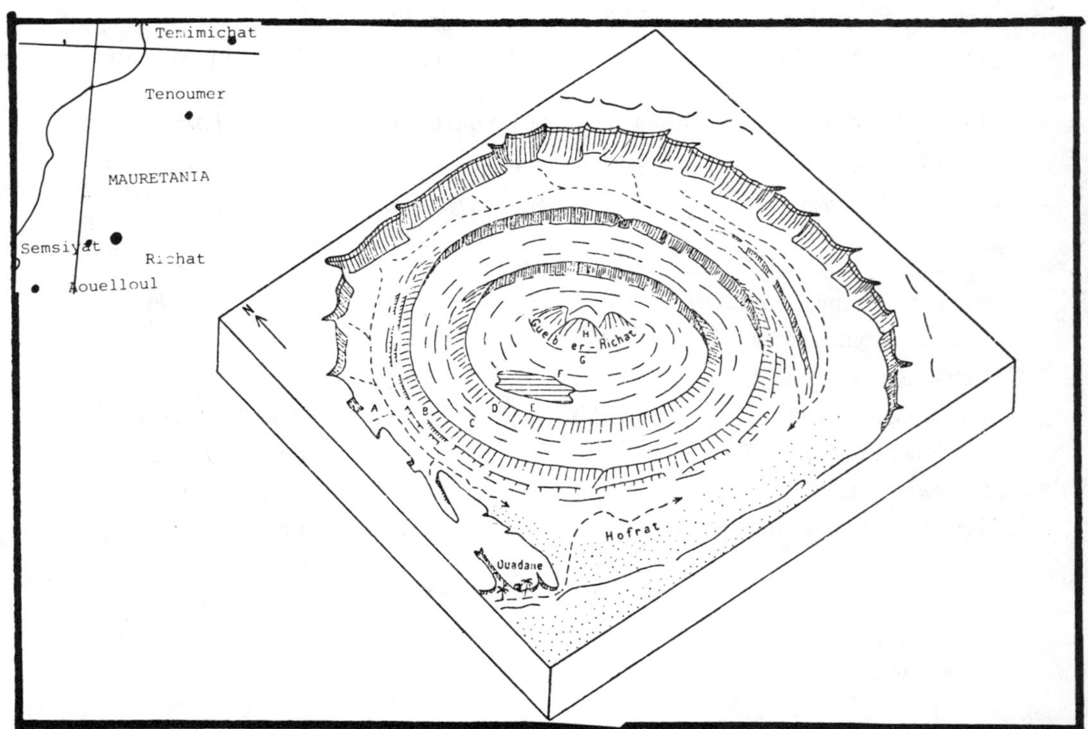

Figure 3. Sketch of the Richart structure, Mauretania. Inset shows the linear arrangement of the Mauretanian structures (after T. Monod, 1952).

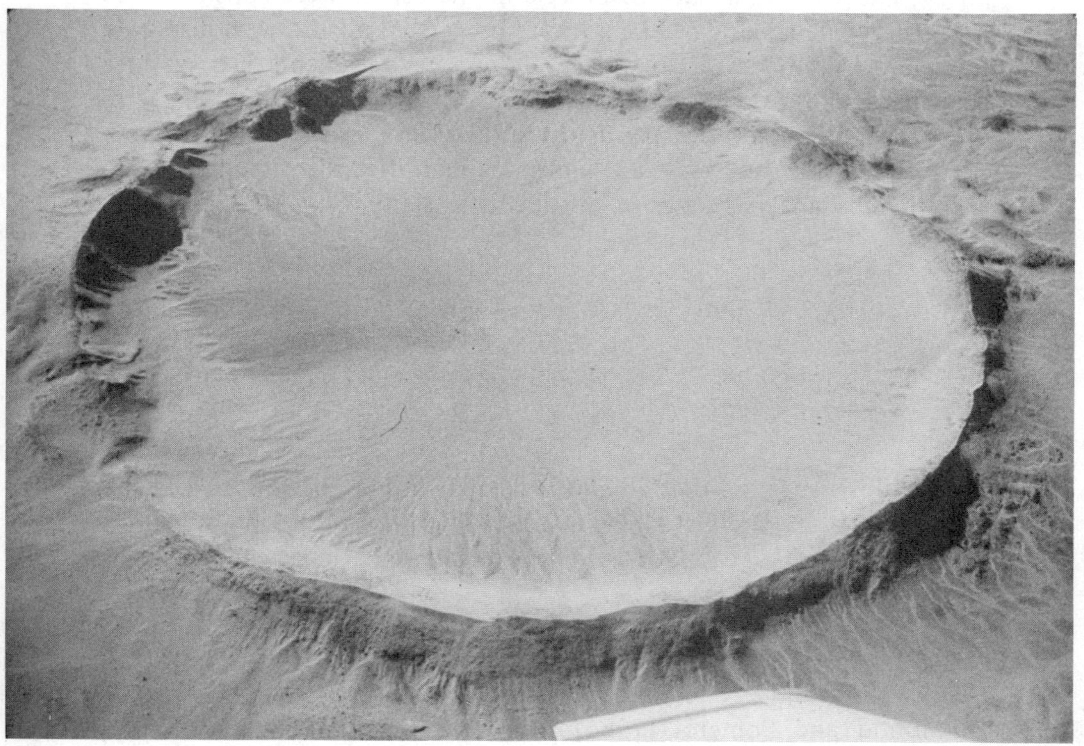

Figure 4. The Tenoumer crater, Mauretania, viewed from the air. (Photo by R. Fudali; reproduced by permission)

with associated basalt flows and disturbance of recent sedimentary cover—an explosion crater of end-Tertiary to Quaternary age, situated astride a northeasterly fracture line.

Richat—"pseudoboutonniere"; a circular, periclinal outburst in Ordovician grits; 50 km diameter; cut by an erosion surface; there is a double seam of gabbro and dolerite, and some evidence of rhyolite activity; post-Ordovician in age—it may be due to a laccolithic push.

Semsiyat—a very flattened dome; 5 km diameter; situated within a plateau of Ordovician grits; faults and dolerite occur nearby, but may be unrelated to the structure, which has a small central depression—Monod favored laccolithic origin, considering the structure to be a smaller, embryonic analog of Richat.

Richat

Cailleux et al. (1964) noted that all five structures are aligned. Richat, the largest, is eroded and appears to be older than the others. The fact that drainage is inward and becomes more incised at the periphery, and the occurrence of breccia at the center were considered to be against igneous origin. The grits are brecciated in situ, with displaced fragments recemented. Brownish microscopic fragments are set in a brownish-red ferruginous matrix. Coesite was identified by diffractometry, but not stishovite. Bardrossy et al. (1963) described analcimolitic quartzites at the focus: analcime was identified diffractometrically, and chemical analysis supports this identification:

SiO_2:	60.15	MnO:	.01	K_2O:	.65
TiO_2:	.90	MgO:	.50	P_2O_5:	.06
Al_2O_3:	18.10	CaO:	1.35	H_2O+:	6.50
Fe_2O_3:	2.15	Na_2O:	6.90	H_2O-:	1.00
FeO:	.20				

This analysis suggests that alkaline magmatism may be involved in the Richat structure. Dietz et al. (1969) could find no coesite in the breccia nor other shock deformations. They considered Richat not to be of impact origin.

Tenoumer

Monod and Pomerol (1966) reassessed the Tenoumer structure, finding the lavas to be rhyodacites and not basalts. They interpreted the structure as a subsidence caldera related to diking. Fudali and Cassidy (1972) reported a negative gravity anomaly: they noted a unique array of concentric dikes forming an outer ring, and contained in these are abundant xenoliths of country rock displaying shock metamorphism. The gravity anomaly suggests that there is an infill of at least 250 m of sediments in the crater, giving a depth/diameter ratio of about 1/8, the same as that of the Arizona crater and consistent with meteoritic origin. There is also a negative gravity anomaly at Temimichat, but this indicates a depth of infill giving a depth/diameter ratio of 1/15, a highly anomalous value in terms of impact theory, as is the case with the El Aouelloul crater.

LIBYAN STRUCTURES

The three Libyan structures are situated near Kufra. The B.P. and Oasis structures display the unique shock metamorphisms. Martin (1969) described the first, a ring of low hills, 2.7 km in diameter, rising above the sand-covered desert surface of south Cyrenaica. Situated amid outcrops of Cretaceous *Nubian Sandstone* in the core of a major syncline, the structure is 280 km from the nearest Tertiary acid igneous rock outcrops to the south. There are three concentric rock outcrops, the inner showing chaotic high dips and standing 15 m above the level of the general desert surface; the middle showing outward dips up to 50°; and the outer showing inward dips up to 15°. The strata of the surrounding desert are flat lying. The inner outcrop has a diameter of 0.5 km, the middle ring a diameter of 1.4–1.9 km, and the outer ring a diameter of 2.7 km. The structure is of probable late Cretaceous age. French et al. (1974) noted that the orthoquartzite is intensely disturbed within the limits of the structure, and displays microscopic shock effects; cleavage planes and lamellations on $00\bar{1}$, $11\bar{2}2$ and $10\bar{1}1$ in quartz grains. As in the case of the Coconino Sandstone in Arizona (McCall, 1977, p. 151), response to shock differs in detail in the case of these slightly cemented sandstones compared with crystalline rocks, developments of lamellations parallel to $10\bar{1}3$ and $10\bar{1}2$ being not well developed. The B.P. Structure is essentially an asymmetric, concentric ring syncline with a central block. Concentric folds within the ring display gentle plunges: dips are up to 70° in the core, which exposed complexly folded, light colored sandstone, the oldest rock unit.

Oasis structure, situated 80 km to the south, 5.1 km in diameter and is a ring of discontinuous hills of *Nubian Sandstone*; it is 100 m high. There is no central block. Dips are mainly outward, and folding is intense; steep and overturned beds are present. Folds have tangential axes and gentle plunges. Shock effects are similar to those in the B.P. structure, and also include microbreccias of quartz and sandstone fragments, including glass, which occurs as flow-banded, dark-brown, partly devitrified fragments; also fragment-bearing glass rims to quartz-grain cores; and as a massive, slightly welded and recrystallised matrix to the breccia. The shock effects are believed to be compatible with pressures in the order of 50–100 kb. These two structures are of the same age and erosional state, and could be due to twin impacts? No shatter cones, megascopic breccias, diaplectic glass, meteoritic material, or Ni/Fe spherules are recorded from either structure. Though Libyan Desert Glass occurs in this region, none is associated with these structures, though its age (28 m yr, Störzer and Wagner, 1971) is compatible with the age for the structures indicated by the field relationships. This highly siliceous, dense glass is significantly different in character to that of tektites (Martin used the term "tektite," but this is not correct). The Libyan Desert Glass has no known source, and remains an enigma.

A third anomalous structure, the East Structure, 160 km north of Kufra Oasis, shows no shock metamorphism, only healed microfractures in quartz. Elliptical (3.5 X 1.8 km), it consists of moderately disturbed beds dipping as much as 10° inward. The relief is 25 m. There is no central block, or ring syncline. There is a low topographic annulus between the central area and outer rim. It is of Class V.

REFERENCES

Bardrossy, C.; T. Monod; and C. Pomerol. 1963. Decouverts d'analcimolites d'origine endogene dans les Richat (Adrar, Mauritanien). *Acad. Sci. Comptes Rendus* **256**:3934-3936.

Cailleux, A.; A. Guillemaut; and C. Pomerol. 1964. Presence de coesite, indice de hautes pressions, dans l'accident circulaire des Richat. *Acad. Sci. Comptes Rendus* **258**:5488.

Dietz, R. S. 1965. Roter Kamm, Southwest Africa: Probable Meteorite Crater. *Meteoritics* **2**(4):311-314.

Dietz, R. S.; R. Fudali; and W. A. Cassidy. 1969. Richat and Semsiyat Domes (Mauretania), Not Astroblemes. *Geol. Soc. America Bull.* **80**:1367-1372.

French, B. M.; J. R. Underwood; and E. P. Fisk. 1974. Shock Metamorphic Features in Two Meteorite Impact Structures, Southeastern Libya. *Geol. Soc. Amer. Bull.* **85**:1425-1428.

Fudali, R. F., and W. A. Cassidy. 1972. Gravity Reconnaissance at Three Mauretanian Craters of Explosive Origin. *Meteoritics* **7**:51-70.

Littler, J.; J. J. Fahey; R. S. Dietz; and E. C. T. Chao. 1961. Coesite from the Lake Bosumtwi Crater, Ashanti, Ghana. *Geol. Soc. America Spec. Paper 68,* p. 218.

McCall, G. J. H., ed. 1977. *Meteorite Craters.* Benchmark Papers in Geology, vol. 36. Stroudsburg, Pa.: Dowden, Hutchinson & Ross.

Maclaren, M. 1931. Lake Bosumtwi, Ashanti. *Geog. Jour.* **78**:270-276.

Martin, A. J. 1969. Possible Impact Structure in Southern Cyrenaica. *Nature* **223**: 940-941.

Milton, D. J., and C. W. Naeser. 1971. Evidence for an Impact Origin of the Pretoria Salt Pan, South Africa. *Nature Phys. Sci.* **229**(7):211-212.

Monod, T. 1952. Les accidents crateriformes ou circulaires du Sahara Occidental. *Bull. Dir. Mines, Gouvt. Gen. Afrique Occidentale, Francaise, Dakar* **15**:169-177.

Monod, T., and C. Pomerol. 1966. Le Cratere de Tenoumer (Mauretanie) et ses laves. *Soc. Geol. France Bull. ser. 7,* **8**:165-172.

Rohleder, H. P. T. 1936. Lake Bosumtwi, Ashanti. *Geol. Jour.* **87**:51-65.

Rohleder, H. P. T. 1937. The Nebiewale Caldera, Gold Coast. *Geol. Jour.* **90**: 546-549.

Storzer, D., and G. A. Wagner. 1971. Fission Track Ages of North American Textites. *Earth and Planetary Sci. Letters* **10**:435-440.

Editor's Comments on Paper 30

30 DIETZ

Vredefort Ring Structure: Meteorite Impact Scar?

The Vredefort Ring (Paper 30), with a diameter of 160 km, is the largest cryptoexplosion structure known in the world. At least a thickness of 13,000 m of strata is affected by the disturbance, the core of Precambrian granitic rocks being uplifted by at least 13 km. Controversial since first recognized, Vredefort has been attributed to tectonic, diapiric, magmatic, and meteoric agency. Dietz (Paper 30) presented the first modern statement of the meteoritic interpretation, following earlier suggestions by Boon and Albritton (1937, 1938) and Daly (1947). The discovery of shatter cones by Hargraves (1961) established the involvement of explosion and shock deformation. Dietz covers the history of meteoritic interpretations and illustrates how he envisages the formation of the structure. It should be noted that the granite core is not intrusive granite, but Archean granitoid basement, mechanically uplifted. The only true granitic intrusions associated with the structure appear to be the granophyres of the marginal ring dikes. Pyroxenitic intrusives have been discovered in restricted outcrops within the core and the marginal rocks of the core have been shown to be overturned like the enveloping sedimentary rocks (Nicolaysen, personal communication).

The meteoritic origin proposed by Dietz has been further considered by Manton (1965), who "unrolled" the structure, revealing the original attitude of the shatter cones in a classic study. While cones are often restricted to the central core, of cryptoexplosion structures, at Vredefort Hargraves found them in the collar rocks, where their development is controlled by the rock material. Hargraves noted that the structure had to be unrolled before focal orientation could be demonstrated. This fact shows that cones formed early and that the explosion took place early, prior to the deformation. Manton concluded:

> An axially symmetric stress field operative over so large an area of the Vredefort ring could only have been brought about by two natural forces: the shock wave radiating out from an explosion or gravity. Were the former true, the cone axes should exhibit spherical symmetry. It has been shown that this occurs when restoration of the sediments to

> the horizontal is made. In other words, the shock occurred before any structural deformation took place, and was directed downwards and outwards from a point which must have been close to the original land surface.* If gravity were the cause, the cone axes should be parallel and normal to the earth's surface. This condition was satisfied when the rocks of the collar were nearly vertical. One can be quite certain that shatter coning can only have occurred at one or the other of these times. The data that have been presented in this paper must be assessed together with the rest appertaining to the origin of the ring. On the one hand its similarities with cryptoexplosion structures suggest an extra-terrestrial origin; on the other, it seems to be closely interwoven in space and time with other events on the geological evolution of the sub-continent (cf. Nicolaysen et al., 1963). Compelling evidence as to which is correct should be given by the age of the shatter coning relative to the formation of the ring. So far this has proved tantalizingly elusive. (Manton, 1965)

Concerning shatter cone distribution, Manton recorded that cones had been discovered in an isolated occurrence in the Black Reef Quartzite, but not (surprisingly) in the dolomite of the Transvaal System. Coning also occurs in Ventersdorp lava, but it only develops in profusion in the Upper Witwatersrand and Jeppestown rocks further into the overturned collar. It is sparingly developed in Government Reef Shales, Hsopital Hill, and Orange Grove Quartzites, but not in the northern part of the Archean core, though schists in the southwest part display cones, and also granite in the southeast part. The distribution of coning is symmetrical to the axis of symmetry of the ring, but is eccentric. Alkali granites and contact metamorphic rocks related to them are coned, anomalously, for these are commonly supposed to be stocks intruded after doming. If there were only a single coning event, according to impact theory, these intrusions must have been emplaced prior to doming. This problem has been discussed by both Manton (1962) and Bischoff (1962). The metamorphics of thermal origin close to the alkali granites show cones, so the metamorphism cannot be impact-generated as Daly (1947) and Dietz (1961) have suggested. Manton recognized two distinct types of pseudotachylite, a tuffisite and an ultramylonite. Shatter coning took place prior to ultramylonization, but the relation of the coning to the tuffisite remains obscure, the two being mutually exclusive. The granophyre dikes are situated beyond the limit of coning, but, because they cut all the faults associated with the doming, should be of postdoming age and therefore also postdate the coning. Manton noted the possibility that coning can be produced under conditions of both violent shock (Shoemaker et al., 1962) and slow stress

*Compare Manton's conclusion with that concerning Popigay (Paper 6) of the same order of size. There penetration to 1.5 km. below the surface is suggested before explosion. There seems to be a considerable degree of discrepancy in depths of penetration before explosion favored by different authorities.

(Bucher, 1963). He discounts the conclusion of Ramsay (1961) that slaty-cleavage in the *Upper Witwatersrand* argillites is older than the cones, and that deformation preceded coning; though he does not completely establish the relationship between cleavage and coning. In a final discussion he notes the problem of explaining shatter cones related apparently to stored residual stress, considering that shatter coning is essentially a mode of tensile failure.

The meteoritic origin proposed has been vigorously disputed by Bucher (1963), McCall (1963) and Nicolaysen (1972). It was defended again by Dietz (1963b). The first two authors stressed the integration of the Vredefort structure in a chain of unusual eruptive manifestations along a lineament, including the Great Dike, Bushveld, and Trompsberg complexes. This regional "interweaving" is also noted by Manton (see above). Dietz (1962) suggested that both the Great Dike and Bushveld Complex are also due to impact, but this attribution appears far fetched—there is little or no evidence to support it. The situation of the Vredefort Dome in the exact center of the Witwatersrand Basin has been noted by Brock (discussion of Manton, 1965). This does appear to be a remarkable coincidence if the structure is an astrobleme, especially, as Manton has observed, in reply, considering that the formation of the Witwatersrand Basin long preceded the Vredefort event. The suggestion has been advanced that the entire basin is a product of impact, but there is sedimentological evidence of a prolonged, stage-by-stage development of the basin in the gold-bearing strata at Welkom, and this suggestion cannot be seriously entertained.

The meteoritic hypothesis must be seriously entertained, despite anomalies in the geological details. The possibility of more than one shatter coning event is, however, real; Simpson (1978) has lately reported shatter cones on fault breccia associated with the downfaulting of already overturned rocks, incompatible with an origin in impact explosion; and the structure is not unlike certain endogenic cauldron complexes such as Glencoe, Scotland (McCall, 1963), at least in certain details. The integration in a linear chain of apparently magmatic complexes remains by far the biggest anomaly. The focal situation to the Witwatersrand Basin is equally significant. The possibility that some form of diapirism related to intense geometric confinement at depth is the key to the endogenic origin of this and some North American structures has been suggested by Nicolaysen (1972)—at the present state of knowledge such ideas should receive at least equal attention as meteoritic hypotheses.

REFERENCES

Bischoff, A. A. 1962. The pseudotachylite of the Vredefort Dome. *Geol. Soc. South Africa Trans.* **65**:205–225.

Boon, J. D., and C. C. Albritton. 1937. Meteorite Scars in Ancient Rocks. *Field and Lab.* **5**:53-64.

Boon, J. D., and C. C. Albritton. 1938. Established and Supposed Examples of Meteorite Craters. *Field and Lab.* **6**:44-56.

Bucher, W. H. 1963. Cryptoexplosion Structures Caused from Within or Without the Earth (Astroblemes or Geoblemes?). *Am. Jour. Sci.* **261**:567-649.

Dietz, R. S. 1962. The Vredefort Ring Structure—A Reply (to Discussion by D. W. Bishopp). *Jour. Geol.* **70**:502-504.

Daly, R. A. 1947. The Vredefort Ring Structure of South Africa. *Jour. Geol.* **55**: 125-145.

Dietz, R. S. 1961. Vredefort Ring Structure: Meteorite Impact Scar? *Jour. Geol.* **69**:499-516.

Dietz, R. S. 1963a. Astroblemes, Ancient Meteorite Impact Structures on Earth. In *The Solar System, Vol. IV: Moon, Meteorites and Comets.* G. P. Kuiper and B. Middlehurst, eds. Chicago: University of Chicago Press, 285-300.

Dietz, R. S. 1963b. Cryptoexplosion Structures: A Discussion. *Am. Jour. Sci.* **261**: 650-664.

Hargraves, R. B. 1961. Shatter Cones in the Rocks of the Vredefort Ring. *Geol. Soc. South Africa Trans.* **64**:147-153.

McCall, G. J. H. 1963. Are Cryptovolcanic structures Due to Meteoric Impact? *Nature* **201**:251-254.

Manton, W. I. 1962. A Discussion of the Pseudotachylite of the Vredefort Dome. *Geol. Soc. South Africa Trans.* **65**:207-225.

Manton, W. I. 1965. The Orientation and Origin of Shatter Cones in the Vredefort Ring. *New York Acad. Sci. Annals* **123** (article 2):1017-1049.

Nicolaysen, L. O. 1972. North American Cryptoexplosion Structures Interpreted as Diapirs Which Obtain Relief from Strong Natural Confinement. *Geol. Soc. America Mem. 132*, pp. 605-620.

Nicolaysen, L. O.; A. J. Burger; and C. B. van Niekerk. 1963. The Origin of the Vredefort Dome Structure in the Light of New Isotopic Data. *Program Abstract,* I.U.G.G. Berkeley Mtg., July 1963.

Ramsay, J. G. 1961. Discussion on Shatter Cones in Rocks of the Vredefort Ring. *Geol. Soc. South Africa Trans.* **64**:156-157.

Shoemaker, E. M.; R. E. Gault; and R. V. Lugn. 1962. Shelter Cones Formed by High Speed Impact in Dolomite. *U.S. Geol. Survey Prof. Paper 477,* pp. 424-426.

Simpson, C. 1978. The Structure of the Rim Synclinorium of the Vredefort Dome. *Geol. Soc. South Africa Trans.* **81**:115-121.

30

Reprinted from pages 499–500 and 502–516 of *Jour. Geology* **69**:499–516 (1961)

VREDEFORT RING STRUCTURE: METEORITE IMPACT SCAR?[1]

ROBERT S. DIETZ
U.S. Navy Electronics Laboratory, San Diego 52, California

ABSTRACT

The peerless Vredefort Ring structure of South Africa defies rational interpretation by tectonic, volcanic, or magmatic forces. Following the heretical opinion of Boon, Albritton, and Daly, it is suggested to be a meteorite impact structure or "astrobleme." Using an impact velocity of 20 km/sec, it was found that an asteroid 2.3 km. in diameter would provide the 6×10^{28} ergs (1.4 million megatons TNT-equivalent) needed to create the original crater 40 km. across and roughly 16 km. deep.

An impact event adequately explains: (1) the apparent impulse direction which was radially outward from the center; (2) the shatter cones as shock-induced; (3) the bilateral symmetry as owing to an oblique impact; (4) the upturned and overtilted 16-km. thick collar as the effect of radial forces spreading out from the explosion focus; (5) the pressure and thermal metamorphism of the sedimentary collar as shock wave–induced; (6) the intensive fracturing and microshearing of rock with little relative displacement as shock wave–induced; (7) the pseudotachylite as a "shock impactite" produced by overpressures in excess of 10^5 bars; and (8) the granophyre dikes as injected crater-lining "impactite."

The uplifted granite plug now filling the original crater was caused by a combination of elastic rebound and isostatic forces. It would seem that, at Vredefort, an impact event transpired equivalent to the formation of Copernicus or Tycho on the moon.

INTRODUCTION

Straddling the border of the Orange Free State and the Transvaal in South Africa lies the peerless but enigmatic Vredefort Dome, or better termed, the Vredefort Ring. In years gone by, the question has been twice asked: Could this structure be the result of a giant meteorite impact?[2] Boon and Albritton (1937, 1938) included it among their possible impact scars but with little discussion. Daly (1947) endorsed this theoretical suggestion and developed it in some detail. This concept seems to have remained unaccepted, however, for, in reply to a recent inquiry, a South African correspondent wrote that, so far as he knows, no South African geologist believes in a cosmic origin of the Vredefort structure. Rather, it is accepted as an integral part of much larger regional structural pattern, all of which is explainable in terms of conventional geological processes.

Recently, I have become convinced (Dietz, 1947, 1959, 1960) that shatter cones are a definitive criterion for identifying meteorite impact scars or "astroblemes" (i.e., star wounds).[3] To further pursue this problem, I corresponded with three South African geologists inquiring about the possible presence of shatter cones at the Vredefort Ring. Subsequently, R. B. Hargraves, of Witwatersrand University, discovered their excellent development in the sedimentary collar of the Vredefort Ring (pl. 1). Shatter coning is a curious type of rock fracturing in which the rock is splintered and shredded with minimum dislocation and without regard for pre-existing lines of weak-

[1] Manuscript received December 16, 1960.

[2] Although the term "meteorite" is mostly used here, a cosmic "cannon ball" of the necessary giant dimensions would be almost certainly an asteroid perturbated from its orbit between Mars and Jupiter.

[3] The validity of shatter cones as a criterion of intense shock has been demonstrated by their artificial production by nuclear detonations and by explosives of high brisance, e.g., shaped-charges. And recently, Gault and Shoemaker (unpublished information) have created them by hypervelocity impact experiments using a light gas gun. Small plastic pellets were fired at 18,000 ft/sec at targets of Kaibab dolomite. At the bottom of the small impact craters, shatter cones were formed which, although minute, are perfect in detail. My efforts to find shatter cones at Meteor Crater were unsuccessful, but recently E. C. T. Chao discovered a small shatter-coned fragment of Coconino sandstone in the fallout debris. Although rare, shatter-coning is therefore present at Meteor Crater. Much of this fracturing is doubtless present in the unexposed rock strata beneath the crater floor.

ness. The cones seem to be explained by the passage of an intense shock wave. The argument of this paper is based not on finding a meteorite but on demonstrating that the Vredefort structure was invaded by an intense shock from a central-point source. This is considered tantamount to proving meteorite impact. Other natural ways of creating such shock, for example, by a natural nuclear detonation, seem to be impossible.

The Vredefort Ring is now the tenth locality in the world from which shatter cones have been reported. Two of these are Quaternary sites: namely, Meteor Crater in Arizona and the Ashanti Crater in Ghana. The others, all of the ancient cryptoexplosion ("cryptovolcanic") type, are Steinheim Basin, Germany; Wells Creek Basin, Tennessee; Flynn Creek, Tennessee; Kentland, Indiana; Sierra Madera, Texas; Serpent Mount, Ohio; and Crooked Creek, Missouri.[4] Shatter cones seem to have a common mode of origin and thus provide an "index fossil" which tie together all these curious "cryptovolcanic" structures. Does Vredefort Ring fit as a probable impact scar?

The evidence from shatter cones bolstered previous suspicions that the Vredefort Ring might be a giant astrobleme. Upon reviewing the literature, the evidence seemed so convincing that I considered it worthwhile to restate the case. The scale of the presumed events at Vredefort is so grand that the general hypothesis for the existence of astroblemes anywhere must stand or fall by the evidence it presents. If this is not an impact scar, then it would seem there are none this side of the moon.

[4] Six structures were described by Bucher (1935) in his classic paper on the "cryptovolcanic" structures of the United States. All of these have been searched for shatter cones either by myself or by colleagues. Wells Creek Basin, Kentland, and Serpent Mound have yielded shatter cones. Upheaval Dome, Decaturville, and Jeptha Knob have not; and for other reasons, as well, they appear probably not to be astroblemes.

GEOLOGIC SETTING OF THE VREDEFORT RING

The center of the remarkable Vredefort Ring is occupied by the Vredefort granite, which has a circular outcrop 40 km. in diameter; a portion of the circle to the south is overlapped by younger Carboniferous Karroo rocks (figs. 1 and 2). Maree (1944), by interpretation of gravity data, concluded that the granite boss may be somewhat pear-shaped, being elongate to the southeast where covered. The granite is a coarse-grained, somewhat gneissic red or gray biotite-granite; and is assigned to the so-called "Older Granite" of the Transvaal. A girdle or collar of sediments some 16,000 meters thick encircles the granite, forming an amphitheater of hard ridges along the NW. quadrant of the perimeter.[5] These sediments, which are entirely unfossiliferous, are of Precambrian age and extend from the base of the Witwatersrand System upward through the Ventersdorp System and as far as the Pretoria Series at the top of the Transvaal System. They are tilted, and for the most part overtilted, beyond the vertical, so that they dip radially into the granite. This inversion locally affects no less than 10,000 meters of sediments.

The central granite is clearly older than the sediments, as shown by the absence of apophyses, by the lack of endomorphic modifications along the granite margins, and by the scale of the metamorphic phenomena. Profound metamorphism has affected the girdle of sediments; in places alterations by heat and pressure are traceable through the

[5] Various figures for the thickness of the sedimentary collar are reported ranging from 13,000 to 16,000 meters. The 16,000-meter thickness used here is taken from Daly (1947).

PLATE 1

Shatter cones, indicating high velocity and intense shock-wave fracturing, from the Vredefort Ring. Hargraves has found that these cones display a preferred orientation toward the center of the structure, indicating that the force arrived from the hub. Photos by R. B. Hargraves. Scale given in inches.

Shatter cones in quartzite from the Vredefort Ring structure of South Africa. From the "crater" collar

entire 3,000-meters thickness of the lower Witwatersrand System. These metamorphic effects are distributed concentrically, and in the area of maximum development, the altered rocks extend 6.5 km. from the granite margin outward.

Evidence of pressure of almost incredible intensity is provided by numerous faults and by an extraordinary profusion of shattered rocks, as well as by the inversion of 6,000–10,000 meters of strata. The crush zone between the central granite and the girdle of sediments was the dominant plane of weakness, as reflected by the distribution of marginal dikes, metamorphism, shattering, etc.

Both the Vredefort granite and the encircling sediments are cut in random fashion by countless veins of black vitreous rock known as pseudotachylite. This seems to be not a true igneous rock but a flinty-crush-rock, the result of intense compression, comminution, and *in situ* fusion. There are also large dikes of enstatite granophyre concentric to the granite periphery recalling the form of ring dikes. Their most characteristic feature is the extraordinary number of xenoliths of crushed quartzite from the lowermost Witwatersrand beds. These granophyre dikes are regarded by Hall and Molengraaff (1925) as flinty-crush-rock on a gigantic scale formed by ultratrituration and fusion.

Diverse views expressed over the years regarding the origin of the Vredefort Ring emphasize that the structure has defied clear understanding in terms of conventional tectonics. Hall and Molengraaff (1925) postulated that centripetal orogenic stresses caused a great updoming closely followed by passive uplift of the granite plug, pushed up, in turn, by deeper basic magma (fig. 3). Bailey (1926) made a questionable comparison between the Vredefort granite and the Arran granite boss in Scotland; he then considered that magmatic upheaval was adequate to account for the uplift and the associated enormous thrusts. Nel (1927) suggested either magmatic upheaval or centripetal pressures. Bishopp (1941), for mechanical reasons, insisted on the need for centrifugal forces causing overturning of the collar by magmatic upheaval and "toggle action." Maree (1944) made an appeal to centripetal and tangential pressures, magma being intruded for isostatic reasons. Daly (1947) reviewed the various concepts of origin of the Vredefort Ring by terrestrial forces and found them all wanting. Not only did the proposed mechanisms seem to him to be doubtful, but they also appeared quantitatively inadequate to explain the magnitude of the event. Brock (1950) proposed purely vertical uplift on the principle of the hydraulic jack owing to liquid basic magma which slowly forced the Vredefort granite plug upward as the surrounding geosyncline subsided. Like Bishopp, he considered the Vredefort cover to have been pierced rather than domed, and hence he uses the term "Vredefort Ring." The unsettled state of the Vredefort problem is emphasized by the vig-

[*Editor's Note:* Figure 1 and Plate 2 have been omitted.]

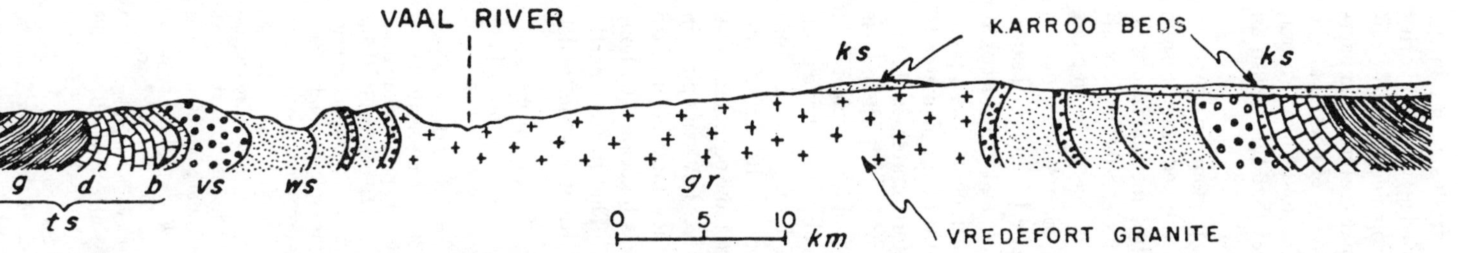

FIG. 2.—Idealized section across the Vredefort Ring structure showing the central granite boss, the sedimentary collar, and part of the great annular syncline. Redrawn and modified from Bishopp (1941). *gr*, Vredefort granite; *ws*, Witwatersrand System; *vs*, Ventersdorp System; *ts*, Transvaal System; *b*, Black Reef Series; *d*, Dolomite Series; *g*, Gatsrand Beds or Pretoria Series; *ks*, Karroo System.

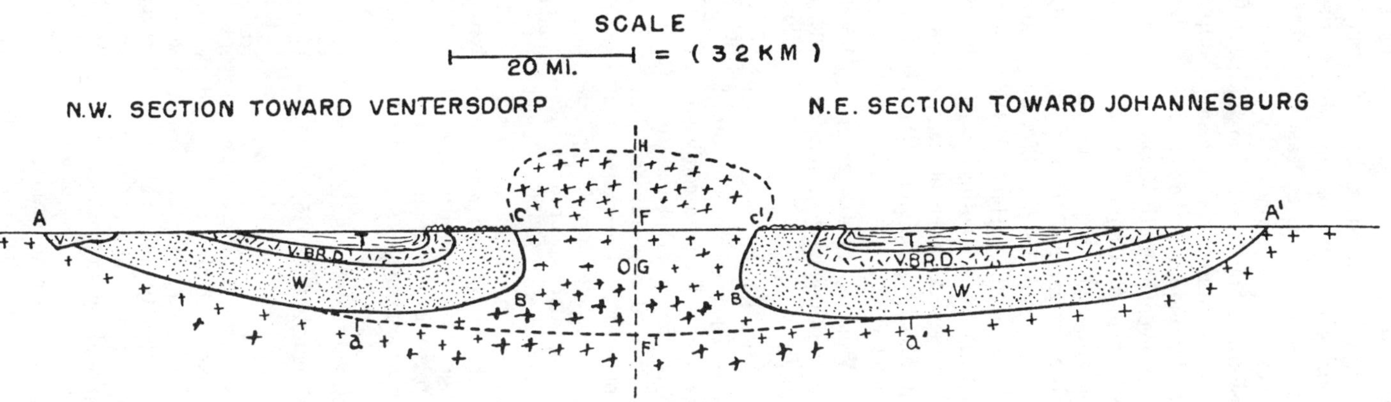

FIG. 3.—Axis of Vredefort Dome. Ideal section across the Vredefort granite and its encircling girdle of sediments. The shape of the hypothetical dome presumed by Hall and Molengraaff is shown, though Bishopp has shown that such a dome could not have existed. Redrawn from Hall and Molengraaff (1925, fig. 1).

$ACFC'A'$ = 132 miles; $ABC + A'B'C'$ = 130 miles; $AaF'a'A'$ = 135 miles; $ABC + A'B'C' + CHC'$ = 170 miles. W = Witwatersrand System; V, BR, D = Ventersdorp, Black Reef, and Dolomite; T = Transvaal System; OG = Old Granite.

orous discussion following Brock's paper which exceeds it in wordage.

Although no volcanic rocks are associated with the Vredefort event, it is commonly supposed that the structure is, at least in part, "cryptovolcanic," i.e., created by a volcanic explosion without extrusion of lava. The great 1888 Bandai-san explosion in Japan is generally quoted as the type example for such an event; but this was the splitting-off of the side of a mountain and occurred in volcanic rock. Explosions related to volcanism appear not to produce shock waves in rock; and the rims of volcanic craters and diatremes (Hack, 1941) are not uptilted. Steam explosions seem quite inadequate by several orders of magnitude to account for the prodigious forces unleashed at Vredefort.

According to Williams (1954), gas pressures involved in eruptions have been calculated most reliably

> by Minakami, from the muzzle velocities and trajectories of ejects. The maximum pressure of explosions from Asama volcano, according to Minakami (1939), is 563 atmospheres, corresponding to muzzle velocities of 212 m/sec. But few volcanic explosions involve pressures of more than 450 atmospheres, which approximates the pressure of water vapour in a critical condition when heated to 1200° C without change in volume. Most major explosions result from pressures between 100 and 300 atmospheres; in eruptions of Strombolian and Kilauean type, the pressures are very much lower and of only a few atmospheres.

NATURE OF THE PRESUMED COSMIC EVENT

Hall and Molengraaff (1925, p. 160) wrote:

> To anyone occupying a suitable point within the central (Vredefort) granite . . . the grand simplicity of design at once suggests a correspondingly broad simplicity in the major cause: almost inevitably one is reminded of the ring-like feature displayed by the craters of the moon . . . or even the geological history of the Rieskessel in Bavaria. The authors do not wish to suggest, however, that such analogies necessarily possess genetic significance in the case of the Vredefort Mountain Land.

This analogy deserves fresh scrutiny in the light of recent knowledge; the similarity may be more than a coincidence. Most scientists now accept lunar craters as impact-created; and the discovery of coesite at the Rieskessel by Shoemaker and Chao (Pecora, 1960) strongly suggests that this is also an impact structure.

Let us attempt to gain some insight into the possible nature of the necessary cosmic collision. Perhaps the best method is to scale up from a subsurface nuclear detonation, although the extrapolation has to be so great that we can only hope for a rough order-of-magnitude estimate. The 1.2 kiloton Teapot Ess crater (292 ft. across and 92 ft. deep) at the Nevada Test Site can be used as a point of departure.[6] The ratio of its diameter to the presumed Vredefort crater is 1:458. Using the scaling law of $W^{1/3.4}$, where W is the yield or "weight" of the explosive in kilotons of TNT-equivalent, we arrive at an energy requirement of 1.4 million megatons of TNT-equivalent, or 6×10^{28} ergs.[7] This is the energy needed to blow out a crater 40 km. across. If superlatives have any place in scientific discussion, they may be used to describe such a natural holocaust without fear of overstatement. The greatest earthquakes of 8.6 on the Richter scale, which

[6] Teapot Ess was a 1.2 kiloton event shot at the Nevada Test Site in 1955 at a depth of 67 ft.; it produced a crater 292 ft. across and 90 ft. deep.

[7] A meteorite impact is primarily a shock event; a nuclear detonation has a total energy distribution of 35 per cent thermal radiation, 15 per cent initial and residual nuclear radiation, and 50 per cent blast and shock (Glasstone, 1957). For this reason, the unknown apparent depth of the explosion, and because of the enormous extrapolation using poorly known scaling laws, the calculation of energy release at 6×10^{28} ergs is, of course, only a very rough approximation. Cratering experiments with explosives suggest that an empirical scaling law of $W^{1/3.4}$ is roughly correct; but for a big event, $W^{1/4}$ is said to be likely (Shelton *et al.*, 1960). This $W^{1/4}$ scaling would make the Vredefort event equivalent to 52 million megatons-TNT. A scaling law of $W^{1/3}$ holds for very small craters, but this breaks down for megaton charges, since the amount of work necessary to lift the material out of the crater is an increasing fraction of the yield; this is the so-called gravity effect.

have occurred only twice in the last century and which seem to represent the maximum ability of the earth's crust to store a stress, release about 2 per cent of this amount. By way of further comparison, the impact which created Meteor Crater, Arizona, is computed to be an event of about 5 megatons of TNT-equivalent.[8]

What size and velocity asteroid is needed to supply this amount of kinetic energy? Let us assume a typical meteorite impact velocity of 20 km/sec;[9] and let us further assume a mixed stony-iron body with a density of 5.[10] From the formula $E = \frac{1}{2}$ mv.^{2}, we can compute that an impacting spherical asteroid would have a diameter of 2.3 km. and a mass of 6×10^{10} metric tons. This is not unreasonable when we recall that astronomers have observed several small asteroids approaching perilously close to the earth. In 1927, Adonis passed by some 900,000 miles away; and in 1937, Hermes passed only 600,000 miles away. Regarding such "cosmic cannonballs," Baldwin (1949) states that, except for Eros which is several times larger, these little bodies are perhaps a mile in diameter.

The Vredefort Ring is more commonly termed the Vredefort Dome. It is abundantly clear that its central granite has been uplifted to the surface from a depth of a least some 16 km. But to term the structure a dome, suggesting that it was arched like a mushroom and pushed upward into the superincumbent strata, is highly questionable. Using reasonable assumptions, Bishopp (1941) showed that, if the now upturned collar beds are restored to their original horizontal position, they would approach within several kilometers of meeting at the bull's eye in the center of the granite. There is then only a small area of missing or eroded strata. In contrast, the Hall and Molengraaff dome concept required the erosional loss of a dome of sediments with a surface trace diameter of 60 km.! It seems better, therefore, to call the structure the Vredefort Ring, a term that is more descriptive and has no genetic implications.

According to the impact concept, the sedimentary collar is the outcrop trace of a giant 40-km.-diameter and 16-km.-deep meteorite explosion crater, into the throat of which a granite plug has been upthrust. In an impact the total energy is released from a small linear source, most of the energy being released several km. below ground zero. Most significantly, this means that the Ring was deformed from an almost point source pushing symmetrically from the center outward. This concept is in sharp contrast with most of the tectonic theories suggesting inwardly directed or tangential stresses. Bishopp (1941), through a consideration of the deformation mechanics, has shown rather well that the great compression was not caused by forces pushing radially inward. Instead, forces of radial compression were transmitted outward from a central focus, and their maximum intensity and accompanying displacement occurred at the sedimentary collar. Bishopp believed that magmatic pressures from below the old granite might account for this. Without at-

[8] A figure of 1.7 megatons was computed by Shoemaker (in press) and a 3-megaton figure is given by Harrison (*in* Merritt, 1959, p. 23). This is probably low as both authors used $W^{1/3}$ scaling, which ignores gravity effects. A.E.C. tables (Glasstone, 1957) indicate that a 30-megaton event detonated at the surface would blow a hole 3,300 ft. across and 250 ft. deep. If buried deeply, a much larger crater would form, but even so, it probably would not be larger than Meteor Crater.

[9] The relative impact velocity of meteorites, which are members of the solar system, may vary from 10 km/sec for a tail-on collision to as much as 70 km/sec for a head-on collision. Hence the mass of the meteorite could be as little as one-twelfth of that suggested here; on the other hand, it could also have weighed four times as much. By way of example, the bolide creating Meteor Crater was probably of an atypically low velocity. On the other hand, the Tunguska 1908 Siberian impact was probably of high velocity as suggested by its arrival at 0700 (at this time, the direction of earth travel is close to the zenith direction), the enormous blast effect with little cratering, and the failure to discover any ponderable meteorite fragments. But possibly this absence is explainable by the bolide being a comet head composed of a free radical ice (NH_3,OH, etc.).

[10] It is assumed here that a body of these dimensions would most likely be composed of mixed stone and nickel-iron.

tempting to point out some dubious aspects of the assumed magmatic force, cannot this centrifugal displacement be better explained by meteorite impact? As is well known, large explosion craters are built more by the outward displacement of rocks than by "spalling," i.e., ejection of debris ballistically through the air.

EVIDENCE OF SHOCK

There are many references by Shand and others to the surprising lack of gross slippage despite the intense powdering and comminution of the rock at Vredefort. Crystals of quartz are commonly cracked and converted into the mosaic of small grains, the twin lamellae of the feldspars are frequently bent and broken; and wavy optical extinction is prevalent in both these minerals. Pulverization of rock is, of course, a micro-shearing; it is macro-shearing that is lacking. Moreover, the geologic map of the Ring shows comparatively few faults when considered in the light of the intense over-all deformation.

Hall and Molengraaff (1925, p. 99) noted that

> the crushing . . . is not accompanied by much displacement or shearing. . . . Small displacements of the micro-faults, however, occur quite regularly. One gets the impression that shattering, pulverization, and mylonisation . . . have been caused by some sort of omnilateral or static pressure.

But, could not this crushing have been caused by a spherically fronted shock wave emanating from an impact source? Micro-shearing is the natural consequence of such a wave; witness the effect of a shaped-charge or other high brisance explosive! The overpressure along such a shock front is uniform; in the absence of free surfaces, crushing would occur, but conditions would be unfavorable for large differential offsets, faulting, and the like.

RING SYNCLINE

The Vredefort deformation does not end at the outer margin of the upturned sedimentary collar. A great ring syncline, about 60 km. across, surrounds the collar so that the over-all diameter of the deformed structure may be as much as 200 km.; that is, if the syncline is considered as part of the structure as Bishopp believed. Alternatively, it might be supposed that the Vredefort deformation originally occurred in the center of the pre-existing dish-shaped geosyncline; in this case, the annular syncline would simply be a passive consequence of the central upheaval rather than a true part of the structure.

One can imagine plausible tectonic reasons why an upheaval might occur in the center of a geosyncline. But an impact is unlikely to be so coincidentally placed. The annular syncline at Vredefort might have been thrown down by the same convulsive spasm that heaved up and overturned the sedimentary girdle.[11] Boon and Albritton (1936) theorized that a deformation created by meteorite impact should have the form of a damped wave, i.e., a central domical uplift encircled by one or more annular synclines and anticlines of rapidly diminishing amplitude. They drew an analogy between an impact deformation and that of any suddenly disturbed elastic medium. Rocks are said to have elasticity of volume if not of shape. Their analysis of the mechanics, however, seems more intuitive than theoretically rigorous. Their assumptions have yet to be confirmed or denied by adequate experiments or by any sophisticated theoretical analysis without doubtful premises. Yet there is adequate field evidence for ring synclines and anticlines at some of the presumed impact sites containing shatter cones. The ring syncline surrounding the Ashanti Crater is a good example (Junner, 1937); around the Wells Creek Basin structure there is another (Wilson, 1953); and we could list others.

[11] There is still another possible explanation. From properly scaled model experiments, it is known that, when the crust is placed under tectonic stresses, it will buckle down rather than up. Less energy is required to displace the substratum than to oppose gravity. If this law is valid for rapidly applied stresses (which seems somewhat doubtful), it might explain the toric syncline.

BILATERAL SYMMETRY

Maree (1945) emphasized that, when the whole structure as well as the gravity anomalies is considered, a remarkable axis of symmetry is revealed striking from 148° to 328°. He writes, "This axis seems to be the key to the solution of the tectonic forces which produced the Vredefort structure." To him, this indicated deformation by some sort of tangential couple. But cannot this bilateral symmetry, with its maximum overturning and disturbance to the north, be better explained by an oblique meteorite impact arriving from the direction of 148°? Boon and Albritton (1936) have already indicated bilateral symmetry of impact structures as one of their basic criteria. Wilson (1953) has also stressed the bilateral symmetry at the Wells Creek Basin structure, which he considers to have been caused by impact. Shock forces from an impact event would be highly vectored so that bilateral structural symmetry would result. Lunar craters provide evidence that angle of impact has little effect on the high circularity of meteorite explosion craters, but the underlying rock structure may well display bilateral symmetry.

METAMORPHISM OF THE COLLAR

The occurrence of metamorphic rocks all around the central granite originally raised the question, does the granite intrude into the sedimentary girdle? But Hall and Molengraaff demonstrated the older granite to be pre-Witwatersrand in age. Also, the granite shows no fine-grained selvages but remains coarse grained along the margins as well as inside its mass. Moreover, there are no chill phenomena or more basic marginal phases, and there are no granitic apophyses. In addition, the absence of endomorphic phenomena negates an intrusive relationship between the granite and its sedimentary collar.

Hall and Molengraaff considered one further aspect to be especially remarkable. Although the granite is only 40 km. across, the metamorphic contact belt is 6.4 km. across. The ratio between the area of granite and the metamorphosed sediments is roughly 3:2. Such a ratio was considered to be out of all reasonable proportion. They concluded that the metamorphism could not have been caused by the central granite.[12]

Metamorphosed rocks occur all around the Vredefort granite; what then is the cause? Hall and Molengraaff suggested polymetamorphism or superposed metamorphism. The great weight of the overlying formations was thought to have been important as well as the unusually great orogenic forces generated by the updoming. But, if the sedimentary collar is visualized as the walls of an impact crater, would not one expect the type and degree of metamorphism observed? Certainly the immediate wall of the crater would be most strongly metamorphosed; and the alterations due to heat and pressure would rapidly decay with distance from the center. In view of the tremendous energy unleashed, alterations to a distance of 6.4 km. do not seem to be unreasonable. The shock wave would have had an intensity of many megabars, releasing great heat as it decayed. Hall and Molengraaff themselves suggest that the crushing of rock could have been the source of the heat.

In some accounts, e.g., Maree's, various intrusive bodies are presumed to have been intruded at different times; but all are said to have a direct connection with the Vredefort deformation. The impact concept implies that the entire structure was instantaneously and completely formed except for isostatic modifications; only solidifaction of the pseudotachylite veins and granophyre dikes would be penecontemporaneous and slightly later than the main spasm. This concept stands in bold contrast to the tectonic concepts which envision slow development punctuated by sudden impulses. Since the alkali granite and nepheline syenite dikes are invaded by pseudotachylite and show the effect of shock, their injection must have preceded the presumed impact event

[12] The great volume of metamorphosed rocks has been explained by some writers as due to unexposed intrusions. Hargraves (personal communication) states that much of the metamorphism is distributed excentrically rather than concentrically and that most of the metamorphism may be pre-event.

and have no genetic connection with it. The mystery of their origin becomes a separate problem.

PSEUDOTACHYLITE

Remarkable features of both the central granite and the sedimentary collar at Vredefort are the anastomozing jet-black veins and stringers of pseudotachylite. This word was coined by Shand in 1916 at Vredefort to identify lustrous veins that superficially resemble both tachylite and trap and also appear to be related somewhat to mylonite and flinty-crush-rock. The veins are irregular in form, direction, and thickness, following sinuous courses as they widen and narrow. They anastomose in a complex manner forming intricate networks. The veins vary in width from 1 mm. to 20 m. and contain many inclusions, becoming in places breccias with a pseudotachylite cement. Externally, the rock has the appearance of a dense, somewhat vitreous, opaque black basalt. Although the effects of strong pressures and crushing are unmistakably present, there is little shearing. Crystals are commonly cut across sharply as though by a knife. All pre-Karroo rocks are invaded by pseudotachylite with the solitary exception of the enstatite granophyre dikes. Many pseudotachylite pockets are completely isolated, indicating their formation *in situ*.

Inclusions in the pseudotachylite show the effects of powerful corrosion by melting (Shand, 1916). High temperatures are indicated, since even the feldspars are fused. Although describing the luster of the pseudotachylites as vitreous, Shand could not determine whether they contain true glass or are entirely microcrystalline. Later, an X-ray analysis by Willemse (1938) showed that the pseudotachylite is microcrystalline, but it may once have been glassy.[13] Chemical analysis shows a close similarity in composition between the pseudotachylites and the immediately surrounding country rock. They appear to have been derived from a wide variety of material, both sedimentary and igneous. Their total bulk greatly surpasses the mass of similar rock known anywhere else in the world.

Shand withdrew his early opinion that the pseudotachylite was injected and concluded that it was generated *in situ* by rock fusion or by gas fluxing. In the absence of obvious shears, he thought that the fusion caused was by shock rather than by slippage friction. He further supposed that the rocks had been invaded by an intense impulse, or series of impulses, of some deep magmatic origin. Hall and Molengraaff (1927) agreed that the pseudotachylites are not true igneous rocks and that the veins are not true igneous veins. They agree that the rocks are crushed without much shearing and presumed that the heating effect of some extraordinary mechanical work raised the temperature of the crushed rocks above their melting point. The fused rocks then solidified as pseudotachylite.

The extreme rarity of pseudotachylite in rock strata suggests that special conditions are needed for its formation. Might it not have been formed by comminution and frictional heat generated by a decaying intense shock wave? Such rock fusion would have been facilitated by the ambient geothermal heat (probably above 300° C.) of the deeply buried Vredefort granite and the lower portions of the sedimentary sequence. Nuclear bomb experiments show that rock *will* fuse at shock pressures in excess of 10^5 bars (Johnson and Violet, 1958). Can instantaneous overpressures of this magnitude be produced by any natural event other than a meteorite impact? How else could superheating of the pseudotachylite melt have occurred in the presence of meltable country rock, unless it was instantaneously induced? Field evidence clearly demonstrates that the pseudotachylite is not a true intrusive igneous rock; but must have formed *in situ*. Rapid cooling in the absence of water and other fluxes, such as are present in true magmas, might account for its microcrystalline or its possible original glassy texture. Formation of the pseudotachylite by the shock following a meteorite impact seems not only

[13] The pseudotachylite, although crystallized now, may have been glassy originally. Glasses are unstable and devitrify. Grout (1932, p. 233) states that there are no fresh glasses of Archean age; indeed, very few are older than Cretaceous.

an adequate but the only possible explanation. It could be termed a "shock-impactite." If it proves to be a shock melted rock, it might usefully be termed a "shocktite" to differentiate it from "impactite"—the term used to describe melted rocks found at some meteor craters. Impactite is also a product of shock melting, of course, but in the immediate vicinity of the striking meteorite.[14]

GRANOPHYRE DIKES

The enstatite granophyre dikes deserve special scrutiny because they are the only rocks affected by the Vredefort event which are *not* invaded by the pseudotachylite, and they are the only ones not shattered. Consequently, they must be younger than the pseudotachylite or were formed at the same time. The dikes run nearly parallel to the periphery of the granite. They are vertical or nearly so; they vary in width, but 35 meters is average. Hall and Molengraaff (p. 111–112) noted certain petrological similarities between the pseudotachylite and the basic granophyre; the apophyses and selvages especially suggested to them that the two rock types are closely related in mode and time of origin. Chilled selvages are crystallized chiefly as spherulites which closely resemble the habit of much of the pseudotachylite. The combination of small phenocrysts of both rhombic and monoclinic pyroxenes, plagioclase, and biotite with the abundant granophyric groundmass were said to make the rock unusual. These phenocrysts are not shattered in contrast to the abundant macerated inclusions.

The granophyre is uniform in composition, but, unlike the pseudotachylite, it seems not to resemble the immediately surrounding country rocks. On the other hand, it is not a normal differentiation product but a magma-type of very abnormal composition (Willemse, 1937). The moisture content is very low, suggesting that the assimilated material was dry. The composition is explained by Willemse as possibly that of an alkali lamprophyric magma which has assimilated quartzite and granite.

Hall and Molengraaff (p. 112) doubted that the basic granophyre was a true igneous rock derived from a deep-seated magma. They suggested that it more likely is "a glorified form of pseudotachylite, i.e., a flinty-crush-rock on a gigantic scale originating from the ultra-trituration and fusion of different rocks."

Is it not possible that the enstatite granophyre may represent dikelike intrusions of the fused rocks which originally lined the cavity of the Vredefort crater at the time of the impact? Fused rocks or "impactites" are well known on a small scale from Quaternary meteorite craters, e.g., Wabar, Meteor Crater, Henbury, and Alloueloul; the coesite-bearing suevite of the Rieskessel may be such an impactite on a large scale. Fusion apparently takes place directly beneath the striking meteorite and in front of the zone of rock pulverization. During an event on the presumed scale of Vredefort, fused rock would be generated in great volume; glazed rocks would line the cavity and a pool of superheated rock would fill the bottom.

Published chemical analyses of the granophyre include no assays for nickel. It would be interesting to know whether traces of this metal, so common in meteorites are present in more than average amounts.[15] Some glassy impactites *do* reveal a high nickel content. It would also be worthwhile to know whether coesite is present.

[14] Shoemaker (personal communication) suggests that the apparent isolation of the pockets of pseudotachylite does not necessarily mean that they were produced *in situ* any more than does the occurrence of isolated blobs of glass in the breccia produced by nuclear explosion. In nuclear explosions the glass apparently is derived from the central very strongly shocked region and is dispersed by some still poorly known processes of explosive injection and mixing. It seems to me, however, by complex focusing effects, differential fusibility of rocks, etc., that hot spots inducing local melting might form far outside of the inner zone of complete fusion.

[15] Nickel-iron or siderite meteorites are the most common in world collections but, for observed falls, "stony" meteorites are much more common. Out of some 600 falls, only 23 are siderites. Therefore, the absence of traces of nickel in "impactites," etc., can hardly be disproof of meteorite impact origin. However, many meteorities other than siderites contain appreciable quantities of nickel.

SHATTER CONES

Hargraves (1960), as already mentioned, has found shatter cones in great abundance in almost all rocks of the Vredefort Ring. The significance of shatter cones as a criterion for meteorite impact has been fully discussed elsewhere (Dietz, 1947, 1959, 1960). They were found in almost all quartzite units of the Witwatersrand System, in hornfels ex Water Tower slates, in Jeppestown amygdaloid, in Ventersdorp amygdaloid, in epidiorite sills, in alkali granite, and in basic schist enclosed in the Vredefort granite. No positive evidence of shatter coning has been found as yet in the Vredefort granite but, as Hargraves points out,

> this may be attributed to the varying susceptibility of different rock types to shatter cone development: fine-grained, homogeneous rock being more favorable. . . . Weathering and erosion of granite takes place primarily by chemically-induced granular disintegration, a process not conducive to the exposure and preservation of shatter cones. Undoubted cones in the alkali granite (much finer-grained than the Vredefort granite) were found only on breaking open large residual boulders.

Alternately, it may be that the Vredefort granite surface now exposed laid below the shatter-coned portion which was eroded away after being uplifted.

Significantly, Hargraves has found *no* shatter coning of the pseudotachylite or the enstatite granophyre dikes. These must have been emplaced post-shock! Also, his study of the orientation of *in situ* shatter cones revealed a common preferred direction of the cone apexes, once the strata are restored to a flattened (preheaved) position. All this fits nicely into the impact concept sequence of events: first, invasion of the rocks by an intense shock wave forming the shatter cones; then, upheaval of the rock by the explosion pulse; and, finally, solidification of the shock-created pseudotachylite and enstatite granophyre. The situation also suggests that a *single* event created the Vredefort Ring rather than a series of explosions, as are characteristic of volcanic phenomena.

The preferred orientation was found by Hargraves to be roughly parallel to the bedding and at right angles to the strike of the upheaved strata. Cones in overturned beds pointed away from the center of the Vredefort Ring and, in strata tilted less than 90°, they pointed toward the bull's eye. If these beds were restored to their original horizontal position the cones would all point roughly toward the center. There is some variation in the average dip of the cone apexes in different beds, but, as Hargraves points out, this is probably a geometrical artifact arising from the preferential exposure of parasitic cones on larger master cones. It is thus apparent that the explosive force came from the central part of the Vredefort Ring.

UPLIFTED GRANITE CORE

Explanations of the Vredefort Ring as an impact structure must account for the central plug of uplifted non-intrusive granite that now chokes the throat of the presumed crater. At a very minimum, assuming that the top of the granite is not deeply eroded, a prodigious uplift of 16 km. is indicated.

Returning once more to the Boon and Albritton thesis, we might appeal to an elastic rebound mechanism. These authors suggested that, under the conditions of a hypervelocity impact, the rocks immediately beneath ground zero are elastically compressed rather than plastically pushed aside. Once a meteorite is fully decelerated, the highly compressed core rebounds elastically and then is damped by tensional fracturing, leaving a dome in the center of the surviving structure. There is some field evidence for such rebound in other presumed impact structures, although it is never quite certain that the central uplift is truly a dome. The Vredefort "dome" is itself a case in point. The Sierra Madera, a presumed impact structure in Texas, originally mapped by King (1930) shows a central dome. Shoemaker (personal communication), however, believes that this is fictitious; rather, the supposed dome is the upturned margin of the impact crater centrally filled with a megabreccia. Lunar craters may show evidence of elastic rebound. Such an effect might account for the central peaks found in many craters, especially for the remarkable central dome of Alpetragius. The mechanics of hypervelocity impact are not well enough understood to provide a definite answer.

Regardless of the validity of the elastic rebound concept, isostatic considerations provide the ultimate control on a structure of the enormous dimensions of the Vredefort Ring.[16] The geophysical principle is well established that large areas of the earth's surface are isostatically compensated; but for how large an area? According to De Sitter (1956), full compensation certainly applies to areas 100 × 100 km. and probably to any area greater than 20 × 20 km. The Vredefort granite, being 40 km. in diameter, should have yielded to isostatic forces. And, when the impact took place, the crust must have been thoroughly fractured down to at least 30 km. The strength of the crust against vertical shear would have been reduced to a low value, permitting the granite plug to rise easily under the isostatic driving force.

The granite is now, and was probably then, lighter than the overlying sediments. Maree (p. 190) gives the average density for the sediments as 2.80 and for the granite as 2.65. Perhaps the hot quasi-plastic granite was forced upward like salt into a salt dome. At the very least, the density differential would have caused the granite plug to rise unusually high before it became isostatically balanced relative to the sedimentary collar. The viscous rock that pushed up the granite plug was presumably the ultrabasic mantle rock at depth. Another important factor promoting the central uplift, in addition to existence of the crater void itself, would be the excessive topographic burden of the uplifted and ejected debris forming the crater rim.

How can one explain the granophyre dikes which are said to pass from the central granite plug without interruption into the sedimentary collar, especially if these dikes are impactites? Isostasy is generally supposed to be a sluggish process involving several tens of thousands of years. It would seem, therefore, that we must invoke extremely rapid isostatic adjustment or accept elastic rebound. On the other hand, the granite periphery in which the granophyre dikes are intruded is more likely a part of the upturned edge of the original crater rather than a central plug; it would be something of a happenstance if the rocks were peeled back exactly at the base of the sedimentary sequence. Presumably, the massive and strongly kneaded nature of the granite makes it difficult, if not impossible, to determine its precise attitudes.

Suppose we accept the idea that the granite plug reached complete isostatic compensation soon after the Vredefort event. Then consider the effects of the subsequent hundreds of millions of years of erosion. The heavy sedimentary collar and the light granite would be eroded more or less evenly. The over-all effect would be to leave the central granite area isostatically heavy relative to the collar. The granite would therefore have a tendency to sink, and it would show a positive anomaly, the amount depending upon the strength of the granite, now rewelded and with its fractures healed, to resist isostatic sinking. Does this account for the 30 milligal positive anomaly which Maree measured over the central granite?

Many questions must be left unanswered about the central granite. However, the existence of this great plug does not negate the impact hypothesis. Hall and Molengraaff themselves state that the central granite acted as a more or less rigid plug or buffer so that a plane of discontinuity or crush zone existed between it and the sedimentary collar. This seems more consistent with the explanation just stated than with the interpretation of the entire structure as a giant mushroom-shaped dome. How could a dome produce overtilting on the observed scale?

THE IMPACT EVENT

We may now attempt to visualize the Vredefort impact event. Some time in the middle of late Precambrian or Paleozoic time an asteroid approaches the earth on collision course (fig. 4). Arriving from the south-

[16] Daly (p. 143) also uses isostasy to explain the central granite plug. He cites evidence that the crust was thinner and the mantle less viscous at the event time, enhancing the isostatic effect.

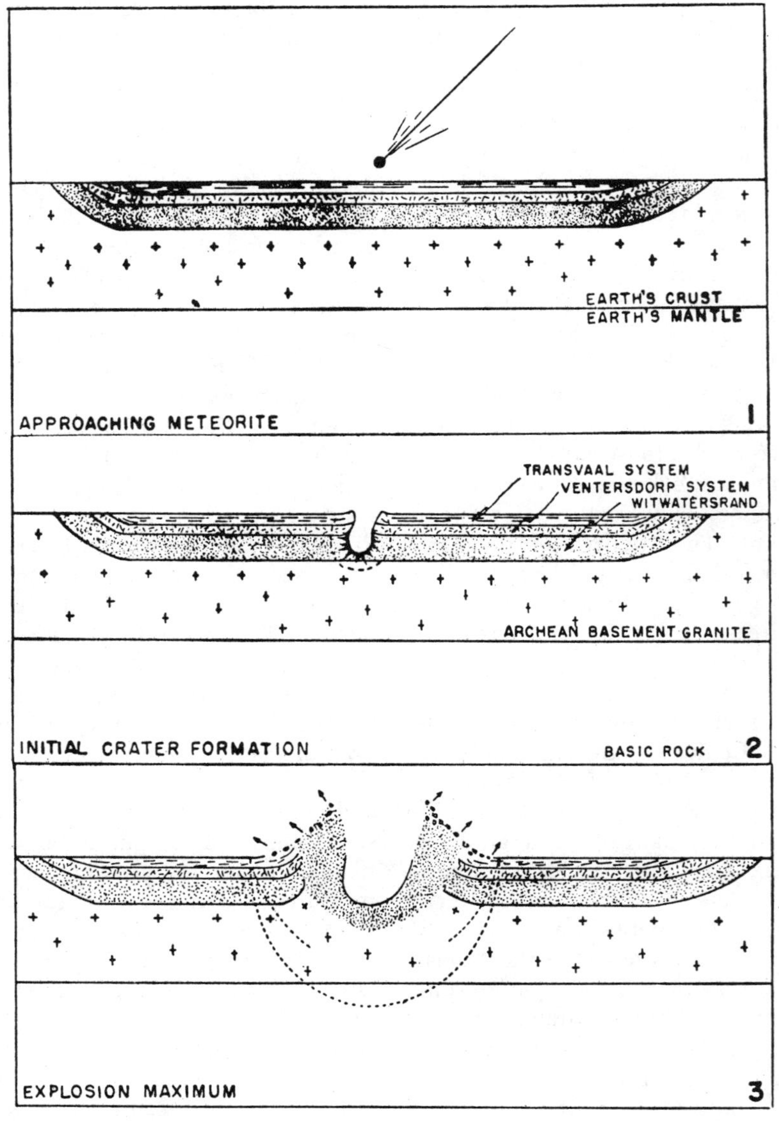

FIG. 4.—A schematic sequence of events which formed the Vredefort Ring structure according to the meteorite impact concept.

1, A giant meteorite approaches from the southwest at Vredefort probably in Precambrian time. 45,000 feet of sediments (the Transvaal System, the Ventersdorp System, and the Witwatersrand System) overlie the Archean granite basement. An initial basin of some type is assumed.

2, Immediately after impact, a crater begins to form. The meteorite is liquefied and gasified, becoming a liner inside cavity. The shock wave spreads out (*dashed line*) followed by fracturing.

3, At maximum crater development, the meteorite is a thin liner (*black line*); debris is spalled off; the formations are thrust upward and even overturned; a zone of intense brecciation is formed (*dotted pattern*); and the shock wave spreads farther out followed by a reflected rarefaction wave (*second dotted line*).

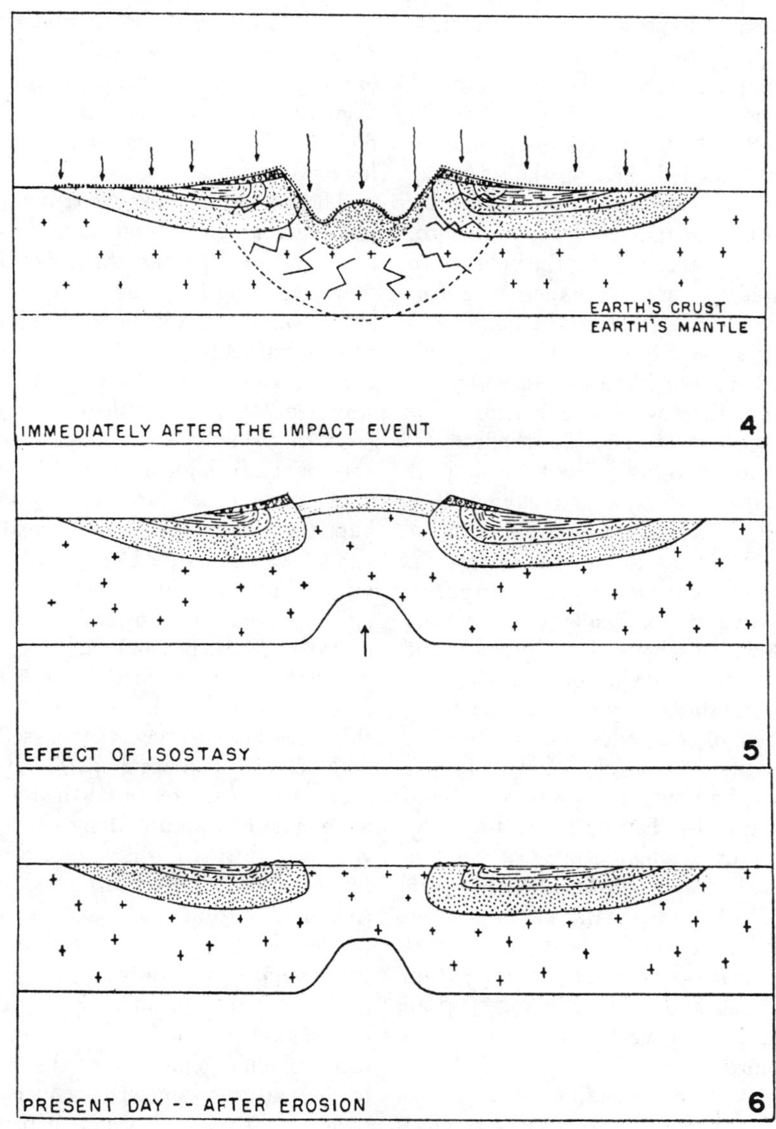

FIG. 4.—*Continued*

4, After the energy is dissipated, the earth recoils to form a central dome in the crater; surrounding the crater is a brecciated zone, and farther out a fracture zone is formed where shatter cones and pockets of pseudotachylite are developed. The fallout settles gradually. Strata to the NW. are overtilted more than those to the SW., owing to the direction of impact being from the SW.

5, After some time, isostatic forces come into play causing the original crater to be destroyed. The plug of Vredefort granite is pushed up by viscous basic rock of the earth's mantle.

6, Vredefort Ring structure at the present time. Erosion has leveled the structure considerably, removing the surficial deposits. The roots of the upturned bed remain as a mountain ring in those areas where the structure has not been overlapped by the Carboniferous Karroo beds.

east, it streams down through the atmosphere obliquely creating an enormous air blast. The impact of the meteorite at ground zero causes a brilliant flash of light.[17] As the front of the bolide penetrates the ground, a spray of fine liquefied particles shoots out at about twice the velocity of impact. Next, an intense shock wave radiates from point zero, compressing and accelerating the bedrock to make room for the onrushing meteorite. Another shock wave backs into the bolide, decelerating it so that the meteorite and bedrock move in unison. Heat generated by the shock waves causes wholesale melting.

The crater now forms. The bedrock is shouldered away rapidly from the impact point opening a dish-shaped cavity. The meteorite itself becomes a liquid liner of the cavity. As the crater grows, the bedrock is upended and even overturned. Meantime, the shock wave rushes ahead; the expanding crater follows closely behind. Between the two is a hemispherical shell of hot rock, compressed by the shock wave and stretched by the expansion of the cavity. The shock wave weakens as its volume of action increases, and it is followed by tension waves reflected from free surfaces. Tensile forces break up the brittle rock throughout a great volume. The smeared out meteorite, like a shaped-charge, drills down for many kilometers before its kinetic energy is spent. As an inert substance, it simply transfers its momentum to the bedrock and continues to set up an intense shock wave vectored somewhat toward the northwest.

The shock wave spreading from the explosion center strikes the surface; a rarefaction wave is reflected back from this free surface. The rarefaction wave, combined with the tail of the outward-moving shock wave, sets up tensile stresses far exceeding the strength of the rocks. Fractured and broken pieces of rock fly into the air with velocities corresponding to the velocity of the shock wave trapped within them. This "spalling" continues as the shock wave spreads until the stress falls below the tensile strength of the rocks. The outward rush of hot gases due to venting of the gas bubble from vaporized rock accelerates the upward flight of the fragments. Debris is scattered for many tens of kilometers.

The explosion pulse centrifugally impels the sedimentary beds outward, peeling them back and overturning them like a gigantic flower opening its petals to the sun. As the shock wave spreads farther, a great toric syncline is thrown down around the upturned beds. Still farther out, the shock waves decay into elastic waves which, attenuating slowly, engulf the entire African continent causing earthquake havoc.[18]

Shock overpressures of several million bars are generated with their greatest intensity along the crater cavity and rapidly diminish outward. Fused rock lines the crater and is forced into tension cracks to cool as enstatite dikes. Beyond this is an aureole of powdered rock. Still farther out, a large fracture zone is formed. The rock is intensely shattered and moves en masse but with little relative displacement. The speed of the shock wave is so great that the rock does not have time to fracture along natural lines of weakness; instead, it is shredded, with shatter cones spreading out from bedding discontinuities. Within the fracture zone, local pockets of rock are fused, eventually to become veins of pseudotachylite.

Finally, the spasm is spent. The elastic stresses around the target relax, and the surfaces rebound somewhat so that the volume of the crater is somewhat reduced. Directly beneath the impacting meteorite, the highly compressed plug of rock backfires and is tensionally damped as a central dome. Fallout debris rains down, and a great base surge of debris, a density current, moves out like a *nuée ardente*.

Sluggishly, isostatic adjustment takes

[17] This generation of a light flash and several other points are taken from the hypervelocity impact experiments described by Charters (1960).

[18] A meteorite impact would not be an efficient quake producer. Because of the short duration of the event, most of the energy would be absorbed in the shock wave before decaying into an elastic seismic wave. But it would seem, nonetheless, that the quake resulting from such impact would far transcend the maximum possible tectonic quakes.

place. The deep basement granite had a pre-impact ambient temperature of more than 300° C.; now, with the added impact heat, it is quasi-plastic. The deep viscous rock of the earth mantle hydrostatically shoves up a vast plug of Vredefort granite, gradually obliterating the crater as a topographic chasm.

Over the subsequent hundreds of millions of years, erosion removes all traces of the ejected meteorite fragments and of the surface rubble as well. But the deep-rooted upturned edges of the crater persist, somewhat denuded so that the hard quartzite ridges stand out but otherwise little changed from their original condition. Indeed, the whole structure survives not greatly different from the form it had shortly after the holocaust.

VREDEFORT RING AND LUNAR CRATERS

Those who espouse the cosmic impact origin of lunar craters, may see in the Vredefort Ring the beginning of an answer to the plaguing question: Why does not the earth show the effect of bombardment as the moon does? There is some logic in relegating most lunar craters to the very early, "pre-readable," geologic time. But there are some comparatively young craters on the moon as well; these are marked by ray systems which apparently are erased with time by the slow, but not completely stilled, lunar erosion. These rayed craters, at least, should have terrestrial equivalents. They may not look the same even disregarding erosion; for one thing, isostasy is a powerful force on the earth but insignificant on the moon for features the size of lunar craters.

Many factors make terrestrial impact craters much smaller than those on the moon. Lunar gravity is only one-sixth that of terrestrial gravity; hence with equivalent work, six times as much material can be excavated on the moon as on the earth. Moreover, the earth's atmosphere absorbs blast and acts as a damper, and the earth's rocks presumably have a greater mass and are better cemented. All in all, we can roughly estimate that, for equivalent impact events, a terrestrial crater would have one-eighth the volume of a lunar crater. In linear terms, such an earth crater would then have half the diameter of its lunar equivalent. The largest of the physiographically youthful craters on the moon are the rayed craters Tycho and Copernicus, each about 70 km. across. The Vredefort Ring, with its crater diameter of 40 km. would seem to record an event equivalent to the formation of a lunar crater 80 km. across. At the Vredefort Ring, then, an impact event may have transpired which was at least equivalent to that which formed the largest of the young craters of the moon.[19]

ACKNOWLEDGMENTS.—It is a pleasure to express my great appreciation to the following persons who critically reviewed the manuscript: Howel Williams (University of California); Francis P. Shepard and H. W. Menard (Scripps Inst. Oceanography); E. L. Hamilton, E. C. Buffington, G. H. Curl, J. L. Stewart, and A. F. Richards (U.S. Navy Electronics Laboratory); H. H. Hess (Princeton University); E. M. Shoemaker (U.S. Geological Survey); and R. B. Hargraves (Witwatersrand University, South Africa). In addition, Hargraves kindly provided detailed information about the shatter cones which he discovered; he also furnished specimens of various Vredefort rocks. Other assistance was provided by W. C. Speidel, I. E. Davies, D. Begnoche, and S. D. Baslee.

REFERENCES CITED

BAILEY, E. B., 1926, Domes in Scotland and South Africa: Geol. Mag., v. 43, p. 481–494.

BALDWIN, R., 1949, The face of the moon: Chicago, Univ. of Chicago Press, 239 p.

BISHOPP, D. W., 1941, The geodynamics of the Vredefort Dome: Geol. Soc. South Africa Trans., v. 44, p. 1–18.

BOON, J. D., and ALBRITTON, C. C., 1936, Meteorite craters and their possible relationship to "cryptovolcanic structures": Field & Lab., v. 5, p. 1–9.

[19] A calculation of the magnitude of the lunar Copernicus event by Shoemaker (unpublished information), using an independent theory of cratering and not involving scaling laws, shows close agreement with the magnitude of the Vredefort event as computed in this paper.

BOON, J. D., and ALBRITTON, C. C., 1937, Meteorite scars in ancient rocks: *ibid.*, p. 53.

——— — ——— 1938, Established and supposed examples of meteoritic craters: *ibid.*, v. 6, p. 44–56.

BROCK, B. B., 1950, The Vredefort Ring: Geol. Soc. South Africa Trans., v. 53, p. 131–157.

BUCHER, W. H., 1935, Cryptovolcanic structures of the United States: Rept. 16th Internat. Geol. Congress, p. 1055–1085.

CHARTERS, A. C., 1960, High speed impact: Sci. American, v. 203, p. 128–140.

DALY, R. A., 1947, The Vredefort Ring structure of South Africa: Jour. Geology, v. 55, p. 125–145.

DE SITTER, L. U., 1956, Structural geology, p. 333: New York, McGraw-Hill Book Co., 552 p.

DIETZ, R. S., 1947, Meteorite impact suggested by orientation of shatter cones at the Kentland, Indiana, disturbance: Science, v. 105, p. 76.

——— 1959, Shatter cones in cryptoexplosion structures (meteorite impact?): Jour. Geology, v. 67, p. 496–505.

——— 1960, Meteorite impact suggested by shatter cones in rock: Science, v. 131, p. 1781–1784.

DU TOIT, A. L., 1926, The geology of South Africa: 2d ed., Edinburgh, Oliver & Boyd.

GLASSTONE, S. (ed.), 1957, The effect of nuclear weapons: Atomic Energy Commission, Supt. Doc., Washington, D.C., 579 p.

GROUT, F. F., 1932, Petrography and petrology, v. 2: New York, McGraw-Hill Book Co., 522 p.

HACK, J. T., 1942, Sedimentation and volcanism in the Hopi Buttes, Arizona: Geol. Soc. America Bull., v. 53, p. 335.

HALL, A. L., and MOLENGRAAFF, G. A. F., 1925, The Vredefort Mountain Land in the southern Transvaal and the northern Orange Free State: Verh. Koninkl. Akad. Wetenschap. Amsterdam, v. 24, 183 p.

HARGRAVES, R. B., 1960, Shatter cones in rocks of the Vredefort Ring (unpublished manuscript).

JOHNSON, G. W., and VIOLET, C. E., 1958, Phenomenology of contained nuclear explosions: Univ. Calif. Rad. Lab. Rev. 1, 27 p.

JUNNER, N. R., 1937, The geology of the Bosumtwi Caldera and surrounding country: Gold Coast Geol. Survey Bull., v. 8, 38 p.

KING, P. B., 1930, The geology of the Glass Mountains, Texas: Univ. of Texas Bull. 3038, p. 123–125.

MACDONALD, G., 1954, Igneous rocks. *In:* EMERY, K. O., TRACEY, J. I., and LADD, H. S., Geology of Bikini and adjacent atolls: U.S. Geol. Survey Prof. Paper 260-A, p. 120–124.

MAREE, B. D., 1944, The Vredefort structure as revealed by a gravimetric survey: Geol. Soc. South Africa Trans., v. 47, p. 183–196.

MERRITT, M. L., *et al.*, 1959, Excavation: Plowshare sym., Univ. Calif. Rad. Lab. Rept. 2, pt. 2, 97 p.

MINAKAMI, T., 1939, Geophysical studies of the Asama volcano: Proc. 6th Pacific Sci. Congress, Berkeley, v. 2, p. 859–73.

NEL, L. T., 1927, The geology of the country around Vredefort: An explanation of the geological map: Geol. Survey South Africa (Pretoria).

PECORA, W. T., 1960, Coesite craters and space geology: Geotimes, no. 9, p. 19.

SHAND, E. J., (1916), The pseudotachylite of Parys: Quart. Jour. Geol. Soc. London, v. 62, p. 198–221.

SHELTON, A. V., NORKDYKE, and GOESKERMANN, 1960: Univ. Calif. Rad. Lab. Rept. 5766.

SHOEMAKER, E. M., Impact mechanics at Meteor Crater, Arizona. *In:* KUIPER, G. P. (ed.), The solar system: Chicago, Univ. Chicago Press.

WILLEMSE, J., 1938, On the Old Granite of the Vredefort region and some of its associated rocks: Geol. Soc. South Africa Trans., v. 40, p. 43–119.

WILLIAMS, H., 1954, Problems and progress in volcanology: Quart. Jour. Geol. Soc. London, v. 109, p. 311–332.

WILSON, C. W., 1953, Wilcox deposits in explosion craters, Stewart County, Tennessee, and their relations to origin and age of Wells Creek Basin structure: Geol. Soc. America Bull., v. 64, p. 753–768.

Part VI

STRUCTURES OF CLASS IV IN AUSTRALIA

Editor's Comments on Papers 31 and 32

31 MILTON et al.
Gosses Bluff Impact Structure, Australia

32 GUPPY, BRETT, and MILTON
Liverpool and Strangways Craters, Northern Territory: Two Structures of Probable Impact Origin

Large cryptoexplosion structures are rare in Australia, and this is probably a real condition and not due to lack of exploration, for the Precambrian shield has lately been subjected to intensive exploration for minerals. Such structures, if they exist in numbers comparable with Canada, the Central United States, Scandinavia, and Russia, would surely have been recognized by now, despite the lack of glacial erosion to emphasize the contrast between circular patterns and glacial grain. The major discoveries, besides Wolf Creek Crater (McCall, 1965; Taylor, 1965) are Gosses Bluff, an immense structure in Central Australia; and the Liverpool and Strangways structures, also in the Northern Territory. The structures in Australia are listed in Table 11.

Table 11

Name	*Class*	*Diameter (km)*	*Age*
Liverpool, Northern Territory	IVa/1	1.6	Cretaceous
Gosses Bluff, Northern Territory	IVa/3	22	Cretaceous
Strangways, Northern Territory	IVa/2 or 3	16	Proterozoic?
Lake Teague, Western Australia	IVb	25	Proterozoic
Dampier Archipelago, Western Australia	V	(elongated basin on similar scale to Sudbury)	Archean?

GOSSES BLUFF

Gosses Bluff, which is described here by Milton et al. in Paper 31, was earlier described by Cook and Crook in two papers (see reference list in Paper 32). Whatever its origin, it ranks among the most remarkable cryptoexplosion structures yet discovered. There are similarities to Sierra Madera (Paper 29), both being immense diapiric piercements of thick Phanerozoic sedimentary sequences extending up to the Cretaceous. Neither has any igneous rock occurrences within the diapiric structure and no clear relationship has even been established between the diapiric structures and neighboring volcanic rock occurrences.

LIVERPOOL

The Liverpool Crater, covered here by Guppy et al. in Paper 32, is of the same size as the Arizona crater. It is infilled with sandstone of likely Cretaceous age, and its wall consists of a ring of breccia derived from the Proterozoic sandstone country rock, exposed both at the foot of the rim mound on the outside and close to the crater in the surrounding country. Physiographic arguments favor an age close to that of the Cretaceous(?) fill. Deformation of quartz includes cleavage, mosaic structure, and decorated multiple planar elements (the decoration being in the form of trains of minute inclusions). This shock metamorphism is believed to indicate shock pressures of more than 100 kb. This crater requires further study, including drilling.

STRANGWAYS

Guppy et al. also discuss Strangways (Paper 32), which is a circular structure 150 km south of the town of Katherine. It is defined by tangentially striking outcrops of Proterozoic sandstone and siltstone, up to 70 m high. The surrounding terrain is underlain by horizontal Lower Cretaceous beds. The structure has a diameter of 16 km, and is cut by faults of probable Precambrian age and may thus be very ancient. Rhombic shatter cleavages (cf. Gosses Bluff, Milton and Brett, 1969), poorly developed shatter cones, and a central area of breccia testify to explosive shock deformation. The breccias consist of angular fragments of quartzite and baked shale together with rounded inclusions of granite gneiss up to 4 m in diameter, set in a fine, chocolate colored matrix. Quartz grains show planar microscopic deformations, which are absent from the sandstone of the outer ring. Diaplectic quartz grains and evidence of liquid flow in the matrix are also recorded. The breccia is not unlike that interpreted as recrystallized shock melt or suevite (French, 1968). It does not contain tridymite (unlike that from Mt. Pyroclast, near Gosses Bluff, Paper 31). Guppy et al. suggest that the Strangways breccia may represent a higher zone in the melt pool, where mixing was more intense and cooling more rapid. At Gosses Bluff, only the lower part is supposed to be preserved. Coesite and stishovite have not been identified at Strangways.

All three of these structures have been attributed to meteorite impact, but their origin remains a matter of controversy. The Lake Teague structure (Butler, 1974) is a large circular feature, with a core of granite reminiscent of the Siljan structure (Paper 2). Deformation is not unlike that at many cryptoexplosion structures in a general way, but lacks shatter cones or other "unique" shock metamorphisms. The Dampier Archipelago, situated on the northwest coast and off it, is an extensive basin of folded basic igneous rocks and felsic granophyres, the latter volumetrically dominant. It is similar in many ways to the Sudbury Basin (Papers 21–24). Radiometric evidence suggests that the age of the structure, situated right on the Archean-Proterozoic unconformity, is late Archean. No shock deformations have been recorded, but little detailed study has been carried out, and there are peculiar quartz-choked breccia dikes radiating from it for scores of kilometers through the Archean rocks. These are not unlike great breccia dikes that radiate out from the Pilansberg alkaline focus within the Bushveld Complex in South Africa. There is a similarity also to quartz-fragment choked lavas of the Ventersdorp Series, also in the Transvaal, south of Johannesburg. Such quartz-choked dikes should be studied in detail, for they could well display the "unique" shock deformation patterns of cryptoexplosion. The Dampier Archipelago also deserves much more detailed study.

REFERENCES

Butler, H. 1974. The Lake Teague Ring Structure, Western Australia. *Search* **10**: 536.

French, B. M. 1968. Shock Metamorphism as a Geological Process. In *Shock Metamorphism of Natural Materials,* B. M. French and N. M. Short, eds. Baltimore, Md.: Mono Book Corp., pp. 1–17.

McCall, G. J. H. 1965. Possible Meteorite Craters—Wolf Creek, Australia and Analogs. *N. Y. Acad. Sci. Annals* **123** (article 2):970–998.

McCall, G. J. H., ed. 1977. *Meteorite Craters.* Benchmark Papers in Geology, vol. 36. Stroudsburg, Pa.: Dowden, Hutchinson & Ross.

Milton, D. J., and R. Brett. 1969. Gosses Bluff Astrobleme, Australia: The Central Uplift. *Geol. Soc. America Spec. Paper 121,* p. 53 (abstr.).

Taylor, S. R. 1965. The Wolf Creek Iron Meteorite. *Nature* **208**:944–945.

31

Reprinted from *Science* **175**:1199-1207 (1972)

Gosses Bluff Impact Structure, Australia

Geological and geophysical techniques establish the origin of an analog of lunar craters.

D. J. Milton, B. C. Barlow, Robin Brett, A. R. Brown, A. Y. Glikson, E. A. Manwaring, F. J. Moss, E. C. E. Sedmik, J. Van Son, G. A. Young

Craters are the dominant surface feature on the Moon and on Mars. Some resemble terrestrial volcanic craters but most appear to have their analogs on Earth among the cryptoexplosion structures (*1*), of which only some sixty are known. A complex of shared features clearly defines cryptoexplosion structures as a class, although no one feature by itself can be relied on as an unfailing index. These features include severe deformation localized within a roughly circular perimeter, an open crater or one filled with undisturbed sediments (unless erosion has reached a level beneath the crater floor), uplifted bedrock in the center, shatter cones, breccias of melted rock fragments, unusual deformation features and anomalous optical properties in mineral grains, and minerals that are formed only at extreme pressures (*2*). Cryptoexplosion structures occur in a wide variety of geologic environments; the apparent randomness of their occurrence may be considered an additional characteristic. It is generally agreed that cryptoexplosion structures were produced by violent events involving intense shock (*3*). It is hotly debated, however, whether the cause is to be sought among processes entirely within the Earth or whether hypervelocity impact of objects from space—meteorites, asteroidal fragments, or comets—is responsible.

The Gosses Bluff structure in central Australia, 160 km west of Alice Springs, is a typical cryptoexplosion structure. In the little over a decade since it was recognized as a geologic anomaly a variety of explanations have been advanced: igneous intrusion (*4*), salt diapirism (*5*), cryptovolcanism (*6*), impact (*6, 7*), and mud volcano activity (*8*). Gosses Bluff is unusually amenable to study for several reasons. It lies in a sequence of nearly flat-lying strata, which simplifies reconstruction of the pre-event geology. Erosion has cut down to just beneath the crater floor and has laid bare the central uplift, and exposures in the semiarid environment are generally good. We have carried out a comprehensive study that includes detailed geologic mapping, and seismic reflection and refraction, gravity, aeromagnetic and ground magnetic surveys. Our conclusion is that the Gosses Bluff structure is indeed an eroded crater formed by a single nearly instantaneous shock event, and that the event can be explained only by impact.

Dr. Milton is with the U.S. Geological Survey, Menlo Park, California; Dr. Brett, with the Survey during field work, is now at the NASA Manned Spacecraft Center, Houston, Texas; the other authors are with the Bureau of Mineral Resources, Canberra, Australia, except Dr. Young, who is now with Anaconda Australia, Inc., Sydney.

Geologic Setting

A photograph taken by the Gemini V astronauts (Fig. 1) shows Gosses Bluff as an isolated circular ridge about 5 km in diameter in Missionary Plain. Rings of slightly differing topography and vegetation show that the structure is larger than the Bluff itself; on the ground, marked structural effects are found as far as 11 km from the center. Missionary Plain, at the northern edge of the Amadeus Basin, is underlain by a flat-floored syncline of Proterozoic and Paleozoic strata which crop out in the MacDonnell Ranges to the north and the James Ranges to the south. Seismic reflection surveys indicate that the sedimentary section in the vicinity of Gosses Bluff is about 8000 meters thick. Although dips near the axis of the Missionary Plain syncline are generally low, a subsurface cross fold, the Gardiner-Tyler anticline (*9*) with dips on the flanks as high as 15° runs northeast-southwest beneath the bluff.

Geologic Structure

The Gosses Bluff structure may be divided into two zones: a lower zone of systematic displacements and an upper zone, now largely stripped by erosion, of more chaotic deformation. The basic pattern of the lower zone at the surface is one of upturned strata facing outward, so that the Brewer Conglomerate, the unit ordinarily at the surface in Missionary Plain, forms an outer ring and increasingly older strata crop out in sequence inward, until at the center strata normally 3000 meters below the plain are exposed.

The resemblance of the hollow ring of Gosses Bluff to a crater is an accident; the form is an expression of lithology. Resistant sandstones of the uppermost unit of the Larapinta Group, the Mereenie Sandstone, and a sandstone member of the Parke Siltstone stand high, rising to about 200 meters above the plains. Outside the bluff are soft siltstones of the Parke, and then poorly lithified Hermannsburg Sand-

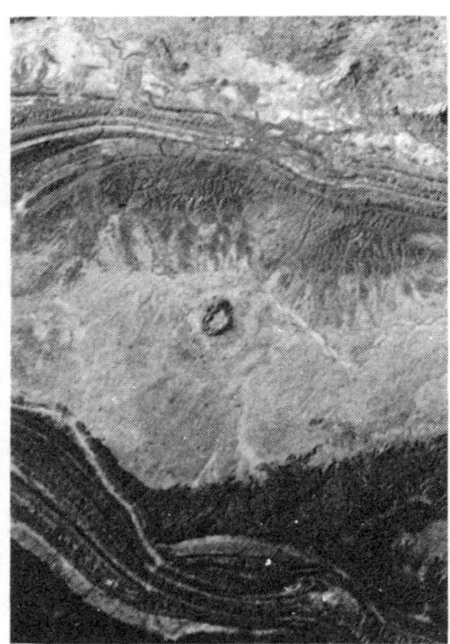

Fig. 1. Gosses Bluff (center) in Missionary Plain, between the MacDonnell Ranges to the north and the James Ranges to the south. Area approximately 60 by 80 km. [Photographed from Earth orbit by G. Cooper and C. Conrad, Jr., Gemini V Spacecraft]

stone and Brewer Conglomerate. Inside the bluff, the lower slopes are underlain by sandstone and siltstone of the Carmichael Sandstone, a low-lying interval with very little outcrop is underlain by Stokes Siltstone, and the low hills in the center are underlain by sandstone beds in the lowermost Stokes and the uppermost Stairway Sandstone.

The map pattern shows that successive strata do not describe continuous belts but occur in discrete plates, each a few hundred meters long (Fig. 3). Some open folding and minor faulting occur within individual plates, but major changes of attitude correspond to the faults which separate plates. Few faults can be traced for long distances; most terminate against another or curve into parallelism with the bedding. En echelon overlapping of plates and lateral wedging between pairs of convergent faults indicate crowding that results from displacement not only upward but inward as well, so that the outcrop band of any stratum is on a perimeter shorter than that which the rocks occupied at depth. Such centripetal displacement is characteristic of the central uplifts of cryptoexplosion structures (*10*). In contrast, around and above diapirs outward displacement is indicated by the characteristic through-going radial fractures (*11*).

The map pattern of the high bluff is a pentagon with a north-south line of symmetry. In the central hills the two shorter sides of the pentagon are suppressed, leaving a triangular pattern. On each side the bedrock plates (or more obviously the faults) are oblique and tend to have one set of orientations in the Stairway-lower Stokes, another in the Carmichael-Mereenie, and yet a third (more or less parallel to the first) in the Parke. The core of the structure, a tight anticline in the Stairway Sandstone, is found on the northwest side of the central triangle. The adjustment to a decreased perimeter requires some systematic pattern, but we offer no explanation why this particular pattern developed rather than any of the multitude of conceivable alternatives. It does not appear to reflect structural predisposition. The north-south line of bilateral symmetry appears to have no relation to the regional geology. One may speculate with no supporting evidence that it marks the direction of entry of the impacting bolide.

Subsurface Structure from Seismic Reflection

An exploratory well drilled for oil in the center of the bluff indicates that steep dips are maintained to a depth of at least 1380 meters. All other subsurface information has been obtained by geophysical techniques.

Two main seismic reflection traverses were run, each 25 km long, crossing at right angles at the center of the bluff (Fig. 4). Reflection quality, normally excellent in the Amadeus Basin, deteriorates markedly within the perimeter of the disturbance. Shot hole and geophone patterns were therefore designed to minimize both coherent and random noise. Multiple coverage, common depth point recording was used within the disturbance, for the most part 3-fold coverage, but as much as 24-fold coverage, inside the bluff. Continuous multiple coverage was obtained under the inaccessible high bluff by offset shooting. Earlier seismic surveys extend the coverage and allow identification of reflectors with horizons penetrated by exploratory wells.

Interpretation of records is complicated by an increase of velocities fror

Fig. 2. Aerial view of Gosses Bluff from the north. [Photograph by C. Zawartko]

south to north, presumably caused by increasing silicification toward the MacDonnell Ranges, and by variable surface conditions close to the bluff which make elevation and weathering corrections deduced by normal methods very unreliable. The velocity in the section under Gosses Bluff does not seem to have been significantly affected by the disruption. The principal feature shown by the traverses is a bowl-shaped zone within which continuous reflections are absent. Migrated seismic horizons along the northwest-southeast traverse are shown in Fig. 5. The limits of disruption are difficult to locate precisely. In the center of the structure the shallowest reflection event was recorded at 2.4 seconds after the shot indicating that disruption does not extend deeper than 5500 meters. On the cross section we have drawn the boundary of disruption as a hemisphere of radius 4.3 km inscribed within the end points to which reflecting horizons can be traced. This hemisphere is surmounted by a shallower saucer of disruption that can be defined by surface geology, the gravity field, the extent of the shallower seismic reflections, and by measurements of near-surface seismic velocity.

The Gardiner-Tyler anticline is evident in the deeper reflections along the northwest-southeast traverse (Fig. 5) and the northeast-southwest traverse shows that the Gosses Bluff disturbance lies above a domal high. This feature, suggested by earlier seismic surveys, led to the hypothesis (*5*) that the Grosses Bluff structure is a diapir with a core of salt rising from the upper Proterozoic Bitter Springs Formation at the crest of the anticline. The continuity of reflecting horizons in and above the Bitter Springs clearly rules out any diapirism. There is, however, an anomalous thickening of the lowest seismic interval (approximately equivalent to the Bitter Springs Formation) beneath the Gosses Bluff disturbance. If the thickening is a result of salt flowage completed in early Paleozoic time, as appears to be the case elsewhere along the anticline, the superposition of an impact structure must be simple coincidence. It is quite possible, however, that the Gosses Bluff event initiated renewed slow movement of salt. A third possibility, that the structure at depth is in some way an immediate response to impact, seems least likely, as the seismic coherence suggests slow deformation.

Shatter Cones

Shatter cones (Fig. 6) are conical fracture surfaces with divergent striations along the length of the cone. They have been found at about 20 cryptoexplosion structures (*12*) and have been produced in craters blasted in granite by multiton charges of TNT at shock pressures of 30 ± 5 kilobars (*13*). Shatter cones apparently form during the compressive phase of a shock event, with axes normal to the shock front and apices at points where the shock wave is diffracted by minor inhomogeneities in the rock (commonly bedding plane discontinuities but occasionally more unusual features; see Fig. 6, right) (*14*).

The orientation of shatter cones records the direction of advance of the shock front and the assemblage of cone axes should point to a focus that corresponds to the center of shock propagation. To use this record, however, displacements of the shatter-coned beds during the subsequent rarefaction phase must be taken into account. Analysis of cone orientation thus requires both finding the axes of the cones and restoring them to their original orientations and positions.

The axial orientation and apical angle of each cone were determined from about 20 measured striations (*15*). Because individual full cones are the exception rather than the rule at Gosses Bluff, striations from several cone segments usually had to be measured (Fig. 6, left). The axis and apical angle of the cone of best fit to the striations and the root mean square deviation of the striations from this cone were computed. Of 95 stations, 87 yielded a well-defined cone (root-mean-square deviation of striations $<9°$), belying the impression of multiple cones or even random orientations which the partial cone segments can give.

To reconstruct the original orientation and location of the shatter cones, it was assumed that bedding was horizontal when shatter coning occurred and that the cones formed directly beneath their present locations at depths corresponding to their stratigraphic positions. The orientation within the horizontal plane must also be specified. Alternate computer programs were run in which (i) the bedding at each station was either simply rotated about the line of strike to be horizontal, or (ii) the bedding was first rotated about a vertical axis to put the strike at right angles to a radius from the center of the structure and then rotated to horizontal. A focus, defined as the point closest to the extended axes of all the cones, was found by least squares anal-

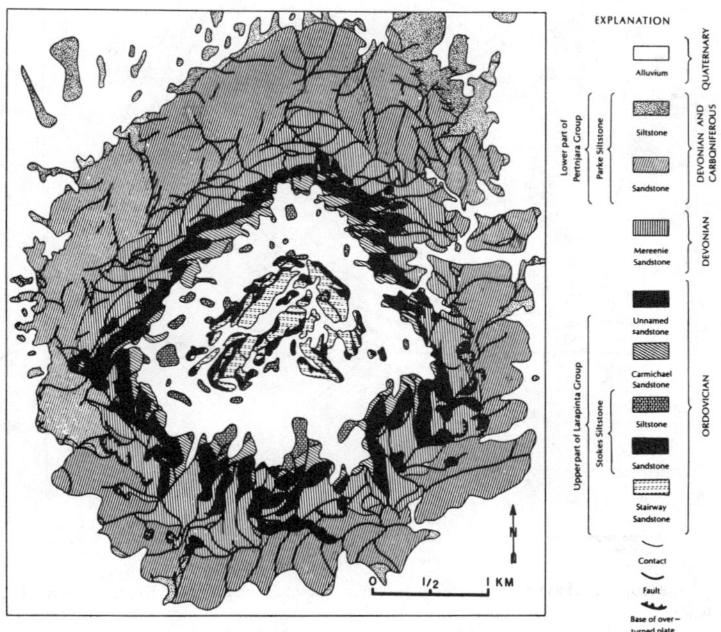

Fig. 3. Geologic map of the central uplift area, Gosses Bluff.

ysis. Foci found by the two methods of reorientation are very close; the simpler method (i) yields a focus with less variance. The map location of this focus is no more than a hundred meters from the center of the bluff.

The deviation of individual cone axes from the direction of the common focus reflects the history of particular rock masses. Stations in the south of the bluff, for example, show less deviation when reoriented by program (i) than (ii), indicating that the plates, despite their marked obliquity to the broader structure, were only rotated about their lines of strike (and translated) during emplacement. Other plates, however, underwent more complex rotations, so in order to treat all stations uniformly, a third program was used in which all cones were rotated so that the bearing of their axes pointed to the center of the structure. This, of course, assumes the horizontal coordinates of the focus, but yields a better estimate of the vertical coordinate.

An overall focus was found about 3000 meters stratigraphically above the lowest exposed bed. More significantly, however, the separate foci obtained when cones in different stratigraphic units are treated separately are spread through a vertical span of 1900 meters in which foci from higher units are progressively higher. A possible interpretation of this spread is that it reflects some nonhemispherical shape of the advancing shock front. We believe, however, that it results from the neglect in our model of the inward displacement that occurred during formation of the central uplift. If the radial distances to the cone stations were greater, the foci would be higher. We assume that the outermost cone stations, in the Hermannsburg Sandstone at about half the radius of the structure, have not moved horizontally. If so, then the foci for the lower units are brought into coincidence with the Hermannsburg focus if the mean horizontal radial distances to the stations have been reduced by the following percentages of the original radial distances: Parke Siltstone, 20 to 30 percent; Mereenie Sandstone, 24 to 37 percent; Carmichael Sandstone, 25 to 38 percent; Stairway and Stokes Formations, 33 to 52 percent. The range of values arises from uncertainties in the stratigraphic thicknesses. We believe that these decreases in radius are the best measure of the centripetal displacement. The Hermannsburg focus is 1500 meters above the mean stratigraphic elevation of the stations in this unit. Neither the stratigraphic thickness in the higher units nor the position of the ground surface at the time of the event are known precisely, but it appears that the focus of energy responsible for the shatter coning at Gosses Bluff was at most 600 meters below the pre-event ground surface and probably much shallower.

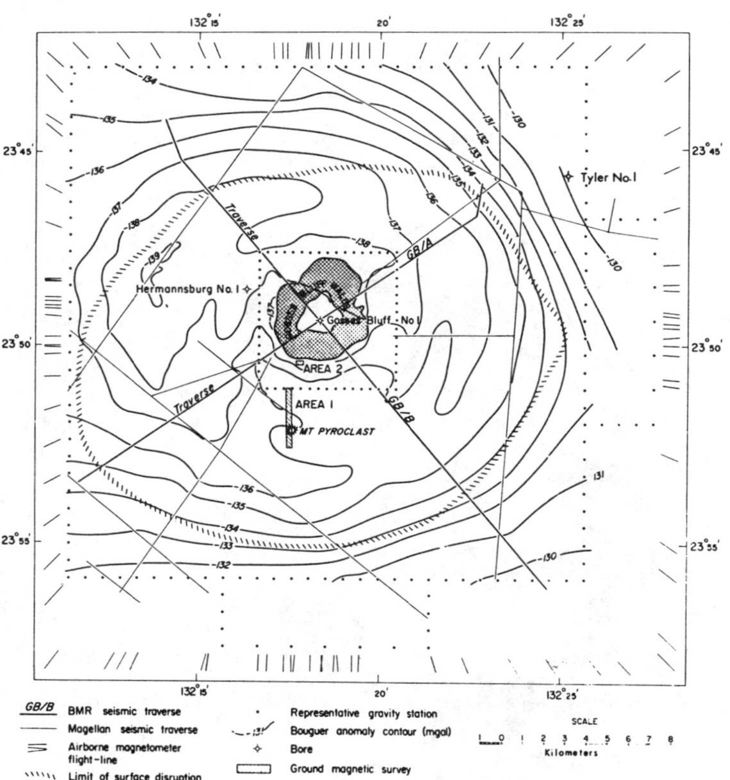

Fig. 4. Geophysical operations and preliminary Bouguer anomalies. Contour interval, 1 mgal.

The Shallow Zone of the Central Uplift

Closer to the focus and the ground surface, where the shock was less attenuated and the rock less constrained, stresses behind the rarefaction front were sufficient not only to break apart and move rock in large plates but to break rock into small fragments and to hurl them upward to fall back as breccia deposits. Erosion has removed much of the breccia but little coherent bedrock, so that the original bedrock-breccia contact is closely approximated by the present ground surface, including the abrupt outer (and probably inner) walls of the bluff and the deep valleys that dissect it. Nevertheless, breccia remnants are found at the summit surface of the bluff, in the valleys, and at the base of the outer wall, as well as on the surrounding plain. Fragmentation is often extreme and clasts are unsorted and randomly oriented, but mixing of clasts from different units is very uncommon.

The depth to which brecciation occurred depended on the strength of the rock, so that resistant strata remained in coherent plates while adjacent weaker strata were brecciated. During the few seconds before the breccia settled, vertical plates of harder strata stood against empty space. In many places the unsupported upper ends toppled outward to lie as overturned plates or detached blocks on the truncated edge of stratigraphically higher units. Some large blocks lie so far from the parent plates that outward propulsion rather than simple toppling must have oc-

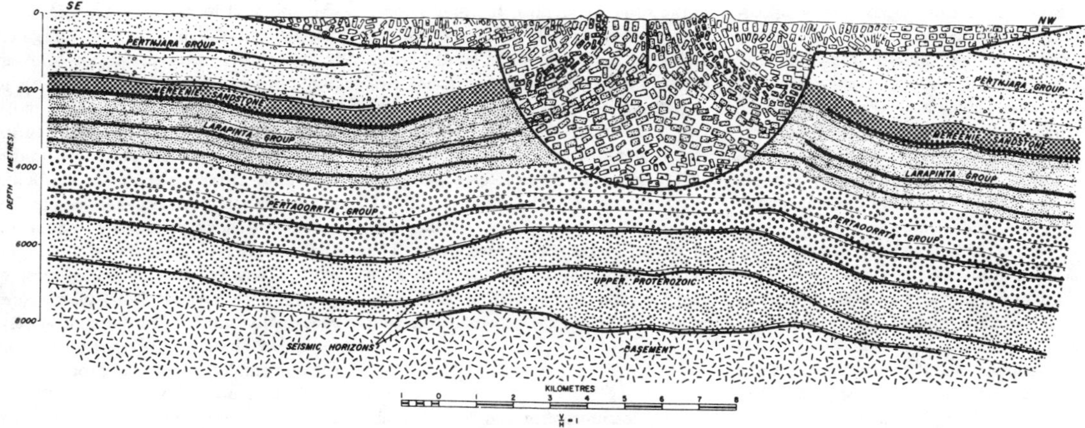

Fig. 5. Migrated seismic horizons and generalized geologic cross section through the Gosses Bluff structure.

curred. Hundred-meter-long blocks of the uppermost Larapinta sandstone lie 350 meters from the Larapinta outcrop; others lie within the narrow valley at the base of the Parke, on edge and facing inward after a 270° rotation from the original horizontal. Breccia is commonly thin or absent beneath overturned blocks, indicating that they were emplaced before the bulk of ejecta settled.

All the displacements observed in the central uplift, as well as the ejection of material that created the topographic crater itself, appear to reflect the flow field established as an immediate response to the rarefaction wave: accelerations dominantly upward, with an inward component at depth and an outward component near the surface. This concept differs significantly from current ideas, which view central uplift as a late phenomenon in an impact event induced by slumping of the crater walls (*16*).

Outer Zones and Gravity

The structural pattern of the bluff continues on the plains outside: steeply dipping, outward-facing bedrock plates, with some plates overturned outward to nearly horizontal downward-facing attitudes; and zones of breccia, both ordinary breccias in which clasts range from meter size to powder, and megabreccias, chaotic assemblages of blocks characteristically tens of meters to a hundred meters in size.

Geophysical data, especially gravity results, partly compensate for the relative lack of outcrop on the plains. Preliminary simple Bouguer anomaly contours (not including terrain corrections) from measurements at about 1550 stations are shown in Fig. 4. The extremely low values are typical of the Amadeus Basin. Removal of the regional gravity field (*17*) and correction for terrain effects indicate a residual gravity low, reasonably circular and centered on the bluff (Fig. 7). The anomaly shows an irregular but basically flat bottom with an average value of about −3 mgal extending to 8 km from the center and a persistent gradient between 8 and 13 km. Gravity models, based on concentric cylinders centered on the bluff, demonstrate that the mass deficiency is not distributed throughout the volume of disturbance as indicated by the seismic data but is primarily due to a near surface disk. A disk of radius 11 km, thickness 2 km, and density contrast -0.04 g/cm^3 gives a reasonable fit to the average gravity profile and is not only the simplest model but also that which gives the estimate of maximum thickness. Figure 5 shows a geologically more acceptable model which also gives a good fit to the profile if a density contrast of -0.1 g/cm^3 is taken for the near surface material. The fine structure of the gravity field is thought to arise from local density contrasts up to 0.1 g/cm^3. Some of this fine structure arises from the different densities of stratigraphic units—for example, the

Fig. 6. (Left) Shatter cone measurement in Mereenie Sandstone. The cone segments near the instrument describe two-thirds of the complete cone; striations on segments in the upper right corner complete the cone. (Right) Shatter-coned "pipe rock" in Mereenie Sandstone. Coning was initiated along Scolithus tubes, casts of burrows normal to the bedding made by organisms dwelling in the sand.

steep-sided annular gravity low apparently associated with the lower density sandstones that underlie the high bluff (Fig. 4). The arcuate gravity lows on the outer plains, however, together with surface geology and shallow seismic refraction, appear to indicate troughs of breccia, one of which has a depth of at least 160 meters at the Hermannsburg No. 1 drill hole.

The maximum gravity gradient suggests a limit to the disturbance at a radius of about 11 km in all directions, and the most remote highly deformed and overturned beds and megabreccia are found at about this radius. In some sectors, however, beds dipping outward at low angles and interrupted at widely spaced intervals by faults or small sharp folds are found out as far as 13 km, while in other sectors flat-lying bedrock crops out as close as 7.5 km to the center. The irregular map pattern and the sharpness of the contact between highly deformed and gently tilted or undeformed rock, together with the general lack of significant associated distortions of the residual gravity field, suggest that these disturbances extend only to shallow depths. It would appear that there is an uptilted crater wall on which lie remnants of the basal part of a structurally complex rim zone, corresponding to the zone of thrust slices and overturned folds mapped at the Ries crater (*18*) or the Henbury craters (*19*). Correlation of structural and topographic features of the crater is difficult, but an original crest to crest diameter of about 20 km seems reasonable. The partial preservation of the rim zone suggests that the present ground level is not far below the pre-crater ground level.

Shock Metamorphism and Melt Breccia

The response to shock of a mineral grain or a rock depends on the entire extremely complicated history during the shock pulse and the relaxation period. Nevertheless, effects can be ranked in a sequence of increasing shock metamorphism (*20*), and the peak pressures associated with these can be determined experimentally (*21*). Quartz is one of the best indicators of progressive shock. The lowest grade of shock metamorphism in quartz—irregular fracturing—is widespread at Gosses Bluff. "Cleavage," widely spaced open fractures along rational planes, most commonly {10$\bar{1}$1}, is well developed in many specimens, particularly in the sandstones from the center of the bluff. In the laboratory cleavage begins to develop at shock pressures of about 50 kilobars (*21*). "Planar features" in quartz, sets of many parallel closely spaced, very narrow closed planes, are perhaps the most diagnostic and widely recognized indicator of shock in rock. Experimentally, they are produced at shock pressures above about 100 kb. They are common at Gosses Bluff (Fig. 8, left); where they are few, {0001} planes dominate; where more abundant, {10$\bar{1}$3}. Planar features are more abundant in heavily silicified sandstone than in adjacent more porous sandstone, suggesting that development is favored by lower rock compressibility. They are also more abundant in breccia fragments than in nearby bedrock outcrops of the same units. Differences in peak shock pressure could not have been great; it may be that confinement during decompression partially suppresses the development of planar features.

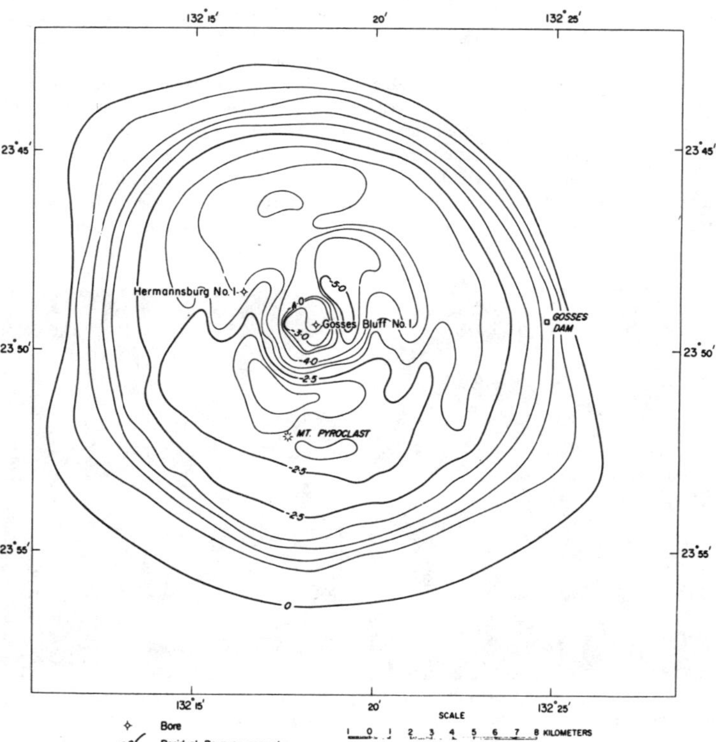

Fig. 7. Bouguer gravity anomaly contours, corrected for regional gradient and terrain effects. Contour interval, 0.5 mgal.

Experimentally, the refractive indices and birefringence of quartz are progressively reduced by higher shock pressures until at pressures between about 200 and 300 kilobars quartz is transformed to an isotropic glass. At about 400 or 500 kilobars the relaxation temperature in shocked sandstone is high enough to melt quartz. These stages of shock metamorphism were not reached in unfragmented bedrock or in the breccia now remaining on the bluff itself. Breccia composed largely of sintered shock-melted fragments ("suevite") is found, however, in isolated occurrences on the exterior plain, remnants of the basal layer of the crater fill. At Mount Pyroclast, a hill 40 m high and 5 km south of the center, a vertical gradation of shock effects is evident. At the base, quartz in breccia fragments shows abundant planar elements, but no evidence of phase change. In sandstone clasts higher on the hill destruction or at least severe disruption of the quartz lattice is indicated by sand grains which retain their outlines but have recrystallized in a diffuse, fine-grained texture. Still higher, actual melting is indicated by flowage features, at first barely detectable but more prominent upward until the breccia at the top is a mass of twisted ropy folds up to several meters long. Much of the quartz in this material recrystallized as felted aggregates of tridymite, the high temperature polymorph of silica, which have since completely reinverted to quartz (Fig. 8, right) (*22*).

Coesite and stishovite, the high-pressure polymorphs of silica, form at shock pressures of several hundred kilobars. Neither has been found at Gosses Bluff, despite extensive search. Their apparent absence, and the absence of any unrecrystallized glass, may be due to reequilibration during slow cooling near the base of the crater fill.

Shale and siltstone clasts in the melt breccia contain zeolites (mostly heulandite and some stilbite and chabazite) and sanidine. In general, no significant change of chemical composition of clasts is evident. Some pumiceous clasts composed largely of fine-grained potassium-rich high sanidine, however, contain 12.6 percent K_2O, more than any possible precursor. These are apparently shock-melted shale fragments enriched in potassium during an essentially hydrothermal stage in the cooling suevite (*23*).

Insofar as can be determined from clasts so highly altered, all material at Mount Pyroclast is Parke Siltstone, bedrock plates of which crop out nearby. The continuous vertical gradation of shock effects contrasts with the sharply marked base of suevite at the Ries Crater (*24*) but resembles the profiles in some Canadian craters (*16*). The clasts are unsorted; the larger ones tend to dip steeply rather than lie flat, although with no consistent strikes. There is no evidence that the material is fallback that was thrown a significant distance upward; it may have been packed against the crater wall after rather slight, largely horizontal transport.

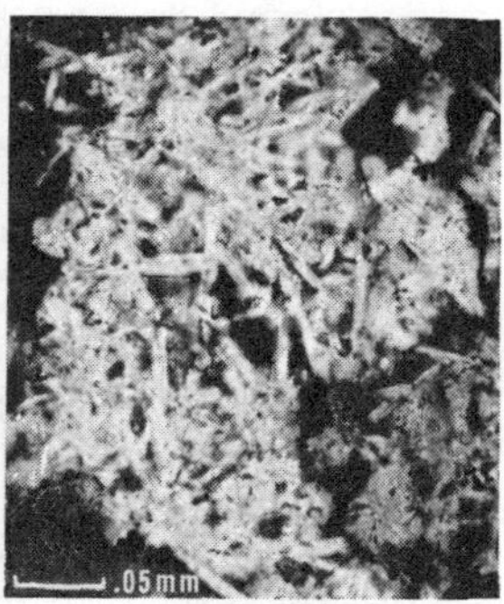

Fig. 8. (Left) Cleavage (darker lines) and planar features (fainter, closely spaced sets) in quartz grains in the basal sandstone of the Stokes Siltstone near the center of Gosses Bluff. (Right) Quartz paramorphous after tridymite in melt breccia, Mount Pyroclast. Crossed nicols.

Age of the Event

An earliest Cretaceous age is indicated by two dating methods. A potassium/argon age of 133 ± 3 million years was measured on the sanidine-rich rock from Mount Pyroclast. Zircon grains from thermally affected sandstones at the bottom of drill holes from beneath Mount Pyroclast have abundant fission tracks indicating Precambrian ages, as would be expected from the provenance of the detritus in the Paleozoic sediments. In zircon grains from the baked zone just beneath the melt breccia the older tracks have been annealed out and the number present would have developed in 130 ± 6 million years.

Magnetic Effects

An earlier regional aeromagnetic survey, flown at 300 meters above ground, showed no anomaly for Gosses Bluff. A survey at about 100 meters, with closely spaced flight lines (Fig. 4), however, showed a local anomaly of −75 gammas at Mount Pyroclast and smaller anomalies south, east, and west of the Bluff. Each of these is associated with an occurrence of highly shocked and thermally affected breccia. Many are nearly concealed by surficial deposits and were found only after being located magnetically. The anomalies are of short wavelength and reduce the intensity of the field; this indicates that they have very shallow sources whose main component of magnetization is in opposition to the Earth's present field, presumably, thermo-remanent magnetization in the direction of the Earth's magnetic field at the time of the Gosses Bluff event (*25*). From the aeromagnetic data alone, the Mount Pyroclast anomaly can be modeled by a dipole inclined at +85° in the magnetic meridian at a depth of 100 meters. The true source of the anomaly, as indicated by drilling, is shallower than calculated.

A detailed ground magnetic survey of the horizontal and vertical field over an area of thermal breccia at the base of the south wall of the Bluff revealed two anomalies which suggest sources similarly inclined at +80°. Measurement of the remanent magnetization of specimens confirmed the conclusions deduced from airborne and ground magnetic interpretation. The mean inclination is $+72° \pm 1°$, the declination 100°, intensities range up to 43×10^{-4} electromagnetic units per cubic centimeter and susceptibilities to 12×10^{-4} emu/cm^3. Australia was an unusually immobile continent during the Permian Period and Mesozoic Era, but the calculated pole position—170°E, 25°S—is satisfactory for an early Cretaceous date.

Origin

We may summarize the principal conclusions from the field evidence, which must be accounted for in any explanatory hypothesis: (i) Shatter cones were formed in undisplaced strata with their axes pointing to a focus near the ground surface. (ii) Displacement occurred after shatter coning, and so rapidly that large blocks were hurled through the air. (iii) Pressures in a large volume near the center of the disturbance reached at the least several hundred kilobars. (iv) Although the structure lies over an anticlinal axis, neither an igneous intrusion nor a salt diapir exists at depth.

The sequence of events accords with a shock process in which shatter coning occurred during the compression phase and displacement during the rarefaction phase. Loosely, the event could be called an "explosion," but this term can refer to processes of two quite distinct types, one exemplified by the deflagration of gunpowder and the other by the detonation of TNT. Deflagration is rapid but ordinary combustion. The pressures that are generated and the work that can be done depend on the thermal expansion of the hot gaseous products and the extent to which these are confined. Volcanic eruptions also operate by the adiabatic expansion of gases, particularly the flashing of water into steam, and are subject to the same limitations. Pressures in processes of these types cannot greatly exceed the 4.85-kilobar breech pressure of Big Bertha, which shelled Paris from 120 km (*26*), or the calculated 3-kilobar pressure of the 1956 eruption of the volcano Bezymianny, which hurled blocks 30 km (*27*).

By contrast, in a detonating explosive the combustion front advances at a supersonic velocity. The combustion front is then a shock front, behind which the energy, transiently in the form of the kinetic energy of motion rather than the potential energy of compression of the combustion products,

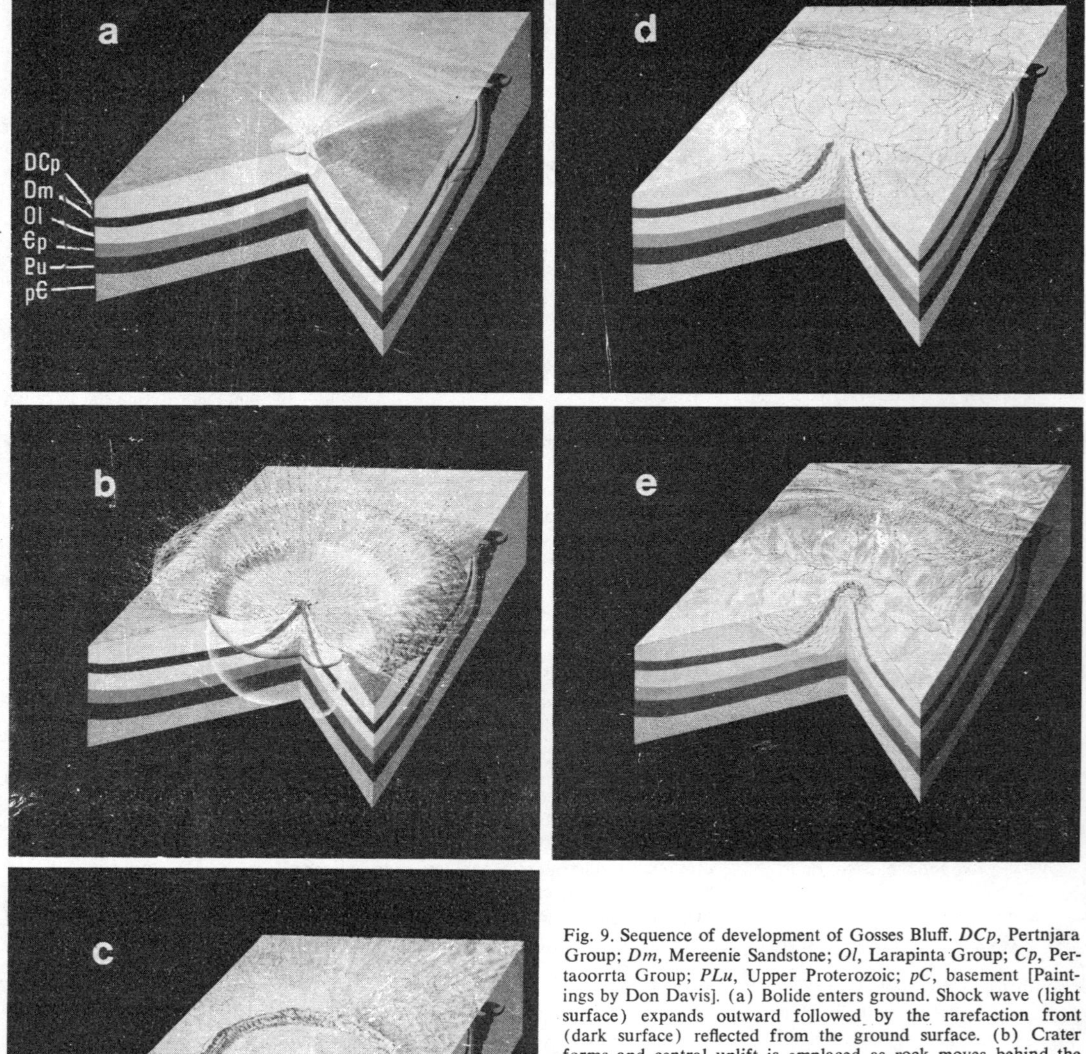

Fig. 9. Sequence of development of Gosses Bluff. *DCp*, Pertnjara Group; *Dm*, Mereenie Sandstone; *Ol*, Larapinta Group; *Cp*, Pertaoorrta Group; *PLu*, Upper Proterozoic; *pC*, basement [Paintings by Don Davis]. (a) Bolide enters ground. Shock wave (light surface) expands outward followed by the rarefaction front (dark surface) reflected from the ground surface. (b) Crater forms and central uplift is emplaced as rock moves behind the rarefaction front. (c) The ancestral Gosses Bluff stands as a central peak surrounded by suevite in the crater floor. (d) The crater rim has been destroyed by erosion and the crater filled by sediments. A plain extends from the present MacDonnell Ranges to the James Ranges (*31*). A stream flows across the site of Gosses Bluff, which is marked only by isolated sandstone hills less than 50 meters high. (e) Present state. Erosion has removed all postimpact sediments and most of the crater floor breccia. Beveled surfaces on the high bluff mark the former plains and notches mark the former watercourse.

can be extremely high. A massive detonation propagating a shock wave outward from the focus provides a model that fits the Gosses Bluff event closely. A large natural detonation seems impossible, but the impact of a bolide moving at many kilometers per second is a phenomenologically equivalent process (*28*) which would produce all the effects seen at Gosses Bluff.

Various scaling relations between crater diameter and energy (*29*) yield energies for the Gosses Bluff event spanning a range of about one order of magnitude with a mean about 10^{27} ergs or 2.4×10^4 megatons of TNT. This would be the kinetic energy of a low-density high-velocity comet (density = 1.3 g/cm^3, velocity = 40 km/sec) or a high-density low-velocity iron meteorite of asteroidal origin (density = 7.85 g/cm^3, velocity = 15 km/sec), each with a diameter of about 600 meters. Penetration of a bolide depends more on its density than on its velocity (*28*). The shallow depth of burst, indicated directly by the shatter cone data and indirectly by the presence of a central peak (*30*), suggests that the bolide was a comet. Whatever the material, it would have been melted or vaporized at impact and so dispersed that relicts are not likely to be found. Stages in the impact and in the subsequent development of Gosses Bluff are shown in Fig. 9.

Craters on Other Planets

Each cryptoexplosion structure that has been studied has its own individuality. Nevertheless, certain general patterns have emerged. Gosses Bluff is representative of its size class. Smaller craters, under about 2 km diameter [for example, Meteor Crater, Arizona (*28*)] have simple bowl-shaped profiles without central uplifts. Craters larger than Gosses Bluff may show multiple annular uplifts and depressions, and, unlike Gosses Bluff, contain igneous rock, either a product of crystallization of rock completely melted by shock [for example, Manicouagan, Quebec (*32*)] or of normal magma from depth, the intrusion of which was triggered by the impact [for example, Sudbury, Ontario (*33*)]. Each of these appears to have analogs on the Moon.

Techniques similar to those used at Gosses Bluff could some day be adapted to exploration of craters on the Moon and Mars. In the more foreseeable future, the findings at Gosses Bluff suggest critical observations that might be made on simpler and briefer missions such as the Apollo landings. The statement at the outset of this article that the terrestrial analogs of large lunar craters are cryptoexplosion structures is not universally accepted—an alternative hypothesis favored by many is that the proper analogs are volcanic calderas (*34*). The nature of the central peaks should be a critical test: Are they constructional volcanic forms contemporaneous with or younger than the floor material, or are they older rock rising from a floor of shock-melted breccia? If the lunar craters are indeed impact craters, the central peaks should furnish material uplifted from considerable depth. A first estimate of the depth could be based on simple analogy to Gosses Bluff and other terrestrial structures. A considerably more refined estimate should be possible, if only one or two orientable shatter cones could be recovered.

References and Notes

1. The term "cryptoexplosion structure" was introduced by R. S. Dietz [*J. Geol.* **67**, 496 (1959)] to avoid the genetic implication of the earlier "cryptovolcanic structure." One may conclude that certain, or all, cryptoexplosion structures are impact structures ("astroblemes") but a nongenetic term is still desirable.
2. B. M. French and N. M. Short, Eds., *Shock Metamorphism of Natural Materials* (Mono Book, Baltimore, 1968) compiles much recent research on cryptoexplosion structures.
3. A shock wave is a pulse of instantaneous compression that propagates away from the source at a supersonic velocity. The energy density or pressure decreases approximately in proportion to the mass of material engulfed by the shock wave. From each point at which the shock front reaches a free surface, a wave of decompression is generated and expands back into the compressed medium as a rarefaction front. Behind the rarefaction front material is accelerated toward the free surface. It is during the rarefaction rather than the compression phase that gross displacement of material may occur. If the solid body is the earth, material accelerated toward the ground surface may be ejected to produce a crater. During the brief duration of a shock pulse, either from a natural hypervelocity impact or in laboratory experiments, pressures may be in the range of hundreds of kilobars or even megabars, far beyond the range of ordinary metamorphism of rocks. Shock compression is essentially isentropic and decompression adiabatic, so that the final temperature of the shocked material is higher than the initial temperature.
4. R. O. Brunnschweiler, unpublished manuscript.
5. D. A. McNaughton, T. Quinlan, R. M. Hopkins, A. T. Wells, *Geol. Soc. Amer. Spec. Pap. 88*, 229 (1968).
6. K. A. W. Crook and P. J. Cook, *J. Geol. Soc. Aust.* **13**, 495 (1966).
7. P. J. Cook, *J. Geology* **76**, 123 (1968), and (*6*).
8. T. S. M. Ranneft, *Amer. Assoc. Petrol. Geol. Bull.* **54**, 517 (1970).
9. A. J. Froelich and E. A. Krieg, *ibid.* **53**, 1968 (1969).
10. Centripetal displacement has been recognized at the Vredefort, South Africa [W. I. Manton, *Ann. N.Y. Acad. Sci.* **123**, 1017 (1965)], Sierra Madera, Texas [H. G. Wilshire and K. A. Howard, *Science* **162**, 259 (1968)], and Wells Creek, Tennessee [C. W. Wilson, Jr., R. G. Stearns, *Tenn. Div. Geol. Bull. 68* (1968)] cryptoexplosion structures. These workers have taken the ratio of mean perimeter to total outcrop length as a measure of the shortening. The accuracy of this method is diminished if beds have rotated about more than one axis, as indicated at Gosses Bluff by shatter cone orientation.
11. G. E. Murray, *Amer. Assoc. Petrol. Geol. Mem.* **8**, 110 (1968); R. Brinkmann and H. Lögters, *ibid.*, p. 284.
12. R. S. Dietz, in *Shock Metamorphism of Natural Materials*, B. M. French and N. M. Short, Eds. (Mono Book, Baltimore, 1968).
13. D. J. Roddy and L. K. Davis, *Trans. Amer. Geophys. Union* **50**, 220 (1969).
14. Theories of shatter coning have been developed for ideal models [P. J. S. Gash, *Nature Phys. Sci.* **230**, 32 (1971)], but as yet have not been tested against actual occurrences. A fact that needs explanation, for example, is the differences in apical angle between cones at Gosses Bluff (mean 80.5°), Sierra Madera (mean 88.5°), and Vredefort (range 90° to 122°).
15. Techniques are basically those developed by Manton (*10*), adapted for computer plotting and calculation.
16. For example, M. R. Dence, in *Shock Metamorphism of Natural Materials*, B. M. French and N. M. Short, Eds. (Mono Book, Baltimore, 1968), p. 182.
17. Based on a regional Bouguer anomaly map compiled from unpublished gravity surveys, including a uniform 11-km grid of the Bureau of Mineral Resources and other data [see (*9*)].
18. E. Preuss, *Geol. Bavarica* **61**, 18 (1969).
19. D. J. Milton, *U.S. Geol. Surv. Prof. Paper 599-C* (1968).
20. E. C. T. Chao, in *Shock Metamorphism of Natural Materials*, N. M. Short and B. M. French, Eds. (Mono Book, Baltimore, 1968), p. 135; W. v. Engelhardt and D. Stöffler, *ibid.*, p. 159; P. B. Robertson, M. R. Dence, M. A. Vos, *ibid.*, p. 433; W. v. Engelhardt and W. Bertsch, *Contrib. Mineral. Petrol.* **20**, 203 (1969).
21. F. Hörz, in *Shock Metamorphism of Natural Materials*, N. M. Short and B. M. French, Eds. (Mono Book, Baltimore, 1968), p. 243.
22. Similar paramorphs after tridymite are found in the Sudbury Basin [J. S. Stevenson, *Can. Mineral.* **7**, 413 (1963)].
23. K. L. Currie and M. Shafiqullah [*Nature* **218**, 457 (1968)] point to potassium enrichment at Canadian craters as evidence for an underlying magmatic source of alkaline solutions. The observation at Gosses Bluff casts doubt on this interpretation.
24. U. Förstner, *Contrib. Mineral. Petrol.* **15**, 281 (1967).
25. An epoch of reversed polarity is indicated at about 130–135 million years [C. E. Helsley and M. B. Steiner, *Earth Planet. Sci. Lett.* **5**, 325 (1969)].
26. See "Ordnance," *Encyclopaedia Britannica* (Benton, Chicago, 1958).
27. G. S. Gorshkov, *Bull. Volcanol.* **20**, 77 (1959).
28. E. M. Shoemaker, in *The Moon, Meteorites, and Comets*, vol. 4, *The Solar System*, B. Middlehurst and G. P. Kuiper, Eds. (Univ. of Chicago Press, Chicago, 1963), p. 301.
29. M. J. S. Innes, *J. Geophys. Res.* **66** 2225 (1961); R. B. Baldwin, *The Measure of the Moon* (Univ. of Chicago Press, Chicago, 1963).
30. Central peaks have not been produced in cratering experiments with buried nuclear devices or high explosives, but have been produced by the detonation of 20- to 500-ton charges of TNT stacked on the ground surface [D. J. Roddy, G. H. S. Jones, C. H. H. Diehl, *Trans. Amer. Geophys. Un.* **50**, 220 (1969)].
31. The summit surface of J. A. Mabbutt, in *Essays in Geomorphology*, C. H. Dury, Ed. (Heinemann, London, 1966), p. 83.
32. J. C. Murtaugh and K. L. Currie, *Que. (Prov.) Dep. of Nat. Resour. Prov. Rep. 583* (1967).
33. B. M. French, *Bull. Volcanol.* **34**, 466 (1970).
34. G. J. H. McCall, *Ann. N.Y. Acad. Sci.* **123**, 843 (1965).

35. Publication authorized by the director, Bureau of Mineral Resources of Australia and the director, U.S. Geological Survey. U.S. Geological Survey work supported by NASA contract R-66.

32

Reprinted from *Jour. Geophys. Research* **76**:5387-5393 (1971)

Liverpool and Strangways Craters, Northern Territory: Two Structures of Probable Impact Origin

D. J. Guppy

Bureau of Mineral Resources, Geology, and Geophysics
Canberra, A.C.T., 2601, Australia

Robin Brett[1]

U.S. Geological Survey
Washington, D.C. 20242

D. J. Milton

U.S. Geological Survey
Menlo Park, California 94025

Two circular structures in the Northern Territory, Australia, are identified as being of probable impact origin. At Liverpool Crater, 1.6 km in diameter, a ring of sandstone breccia with tangentially striking sandstone at the base is exposed. Quartz grains in the breccia exhibit planar features, mosaic structure, and cleavage, indicating moderate shock deformation. At Strangways crater, tangentially striking sandstone and siltstone units define a circular pattern about 16 km in diameter. In the interior are large outcrops of melt breccia with highly shocked inclusions and an outcrop of basement rock, which may mark a central uplift.

An expedition through northern Australia in March 1968 by Guppy and Brett to investigate structures of possible impact origin resulted in two noteworthy finds, which we name the Liverpool crater and the Strangways crater. This paper is based on very brief reconnaissance of the two localities and petrographic examination of specimens by Milton.

Liverpool Crater

Location and geology. The Liverpool Crater (latitude 12°24′S, longitude 134°03′E) is situated approximately 3.5 km west of the Liverpool River in northern Arnhem Land, Northern Territory (Figure 1). *Rix* [1965] noted the structure on the Milingimbi map sheet as a possible meteorite crater. The swampy terrain and meandering rivers inhabited by crocodiles and the high summer rainfall make the crater accessible by land only in very dry years. We therefore visited the structure by helicopter, and our examination was restricted to several hours.

The crater is at the edge of the flood plain of the Liverpool River. A few hundred meters southwest, flat-lying Proterozoic sandstone crops out, dissected along joints into low, very rugged hills. The sandstone contains conglomeratic facies with rounded white quartz pebbles up to 5 cm in diameter, and minor red beds. The crater, 1.6 km in diameter and nearly circular, is marked by a ring of sandstone breccia standing as much as 50 meters above the surrounding plain and up to 300 meters wide (Figures 2 and 3). The breccia consists of angular blocks of sandstone from 7 meters to a few millimeters across. Dikes of pulverized sandstone up to 10 cm wide cross cut the breccia blocks. Matching opposite walls indicate simple dilation during the emplacement of the dikes. Small exposures of Proterozoic sandstone bedrock were seen at the exterior base of the breccia ring, in which strikes are tangential to the structure and dips are at moderate angles both inward and outward.

[1] Now at NASA Manned Spacecraft Center, Houston, Texas 77058.

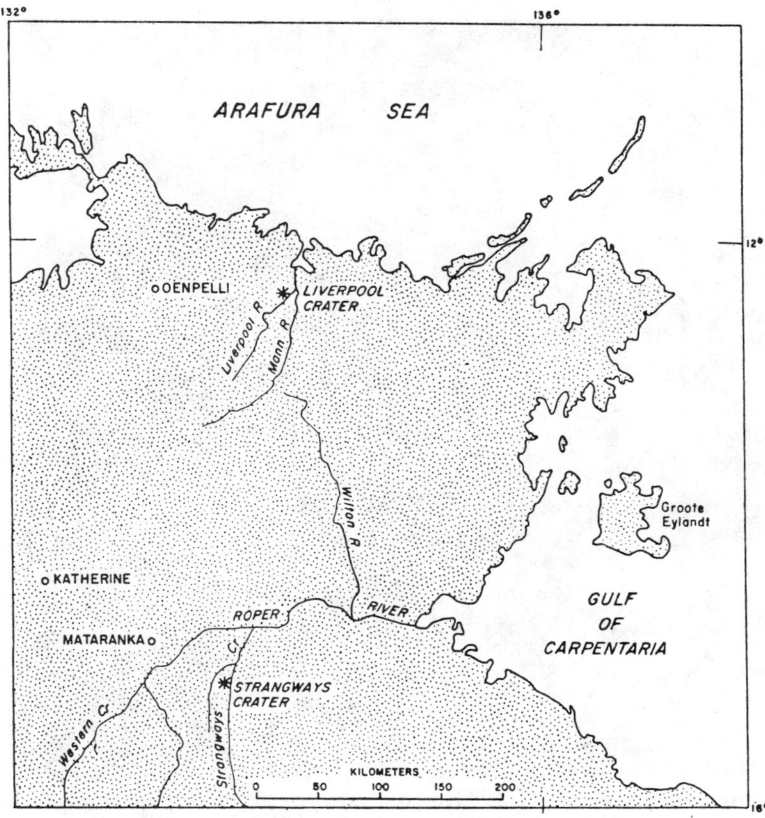

Fig. 1. Locality map of Liverpool and Strangways structures.

The interior of the crater contains friable sandstone dipping radially inward at 5–7 degrees that was evidently deposited subsequent to the formation of the crater. This resembles sandstone of the Cretaceous Mullamen group which is widespread in coastal Arnhem Land, although the closest outcrops are about 30 km distant from the crater [*Rix*, 1965]. If the correlation is correct, the age of the structure lies between Proterozoic and Cretaceous. The excellent preservation suggests no more than a single cycle of burial and exhumation, so that the age is probably close to the latter limit.

Petrographic examination of the breccia shows deformation ranging from slight to intense. In some specimens, the sandstone has been completely disaggregated and many quartz grains have been broken, yet within individual grains or grain fragments no deformation is visible. Other specimens consist of multigrain clasts veined by finely crusted quartz. Quartz grains of the larger clasts commonly show fractures parallel to the nearby veins. In one specimen, every quartz grain shows mosaic structure, cleavage, and abundant multiple planar elements 'decorated' by minute inclusions. This specimen is equivalent to the most intensely shocked material found in the central uplift of Gosses Bluff, Northern Territory, and indicates shock pressure of at least 100 kb (*Cook* [1968]; D. J. Milton and others, work in preparation).

Coesite and stishovite, the high-pressure polymorphs of silica, are ordinarily found only in rock shocked to the point of incipient melting, a stage more intense than represented by our specimens. Nevertheless, J. J. Fahey kindly attempted extractions on small breccia chips, with negative results.

The circularity of the structure, the attitude of the bedrock, the brecciation, and the shock deformation of the quartz indicate hypervelocity impact as the cause of the structure. The exposed ring apparently corresponds to the

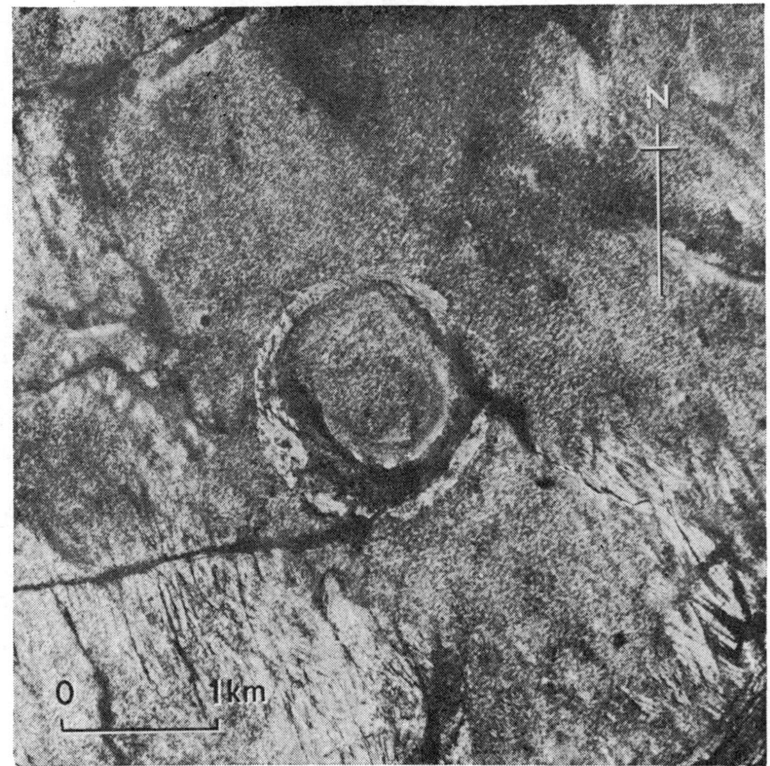

Fig. 2. Vertical air photograph of Liverpool crater and its surroundings (courtesy of Australian Department of National Development).

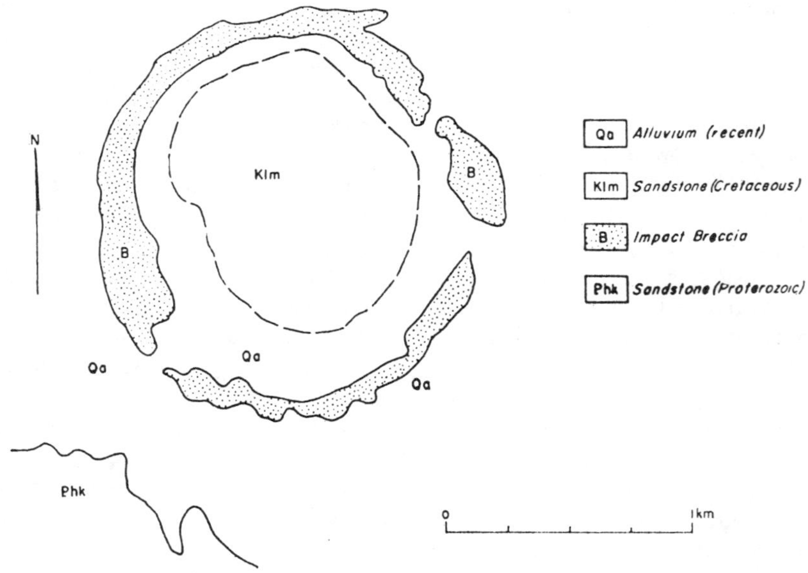

Fig. 3. Geological sketch map of Liverpool crater and surroundings. Map based mainly on air photo interpretation with some field checking.

raised rims of Meteor Crater, Arizona, and Wolf Creek crater, Western Australia, although breccia apparently conceals uplifted bedrock. Further evidence of impact would be shock-melted quartzite or actual fragments of the projectile. At Meteor Crater, such material is found at the surface only as isolated fragments, but large masses of melted sandstone and abundant nickel-iron particles have been encountered in drill holes and shafts sunk in the center floor. At Liverpool crater such material is unlikely to be found at the surface

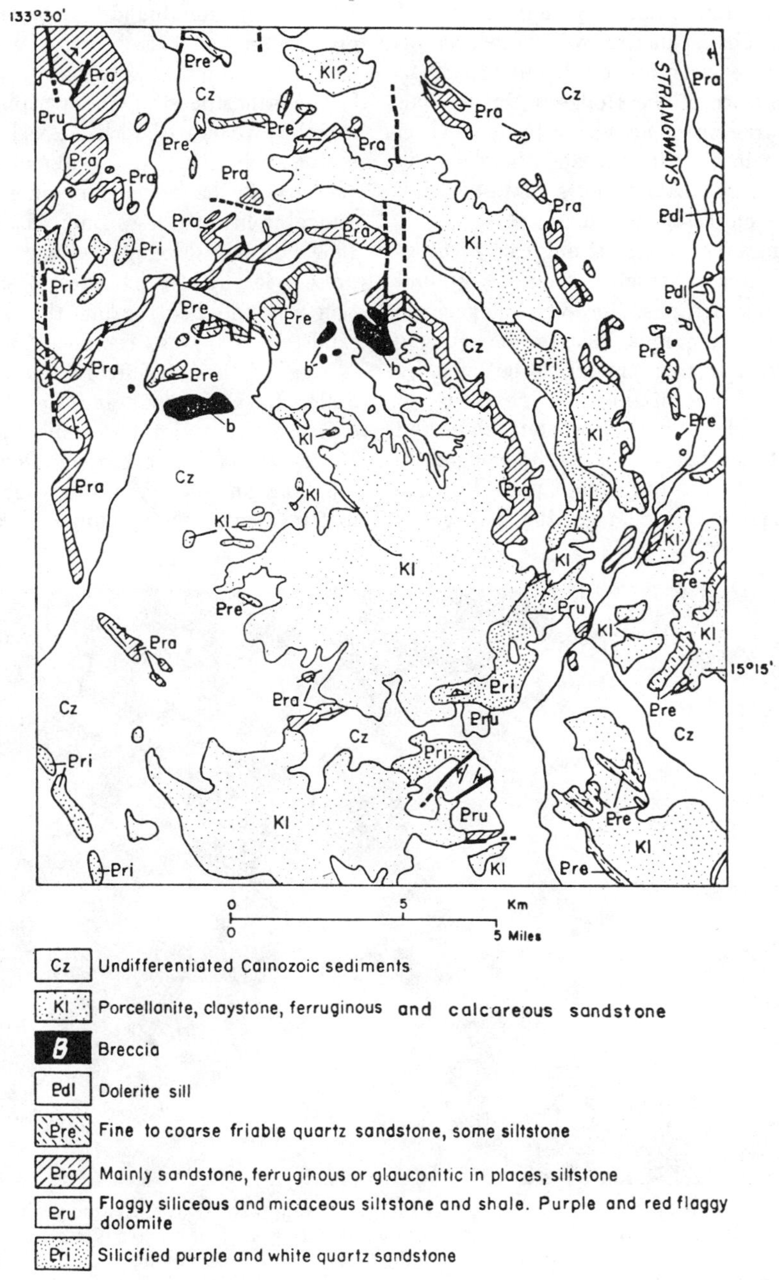

Fig. 4. Geological map of Strangways crater, after *Dunn* [1963].

owing to the greater age and more complicated history of this crater compared to Meteor Crater.

STRANGWAYS CRATER

The Strangways crater (latitude 15°12′S, longitude 133°35′E) lies west of the Strangways River, approximately 75 km southeast of Elsey station, near Mataranka, Northern Territory (Figure 1). The structure was noted as an igneous collapse structure by *Dunn* [1963] in his reconnaissance of the Hodgson Downs map sheet. The structure is accessible by four-wheel drive vehicle in the dry season; there are no roads nearby, so that it takes almost a full day's travel from Elsey station.

The Strangways crater is not easily recognizable as a circular structure from the ground or from air photographs. However, the geologic map of Dunn (Figure 4) shows the circular form clearly. The ring pattern is outlined by approximately tangentially striking beds of sandstone in ridges up to 70 meters high, and siltstone in broad flat valleys, both belonging to the Roper group of probable Upper Proterozoic age. The outer boundary of the disturbance is difficult to locate owing both to lack of outcrop and to the presence of horizontal sediments, mainly sandstone of Lower Cretaceous age, which partially cover the structure, but the diameter is at least 16 km.

An early age for the disturbance is suggested by the north–south faults that are shown cutting the structure, which belong to a regional set that Dunn found to have been active mainly in the Precambrian, with movement continuing into the Paleozoic.

The sandstones of the periphery exhibit a rhombic pattern of rock cleavage with a spacing of about 5 cm, which resembles the shatter cleavage we have found in quartzite from the central uplift of Gosses Bluff [*Milton and Brett,* 1969]. Poorly developed shatter cones are also present locally. No evidence of shock deformation was noted in the few thin sections of the sandstone that were examined.

Most of the interior of the structure is concealed by Cretaceous and Cenozoic sediments, except for several outcrops mapped by Dunn as Proterozoic volcanic breccia. One of these is 2 km long and nearly 1 km wide and rises 50 meters above the surrounding plain. The bulk

Fig. 5. Three quartz inclusions in melt breccia, Strangways crater. Center grain has irregular fractures only, indicating mild shock or none; left grain shows five directions of 'cleavage' and (more apparent at high magnifications) three of 'planar features,' indicating moderate shock; rich grain has entirely recrystallized, indicating shock intense enough to destroy crystal lattice. Crossed Nicols.

of the rock is extremely fine-grained, chocolate brown, and contains numerous angular inclusions of quartzite and baked shale, and rounded inclusions of granite gneiss up to 4 meters in diameter.

Thin sections show angular grains of quartz and feldspar in a fine-grained matrix. Some grains show no evidence of shock, but most show cleavage and abundant planar features (Figure 5). Some grains retain their angular outline but have completely recrystallized internally, indicating a higher degree of shock, in which the crystal lattice was completely or largely destroyed without reaching temperatures of true melting. Some of these diaplectic grains show slight flowage and are thus transitional to the matrix, in which swirled trains of dust inclusions indicate liquid flow. There are two distinct but intermingled materials in the matrix: a colorless material composed of completely intergrown fine granular quartz, and a lower index brown material composed of fine feathery and spherulitic growths, which is apparently alkali feldspar according to qualitative microprobe analysis. The crystallite patterns have no relation to the flowage pattern, and indicate later crystallization. Isotropic patches are apparently original glass, but they are too small and scarce and too largely altered to clay(?) minerals for positive identification.

The mingling of liquids of grossly different compositions and the absence of eutectic crystallization distinguish this material from volcanic rock and indicate that the matrix is rock melted by extreme shock. The breccia closely resembles those from some of the Canadian craters, which have been interpreted as recrystallized shock-melt breccias or suevite, for example the Onaping formation in the Sudbury basin [*French*, 1968]. The recrystallized melt breccia of Mount Pyroclast at Gosses Bluff crater is similar, except that most of the silica initially crystallized as tridymite, of which there is no evidence at Strangways. Only the basal layer of the melt breccia remains at Gosses Bluff, whereas the outcrop at Strangways may be part of a higher zone where mixing was more intense and cooling more rapid. Attempts at extraction of coesite and stishovite by J. J. Fahey again had negative results.

A chemical analysis of the matrix of the breccia done by the rapid rock analysis technique of the U.S. Geological Survey is given in Table 1, and the trace-element content is given in Table 2. The composition of the breccia matrix is consistent with a mixture of quartzite, shale, and gneiss, which are the rock types observed in unmelted clasts. The larger quartz-rich areas in the thin sections studied apparently represent melted quartzite; it has not been determined whether the alkali feldspar glasses are derived from shale or the feldspar of the gneiss. The nickel content of the breccia matrix is anomalously high for a nonultramafic rock, as is the Ni/Co ratio, suggesting possible

TABLE 1. Major-Element Analysis of Breccia Matrix from Strangways Crater

Oxide	Content, wt %	Oxide	Content, wt %
SiO_2	59.0	K_2O	7.3
Al_2O_3	17.1	H_2O^-	1.3
Fe_2O_3	6.6	H_2O^+	3.0
FeO	0.36	TiO_2	0.74
MgO	2.9	P_2O_5	0.06
CaO	0.50	MnO	0.10
Na_2O	0.44	CO_2	0.05
Total oxide content, 99.4			

Analysis by Rapid Rock Analysis Laboratory, U. S. Geological Survey.

TABLE 2. Optical Emission Spectrographic Analysis of Trace Elements in Breccia Matrix

Element	Content, ppm
Pb	10
Ag	<1
Cu	5
Ga	22
Sn	<10
Mn	300
Cr	68
Mo	<2
B	18
Co	20
Ni	247
Ba	1650
Sr	83
V	60
Be	<2
Nb	<10
Sc	16
La	83
Y	40
Zr	248

Analysis by C. Annell, U. S. Geological Survey.

meteoritic contamination. Such contamination is also suspected in some of the large Canadian craters [*Dence*, 1970].

One outcrop of basement rock 20 meters long was found within the structure, a quartz-orthoclase-biotite gneiss. The plagioclase is highly altered, and the biotite is slightly chloritzed. Multiple planar elements occur in some quartz grains although most show no shock damage. The closest outcrop of basement rock outside the structure is an adamellite porphyry at Maranboy, one hundred kilometers distant. The thickness of the sedimentary section in the area and the distance to the nearest basement outcrop suggest that the gneiss outcrop at Strangways crater indicates significant uplift, whether the outcrop is part of a coherently uplifted central peak or a large block in the breccia.

The form of the structure, the presence of probable shatter cleavage and possible shatter cones, the evidence of central uplift, the shock deformation, and the presence of large outcrops of melt breccia indicate that the Strangways was formed by hypervelocity impact.

The identification of two new structures of probable impact origin raises the number known in Australia to seven. Aside from the smallest, Dalgaranga, they fall into pairs in size and general character: Strangways and Gosses Bluff; Liverpool and Wolf Creek; Henbury and Boxhole. The two large structures complement each other particularly well. At Gosses Bluff the rim has been destroyed by erosion and only remnants of the basal layer of highly shocked breccia of the crater floor are preserved, while the central peak is excellently exposed. At Strangways, on the other hand, while the central uplift is obscure, more of the crater fill is preserved by a nearly complete ring of outcrop.

Acknowledgments. We thank P. R. Dunn (Bureau of Mineral Resources) for pointing out the structures to us on aerial photographs, C. Annell and Frank Cuttitta (U.S.G.S.) for chemical analyses, J. J. Fahey (U.S.G.S.) for coesite and stishovite separations, L. C. Calk (U.S.G.S.) for microprobe analyses, and S. R. Taylor (Australian National University) for helpful discussion. We also thank the staff of Maningrida Aboriginal Settlement for providing accomodation during our stay in Arnhem Land, and the Welfare Branch of the Northern Territory Administration for granting us permission to visit the area.

Publication authorized by the Director, Bureau of Mineral Resources, Geology and Geophysics, Australia, and the Director, U.S. Geological Survey.

REFERENCES

Cook, P. J., The Gosses Bluff cryptoexplosion structure, *J. Geol.*, *76*, 123, 1968.

Dence, M. R., Terrestrial and lunar impact melts (abstract), *Meteoritics*, *5*, 190, 1970.

Dunn, P. R., *1:250,000 Geological Series, Hodgson Downs, N.T., Sheet SD 53-14* (explanatory notes), Bureau of Mineral Resources, Geology, and Geophysics, Canberra, Australia, 1963.

French, B. M., Sudbury structure, Ontario: Some petrographic evidence for an origin by meteorite impact, in *Shock Metamorphism of Natural Materials*, edited by B. M. French and N. M. Short, p. 383, Mono, Baltimore, Md., 1968.

Milton, D. J., and Robin Brett, Gosses Bluff astrobleme, Australia—the central uplift (abstract), *Geol. Soc. Amer., Spec. Paper 121*, 531, 1969.

Rix, P., *1:250,000 Geological Series, Milingimbi, N.T., Sheet SD 53-2* (explanatory notes), Bureau of Mineral Resources, Geology, and Geophysics, Canberra, Australia, 1965.

(Received January 13, 1971;
revised March 4, 1971.)

Part VII

STRUCTURES OF CLASS IV IN SOUTH AMERICA

Editor's Comments on Paper 33

33 DIETZ and FRENCH
Two Probable Astroblemes in Brazil

It is only recently (Dietz and French, 1973, Paper 33) that possible astroblemes have been recognized in South America, both the structures being in Brazil.

The Araguainha Dome, a class IVa structure at Matte Grosso in central Brazil, was formerly supposed to be an alkaline intrusive complex, similar to numerous such complexes in Brazil. Here, the Devonian *Furnas Standstone* is cataclastically deformed in a central uplift (10 km in diameter) enclosed within a ring syncline in which Permian-Carboniferous sediments are downdropped within a graben. The outer margin of the syncline is situated 20 km out from the focus, but ERTS imagery shows that the structure is 40 km in diameter. Unusual igneous breccias outcrop in the central uplift. Shock lamellations in quartz grains suggest possible shock pressures exceeding 100 kb. The matrix of the breccia is heterogeneous and contains partly melted, shocked rock and mineral fragments. The age of the structure is Carboniferous or younger. Its general character resembles other cryptoexplosion structures, but more detailed study is required—particularly in the light of the geographic association with alkaline central complexes, which incorporate manifestations of explosive eruptivity.

Serrada Canghala is remotely situated in northeast Brazil and has a damped-wave form, comprising central uplift 5 km in diameter, and a downdropped annulus extending to 12 km from the focus. The uplift consists of an outer cricular wall of steeply dipping Carboniferous *Poti Sandstone* and a central basin of uplifed Devonian *Longa Shale.* This basin is erosionally produced by streams that breach the sandstone wall and is due to differential erosion of the softer central rocks. Diamond drilling in search of kimberlite—three holes to 200 m depth—only encountered steeply dipping, deformed *Longa Shale* beneath the central depression. Such very shallow drilling must, however, be regarded as inconclusive. Of Carboniferous age or younger, the structure appears to be an eviscerated dome and the disturbance increases toward its focus. It is not unlike Gosses Bluff (Paper 31). The rocks exposed are all clastic sediments—carbonates and saline evaporites are absent. Large sandstone dikes cross-cut the bedding, but igneous rocks are absent. No shatter cones or metamorphism indicating intense explosive shock have been recorded. Unusual fracturing in quartzites, occurring as pebbles in the *Poti* conglomerate from the central uplift, resemble weak shock deformations in other cryptoexplosion structures, but are in no way diagnostic of extreme shock. The structure is therefore classified as IVb.

33

Reprinted from *Nature* **244**:561–562 (1973)

TWO PROBABLE ASTROBLEMES IN BRAZIL

R. S. Dietz and B. M. French

More than forty structures have now been identified as astroblemes, ancient geological scars produced by the impacts of large cosmic bodies on the Earth[1]. We report here some preliminary results on the study of two new astroblemes in Brazil, the Araguainha Dome and the Serra da Canghala structure. Although relatively young meteorite impact craters have been identified in Argentina[2] and Chile[3], these two Brazilian structures are the first ancient astroblemes to be reported from South America. The impact origin of the Araguainha Dome has been established by the discovery of diagnostic shock-metamorphic effects[4] in its rocks. No shock effects have yet been identified in rocks from the Serra da Canghala structure.

The Araguainha Dome, centred at 16° 46′ S, 52° 59′ W, is located near the town of Araguainha in the state of Matto Grosso, Brazil. The structure has been regarded as an alkaline intrusive, of which there are many examples in Brazil. It consists of a central uplift about 10 km across, in which cataclastically deformed Devonian Furnas Sandstone is exposed. Around this uplift is an annular ring syncline in which Permo-Carboniferous sediments, chiefly sandstones, are exposed in a series of downdropped grabens. The diameter of the structure estimated from ground studies is about 20 km. But a photograph of the structure taken from the ERTS-1 satellite (ERTS-1 photograph 1089-13005; October 20, 1972) shows that the circular pattern extends across a diameter of about 40 km (Fig. 1). If this pattern represents the true size of the structure, then the Araguainha Dome is one of the larger known astroblemes.

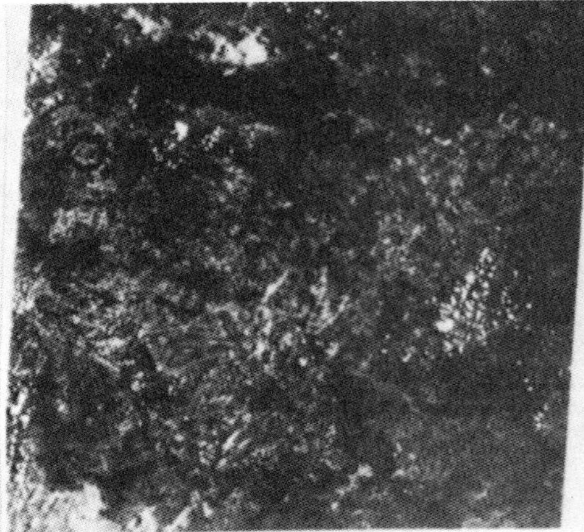

Fig. 1 ERTS-1 photograph of the Araguainha Dome. The structure appears at upper left as a dark inner ring and an outer circular pattern about 40 km across. NASA photograph, courtesy N. M. Short.

The central part of the central uplift is occupied by an unusual igneous breccia which contains inclusions of buried quartzofeldspathic Precambrian crystalline rocks in an igneous matrix. Thin sections of this breccia and the inclusions, made available to us by officials of Petróleo Brasiliero SA (Petrobras), contain a wide variety of distinctive shock-metamorphic effects, which are diagnostic of the intense shock pressures ($\geqslant$100 kbar) produced by meteorite impact. The inclusions show extensive development of planar features (shock lamellae) in quartz, often as many as six to eight distinct sets of lamellae per grain, with orientations corresponding to the distinctive $\{10\bar{1}3\}$, $\{10\bar{1}2\}$, $\{11\bar{2}2\}$, and $\{11\bar{2}1\}$ planes[5,6]. The igneous matrix of the breccia exhibits a heterogeneous texture and contains numerous shocked and partly melted rock and mineral inclusions. The breccia from the Araguainha Dome is identical in appearance to impact produced breccias observed at other astroblemes.

Fig. 2 Aerial photograph of Serra da Canghala structure, showing high wall of uplifted Poti Sandstone around the central uplift. Part of the downdropped ring syncline is in the foreground. White streak at left is an aircraft landing strip.

The Serra da Canghala structure ("Pack-Saddle Mountain") is located in the remote northern corner of the state of Goiaz (8° 05′ S, 46° 52′ W) in north-eastern Brazil, on the southern limb of the Amazon basin. The structure (diameter 12 km) has a damped-wave form with a central uplift of diameter 5 km surrounded by a downdropped annulus at least 3.5 km across. Serra da Canghala is a remarkably impressive feature of the region. In the central uplift, a circular wall, 250 m high, of steeply dipping Poti (Carboniferous) Sandstone surrounds a central basin of uplifted Longa (Devonian) Shale (Fig. 2). The central part of the structure thus has a crater-like apearance resulting from the preferential erosion of the Longa Shale by streams which flow through a breach in the Poti Sandstone wall.

In 1971 the Company for Research on Mineral Resources (CRPM) drilled three diamond drill holes, each 200 m deep, within the centre of the structure to test the possibility that the structure might be a kimberlite intrusion. No igneous rocks were found, and the drilling encountered only highly deformed and steeply dipping Longa Shale. The drilling results and the surface outcrop pattern indicate that the central uplift becomes progressively more disturbed towards the centre of the structure and displays the forms of an eviscerated dome. The intensely disturbed central area, the down thrown peripheral annulus, and the ratio of the central uplift diameter to total diameter are consistent with the structural features found in astroblemes. The overall structural pattern and geomorphology are strongly similar to those at the Gosses Bluff astrobleme in central Australia[7,8].

The rock units underlying the Serra da Canghala structure down to the Precambrian basement are entirely clastic and do not contain significant carbonate or salt units ; the structure cannot be a salt diapir. Igneous rocks are entirely absent from the structure, and large cross-cutting sandstone dikes are the only anomalous features.

One of us (R. S. D.) made a three-day reconnaissance and collection at the Serra da Canghala structure in October 1972. Shatter coning, a useful field criterion for meteorite impact[9], was not observed. The rocks collected were almost entirely fine-grained siltstones which did not show any shock-metamorphic effects. One specimen of coarser grained quartzite, collected as a cobble in a conglomeratic unit of the Poti Sandstone in the central part of the structure, showed small areas characterized by unusual fracturing of the quartz grains. Similar fracturing in quartz has been reported in mildly-shocked rocks from known meteorite impact craters[10,11], but the features observed in the Poti Sandstone are not distinctive enough to be considered diagnostic for meteorite impact. At present, the structural deformation in the Serra da Canghala structure, combined with the apparent absence of igneous activity, favoure the theory that the structure is an astrobleme. But definite proof will require the identification of unquestionable shock-metamorphic effects in its rocks.

We thank the Company for Research on Mineral Resources (CRPM) for making possible the field work at Serra da Canghala and for providing the assistance of J.-M. Coehlo in the field. M. Antonio M. de Oliveira of Petróleo Brasiliero SA (Petrobras) provided the thin sections and information on the Araguainha Dome. Petrographic studies were carried out at NASA, Goddard Space Flight Center, courtesy of C. C. Schnetzler. N. M. Short, also of NASA, Goddard, provided information and ERTS photographs of the Araguainha Dome. Partial support for R. S. D.'s travel to South America was provided by the Barringer Crater Co., Philadelphia, Pennsylvania.

1 Dence, M. R., *Rep. Twenty-fourth Int. geol. Cong., Montreal, Canada*, part 15, 77 (1972).

2 Cassidy, W. A., Villar, L. M., Bunch, T. E., Kohman, T. P., and Milton, D. J., *Science, N.Y.*, **149**, 1055 (1965).

3 Sanchez, J., and Cassidy, W., *J. geophys. Res.*, **71**, 4891 (1966).

4 *Shock Metamorphism of Natural Materials* (edit. by French, B. M., and Short, N. M.) (Mono Book Corp., 1968).

5 Robertson, P. B., Dence, M. R., and Fos, M. A., in *Shock Metamorphism of Natural Materials* (edit. by French, B. M., and Short, N. M.), 433 (Mono Book Corp., 1968).

6 Engelhardt, W. v., and Bertsch, W., *Contrib. Mineral. Petrol.*, **20**, 203 (1969).

7 Dietz, R. S., *Nature*, **216**, 1082 (1967).

8 Milton, D. J., Barlow, B. C., Brett, R., Brown, A. R., Glikson, A. Y., Mainwaring, E. A., Moss, F. J., Sedmik, E. C. E., Van Son, J., and Young, G. A., *Science, N.Y.*, **175**, 1199 (1971).

9 Dietz, R. S., *Rep. Twenty-fourth Int. geol. Cong., Montreal, Canada*, Part 15, 112 (1972).

10 Bunch, T. E., in *Shock Metamorphism of Natural Materials* (edit. by French, B. M., and Short, N. M.), 413 (Mono Book Corp., 1968).

11 Kieffer, S. W., *J. geophys. Res.*, **76**, 5449 (1971).

Part VIII
STRUCTURES OF CLASSES V AND VI

CLASS V

Structures of Class V, about which too little is known for them to be referred to Classes I–IV, are listed in Table 2. They include no less than eleven structures situated in the former French West African Territories. This remote region has revealed a number of interesting and unusual cryptoexplosion structures (Talemzane Crater, Algeria—Karpoff 1963; Aouelloul Crater—Monod 1952; and the other Mauretanian structures, see editorial comments Part V of this book). These structures call for more detailed study. The Nebiewale and Dampier Archipelago structures have already been mentioned (editorial comments Parts V and VI). Other interesting structures of class V include:

Al Umchaimin, Iraq

Merriam and Holwerda (1957) described a 3.2 km diameter crater situated 145 km southwest of Rutba, amid near-horizontal Eocene limestones that show no disturbance at the crater. The floor lies 25 m below the rim. No meteoritic material has been recovered. Neither solution nor wind erosion seems to be favored by the evidence, and the nearest igneous rocks are 100 km distant. The authors favor meteoritic origin, but the lack of any visible deformation would seem to rule this out. The origin remains obscure. Drilling might be informative.

Herault, France

Gèze and Cailleux (1950) describe craters situated amid Viséean "schistes" and argillites. There is a larger structure at Cabrérolles; and smaller ones are situated at Faugères in the Montagne Noir. The largest, le Clot, has a diameter of 200–220 m and is 100 m deep, possessing a flat floor. The smaller is situated 5 km to the northwest, has a diameter of 50–60 m and a depth of 23 m. Some smaller craters may be associated with the Faugères crater. The whole group appears to possess a NNE–SSW alignment. The largest, at Cabrérolles, may have been the last to form (down-approach); the mass being the last to arrive in a shower. These structures are all Quaternary or Holocene. They are not karst phenomena. A magnetic anomaly is associated with each of them. A comparison drawn by the authors with Meteor Crater, Arizona, is not entirely convincing, and the origin of these craters remains obscure.

Kalkkop, South Africa

Situated in the Cape Province, this is a circular, saucer-shaped depression 640 m across. The rim is 10 m and the floor 6 m above the surrounding country, the form being that of a raised pan or crater (Blignaut et al., 1948). It lies within gently flexed Karroo (Permian-Carboniferous) Beds, and is situated on a small anticline. The rim is formed of white limestone. Dips are outward and steep. The limestone contains recent fossils. A 300 m deep central drill hole penetrated limestones and recent fossiliferous sediments (gravels etc.) down to 290 m. At that depth breccias were encountered, of angular sandstone and shale, embedded in limestone. Megabreccias appeared below, then cracked rock. The crater was clearly formed by some sort of violent upheaval, which produced the brecciation. The breccia was later cemented by limestone and deposited in the crater. The authors leave the question of origin open: it remains obscure.

Lakes Michikamau, Mecatina, etc., Canada

Beals and Hitchen (1970) refer to these two occurrences and eight others on the Canadian Shield, in the eastern region, suggesting possible meteoritic origin. They do not appear to be true craters, but have the character of stratified, circular structures. Their origin remains obscure, and they deserve detailed study.

Ramgarh, India

Crawford (1972) described this isolated, rectangular hill, with rounded corners (in shape not unlike the Arizona crater). The external diameter is 3 km. The structure is set amid flat-lying, undisturbed sandstones of the *Vindhayan System,* much obscured by alluvium. The hill rises 200 m above the plain, and there is a low central hill within the ring, which is breached on the southwest. Brecciation without faulting and possible shatter cones are reported. Dips are steep and steepen up toward the focus. The origin remains obscure, more detailed field and laboratory study being called for.

CLASS VI

Structures of Class VI, those for which meteoritic origin has been discounted, are listed in Table 2; they number 45. They include Ellef Ringnes Island and Upheaval Dome, both known to be salt diapirs. Others are simply solution cavity phenomena—for example those of the Eyre Peninsula. Some, such as Panamint, have been shown to be volcanic craters. Winkler Crater (McCall, 1977, Part III) has recently been shown by drilling to be underlain by kimberlite (pers. comm., A. J. A. Janse). Some of the structures listed were attributed to impact on the basis of very meagre evidence. Many, such as Hudsons Bay, owing the attribution to arcuate coastline alone. The sources of these attributions are listed by Hey (1966), Freeberg (1966, 1969), Monod (1965), O'Connell (1965), and Zotkin (1969).

REFERENCES

Beals, C. S., and A. Hitchen. 1970. On Deposition of Sediments in Craters. *Dominion Observatory Ottawa Pubs.* **39**:105–118.

Blignaut, J. J. G.; P. J. Roussow; J. de Villiers; and H. D. Russell. 1948. The Geology of the Schoorteensberg Area, Cape Province. *South Africa Dept. of Mines.* 48 pp.

Crawford, A. R. 1972. Possible Impact Structure in India. *Nature* **237**:96.

Freeberg, J. H. 1966, 1969. Terrestrial Impact Structures: A Bibliography. *U.S. Geol. Survey Bull. 1220,* 91 pp; *1320,* 39 pp.

Gèze, B., and A. Cailleux. 1950. Probable Existence of Meteorite Craters at Cabrérolles and Faugeres (Herault). *Acad. Sci. Comptes Rendus* **230**:1534–1536.

Hey, M. H. 1966. *Catalogue of Meteorites.* London: British Museum (Natural History).

Karpoff, R. 1963. The Meteorite Crater of Talemzane in Southern Algeria. *Meteoritics* **1**:31–38.

McCall, G. J. H., ed. 1977. *Meteorite Craters.* Benchmark Papers in Geology, vol. 36. Stroudsburg, Pa.: Dowden, Hutchinson & Ross.

Merriam, R., and J. G. Holwerda. 1957. Al Umchaimin, A Crater of Possible Meteoritic Origin in Western Iraq. *Geol. Jour.* **123**:231–233.

Monod, T. 1952. Les accidentes cratiformes ou circulaire du Sahara Occidentale. *Bull. Dir. Mines, Gouvt. Gen. Afrique Occidentale, Francaise, Dakar* **15**:169–177.

Monod, T. 1965. *Contributions a l'establissement d'une liste d'accidents circulaire d'origine meteoritique (reconnue, possible ou supposé), cryptoexplosive.* Université de Dakar, Ifan, Dakar.

O'Connell, E. 1965. *A Catalogue of Meteorite Craters and Related Features with a Guide to the Literature.* RAND Corp., Santa Monica, California.

Zotkin, I. T. 1969. The Moon Craters on Earth. *Priroda (USSR)* **9**:95–105.

AUTHOR CITATION INDEX

SUBJECT INDEX

About the Editor

GERALD JOSEPH HOME McCALL was most recently senior consultant geologist for Paragon-Contech, Teheran, where he engaged in a major regional mapping program for the Geological Survey of Iran. In addition to having served as an independent consultant to various other organizations, he was director of exploration at Research and Exploration Management Pty Ltd. in Melbourne from 1970 to 1972 and spent ten years prior to this as Reader in Geology at the University of West Australia and ten as geologist in the Kenya Geological Survey.

Dr. McCall earned his Bachelor of Science degree in geology from London University in 1949 and also holds Ph.D. (1951) and D.Sc. (1968) degrees in geology from that institution. The author of over 90 scientific papers, of *Meteorites and Their Origins* (David & Charles Ltd.), of *The Archean: Search for the Beginning* (Dowden, Hutchinson & Ross, Inc., 1977), and of *Meteorite Craters* (Dowden, Hutchinson & Ross, Inc., Benchmark Papers in Geology, Vol. 36, 1977), he is active in a number of scientific societies and has just completed a term as vice president of the International Association of Planetology. As of December 1968, Dr. McCall became a Fellow of Liverpool University.